The series "Studies in Computational Intelligence" (SCI) publishes new developments and advances in the various areas of computational intelligence—quickly and with a high quality. The intent is to cover the theory, applications, and design methods of computational intelligence, as embedded in the fields of engineering, computer science, physics and life sciences, as well as the methodologies behind them. The series contains monographs, lecture notes and edited volumes in computational intelligence spanning the areas of neural networks, connectionist systems, genetic algorithms, evolutionary computation, artificial intelligence, cellular automata, self-organizing systems, soft computing, fuzzy systems, and hybrid intelligent systems. Of particular value to both the contributors and the readership are the short publication timeframe and the world-wide distribution, which enable both wide and rapid dissemination of research output.

More information about this series at http://www.springer.com/series/7092

Alireza Rezvanian · Ali Mohammad Saghiri
S. Mehdi Vahidipour · Mehdi Esnaashari
Mohammad Reza Meybodi

Recent Advances in Learning Automata

 Springer

Alireza Rezvanian
School of Computer Science
Institute for Research in Fundamental
  Sciences (IPM)
Tehran
Iran

and

Computer Engineering and Information
  Technology Department
Amirkabir University of Technology
  (Tehran Polytechnic)
Tehran
Iran

Ali Mohammad Saghiri
Computer Engineering and Information
  Technology Department
Amirkabir University of Technology
  (Tehran Polytechnic)
Tehran
Iran

S. Mehdi Vahidipour
Faculty of Electrical and Computer
  Engineering, Computer Engineering
  Department
University of Kashan
Kashan
Iran

Mehdi Esnaashari
Faculty of Computer Engineering
K.N.Toosi University of Technology
Tehran
Iran

Mohammad Reza Meybodi
Soft Computing Laboratory
Amirkabir University of Technology
  (Tehran Polytechnic)
Tehran
Iran

ISSN 1860-949X   ISSN 1860-9503   (electronic)
Studies in Computational Intelligence
ISBN 978-3-319-89182-8   ISBN 978-3-319-72428-7   (eBook)
https://doi.org/10.1007/978-3-319-72428-7

Printed on acid-free paper

This Springer imprint is published by Springer Nature
The registered company is Springer International Publishing AG
The registered company address is: Gewerbestrasse 11, 6330 Cham, Switzerland

To my lovely wife, my beloved dad, my merciful mom, and my dear sisters for their love and supports

Alireza

To my late mother, Dr. H. Afsar Lajevardi. I will never forget her kindness and support. To my brother, Dr. Mohammad Ali Saghiri and my father for their support during difficult days of my life

Ali Mohammad

To my family

S. Mehdi

To my love, my life, my soulmate, my ONE and only, Najmeh

Mehdi

Preface

This book is written for computer engineers, scientists, and students studying/working in reinforcement learning and artificial intelligence domains. The book collects recent advances in learning automaton theory as well as its applications in different computer science problems and domains. The book, in detail, describes the distributed learning automata and the cellular learning automata models for solving a variety of complex problems in wireless sensor networks, complex social networks, cognitive peer-to-peer networks, and adaptive Petri nets. Validation of the given learning automata-based methodologies is provided through extensive computer simulations. In addition, the book presents detailed mathematical and theoretical aspects of recent developments of cellular learning automata in real-world problems. The mathematical level in all chapters is well-suited within the grasp of the scientists as well as the graduate students from the engineering and computer science streams. The reader is encouraged to have basic understanding of probability, stochastic processes, and related mathematical analyses.

This book consists of six chapters dedicated toward using recent models of learning automata for computer science applications. Chapter 1 provides the necessary background about learning automata theory and distributed learning automata. Chapter 2 gives a brief introduction about recent cellular learning automata models including irregular cellular learning automata and dynamic cellular learning automata models. Chapter 3 introduces recent applications of learning automata for wireless sensor networks. Chapter 4 is devoted to applications of learning automata in cognitive peer-to-peer networks. Chapter 5 discusses about the applications of learning automata for social network analyses when the underlying graph model is assumed to be stochastic. Finally, Chap. 6 provides new models of adaptive Petri nets based on learning automata.

The authors would like to thank Dr. Thomas Ditzinger, Springer, Executive Editor, Interdisciplinary and Applied Sciences and Engineering and Mr. Ramamoorthy Rajangam, Project Coordinator, Books Production, Springer Verlag, Heidelberg, for the editorial assistance and excellent cooperative collaboration to produce this important scientific work. We hope that readers will share our pleasure to present this book on recent advances in learning automata and will find it useful in their careers.

Tehran, Iran   Alireza Rezvanian
Tehran, Iran   Ali Mohammad Saghiri
Kashan, Iran   S. Mehdi Vahidipour
Tehran, Iran   Mehdi Esnaashari
Tehran, Iran   Mohammad Reza Meybodi

Acknowledgements

We are grateful to many people who have helped us during the past few years, who have contributed to work presented here, and who have offered critical reviews of prior publications. We thank Springer for their assistance in publishing the book. We are also grateful to our academic supervisor, our family, our parents, and all our friends for their love and support.

Contents

About the Authors

Alireza Rezvanian was born in Hamedan, Iran, in 1984. He received his B.Sc. from Bu-Ali Sina University of Hamedan, Iran, in 2007, M.Sc. in Computer Engineering with honors from Islamic Azad University of Qazvin, Iran, in 2010, and Ph.D. in Computer Engineering at the Computer Engineering and Information Technology Department from Amirkabir University of Technology (Tehran Polytechnic), Tehran, Iran, in 2016. Currently, he works as a researcher in School of Computer Science, Institute for Research in Fundamental Sciences (IPM), Tehran, Iran. Prior to the current position, he joined the Department of Computer Engineering and Information Technology at Hamedan University of Technology, Hamedan, Iran as a lecturer. He has authored or coauthored more than 70 research publications in reputable peer-reviewed journals and conferences including IEEE, Elsevier, Springer, Wiley and Taylor & Francis. He has been Guest Editor of special Issue on new applications of learning automata-based techniques in real-world environments for Journal of Computational Science (Elsevier). He is an associate editor of the Human-centric Computing and Information Sciences (Springer). His research activities include soft computing, evolutionary algorithms, complex social networks, and learning automata.

Ali Mohammad Saghiri received his B. Sc. and M. Sc. in computer engineering in Iran, in 2008 and 2010, respectively. He received Ph.D. in computer engineering at the Computer Engineering and Information Technology Department from Amirkabir University of Technology (Tehran Polytechnic), Tehran, Iran, in 2017. His research interests include peer-to-peer networks, distributed systems, artificial intelligence, learning automata, reinforcement learning, parallel algorithms, and soft computing.

S. Mehdi Vahidipour received his B.Sc. and M.Sc. in computer engineering in Iran, in 2000 and 2003, respectively. He received Ph.D. in computer engineering at the Computer Engineering and Information Technology Department from Amirkabir University of Technology (Tehran Polytechnic), Tehran, Iran, in 2016. Currently, he is an Assistant Professor in Electrical and Computer Engineering department, University of Kashan. His research interests include distributed artificial intelligence, learning automata, and Adaptive Petri nets.

Mehdi Esnaashari received his B.S., M.S., and the Ph.D. in computer engineering, all from the Amirkabir University of Technology, Tehran, Iran, in 2002, 2005, and 2011, respectively. Prior to his current position, he was an Assistant Professor with Iran Telecommunications Research Center, Tehran. He is currently an Assistant Professor with faculty of computer engineering, K. N. Toosi University of Technology, Tehran, Iran. His current research interests include computer networks, learning systems, learning automata, soft computing, and information retrieval.

Mohammad Reza Meybodi received B.S. and M.S. in Economics from the Shahid Beheshti University in Iran, in 1973 and 1977, respectively. He also received M.S. and Ph.D. from the Oklahoma University, USA, in 1980 and 1983, respectively, in Computer Science. Currently, he is a Full Professor in Computer Engineering Department, Amirkabir University of Technology, Tehran, Iran. Prior to current position, he worked from 1983 to 1985 as an Assistant Professor at the Western Michigan University, and from 1985 to 1991 as an Associate Professor at the Ohio University, USA. His current research interests include, learning systems, cloud computing, soft computing, and social networks.

Abstract

Learning automaton (LA) as a promising field of artificial intelligence is a self-adaptive decision-making device that interacts with an unknown stochastic environment and progressively is able to find the optimal action even provided with probabilistic wrong hints. LA has made a significant impact in many areas of computer science and engineering problems. In the past decade, a wide range of learning automata theories, models, and paradigms have been published by researchers in vast areas of computer science domain such as resource allocation, pattern recognition, image processing, task scheduling, data mining, computer networks, communication networks, distributed adaptive systems, cognitive networks, vehicular sensor networks, grid computing, cloud computing, adaptive Perti-nets, complex social networks, optimization, and so on. Learning automata are extremely suitable for modeling, learning, controlling, and solving real-world problems, especially when the information is incomplete; that is, when the environment is noisy or has a high degree of uncertainty. This book is intended to collect recent advances in learning automata including research results that address key issues and topics related to learning automata theories, architecture, models, algorithms, and their applications.

Part I
Models

Chapter 1
Learning Automata Theory

1.1 Learning Automata

Reinforcement learning or learning with a critic is defined by characterizing a learning problem, instead of characterizing the learning methods (Sutton and Barto 1998). Any method that is well suited to solve that problem is considered as a reinforcement learning method. The reinforcement learning problem is the problem of learning from interactions in order to achieve a certain goal. Considering this specification for the reinforcement learning problem, there must be an agent capable of learning, called the learner or the decision-maker. The learner must interact with a so-called environment or teacher, comprising everything outside of the learner, which provides evaluative responses to the actions performed by the learner. The learner and the environment interact continually; the learner selects actions and the environment responds to the selected actions, presenting new situations to the learner. In short, reinforcement learning is a framework of the learning problems in which the environment does not indicate the correct action, but provides only a scalar evaluative response to the selection of an action by the learner.

Learning automaton (LA) as one of computational intelligence techniques have been found very useful tool to solve many complex and real world problems in networks where a large amount of uncertainty or lacking the information about the environment exists (Torkestani 2012; Mousavian et al. 2013, 2014; Mahdaviani et al. 2015; Zhao et al. 2015; Khomami et al. 2016b; Mirsaleh and Meybodi 2016a). A learning automaton is a stochastic model operating in the framework of the reinforcement learning (Narendra and Thathachar 1989; Thathachar and Sastry 2004). This model can be classified under the reinforcement learning schemes in the category of the temporal-difference (*TD*) learning methods. *TD* learning is a combination of the Monte Carlo ideas and the dynamic programming ideas. Like Monte Carlo methods, *TD* methods can learn directly from raw experience without a model of the environment's dynamics. Like the dynamic programming, *TD* methods update estimates based in part on the other learned estimates, without

© Springer International Publishing AG 2018
A. Rezvanian et al., *Recent Advances in Learning Automata*, Studies in
Computational Intelligence 754, https://doi.org/10.1007/978-3-319-72428-7_1

waiting for a final outcome (Sutton and Barto 1998). Sarsa (Rummery and Niranjan 1994), Q-learning (Watkins 1989), Actor-Critic methods (Barto et al. 1983) and R-learning (Schwartz 1993) are other samples of *TD* methods. Learning automata differ from other *TD* methods in the following two ways; the representation of the internal states and the updating method of the internal states.

The automata approach to learning can be considered as the determination of an optimal action from a set of actions. A learning automaton can be regarded as an abstract object that has a finite number of actions. It selects an action from its finite set of actions and applies it to an environment. The environment evaluates the applied action and sends a reinforcement signal to the learning automaton (Fig. 1.1). The reinforcement signal provided by the environment is used to update the internal state of the learning automaton. By continuing this process, the learning automaton gradually learns to select the optimal action, which leads to favorable responses from the environment.

A learning automaton is a quintuple $\langle \alpha, \Phi, \beta, F, G \rangle$ where $\alpha = \{\alpha_1, \alpha_2, \ldots \alpha_r\}$ is the set of actions that it must choose from, $\Phi = (\Phi_1, \Phi_2, \ldots, \Phi_s)$ is the set of states, $\beta = \{\beta_1, \beta_2, \ldots, \beta_q\}$ is the set of inputs, $G: \Phi \rightarrow \alpha$ is the output map and determines the action taken by the learning automaton if it is in the state Φ_j, and $F: \Phi \times \beta \rightarrow \Phi$ is the transition map that defines the transition of the state of the learning automaton upon receiving an input from the environment.

The selected action at the time instant k, denoted by $\alpha(k)$, serves as the input to the environment which in turn emits a stochastic response, $\beta(k)$, at the time instant k, which is considered as the response of the environment to the learning automaton. Based upon the nature of β, environments could be classified in three classes: P-, Q-, and S-models. The output of a P-model environment has two elements of success or failure. Usually in P-model environments, a failure (or unfavorable response) is denoted by 1 while a success (or a favorable response) is denoted by 0. In Q-model environments, β can take a finite number of values in the interval [0, 1] while in S-model environments, β lies in the interval [0, 1]. Based on the response $\beta(k)$, the state of the learning automaton $\Phi(k)$ is updated and a new action is chosen at the time instant $(k + 1)$. Learning automata can be classified into two main classes: fixed and variable structure learning automata (VSLA) (Narendra and Thathachar 1989).

A simple pseudo-code for the behavior of an r-action learning automaton within a stationary environment with $\beta \in \{0, 1\}$ is presented in Fig. 1.2.

Fig. 1.1 The interaction of a learning automaton and its environment

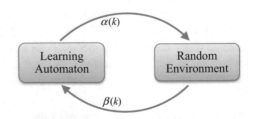

Algorithm 1-1. Learning automaton

Input: Action-set α

Output: Action probability vector p

Assumptions

   Initialize r-dimensional action-set: $\alpha = \{\alpha_1, \alpha_2, ..., \alpha_r\}$ with r actions

   Initialize r-dimensional action probability vector $p(k) = \{p_1, p_2, ..., p_r\} = \left\{\frac{1}{r}, \frac{1}{r}, ..., \frac{1}{r}\right\}$ at instant k

   Let i denotes the selected action by the automaton

   Let j denotes the current checking action

Begin

   While (automaton does not converge to any action)

      The learning automaton selects an action based on the action probability vector p.

      The environment evaluates the selected action and gives the reinforcement signal $\beta \in \{0, 1\}$ to the learning automaton.

      //The learning automaton updates its probability vector $p(k)$ using the provided reinforcement signal β

      For each action $j \in \{1, ..., r\}$ **Do**

         If ($\beta=0$) // favorable response

            The selected action by learning automaton is rewarded according to equation (1-1)

         Else If ($\beta=1$) // unfavorable response

            The selected action by learning automaton is penalized according to equation (1-2)

         End If

      End For

   End While

End Algorithm

Fig. 1.2 Pseudo-code of the behavior of a learning automaton

Learning automata have several good features, which make them suitable for using in many applications. Main features of learning automata are stated.

1. Learning automata can be used without any priori information about the underlying application.
2. Learning automata are very useful for applications with large amount of uncertainty.
3. Unlike traditional hill-climbing algorithms, hill-climbing in learning automata is done in expected sense in a probability space.
4. Learning automata require a very little and simple feedback from their environment.
5. Learning automata are very useful in multi-agent and distributed systems with limited intercommunication and incomplete information.
6. Learning automata are very simple in structure and can be implemented easily in software or hardware.
7. The action set of learning automata can be a set of symbolic or numeric values.
8. Optimization algorithms based on learning automata don't need the objective function to be an analytical function of adjustable parameters.
9. Learning automata can be analyzed by powerful mathematical methodologies. It has been shown that learning automata are optimal in single, hierarchical, and distributed structures.
10. Learning automata require a few mathematical operations at each iteration so it can be used in real-time applications.
11. Learning automata have flexibility and analytical tractability needed for most applications.

1.1.1 Fixed-Structure Learning Automata

The learning automaton is called fixed-structure if the followings are fixed: 1. the probability of the transition from one state to another state and 2. the action probability of any action in any state. Examples of the fixed-structure *LA* are L_2N_2, G_2N_2, Krylov, and Krinsky (Narendra and Thathachar 1989) which has been used in many applications such as adaptation of the back-propagation algorithm parameters.

1.1.2 Variable-Structure Learning Automata (VSLA)

A VSLA is represented by a 6-tuple $\langle \beta, \phi, \alpha, P, G, T \rangle$, where β denotes the set of inputs, ϕ is a set of internal states, α is the action sets as output, P is the state probability vector governing the choice of state as each instant k, G is the output mapping, and T is the learning algorithm (also known as learning scheme). The learning algorithm T refers to a recurrence relation which is used to update the action probability vector. Let $\alpha_i(k) \in \alpha$ be the action that is chosen by a learning automaton and $p(k)$ is the probability vector defined over the set of action at instant k. At each instant k, the action probability vector $p(k)$ is updated by the linear learning algorithm given in Eq. (1.1) if the chosen action $\alpha_i(k)$ is rewarded by the random environment ($\beta = 0$) and it is updated according to Eq. (1.2) if the chosen action is penalized ($\beta = 1$).

$$p_j(k+1) = \begin{cases} p_j(k) + a[1 - p_j(k)] & j = i \\ (1-a)p_j(k) & \forall j \neq i \end{cases} \tag{1.1}$$

$$p_j(k+1) = \begin{cases} (1-b)p_j(k) & j = i \\ \left(\frac{b}{r-1}\right) + (1-b)p_j(k) & \forall j \neq i \end{cases} \tag{1.2}$$

where a denotes reward parameter which determines the amount of increases of the action probability values, b is the penalty parameter determining the amount of decrease of the action probabilities values and r is the number of actions that learning automaton can take. If $a = b$, the learning algorithm is called linear reward-penalty (L_{RP}) algorithm; if $b = \varepsilon a$ where $0 < \varepsilon < 1$, then the learning algorithm is called linear reward-ε-penalty ($L_{R\varepsilon P}$) algorithm; and finally if $b = 0$, the learning algorithm is called linear reward-Inaction (L_{RI}) algorithm which is perhaps the earliest scheme considered in mathematical psychology.

1.1.3 Learning Automata with Variable Number of Actions

In variable action-set learning automata (also called a learning automaton with variable number of actions) the number of available actions that can be taken by the learning automaton at every instant can be changed with time (Thathachar 1987). In variable action set learning automata, at each instant k, a learning automaton selects an action based on the scaled action probability vector $\hat{p}(k)$ defined as

$$\hat{p}_i(k) = \frac{p_i(k)}{K(k)} \tag{1.3}$$

where $K(k)$ is the sum of the probabilities of the available actions at instant k, and $p_i(k) = Prob[\alpha(k) = \alpha_i]$. And then the selected action performed on the environment. After receiving the reinforcement signal, the LA will similarly update probability vector of available actions via learning algorithm given in Eqs. (1.1) and (1.2). Finally, the probability vector of available actions $p(k+1)$ is rescaled as follows

$$p_i(k+1) = \hat{p}_i(k+1).K(k) \tag{1.4}$$

The pseudo-code of the behavior of a variable action set learning automaton is shown in Fig. 1.3.

1.1.4 Estimator Algorithms

The slow convergence speed of the *LA* is the most important problem with this learning model. Estimator learning algorithms (Thathachar and Sastry 1985) are a class of the learning schemes which improve the convergence speed of the *LA*. In such algorithms, additional information about the characteristics of the environment is maintained and is used to increase the speed of convergence of the *LA*. Unlike other learning algorithms, in which the action probability vector is updated based solely on the current response received from the environment, in estimator learning algorithms, an estimation of the reward strength is maintained for each action of the *LA* and is used along with the current response of the environment to update the action probability vector. The estimation of the reward strength for any action α_i is computed as the ratio of the number of times α_i is rewarded to the number of times α_i is selected. Different estimator learning algorithms have been proposed in literature thus far, such as stochastic (Papadimitriou 1994), absorbing stochastic (Papadimitriou et al. 2002), Discretized TS (Lanctot and Oommen 1992), relative strength (Simha and Kurose 1989), and S-model ergodic discretized (Paximadis and Paximadis 1994).

Algorithm 1-2. Variable action-set learning automata

Input: Action-set α

Output: Action probability vector p

Assumptions

    Initialize r-dimensional action-set: $\alpha = \{\alpha_1, \alpha_2, \dots, \alpha_r\}$ with r actions

    Initialize r-dimensional action probability vector $p(k) = \{p_1, p_2, \dots, p_r\} = \left\{\frac{1}{r}, \frac{1}{r}, \dots, \frac{1}{r}\right\}$ at instant k

    Let i denotes the selected action by the automaton

    Let j denotes the current checking action

Begin

    While (automaton does not converge to any action)

        Calculate available actions of learning automaton

        Calculate the sum of the probability of available actions

        For each action $j \in \{1, \dots, r\}$ **Do**

            If (α_j is available action)

                Scale action probability vector $\hat{p}_i(k)$ according to equation (1-3)

            End if

        End for

        Generate map function between available actions and all actions of learning automaton

        The learning automaton selects an action according to the probability vector of available actions $\hat{p}(k)$

        The environment evaluates the selected action and gives the reinforcement signal $\beta \in \{0, 1\}$ to the learning

            automaton.

        For each available action $j \in \{1, \dots, m\}$ **Do**

            If ($\beta = 0$) // favorable response

                The selected action by learning automaton is rewarded according to equation (1-1)

            Else If ($\beta = 1$) // unfavorable response

                 The selected action by learning automaton is rewarded according to equation (1-2)

            End if

        End for

        For each action $j \in [1, \dots, r]$ do

            If (α_j is available action)

                Rescale the probability vector of selected available action by equation (1-4)

            End

        End for

    End While

End Algorithm

Fig. 1.3 Pseudo-code of the behavior of a variable action-set learning automaton

1.1.5   Pursuit Algorithms

A pursuit learning algorithm is a simplified version of the estimator learning algorithms, in which the action probability vector chases after the currently optimal action. At each iteration of a pursuit learning algorithm, the action probability of the action with the maximum rewarding estimation is increased. By this updating method, learning algorithm always pursues the optimal action. Several different pursuit algorithms have been proposed in literature, namely reward-penalty (Thathachar and Sastry 1986), discretized reward-penalty (Oommen and Lanctot 1990), reward-inaction (Agache and Oommen 2001), discretized reward-inaction (Agache and Oommen 2001), generalized (Agache and Oommen 2002), and discretized generalized (Agache and Oommen 2002).

1.1.6 Continuous Action-Set Learning Automata

In a continuous action-set learning automaton (*CALA*), the action-set is defined as a continuous interval over the real numbers. This means that each automaton chooses its actions from the real line. In such a learning automaton, the action probability of the possible actions is defined as a probability distribution function. All actions are initially selected with the same probability, that is, the probability distribution function under which the actions are initially selected has a uniform distribution. The probability distribution function is updated depending upon the responses received from the environment.

In (Santharam et al. 1994), a continuous action-set learning automaton has been proposed by Santharam et al. In this *CALA*, at any time instant k, the probability distribution function ξ has a normal distribution with mean μ_k and standard deviation σ_k. The action α_k and the mean μ_k are two parameters using which the *CALA* interacts with its environment. *CALA* applies these parameters as two actions to the environment and gets two reinforcement signals from the environment, one for each of them. Denote these reinforcement signals as $\beta(\mu)$ and $\beta(\alpha)$. Based on the supplied reinforcement signals, *CALA* updates the probability distribution function of its actions using the following equations.

$$\mu_{k+1} = \mu_k + af_1[\mu_k, \sigma_k, \alpha_k, \beta(\alpha), \beta(\mu)]$$
$$\sigma_{k+1} = \sigma_k + af_2[\mu_k, \sigma_k, \alpha_k, \beta(\alpha), \beta(\mu)] - Ca[\sigma_k - \sigma_L] \tag{1.5}$$

where C, σ_L, and a are the parameters of the *CALA* and f_1 and f_2 are defined as given below.

$$f_1[\mu, \sigma, \alpha, \beta(\alpha), \beta(\mu)] = \left[\frac{\beta(\alpha) - \beta(\mu)}{\phi(\sigma)}\right]\left[\frac{\alpha - \mu}{\phi(\sigma)}\right]$$

$$f_2[\mu, \sigma, \alpha, \beta(\alpha), \beta(\mu)] = \left[\frac{\beta(\alpha) - \beta(\mu)}{\phi(\sigma)}\right]\left[\left(\frac{\alpha - \mu}{\phi(\sigma)}\right)^2 - 1\right] \tag{1.6}$$

$$\phi(\sigma) = (\sigma - \sigma_L)I\{\sigma > \sigma_L\} + \sigma_L.$$

The *CALA*, by using the above updating equations, converges to a normal distribution $N(\alpha^*, \sigma_L)$, where α^* is the optimum action. One another continuous action-set learning automaton is given by Howell et al. (1997).

1.2 Interconnected Learning Automata

It seems that the full potential of learning automaton is realized when multiple automata interact with each other. It is shown that a set of interconnected learning automata is able to describe the behavior of an ant colony capable of finding the shortest path from their nest to food sources and back (Verbeeck and Nowe 2002). In this section, we study the interconnected learning automata. The interconnected learning automata techniques based on activation type of learning automata for taking an action can be classified into three classes: synchronous, sequential, and asynchronous, as follows

- Synchronous Model of Interconnected Automata: In synchronous model of interconnected automata, at any time instant, all automata are activated simultaneously, choose their actions, then apply their chosen actions to the environment, and finally update their states. Two models of synchronous interconnected learning automata have been reported in the literature: game of automata and synchronous cellular learning automata.
- Asynchronous Model of Interconnected Automata: In asynchronous model of interconnected automata, at any time instant only a group of automaton is activated, independently. The only proposed model for asynchronous model is asynchronous cellular learning automata. An asynchronous cellular learning automaton is a cellular automaton in which an automaton (or multiple automata) is assigned to its every cell. The learning automata residing in a particular cell determines its state (action) on the basis of its action probability vector. Like cellular automata, there is a rule that cellular learning automata operate under it. The rule of cellular learning automata and the state of neighboring cell of any particular cell determine the reinforcement signal to the learning automata residing in that cell. In cellular learning automata, the neighboring cells of any particular cell constitute its environment because they produce the reinforcement signal to the learning automata residing in that cell. This environment is a nonstationary environment because it varies as action probability vectors of cell vary and called local environment because is local to every cell. Krishna proposed an asynchronous cellular learning automata in which the order to which learning automata is determined is imposed by the environment (Krishna 1993).
- Sequential Model of Interconnected Automata: In sequential model of interconnected automata, at any time instant only a group of automaton is activated and the actions chosen by currently activated automata determine next automata to be activated. The hierarchical structure learning automata, network of learning automata, distributed learning automata, and extended distributed learning automata are examples of sequential model.

In following subsections, we focus on sequential interconnected learning automata.

1.2.1 Hierarchical Structure Learning Automata

When the number of actions for a learning automaton becomes large (more than 10 actions), the time taken for the action probability vector to converges it also increases. Under such circumstances, a hierarchical structure of learning automata (HSLA) can be used. A hierarchical system of automata is a tree structure with depth of M where each node corresponds to an automaton and the arcs emanating from that node corresponds to actions of that automaton. In HSLA, an automaton with r actions is in the first level (root of tree) and kth level has r^{k-1} automata each with r actions. The root node corresponds to an automaton which will be referred to as the first-level or top-level automaton. Selection of each action of this automaton leads to activate an automaton at the second level. In this way, the structure can be extended to an arbitrary number of levels. A three-level hierarchy with three actions per automaton is shown in Fig. 1.4.

The operation of hierarchical structure learning automata can be described as follows: initially, root automaton selects one action, say action α_{i_1}. Then the i_1^{th} automaton at the second level will be activated. The action selected by i_1^{th} automaton at the second level (say $\alpha_{i_1 i_2}$) activates an automaton at the third level. This process is continued until a leaf automaton is activated. The action of this automaton is applied to the environment. The response from the environment is used to update the action probability vectors of the activated automata at the path from root to the selected leaf node. The basic idea of learning algorithm is to increase the probability of selecting good action and to decrease the probability of selecting other actions. In HSLA, set of actions $\{\alpha_{i_1}, \alpha_{i_1 i_2}, \ldots, \alpha_{i_1 i_2 \ldots i_M}\}$ is said to be on optimal path if the product of their respective reward probabilities are maximum. The HSLA can be classified into three types I, II and III (Thathachar and Sastry 1997). A HSLA is said to be of type I if actions constituting the optimal path are also individual optimal at their respective levels. A HSLA is said to be of type II if the actions constituting the optimal path are also individual optimal at their respective automata. Any general hierarchy is said to be of type III.

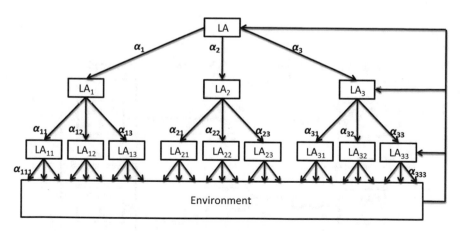

Fig. 1.4 Hierarchical structure learning automata

1.2.2   Multi-Level Game of Learning Automata

Multi-level game of *LA*s can be effectively used for solving the multi-criteria optimization problems in which several objective functions are required to be optimized. In multi-level game of *LA*s, multiple games are played in different levels (Billard 1994, 1996). The game, which is being played at each level of the multi-level game of *LA*s, decides the game, which has to be played in the next level. Figure 1.5 shows a two-level game of *LA*s with four players.

1.2.3   Network of Learning Automata

A network of *LA*s (Williams 1988) is a collection of *LA*s connected together as a hierarchical feed-forward layered structure. In this structure, the outgoing link of the *LA*s in the preceding layers is the input of the *LA*s of the succeeding layers. In this model, the *LA*s (and consequently the layers) are classified into three separate groups. The first group includes the *LA*s located at the first level of the network, called the input *LA*s (input layer). The second group is composed of the *LA*s located at the last level of the network, called the output *LA*s (output layer). The third group includes the *LA*s located between the first and the last layers, called the hidden *LA*s (hidden layer). In a network of *LA (NLA)*, the input *LA*s receive the context vectors as external inputs from the environment, and the output *LA*s apply the output of the network to the environment. The difference between the feed-forward neural networks and NLA is that units of neural networks are deterministic while units of NLA are stochastic and the learning algorithms used in two networks are different. Since units are stochastic, the output of a particular unit i is drawn from a

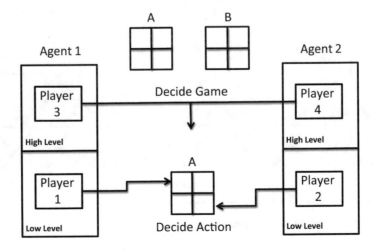

Fig. 1.5 Multi-level game of learning automata

distribution depending its input weight vector and output of the units in the preceding layers. This model operates as follows. The context vector is applied to the input *LA*s. Each input *LA* selects one of its possible actions on the basis of its action probability vector and the input signals it receives from the environment. The chosen action activates the *LA*s of the next level, which are connected to this *LA*. Each activated *LA* selects one of its actions as stated before. The actions selected by the output *LA*s are applied to the random environment. The environment evaluates the output action in comparison with the desired output and generates the reinforcement signal. This reinforcement signal is then used by all *LA*s for updating their states. The structure of such a network is shown in Fig. 1.6.

1.2.4 Distributed Learning Automata (DLA)

The hierarchical structure learning automata has a tree structure, in which there exists a unique path between the root of the tree and each of its leaves. However, in some applications, such as routing in computer networks, there may be multiple paths between the source and destination nodes. This system is a generalization of HSLA, which referred to as distributed learning automata (DLA). A Distributed learning automata (DLA) (Beigy and Meybodi 2006) shown in Fig. 1.7 is a network of interconnected learning automata which collectively cooperate to solve a particular problem. The number of actions for a particular LA in DLA is equal to the number of LA's that are connected to this LA. Selection of an action by a LA in DLA activates another LA which corresponds to this action. Formally, a DLA can

Fig. 1.6 Network of learning automata

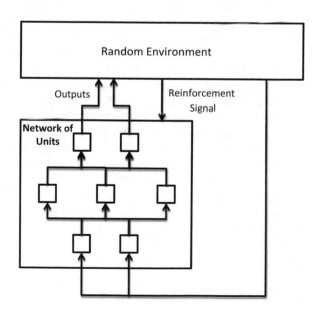

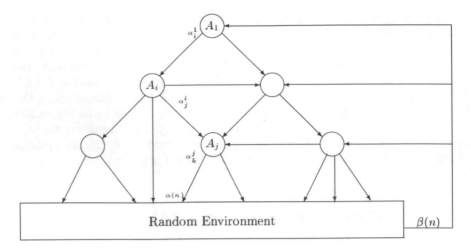

Fig. 1.7 Structure of distributed learning automata

be defined by a quadruple $\langle A, E, T, A_0 \rangle$, where $A = \{A_1, A_2, ..., A_n\}$ is the set of learning automata, $E \subset A \times A$ is the set of edges where edge (v_i, v_j) corresponds to action α_i^j of automaton A_i, T is the set of learning algorithms with which learning automata update their action probability vectors, and A_l is the root automaton of DLA at which activation of DLA starts.

The operation of a DLA can be described as follows: At first, the root automaton A_0 randomly chooses one of its outgoing edges (actions) according to its action probabilities and activates the learning automaton at the other end of the selected edge. The activated automaton also randomly selects an action which results in activation of another automaton. The process of choosing actions and activating automata is continued until a leaf automaton (an automaton which interacts with the environment) is reached. The chosen actions, along the path induced by the acti- vated automata are applied to the random environment. The environment evaluates the applied actions and emits a reinforcement signal to DLA. The activated learning automata along the chosen path update their action probability vectors on the basis of the reinforcement signal according to the learning algorithms. The paths from the unique root automaton to one of the leaf automata are selected until the probability with which one of the chosen paths is close enough to unity. Each DLA has exactly one root automaton which is always activated, and at least one leaf automaton which is activated probabilistically. For example in Fig. 1.8, every automaton has two actions. If automaton LA_1 selects α_2^1 from its action set, then it will activate automaton LA_2. Afterward, automaton LA_2 chooses one of its possible actions and so on. At any time, only one LA in the network is active.

In (Sato 1999), a restricted version of above DLA is introduced. In this model, the underlying graph, in which the DLA is embedded, is a finite directed acyclic graph (DAG). Sato used the L_{R-I} learning algorithm with decaying reward parameter. It is shown that every edge of DAG is selected infinitely often and thus

Fig. 1.8 Example of
distributed learning automata

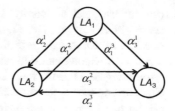

every learning automaton is activated infinitely often. Also, it is shown that when every learning automaton has unique best action, the DLA converges to its best action with probability 1 (Sato 1999). Meybodi and Beigy introduced a DLA in which the underlying graph is not restricted to be a DAG (Beigy and Meybodi 2006). But in order to restrict a learning automaton to appear more than once in any path, the learning automaton with changing number of actions are used. The DLA was used for solving the several stochastic graph problems (Torkestani and Meybodi 2010a, 2012a; Mollakhalili Meybodi and Meybodi 2014; Rezvanian and Meybodi 2015a, b).

1.2.5 Extended Distributed Learning Automata (eDLA)

An extended distributed learning automata (eDLA) (Mollakhalili Meybodi and Meybodi 2014) is a new extension of DLA supervised by a set of rules governing the operation of the LAs. *Mollakhalili-Meybodi* et al. presented a framework based on eDLA for solving stochastic graph optimization problems such as stochastic shortest path problem and stochastic minimum spanning tree problem. Here, we provide a brief introduction to the eDLA. In general in eDLA, the ability of a DLA is improved by adding communication rules and changing the activity level of each LA depends on a problem to be solved by an eDLA. An eDLA similar to a DLA can be modeled by a directed graph in which the node-set of graph constructs the set of LA and the number of actions for each LA in eDLA equals to the number of LAs that are connected to that LA. In eDLA, at any time, not only each LA can be in one mode of activity level but also each LA with a high activity level can be performed an action according to its probabilities on the random environment. Formally, an eDLA can be described by a 7-tuple $\langle A, E, S, P, S^0, F, C \rangle$, where A is the set of LA, $E \subseteq A \times A$ is the edge-set of communication graph $G = \langle V, E \rangle$ and $S = \{s_1, s_2, \ldots, s_n\}$ is a set of activity levels corresponding to each LA in eDLA; specially s_i indicates the activity level for learning automaton A_i in which $s_i \in \{Pa, Ac, Fi, Of\}$ consists of one of the following activity levels: *Passive* (initial level of each LA and can be changed to *Active*), *Active* (activity level for set of available LAs and its level can be upgraded to *Fire*), *Fire* (the highest level of activity, LA can be performed and its level can be changed to *Off*) and *Off* (the lowest level of activity, LA is disabled and its level stay unchanged), represented briefly by *Pa*, *Ac*, *Fi*, and *Of* respectively. As mentioned, at any time only one LA in *e*DLA can be in the *Fi* level of activity and can be determined by *fire function C* which randomly selects a LA from a set of LAs with activity level of *Ac*. *Governing rule P* is the finite set of rules that governs the activity levels of each LA. *P* according to the current activity level of each LA, its adjacent LA or depending on the particular problem which *e*DLA is designed is defined. $S^0 = \left(s_1^0, s_2^0, \ldots, s_n^0\right)$ and $F = \left\{S^F | S^F = \left(s_1^F, s_2^F, \ldots, s_n^F\right)\right\}$ are the *initial state* and *final conditions* of *e*DLA.

The operation of *e*DLA can be described as follows. In *e*DLA, at first at initial state S^0, a starting LA is randomly selected by firing function *F* to fires, selects one of outgoing edges (actions) according to its action probabilities and performs it on the random environment and at the same time the activity level of fired LA and neighboring LAs are changed to *Of* and *Ac* respectively. Changing activity levels of LAs result in the state of *e*DLA transferred from state S^k to state S^{k+1} at instant *k* by governing rule *P*. Then the firing function *C* fires one LA from the set of LA with activity level of *Ac* to selects an action and then changes its activity level and neighboring LAs. The process of firing one LA by firing function, performing an action by fired LA, changing the activity level of fired LA and its neighbors by governing rule *P* is continued until the final condition of *e*DLA *F* is reached. *F* can be defined based on a set of criteria in terms of activity levels of LAs such that, if one of them is satisfied, the final condition of *e*DLA is realized. The environment evaluates the performed actions by fired LA and generates a reinforcement signal to *e*DLA. The action probabilities of fired LA along the visited nodes or LA of the nodes which are part of a solution to the problem of graph are then updated on the basis of the reinforcement signal according to the learning algorithm. Firing LAs of *e*DLA by starting from randomly LA is repeated predefined number of times until the solution of the problem for which *e*DLA is designed is obtained.

1.3   Cellular Learning Automata

Cellular learning automaton (*CLA*) (Meybodi et al. 2003) is a combination of cellular automaton (*CA*) (Packard and Wolfram 1985) and learning automaton (*LA*) (Narendra and Thathachar 1989). The basic idea of *CLA* is to use *LA* for adjusting the state transition probability of a stochastic *CA*. This model, which opens a new learning paradigm, is superior to *CA* because of its ability to learn and is also superior to single *LA* because it consists of a collection of *LAs* interacting with each other (Beigy and Meybodi 2004). A *CLA* is a *CA* in which a number of *LAs* is assigned to every cell. Each *LA* residing in a particular cell determines its action (state) on the basis of its action probability vector. Like *CA*, there is a local rule that the *CLA* operates under. The local rule of the *CLA* and the actions selected by the neighboring *LAs* of any particular *LA* determine the reinforcement signal to that *LA*.

The neighboring *LA*s (cells) of any particular *LA* (cell) constitute the local environment of that *LA* (cell). The local environment of an *LA* (cell) is non-stationary due to the fact that the action probability vectors of the neighboring *LA*s vary during the evolution of the *CLA*. The operation of a *CLA* could be described as the following steps (Fig. 1.9): At the first step, the internal state of every cell is determined on the basis of the action probability vectors of the *LA*s residing in that cell. In the second step, the local rule of the *CLA* determines the reinforcement signal to each *LA* residing in that cell. Finally, each *LA* updates its action probability vector based on the supplied reinforcement signal and the chosen action. This process continues until the desired result is obtained.

CLA can be either *synchronous* or *asynchronous*. In a synchronous *CLA* (Meybodi et al. 2003), *LA*s in all cells are activated at the same time synchronously using a global clock whereas in an asynchronous *CLA* (*ACLA*) (Beigy and Meybodi 2008), *LA*s in different cells are activated asynchronously. The *LA*s may be activated in either time-driven or step-driven manner. In a time-driven *ACLA*, each cell is assumed to have an internal clock which wakes up the *LA*s associated to that cell. In a step-driven *ACLA*, a cell is selected for activation in either a random or a fixed sequence. From another point of view, *CLA* can be either *close* or *open*. In a close *CLA*, the action selection of any particular *LA* in the next iteration of its evolution only depends on the state of the local environment of that *LA* (actions of its neighboring *LA*s) whereas in an open *CLA* (Beigy and Meybodi 2007), this not only depends on the local environment, but also on the external environments. In (Beigy and Meybodi 2010), a new type of *CLA*, called *CLA* with multiple *LA*s in each cell, has been introduced. This model is suitable for applications such as channel assignment in cellular networks, in which it is needed that each cell is equipped with multiple *LA*s. In (Beigy and Meybodi 2004), a mathematical framework for studying the behavior of the *CLA* has been introduced. It was shown that, for a class

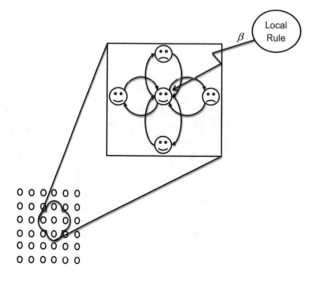

Fig. 1.9 Operation of the cellular learning automaton

of local rules called commutative rules, different models of *CLA* converge to a globally stable state (Beigy and Meybodi 2004, 2007, 2008, 2010).

1.3.1   CLA-EC: CLA-Based Evolutionary Computing

CLA-EC model, proposed in (Rastegar et al. 2004), is a combination of *CLA* and the evolutionary computing model (Fig. 1.10). In this model, each cell of the *CLA* is equipped with an m-bit binary genome. Each genome has two components; model genome and string genome. Model genome is composed of m learning automata. Each *LA* has two actions; 0 and 1. The set of actions selected by the set of *LA*s of a particular cell concatenated to each other to form the second component of the genome, i.e. the string genome. The operation of a CLA-EC, in any particular cell c_i, takes place as follows. Each *LA* residing within the cell c_i randomly selects one of its actions according to its action probability vector. The actions selected by the set of *LA*s of the cell c_i are concatenated to each other to form a new string genome for that cell. The fitness of this newly generated genome is then evaluated. If the newly generated genome is better than the current genome of the cell, then the current genome of the cell c_i is replaced by the newly generated genome. Next, a number of the neighboring cells of the cell c_i, according to the fitness evaluation of their corresponding genomes, are selected for mating. Note that the mating in this context is not reciprocal, i.e., a cell selects another cell for mating but not necessarily vice versa.

The results of the mating process in the cell c_i are a number of reinforcement signals, one for each *LA* of the cell. The process of computing the reinforcement signal for each *LA* is described in Fig. 1.11. Each *LA* updates its action probability

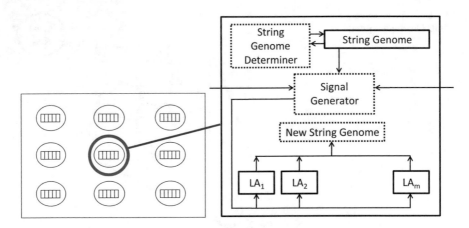

Fig. 1.10   The structure of the CLA-EC model

Fig. 1.11 The process of computing the reinforcement signal for each *LA*

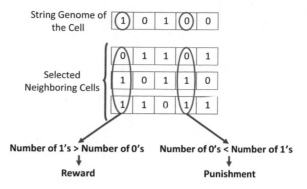

vector based on the supplied reinforcement signal and its selected action. This process continues until a termination condition is satisfied.

1.4 Applications of Learning Automata

In the recent years, learning automata have been used as optimization tools in complex and dynamic environments where a large amount of uncertainty or lacking the information about the environment exists. In the recent years, learning automata have been successfully applied to a wide variety of applications such as optimization (Rezvanian and Meybodi 2010a, b; Hasanzadeh et al. 2013; Moradabadi and Beigy 2014; Mahdaviani et al. 2015; Mirsaleh and Meybodi 2015; Moradabadi et al. 2016; Vafashoar and Meybodi 2016), (Moradabadi and Beigy 2014), image processing (Mofrad et al. 2015; Damerchilu et al. 2016), graph problems (Soleimani-Pouri et al. 2012; Mousavian et al. 2013, 2014; Khomami et al. 2016a; Mirsaleh and Meybodi 2016b; Vahidipour et al. 2017a), data clustering (Ahangaran et al. 2017; Hasanzadeh-Mofrad and Rezvanian 2017), community detection (Amiri et al. 2013; Khomami et al. 2016b, 2017b; Mirsaleh and Meybodi 2016a; Elyasi et al. 2016.), link prediction (Moradabadi and Meybodi 2016, 2017), grid computing (Hasanzadeh and Meybodi 2014, 2015; Mofrad et al. 2016), stochastic social networks (Rezvanian and Meybodi 2015a, b, 2016a, c, 2017a), network sampling (Rezvanian et al. 2014; Jalali et al. 2015, 2016; Rezvanian and Meybodi 2015c, 2016b; Ghavipour and Meybodi 2017a), information diffusion (Daliri Khomami et al. 2014; Khomami et al. 2017a, c), recommender systems (Ghavipour and Meybodi 2016), wireless sensor networks (Safavi et al. 2014; Nicopolitidis 2015), WiMAX networks (Misra et al. 2015), network security (Krishna et al. 2014), wireless mesh networks (Kumar and Lee 2015), mobile video surveillance (Kumar et al. 2015a), vehicular environment (Kumar et al. 2014, 2015b), Peer-to-Peer networks (Saghiri and Meybodi 2016a, 2017a), and cloud computing (Morshedlou and Meybodi 2014, 2017).

Chapter 2
Cellular Learning Automata

2.1 Introduction

In recent years, *Cellular Learning Automata* (*CLAs*) which are hybrid models based on *Cellular Automata* (*CAs*) and *Learning Automata* (*LAs*) have received much attentions (Thathachar and Sastry 2004; Beigy and Meybodi 2010; Esnaashari and Meybodi 2015). *CLAs* are distributed learning models which bring together the computational power of *CAs* and also the learning capability of *LAs* in unknown environments. These models are superior to *CAs* because of their ability to learn and are also superior to single *LA* because they consist of a collection of *LAs* interacting with each other. *CLAs* have been used in wide range of applications such as computer networks (Esnaashari and Meybodi 2008, 2010a, 2011, 2013; Beigy and Meybodi 2010; Saghiri and Meybodi 2016a, b, 2017a), social networks (Zhao et al. 2015; Ghavipour and Meybodi 2017b; Khomami et al. 2017c; Rezvanian and Meybodi 2017b), evolutionary computing (Rastegar and Meybodi 2004; Rastegar et al. 2006), and optimization (Mozafari et al. 2015). The *CLAs* can be classified from different perspectives, some of which are described as given bellow.

- *Closed* versus *Open CLAs*: In *closed CLAs*, states of neighboring cells of each cell, called local environment, affect the action selection process of the *LA* of that cell, whereas in open *CLAs*, the local environment of each cell, a global environment, and an exclusive environment affect the action selection process of the *LA* of that cell. In an *open CLA*, each cell has its own exclusive environment and one global environment is defined for the whole *CLA* (Beigy and Meybodi 2007; Saghiri and Meybodi 2017b).
- *Regular* versus *Irregular CLAs*: Depending on their structures, *CLAs* can be also classified as regular (Beigy and Meybodi 2004) or irregular (Esnaashari and Meybodi 2015). In irregular *CLAs*, the structure regularity assumption is relaxed.

© Springer International Publishing AG 2018
A. Rezvanian et al., *Recent Advances in Learning Automata*, Studies in
Computational Intelligence 754, https://doi.org/10.1007/978-3-319-72428-7_2

- *Synchronous* versus *Asynchronous CLAs*: In synchronous *CLAs*, all *LAs* in different cells are activated synchronously, whereas in asynchronous *CLAs*, *LAs* in different cells are activated asynchronously (Beigy and Meybodi 2007, 2008).
- *Static* versus *Dynamic CLAs*: In a *static CLA*, the structure of the cells remains fixed during the evolution of the *CLA* (Beigy and Meybodi 2007, 2008, 2010; Esnaashari and Meybodi 2015; Mozafari et al. 2015; Zhao et al. 2015). Due to the dynamic nature of many real world problems, their structural properties are time varying and for this reason, using fixed graphs for modeling them is too restrictive. In a *dynamic CLA*, one aspect of the model such as structure, local rule, or neighborhood radius may change over time (Esnaashari and Meybodi 2011, 2013; Saghiri and Meybodi 2016a). Dynamic *CLAs* can be used to solve problems that can be modeled as dynamic graphs. Dynamic *CLAs* have higher capability than static *CLAs* in adapting themselves to the environment under which they operate. These models are used to solve complicated learning problems, which arise in many real world situations.
- *CLAs with fixed number of LAs in each cell* versus *CLAs with varying number of LAs in each cell*: in *CLAs* with varying number of *LAs* in each cell, the number of *LAs* of each cell changes over time (Saghiri and Meybodi 2017b).
- *CLAs with one LAs in each cell* versus *CLAs with multiple LAs in each cell*: in *CLAs* with multiple *LAs* in each cell, each cell is equipped with multiple *LAs* (Beigy and Meybodi 2010).
- *CLAs with fixed structure LAs* versus *CLAs with variable structure LAs*: *LAs* can be classified into two main families; fixed and variable structure (Narendra and Thathachar 1989; Thathachar and Sastry 2004). In *CLAs* with fixed structure *LAs*, constituting *LAs* are of fixed structure type, whereas in *CLAs* with variable structure *LAs*, *LAs* are of variable structure type.

In the rest of this section, we focus on irregular and dynamic models of *CLAs*.

2.2 Irregular Cellular Learning Automata

Irregular cellular learning automaton (Fig. 2.1) (Esnaashari and Meybodi 2015) is a generalization of *CLA* in which the restriction of regular structure is removed. An *ICLA* is defined as an undirected graph in which, each vertex represents a cell and is equipped with an *LA*, and each edge induces an adjacency relation between two cells (two *LAs*). The *LA* residing in a particular cell determines its state (action) according to its action probability vector. Like *CLA*, there is a rule that the *ICLA* operates under. The local rule of the *ICLA* and the actions selected by the neighboring *LAs* of any particular *LA* determine the reinforcement signal to that *LA*. The neighboring *LAs* of any particular *LA* constitute the local environment of that *LA*. The local environment of an *LA* is non-stationary because the action probability vectors of the neighboring *LAs* vary during the evolution of the *ICLA*.

Fig. 2.1 Irregular cellular
learning automaton

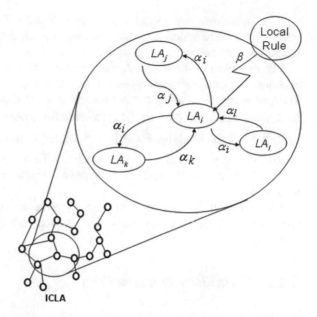

The operation of the *ICLA* is similar to the operation of the *CLA*. At the first step, the internal state of each cell is specified on the basis of the action probability vector of the *LA* residing in that cell. In the second step, the rule of the *ICLA* determines the reinforcement signal to the *LA* residing in each cell. Finally, each *LA* updates its action probability vector on the basis of the supplied reinforcement signal and the internal state of the cell. This process continues until the desired result is obtained. Formally, an *ICLA* is defined as given below.

Definition   2.1 Irregular   cellular   learning   automaton   is   a   structure $A = (G<E, V>, \Phi, A, F)$, where:

- G is an undirected graph, with V as the set of vertices (cells) and E as the set of edges (adjacency relations).
- Φ is a finite set of states and Φ_i represents the state of the cell c_i.
- A is the set of *LA*s each of which is assigned to one cell of the *ICLA*.
- $F: \underline{\varphi}_i \to \underline{\beta}$ is the local rule of the *ICLA* in each cell c_i, where $\underline{\varphi}_i = \{\varphi_j | \{i,j\} \in E\} \cup \{\varphi_i\}$ is the set of states of all neighbors of c_i and $\underline{\beta}$ is the set of values that the reinforcement signal can take. Local rule gives the reinforcement signal to each *LA* from the current actions selected by the neighboring *LA*s of that *LA*.

Note that in the definition of the *ICLA*, no explicit definition is given for the neighborhood of each cell. It is implicitly defined in the definition of the graph G.

In what follows, we consider *ICLA* with N cells. The learning automaton LA_i which has a finite action set $\underline{\alpha}_i$ is associated to cell c_i (for $i = 1, 2, \ldots, N$) of the *ICLA*. Let the cardinality of $\underline{\alpha}_i$ be m_i.

The operation of the *ICLA* takes place as the following iterations. At iteration k, each learning automaton selects an action. Let $\alpha_i \in \underline{\alpha}_i$ be the action selected by LA_i. Then all learning automata receive a reinforcement signal. Let $\beta_i \in \underline{\beta}$ be the reinforcement signal received by LA_i. This reinforcement signal is produced by the application of the local rule $F^i\left(\underline{\varphi}_i\right) \rightarrow \underline{\beta}$. Higher values of β_i mean that the selected action of LA_i will receive higher penalties. Then, each LA_i updates its action probability vector on the basis of the supplied reinforcement signal and its selected action α_i.

Like *CLA*, *ICLA* can be either synchronous or asynchronous and an asynchronous *ICLA* can be either time-driven or step-driven.

2.2.1 Definitions and Notations

In this section, we give some definitions, which will be used later for the analysis of the *ICLA*.

Definition 2.2 A configuration of the *ICLA* at step k is denoted by $\underline{p}(k) = \left(\underline{p}_1, \underline{p}_2, \ldots, \underline{p}_N\right)^T$, where $\underline{p}_i$ is the action probability vector of the learning automaton LA_i and T denotes the transpose operator.

Definition 2.3 A configuration p is called deterministic if the action probability vector of each learning automaton is a unit vector, otherwise it is called probabilistic. Hence, the set of all deterministic configurations, K^*, and the set of probabilistic configurations, K, in *ICLA* are

$$\mathsf{K}^* = \left\{ \underline{p} \,\middle|\, \begin{array}{l} \underline{p} = \left(\underline{p}_1, \underline{p}_2, \ldots, \underline{p}_N\right)^T, \; \underline{p}_i = (p_{i1}, \ldots, p_{im_i})^T, \\ \quad p_{iy} = 0 \, or \, 1 \, \forall y, \, i, \; \sum_y p_{iy} = 1 \, \forall i \end{array} \right\}, \quad (2.1)$$

and

$$\mathsf{K} = \left\{ \underline{p} \,\middle|\, \begin{array}{l} \underline{p} = \left(\underline{p}_1, \underline{p}_2, \ldots, \underline{p}_N\right)^T, \; \underline{p}_i = (p_{i1}, \ldots, p_{im_i})^T, \\ \quad 0 \leq p_{iy} \leq 1 \, \forall y, \, i, \; \sum_y p_{iy} = 1 \, \forall i \end{array} \right\}, \quad (2.2)$$

respectively.

Lemma 2.1 K *is the convex hull of* K^*.

Proof Proof of this lemma is given in (Beigy and Meybodi 2004). ∎

The application of the local rule to every cell allows transforming a configuration to a new one.

Definition 2.4 The global behavior of an *ICLA* is a mapping G: $\mathsf{K} \to \mathsf{K}$ that describes the dynamics of the *ICLA*. The evolution of the *ICLA* from a given initial configuration $\underline{p}(0) \in \mathsf{K}$ is a sequence of configurations $\left\{ \underline{p}(k) \right\}_{k \geq 0}$, such that $\underline{p}(k+1) = \mathsf{G}\left(\underline{p}(k) \right)$.

Definition 2.5 Neighborhood set of any particular LA_i, denoted by $N(i)$, is defined as the set of all learning automata residing in the adjacent cells of the cell c_i, that is,

$$N(i) = \left\{ LA_j | \{i,j\} \in E \right\}. \tag{2.3}$$

Let N_i be the cardinality of $N(i)$.

Definition 2.6 The average penalty for action r of learning automaton LA_i for configuration $\underline{p} \in \mathsf{K}$ is defined as

$$d_{ir}\left(\underline{p} \right) = \mathrm{E}\left[\beta_i | \underline{p}, \alpha_i = r \right] = \sum_{y_{j_1}, \ldots, y_{j_{\mathsf{N}_i}}} \mathsf{F}^i \left(y_{j_1}, \ldots, y_{j_{\mathsf{N}_i}}, r \right) \prod_{LA_l \in N(i)} p_{l y_{j_l}}, \tag{2.4}$$

and the average penalty for the learning automaton LA_i is defined as

$$\begin{aligned} D_i\left(\underline{p} \right) &= \mathrm{E}\left[\beta_i | \underline{p} \right] \\ &= \sum_y d_{iy}\left(\underline{p} \right) p_{iy}. \end{aligned} \tag{2.5}$$

The above definition implies that if the learning automaton LA_j is not a neighboring learning automaton for LA_i, then $d_{ir}\left(\underline{p} \right)$ does not depend on $\underline{p}_j$. We assume that $d_{ir}\left(\underline{p} \right) \neq 0$ for all i, r and $\underline{p}$, that is, in any configuration, any action has a non-zero chance of receiving penalty.

Definition 2.7 The total average penalty for the *ICLA* at configuration $\underline{p} \in \mathsf{K}$ is the sum of the average penalties for all learning automata in the *ICLA*, that is,

$$D\left(\underline{p} \right) = \sum_i D_i\left(\underline{p} \right). \tag{2.6}$$

2.2.2 *Behavior of ICLA*

In this section, we will study the asymptotic behavior of the *ICLA*, in which all
learning automata use the L_{RP} learning algorithm (Narendra and Thathachar 1989),
when operating within an S-model environment. We refer to such an *ICLA* as the
ICLA with SL_{RP} learning automata hereafter. The process $\left\{\underline{p}(k)\right\}_{k \geq 0}$ which
evolves according to the L_{RP} learning algorithm can be described by the following
difference equation:

$$\underline{p}(k+1) = \underline{p}(k) + \underline{a} \cdot \underline{g}\left(\underline{p}(k), \underline{\beta}(k)\right), \tag{2.7}$$

where $\underline{\beta}(k)$ is composed of components $\beta_{iy}(k)$ (for $1 \leq i \leq n, 1 \leq y \leq m_i$, and
$\beta_{iy_1}(k) = \beta_{iy_2}(k)$ for $\forall\, 1 \leq y_1, y_2 \leq m_i$), which are dependent on $\underline{p}(k) \cdot \underline{a}$ is an $n \times n$
diagonal matrix with $a_{ii} = a_i$ and a_i represents the learning parameter for learning
automaton LA_i. $\underline{g}$ represents the learning algorithm, whose components can be
obtained using L_{RP} learning algorithm in S-model environment as follows:

$$\underline{g}_{ir}(p_{ir}, \beta_{ir}) = \begin{cases} (1 - p_{ir} - \beta_{ir}); & \alpha_i = r \\ \left(\frac{\beta_{ir}}{m_i-1} - p_{ir}\right); & \alpha_i \neq r \end{cases} . \tag{2.8}$$

From Eq. (2.7) it follows that $\left\{\underline{p}(k)\right\}_{k \geq 0}$ is a discrete-time Markov process
(Isaacson and Madsen 1976) defined on the state space K [given by Eq. (2.2)]. Let
(K, d) be a metric space, where d is the metric defined according to Eq. (2.9) as
given below.

$$\mathsf{d}\left(\underline{p}, \underline{q}\right) = \sum_i \left\|\underline{p}_i - \underline{q}_i\right\|, \tag{2.9}$$

where $\|\underline{X}\|$ stands for the norm of the vector $\underline{X}$.

Lemma 2.2 and Lemma 2.3, given below, state some properties of the
Markovian process given by Eq. (2.7).

Lemma 2.2 *The Markovian process given by Eq. (2.7) is strictly distance
diminishing.*

Proof To prove this lemma, we will show that the Markovian process given by
Eq. (2.7) follows the definition of the strictly distance diminishing processes given
by Norman (1968). Let $ES = \left\{\underline{\alpha} \middle| \underline{\alpha} = \left(\underline{\alpha}_1^T, \underline{\alpha}_1^T, \ldots, \underline{\alpha}_n^T\right)^T\right\}$ be the event set which
causes the evolution of the state $\underline{p}(k)$. The evolution of the state $\underline{p}(k)$ is dependent
on the occurrence of an associated event $ES(k)$. Thus

$$\underline{p}(k+1) = f_{ES(k)}\left(\underline{p}(k)\right). \tag{2.10}$$

where $f_{ES(k)}$ is defined according to Eq. (2.7). Let $\Pr\left[ES(k) = e \middle| \underline{p}(k) = \underline{p}\right] = \phi_e\left(\underline{p}\right)$ where $\phi_e\left(\underline{p}\right)$ is a real-valued function on $ES \times \mathsf{K}$. Now define $m(\phi_e)$ and $\mu(f_e)$ as

$$m(\phi_e) = \sup_{\underline{p} \neq \underline{p}'} \frac{\left|\phi_e\left(\underline{p}\right) - \phi_e\left(\underline{p}'\right)\right|}{\mathsf{d}\left(\underline{p},\underline{p}'\right)} \tag{2.11}$$

and

$$\mu(f_e) = \sup_{\underline{p} \neq \underline{p}'} \frac{\mathsf{d}\left(f_e\left(\underline{p}\right), f_e\left(\underline{p}'\right)\right)}{\mathsf{d}\left(\underline{p},\underline{p}'\right)}, \tag{2.12}$$

whether or not these are finite. Following propositions are hold:

- ES is a finite set.
- (K, d) is a metric space.
- (K, d) is compact.
- $m(\phi_e) < \infty$ for all $e \in ES$.
- $\mu(f_e) < 1$ for all $e \in ES$. To see this, consider $\underline{p}$ and $\underline{q}$ as two states of the process $\left\{\underline{p}(k)\right\}$. From Eqs. (2.8) and (2.9) we have

$$\mu(f_e) = \frac{\mathsf{d}\left(f_e\left(\underline{p}\right), f_e\left(\underline{q}\right)\right)}{\mathsf{d}\left(\underline{p},\underline{q}\right)} = \frac{\sum_i \left\|f_{e_i}\left(\underline{p}_i\right) - f_{e_i}\left(\underline{q}_i\right)\right\|}{\sum_i \left\|\underline{p}_i - \underline{q}_i\right\|}$$

$$= \frac{\sum_i (1 - a_i) \cdot \left\|\underline{p}_i - \underline{q}_i\right\|}{\sum_i \left\|\underline{p}_i - \underline{q}_i\right\|}. \tag{2.13}$$

Since $0 < a_i < 1$, $\forall i$ it follows that $\mu(f_e) < 1$.

Therefore, and according to the definition of the distance diminishing processes given by Norman (1968), Markovian process given by Eq. (2.7) is strictly distance diminishing. ∎

Corollary 2.1 *Let $\underline{p}^{(h)}$ denotes $\underline{p}(k+h)$ when $\underline{p}(k) = \underline{p}$ and $\underline{q}^{(h)}$ denotes $\underline{p}(k+h)$ when $\underline{p}(k) = \underline{q}$, then $\underline{p}^{(h)} \to \underline{q}^{(h)}$ as $h \to \infty$ irrespective of the initial configurations $\underline{p}$ and $\underline{q}$.*

Proof From Lemma 2.2 It follows that

$$\mathsf{d}\left(\underline{p}^{(h)}, \underline{q}^{(h)}\right) = \sum_i (1 - a_i)^h \cdot \left\| \underline{p}_i - \underline{q}_i \right\|. \tag{2.14}$$

Right hand side of Eq. (2.14) tends to zero as $h \to \infty$. Hence $\underline{p}^{(h)} \to \underline{q}^{(h)}$ irrespective of the initial configurations $\underline{p}$ and $\underline{q}$.   ∎

Lemma 2.3 *The Markovian process given by* Eq. (2.7) *is ergodic.*

Proof To prove the lemma we can see that the Markovian process given by Eq. (2.7) has the following two properties:

- There are no absorbing states for $\left\{ \underline{p}(k) \right\}$, since there is no $\underline{p}$ that satisfies $\underline{p}(k+1) = \underline{p}(k)$.
- The proposed process is strictly distance diminishing (Lemma 2.2).

From the above two properties and considering the results given in Corollary 2.1, we can conclude that the Markovian process $\left\{ \underline{p}(k) \right\}_{k \geq 0}$ is ergodic.   ∎

Now define

$$\Delta \underline{p}(k) = \mathrm{E}\left[\underline{p}(k+1) \middle| \underline{p}(k) \right] - \underline{p}(k). \tag{2.15}$$

Since $\left\{ \underline{p}(k) \right\}_{k \geq 0}$ is Markovian and $\underline{\beta}(k)$ depends only on $\underline{p}(k)$ and not on k explicitly, then $\Delta \underline{p}(k)$ can be expressed as a function of $\underline{p}(k)$. Hence we can write

$$\Delta \underline{p} = a \underline{f}\left(\underline{p} \right). \tag{2.16}$$

The components of $\Delta \underline{p}$ can be obtained as follows:

$$\Delta p_{ir} = a_i p_{ir} \cdot [1 - p_{ir} - \mathrm{E}[\beta_{ir}]] - a_i \sum_{y \neq r} p_{iy} \cdot \left[\frac{1}{m_i - 1} \mathrm{E}[\beta_{iy}] - p_{ir} \right]$$

$$= a_i \cdot \left[\frac{1}{m_i - 1} \sum_{y \neq r} p_{iy} \mathrm{E}[\beta_{iy}] - p_{ir} \mathrm{E}[\beta_{ir}] \right] \tag{2.17}$$

$$= a_i \cdot \left[\frac{1}{m_i - 1} \sum_{y \neq r} p_{iy} d_{iy}\left(\underline{p}\right) - p_{ir} d_{ir}\left(\underline{p}\right) \right]$$

$$= a_i f_{ir}\left(\underline{p}\right),$$

where

$$f_{ir}\left(\underline{p}\right) = \frac{1}{m_i - 1} \sum_{y \neq r} p_{iy} d_{iy}\left(\underline{p}\right) - p_{ir} d_{ir}\left(\underline{p}\right)$$

$$= \frac{1}{m_i - 1} \sum_{y} p_{iy} d_{iy}\left(\underline{p}\right) - \left(1 + \frac{1}{m_i - 1}\right) \cdot \left[p_{ir} d_{ir}\left(\underline{p}\right) \right] \tag{2.18}$$

$$= \frac{1}{m_i - 1} \cdot \left[D_i\left(\underline{p}\right) - m_i p_{ir} d_{ir}\left(\underline{p}\right) \right].$$

Lemma 2.4 *Function* $\underline{f}\left(\underline{p}(k)\right)$ *whose components are given by* Eq. (2.18) *is Lipschitz continuous over the compact space* $\mathbf{K}$.

Proof Function $\underline{f}\left(\underline{p}\right)$ has compact support (it is defined over $\mathbf{K}$), is bounded because $-1 \leq f_{ir}\left(\underline{p}\right) \leq 1$ for all $\underline{p}$, i, and is also continuously differentiable with respect to $\underline{p}$ over $\mathbf{K}$. Therefore, its first derivative with respect to $\underline{p}$ is also bounded. Thus, using the Cauchy's mean value theorem, it can be concluded that $\underline{f}\left(\underline{p}\right)$ is Lipschitz continuous over the compact space $\mathbf{K}$ with Lipschitz constant $\mathrm{K} = \sup_{\underline{p}} \left\| \nabla_{\underline{p}} \underline{f}\left(\underline{p}\right) \right\|$. ∎

For different values of $\underline{a}$, Eq. (2.7) generates a different process and we shall use $\underline{p}^{\underline{a}}(k)$ to denote this process whenever the value of $\underline{a}$ is to be specified explicitly. To find the approximating ODE for the learning algorithm given by (2.7), we define a sequence of continuous-time interpolation of (2.7), denoted by $\tilde{\underline{p}}^{\underline{a}}(t)$ and called an *interpolated process*, whose components are defined by:

$$\tilde{p}_i^a(t) = \underline{p}_i(k), \quad t \in [ka_i, (k+1)a_i), \tag{2.19}$$

where a_i is the learning parameter of the L_{RP} algorithm for learning automaton LA_i. The interpolated process $\left\{ \tilde{\underline{p}}^{\underline{a}}(t) \right\}_{t \geq 0}$ is a sequence of random variables that takes values from $\mathsf{R}^{m_1 \times \cdots \times m_N}$, where $\mathsf{R}^{m_1 \times \cdots \times m_N}$ is the space of all functions that, at each point, are continuous on the right and have a limit on the left over $[0, \infty)$ and take values in K, which is a bounded subset of $\mathsf{R}^{m_1 \times \cdots \times m_n}$. Consider the following ordinary differential equation (ODE):

$$\dot{\underline{p}} = \underline{f}\left(\underline{p}\right), \tag{2.20}$$

where $\dot{\underline{p}}$ is composed of the following components:

$$\frac{dp_{ir}}{dt} = \frac{1}{m_i - 1} \cdot \left[D_i\left(\underline{p}\right) - m_i p_{ir} d_{ir}\left(\underline{p}\right) \right]. \tag{2.21}$$

In the following theorem, we will show that Eq. (2.7) is the approximation to the ODE (2.20). This means that if we have the solution to (2.20), then we can obtain information regarding the behavior of $\underline{p}(k)$.

Theorem 2.1 *Using the learning algorithm* (2.8) *and considering* $\max\{\underline{a}\} \to 0, \underline{p}(k)$ *is well approximated by the solution of the ODE* (2.20).

Proof Following conditions are satisfied by the learning algorithm given by Eq. (2.8):

- $\left\{\underline{p}(k)\right\}_{k \geq 0}$ is a Markovian process.
- Given $\underline{p}(k), \underline{\alpha}(k)$ and $\underline{\beta}(k)$ are independent of $\underline{\alpha}(k-1)$ and $\underline{\beta}(k-1)$.
- $\underline{f}\left(\underline{p}(k)\right)$ is independent of k.
- $\underline{f}\left(\underline{p}(k)\right)$ is Lipshitz continuous over the compact space K (Lemma 2.4) .
- $E\left\|\underline{g}\left(\underline{p}\right) - \underline{f}\left(\underline{p}\right)\right\|^2$ is bounded since $\underline{g}\left(\underline{p}\right) \in [-1, 1]^{m_1 \times \cdots \times m_n}$.
- Learning parameters a_i, $i = 1, \ldots, n$ are sufficiently small since $\max\{\underline{a}\} \to 0$.

Therefore, using theorem (A.1) in Thathachar and Sastry (2004), we can conclude the theorem.   ∎

Equation (2.7) is the so called Euler approximation to the ODE (2.20). Specifically, if $\underline{p}(k)$ is a solution to (2.7) and $\tilde{\underline{p}}(t)$ is a solution to (2.20), then for any $T > 0$, we have

$$\lim_{a_i \to 0} \sup_{0 \leq k \leq T/a_i} \left\| \underline{p}_{ir}(k) - \tilde{\underline{p}}(ka_i) \right\| = 0. \tag{2.22}$$

What Theorem 2.1 says is that $\underline{p}(k)$, given by Eq. (2.7), will closely follow the solution of the ODE (2.20), that is, $\underline{p}(k)$ can be made to closely approximate the solution of its approximating ODE by taking $\max\{\underline{a}\}$ sufficiently small. Thus, if the ODE (2.20) has a globally asymptotically stable equilibrium point, then we can conclude that (by taking $\max\{\underline{a}\}$ sufficiently small), $\underline{p}(k)$, for large k, would be close to this equilibrium point irrespective of its initial configuration $\underline{p}(0)$. Therefore, the analysis of process $\left\{\underline{p}(k)\right\}_{k \geq 0}$ is done in two stages. In the first stage, we solve ODE (2.20) and in the second stage, we characterize the solution of this ODE.

In the following subsections, we first find the equilibrium points of ODE (2.20), then study the stability property of these equilibrium points, and finally state some theorems about the convergence of *ICLA*.

2.2.2.1   Equilibrium Points

To find the equilibrium points of ODE (2.20), we first show that this ODE has at least one equilibrium point and then specify a set of conditions which must be satisfied by a configuration $\underline{p}$ to be an equilibrium point of the ODE (2.20).

Lemma 2.5 *ODE (2.20) has at least one equilibrium point.*

Proof To prove this lemma, we first propose a continuous mapping $\zeta\left(\underline{p}\right)$ from **K** to **K**. Then, using the Brouwer's fixed point theorem, we will show that any continuous mapping from **K** to **K** has at least one fixed point. Finally, we will show that the fixed point of $\zeta\left(\underline{p}\right)$ is the equilibrium point of the ODE (2.20). This indicates that ODE (2.20) has at least one equilibrium point. Let $\zeta\left(\underline{p}\right) = \underline{a}f\left(\underline{p}\right) + \underline{p}$ where matrix $\underline{a}$ is equivalent to the one given in Eq. (2.7). Components of $\zeta\left(\underline{p}\right)$ can be obtained as follows:

$$\zeta_{ir}\left(\underline{p}\right) = \frac{a_i}{m_i - 1} \cdot \left[D_i\left(\underline{p}\right) - m_i p_{ir} d_{ir}\left(\underline{p}\right)\right] + p_{ir}. \tag{2.23}$$

It is easy to verify that $0 \leq \zeta_{ir}\left(\underline{p}\right) < 1$ for all i and r. Thus, $\zeta\left(\underline{p}\right)$ is a continuous mapping from **K** to **K**. Since **K** is closed, bounded, and convex (Lemma 2.1), we can use the Brouwer's fixed point theorem to show that $\zeta\left(\underline{p}\right)$ has at least one fixed point. Let $\underline{p}^*$ be a fixed point of $\zeta\left(\underline{p}\right)$, thus we have

$$\zeta\left(\underline{p}^*\right) = \underline{p}^*, \tag{2.24}$$

or equivalently

$$\underline{a}f\left(\underline{p}^*\right) + \underline{p}^* = \underline{p}^*. \tag{2.25}$$

Since $\underline{a}$ is a diagonal matrix with no zero elements on its main diagonal, it can be concluded from (2.25) that

$$\underline{f}\left(\underline{p}^*\right) = \underline{0}. \tag{2.26}$$

Since every point $\underline{p}^*$, that satisfies $\underline{f}\left(\underline{p}^*\right) = \underline{0}$, is an equilibrium point of ODE (2.20), we can conclude that ODE (2.20) has at least one equilibrium point.   ∎

Theorem 2.2 *The equilibrium points of ODE* (2.20) *are the set of configurations* $\underline{p}^*$ *which satisfy the set of conditions* $p_{ir}^* = \dfrac{\prod_{y \neq r} d_{iy}\left(\underline{p}^*\right)}{\sum_{y_1}\left(\prod_{y_2 \neq y_1} d_{iy_2}\left(\underline{p}^*\right)\right)}$ *for all* i, r.

Proof To find the equilibrium points of ODE (2.20), we have to solve equations of the from:

$$\frac{dp_{ir}^*(t)}{dt} = 0 \quad \text{for all } i, r. \tag{2.27}$$

Using Eq. (2.21), (2.27) can be rewritten as

$$\frac{1}{m_i - 1} \cdot \left[D_i\left(\underline{p}^*(t)\right) - m_i p_{ir}^*(t) d_{ir}\left(\underline{p}^*(t)\right)\right] = 0, \quad \text{for all } i, r. \tag{2.28}$$

These equations have solutions of the form

$$p_{ir}^*(t) = \frac{D_i\left(\underline{p}^*(t)\right)}{m_i d_{ir}\left(\underline{p}^*(t)\right)}, \quad \forall i, r, \tag{2.29}$$

which after some algebraic manipulations, can be rewritten as

$$p_{ir}^* = \frac{\prod_{y \neq r} d_{iy}\left(\underline{p}^*\right)}{\sum_{y_1}\left(\prod_{y_2 \neq y_1} d_{iy_2}\left(\underline{p}^*\right)\right)}, \quad \text{for all } i, r, \tag{2.30}$$

and hence the theorem.   ∎

It follows from Theorem 2.2 that the difference equation given by (2.16) has equilibrium points $\underline{p}^*$ that satisfy the set of conditions $p_{ir}^*(k) =$

$$\frac{\prod_{y \neq r} d_{iy}(\underline{p}^*(k))}{\sum_{y_1}\left(\prod_{y_2 \neq y_1} d_{iy_2}(\underline{p}^*(k))\right)} \text{ for all } i, r.$$

Lemma 2.6 *If at an equilibrium point $\underline{p}^*$ of the ODE (2.20), the average penalties for all actions of learning automaton LA_i are equal, then LA_i becomes a pure-chance automaton.*

Proof By Theorem 2.2, if $\underline{p}^*$ is an equilibrium point of ODE (2.20), then $p_{ir}^* =$

$$\frac{\prod_{y \neq r} d_{iy}(\underline{p}^*)}{\sum_{y_1}\left(\mathrm{II}_{y_2 \neq y_1} d_{iy_2}(\underline{p}^*)\right)} \text{ for all } i, r. \text{ Thus, if the average penalties for all actions of the}$$

learning automaton LA_i are equal, that is, $d_{ir}\left(\underline{p}^*\right) = d_{ir'}\left(\underline{p}^*\right), \forall r, r'$, then $p_{ir}^* = \frac{1}{m_i}, \forall r$ which implies that the learning automaton LA_i is a pure-chance automaton. ∎

Lemma 2.7 *If $\underline{p}^*$ is an equilibrium point of ODE (2.20), then its components p_{ir}^* are bounded from below to $\frac{D_i(\underline{p}^*)}{m_i}$ and from above to $\frac{D_i(\underline{p}^*)}{m_i} + \left(1 - D_i\left(\underline{p}^*\right)\right)$.*

Proof According to Eq. (2.29), if $\underline{p}^*$ is an equilibrium point of ODE (2.20), then $p_{ir}^* = \frac{D_i(\underline{p}^*)}{m_i d_{ir}(\underline{p}^*)}$, for all i, r. p_{ir}^* attains its minimum value when $d_{ir}\left(\underline{p}^*\right) = 1$, that is, p_{ir}^* is bounded from below to $\frac{D_i(\underline{p}^*)}{m_i}$. Since for every i, $\sum_y p_{iy}^* = 1, p_{ir}^*$ attains its maximum value when $p_{iy}^* = \frac{D_i(\underline{p}^*)}{m_i}$, for every $y \neq r$, that is,

$$\max\left(p_{ir}^*\right) = 1 - \sum_{y \neq r} \frac{D_i\left(\underline{p}^*\right)}{m_i}$$

$$= \frac{D_i\left(\underline{p}^*\right)}{m_i} + \left(1 - D_i\left(\underline{p}^*\right)\right),$$

(2.31)

and hence the lemma. ∎

2.2.2.2   Stability Property

In this subsection, we characterize the stability of the equilibrium configurations of
ICLA, that is, the equilibrium points of ODE (2.20). To do this, the origin is first
transferred to an equilibrium point $\underline{p}^*$, and then a candidate for a Lyapunov function
is introduced for studying the stability of this equilibrium point. Consider the
following transformation

$$\hat{p}_{ir} = p_{ir} - \frac{D_i\left(\underline{p}^*\right)}{m_i d_{ir}\left(\underline{p}^*\right)} \quad \text{for all } i, r. \tag{2.32}$$

Using this transformation, the origin is transferred to $\underline{p}^*$.

Lemma 2.8 *Derivative of $\hat{\underline{p}}$ with respect to time has components of the following
form*:

$$\frac{d\hat{p}_{ir}}{dt} = -d_{ir}\left(\underline{p}\right)\hat{p}_{ir} \quad \text{for all } i, r. \tag{2.33}$$

Proof Let

$$\delta_{ir}^* = \frac{D_i\left(\underline{p}^*\right)}{m_i d_{ir}\left(\underline{p}^*\right)}. \tag{2.34}$$

Using Eqs. (2.32) and (2.34), components of the derivative of $\hat{\underline{p}}$ with respect to
time can be given as

$$\frac{d\hat{p}_{ir}}{dt} = \frac{dp_{ir}}{dt} + \frac{d\delta_{ir}^*}{dt} = \frac{dp_{ir}}{dt} \quad \text{for all } i, r. \tag{2.35}$$

Using Eqs. (2.21) and (2.34), Eq. (2.35) can be rewritten as

$$\frac{d\hat{p}_{ir}}{dt} = \frac{1}{m_i - 1} \cdot \left[D_i\left(\underline{p}\right) - m_i \cdot \left(\hat{p}_{ir} + \delta_{ir}^*\right) \cdot d_{ir}\left(\underline{p}\right)\right]. \tag{2.36}$$

Equation (2.36) is valid for all configurations $\hat{\underline{p}}$ including $\hat{\underline{q}}$ in which $\hat{q}_{ir} = 0$ for
a particular action r of the ith learning automaton. For this configuration, it is easy
to verify that $\frac{d\hat{q}_{ir}}{dt} = 0$.

Next, use Eq. (2.5) to rewrite Eq. (2.36) as follows:

$$\frac{d\hat{p}_{ir}}{dt} = \frac{1}{m_i - 1} \cdot \left[d_{ir}\left(\underline{p}\right) \cdot (1 - m_i) \cdot \left(\hat{p}_{ir} + \delta_{ir}^*\right) + \sum_{y \neq r} \left(\hat{p}_{iy} + \delta_{iy}^*\right) d_{iy}\left(\underline{p}\right) \right] \quad \forall i, r. \tag{2.37}$$

Equation (2.37) is also valid for all configurations $\hat{p}$ including $\hat{q}$. Thus, we have

$$\frac{d\hat{q}_{ir}}{dt} = \frac{1}{m_i - 1} \cdot \left[d_{ir}\left(\underline{q}\right) \cdot (1 - m_i) \cdot \delta_{ir}^* + \sum_{y \neq r} \left(\hat{q}_{iy} + \delta_{iy}^*\right) d_{iy}\left(\underline{q}\right) \right] = 0, \tag{2.38}$$

from which it immediately follows that

$$d_{ir}\left(\underline{q}\right) \cdot (1 - m_i) \cdot \delta_{ir}^* + \sum_{y \neq r} \left(\hat{q}_{iy} + \delta_{ir}^*\right) d_{iy}\left(\underline{q}\right) = 0. \tag{2.39}$$

Since none of the terms in Eq. (2.39) depends on the value of $\hat{q}_{ir}$, this equation is valid for all configurations $\hat{p}$, that is,

$$d_{ir}\left(\underline{p}\right) \cdot (1 - m_i) \cdot \delta_{ir}^* + \sum_{y \neq r} \left(\hat{p}_{iy} + \delta_{iy}^*\right) d_{iy}\left(\underline{p}\right) = 0, \quad \text{for all } \hat{p}, i, r. \tag{2.40}$$

Using Eq. (2.40) in Eq. (2.37) we get

$$\frac{d\hat{p}_{ir}}{dt} = -d_{ir}\left(\underline{p}\right)\hat{p}_{ir} \quad \text{for all } i, r, \tag{2.41}$$

and hence the lemma.   ∎

Corollary 2.2 $\hat{p}_{ir}$ *and its time derivative* $\frac{d\hat{p}_{ir}}{dt}$ *have different signs.*

Proof Considering the fact that $d_{ir}\left(\underline{p}\right) \geq 0$ for all configurations $\underline{p}$ and for all i and r, the proof is an immediate result of Eq. (2.33).   ∎

Theorem 2.3 *A configuration* p^* *whose components satisfy the set of equations*

$$p_{ir}^* = \frac{\prod_{y \neq r} d_{iy}(\underline{p}^*)}{\sum_{y_1} \left(\prod_{y_2 \neq y_1} d_{iy_2}(\underline{p}^*) \right)} \text{ for all } i, r, \text{ is an asymptotically stable equilibrium point}$$

of ODE (2.20) over **K**.

Proof Apply the transformation (2.32) to transfer the origin to p^*. Now consider the following positive definite Lyapunov function:

$$V\left(\hat{\underline{p}}\right) = -\sum_i \sum_y \hat{p}_{iy} \cdot \ln\left(1 - \hat{p}_{iy}\right). \qquad (2.42)$$

$V\left(\hat{\underline{p}}\right) \geq 0$ for all configurations $\hat{\underline{p}}$ and is zero only when $\hat{p}_{ir} = 0$ for all i and r. Time derivative of $V(.)$ can be expressed as

$$\dot{V}\left(\hat{\underline{p}}\right) = -\sum_i \sum_y \frac{d\hat{p}_{iy}}{dt} \cdot \upsilon\left(\hat{p}_{iy}\right), \qquad (2.43)$$

where

$$\upsilon\left(\hat{p}_{iy}\right) = \left[\frac{\ln\left(1 - \hat{p}_{iy}\right) \cdot \left(1 - \hat{p}_{iy}\right) - \hat{p}_{iy}}{1 - \hat{p}_{iy}}\right]. \qquad (2.44)$$

Considering the value of $\hat{\underline{p}}_{iy}$, following three cases may arise for each term of this derivative:

$$\begin{cases} 0 < \hat{p}_{iy} \leq 1; & Case\,h_1 \\ -1 \leq \hat{p}_{iy} < 0; & Case\,h_2 \\ \hat{p}_{iy} = 0; & Case\,h_3 \end{cases}$$

- *Case h_1*: If $0 < \hat{p}_{iy} \leq 1$, then $\upsilon\left(\hat{p}_{iy}\right) < 0$ and $\frac{d\hat{p}_{iy}}{dt} < 0$ (Corollary 2.2). Therefore, $\frac{d\hat{p}_{iy}}{dt} \cdot \upsilon\left(\hat{p}_{iy}\right) > 0$.
- *Case h_2*: If $-1 \leq \hat{p}_{iy} < 0$, then $\upsilon\left(\hat{p}_{iy}\right) > 0$ and $\frac{d\hat{p}_{iy}}{dt} > 0$ (Corollary 2.2). Therefore, $\frac{d\hat{p}_{iy}}{dt} \cdot \upsilon\left(\hat{p}_{iy}\right) > 0$.
- *Case h_3*: If $\hat{p}_{iy} = 0$, then $\frac{d\hat{p}_{iy}}{dt} \cdot \upsilon\left(\hat{p}_{iy}\right) = 0$.

Thus, $\dot{V}\left(\hat{\underline{p}}\right) \leq 0$, for all configurations $\hat{\underline{p}}$ and is zero only when $\hat{p}_{ir} = 0$ for all i and r. Therefore, using the Lyapunov theorems for autonomous systems, it can be proved that p^* is an asymptotically stable equilibrium point of ODE (2.20) over K. ∎

Corollary 2.3 *The equilibrium point of ODE* (2.20) *is unique over* K.

Proof Let p^*, q^* be two equilibrium points of ODE (2.20). Theorem 2.3 proves that any equilibrium point of ODE (2.20), including p^*, is asymptotically stable over K. This means that all initial configurations within the state space K converge to p^*. Using a similar approach for q^*, one can conclude that all initial configurations within the state space K converge to q^*. This implies $p^* = q^*$, and thus, the equilibrium point of ODE (2.20) is unique over K.   ∎

2.2.2.3 Convergence Results

In this section, we summarize the main results specified in the above lemmas and theorems in a main theorem (Theorem 2.4).

Theorem 2.4 *An ICLA with SL_{RP} learning automata, regardless of its initial configuration, converges in distribution to a random configuration, in which the mean value of the action probability of any action of any LA is inversely proportional to the average penalty received by that action.*

Proof The evolution of an *ICLA* with SL_{RP} learning automata is described by Eq. (2.7). From this equation, it follows that $\left\{\underline{p}(k)\right\}_{k \geq 0}$ is a discrete-time Markov process. Lemma 2.3 states that this Markovian process is ergodic, and hence, it converges in distribution to a random configuration $\underline{p}^*$, irrespective of its initial configuration. Lemma 2.5 shows that such a configuration exists for the *ICLA*, Corollary 2.3 states that it is unique, and Theorem 2.3 proves that it is asymptotically stable. Theorem 2.2 specifies the properties of the configuration $\underline{p}^*$. It shows that $\underline{p}^*$ satisfies the set of conditions given by (2.30). According to Eq. (2.30), in configuration $\underline{p}^*$, the action probability of any action of any *LA* is inversely proportional to the average penalty received by that action. ∎

2.2.3 Expediency of ICLA

In this subsection, we introduce the concept of expediency of *ICLA* and specify the set of conditions under which an *ICLA* becomes expedient or totally expedient.

Definition 2.8 A pure-chance automaton is an automaton that chooses each of its actions with equal probability i.e., by pure chance, that is, an m-action automaton is pure-chance if $p_i = \frac{1}{m}$, $i = 1, 2, \ldots, m$.

Definition 2.9 A pure-chance *ICLA* is an *ICLA*, for which every cell contains a pure-chance automaton rather than a learning automaton. The configuration of a pure-chance *ICLA* is denoted by $\underline{p}^{pc}$.

Definition 2.10 An *ICLA* is said to be expedient with respect to the cell c_i if $\lim_{k \to \infty} \underline{p}(k) = \underline{p}^*$ exists and the following inequality holds:

$$\lim_{k \to \infty} E\left[D_i\left(\underline{p}(k)\right)\right] < \frac{1}{m_i} \sum_y d_{iy}\left(\underline{p}^*\right). \tag{2.45}$$

In other words, an *ICLA* is expedient with respect to the cell c_i if, in the long run, the ith learning automaton performs better (receives less penalty) than a pure-chance automaton.

Definition 2.11 An *ICLA* is said to be expedient if it is expedient with respect to every cell in *ICLA*.

Theorem 2.5 *An ICLA with SL_{RP} learning automata, regardless of the local rule being used, is expedient.*

Proof We have to show that an *ICLA* with SL_{RP} learning automata is expedient with respect to all of its cells, that is, $\lim_{k\to\infty} \underline{p}(k) = \underline{p}^*$ exists and the following inequality holds:

$$\lim_{k\to\infty} \mathrm{E}\left[D_i\left(\underline{p}(k)\right)\right] < \frac{1}{m_i}\sum_y d_{iy}\left(\underline{p}^*\right), \quad \text{for every } i \tag{2.46}$$

As it was shown in Theorem 2.4, for an *ICLA* with SL_{RP} learning automata, $\lim_{k\to\infty} \underline{p}(k) = \underline{p}^*$ exists. Thus, we only have to show that the inequality (2.47) holds. Using Eq. (2.5), the left hand side of this inequality can be rewritten as

$$\begin{aligned}
\lim_{k\to\infty} \mathrm{E}\left[D_i\left(\underline{p}(k)\right)\right] &= \lim_{k\to\infty} \mathrm{E}\left[\sum_y d_{iy}\left(\underline{p}(k)\right)p_{iy}(k)\right] \\
&= \sum_y \lim_{k\to\infty} \mathrm{E}\left[d_{iy}\left(\underline{p}(k)\right)p_{iy}(k)\right].
\end{aligned} \tag{2.47}$$

Since $d_{iy}\left(\underline{p}(k)\right)$ and $p_{iy}(k)$ are independent, (2.47) can be simplified to

$$\lim_{k\to\infty} \mathrm{E}\left[D_i\left(\underline{p}(k)\right)\right] = \sum_y \left(\lim_{k\to\infty} \mathrm{E}\left[d_{iy}\left(\underline{p}(k)\right)\right] \cdot \lim_{k\to\infty} \mathrm{E}[p_{iy}(k)]\right). \tag{2.48}$$

From Theorem 2.4, we have $\lim_{k\to\infty} \mathrm{E}\left[\underline{p}(k)\right] = \underline{p}^*$ and hence $\lim_{k\to\infty} \mathrm{E}[p_{ir}(k)] = p_{ir}^*$ for all i and r. Using Eq. (2.4), $\lim_{k\to\infty} \mathrm{E}\left[d_{ir}\left(\underline{p}(k)\right)\right]$ can be computed as follows:

$$\lim_{k \to \infty} \mathrm{E}\left[d_{ir}\left(\underline{p}(k)\right)\right]$$

$$= \lim_{k \to \infty} \mathrm{E}\left[\sum_{y_{j_1}, \dots, y_{j_{N_i}}} \mathsf{F}^i\left(y_{j_1}, y_{j_2}, \dots, y_{j_{N_i}}, r\right) \prod_{LA_l \in N(i)} p_{ly_{j_l}}(k)\right]$$

$$= \sum_{y_{j_1}, \dots, y_{j_{N_i}}} \mathsf{F}^i\left(y_{j_1}, y_{j_2}, \dots, y_{j_{N_i}}, r\right) \prod_{LA_l \in N(i)} \lim_{k \to \infty} \mathrm{E}\left[p_{ly_{j_l}}(k)\right]$$

$$= d_{ir}\left(\underline{p}^*(k)\right).$$

(2.49)

Thus, we get

$$\lim_{k \to \infty} \mathrm{E}\left[D_i\left(\underline{p}(k)\right)\right] = \sum_y \left(d_{iy}\left(\underline{p}^*\right)p_{iy}^*\right). \tag{2.50}$$

Now, we have to show that:

$$\sum_y d_{iy}\left(\underline{p}^*\right)p_{iy}^* < \frac{1}{m_i}\sum_y d_{iy}\left(\underline{p}^*\right) \quad \text{for every } i. \tag{2.51}$$

Each side of this inequality is a convex combination of $d_{iy}\left(\underline{p}^*\right), y = 1, \dots, m_i$. In the convex combination given on the right hand side of (2.51), weights of all $d_{ir}\left(\underline{p}^*\right)$ are equal to $\frac{1}{m_i}$ whereas in the convex combination given on the left hand side, weight of each $d_{ir}\left(\underline{p}^*\right)$ is inversely proportional to its value, that is, the larger $d_{ir}\left(\underline{p}^*\right)$, the smaller its weight is [considering Eq. (2.29)]. Therefore, the convex combination given on the left hand side of inequality (2.51) is smaller than the one given on the right hand side, and hence the theorem. ∎

2.3 Dynamic Models of Cellular Learning Automata

As it was mentioned before, *ICLA* is suitable for modeling problems which are not regular in nature, such as problems in the area of sensor networks, web mining, and grid computing. The structure of *ICLA* remains fixed during its operation, and therefore, it is not suitable for problem solving in areas with dynamic structures, such as mobile ad hoc and sensor networks. Therefore, in this section, we introduce dynamic models of *CLAs* (Esnaashari and Meybodi 2011, 2013; Saghiri and Meybodi 2017b, c, d) which are better suited for problem solving in environments with dynamic structures.

Dynamic models of *CLAs* are of two main classes, *DICLAs* and *DCLAs*. In *DICLAs,* a set of interests is defined for describing changes in the structure of the *CLA*. In these models, the rule which changes the cellular structure is fixed. Several algorithms based on these models were reported in (Esnaashari and Meybodi 2011, 2013). However, there are applications in which changes in the structure cannot be adequately modeled by a set of interests. Instead, we need to define problem specific rules for changing the cellular structure. In *DCLA*, each cell has a set of attributes and there is a rule called structure-updating rule, which can be used to define problem specific changes in the cellular structure. *DICLAs* and *DCLAs* are defined with more details in the rest of this section.

The rest of this section is organized as follows. In Sects. 2.3.1 and 2.3.2 Dynamic *Irregular CLA* (*DICLA*), and Heterogeneous *DICLA* (*HDICLA*) models will be introduced. Sections 2.3.3 and 2.3.4 will include the definition of the *Closed Asynchronous Dynamic CLA* (*CADCLA*) and *CADCLA* with varying number of LAs in each cell (*CADCLA-VL*).

2.3.1 Dynamic Irregular Cellular Learning Automata

A *DICLA* is defined as an undirected graph in which, each vertex represents a cell and a learning automaton is assigned to every cell (vertex) (Esnaashari and Meybodi 2017). A finite set of interests is defined for *DICLA*. For each cell of *DICLA* a tendency vector is defined whose *j*th element shows the degree of tendency of that cell to the *j*th interest. In *DICLA*, the state of each cell consists of two parts; the action selected by the learning automaton and the tendency vector. Two cells are neighbors in *DICLA* if the distance between their tendency vectors is smaller than or equal to the neighborhood radius.

Like *ICLA*, there is a local rule that *DICLA* operates under. The local rule of *DICLA*, the actions selected by the neighboring *LAs* of any particular learning automaton LA_i determine the followings: (1) the reinforcement signal to the learning automaton LA_i, and (2) the restructuring signal to the cell in which LA_i resides. Restructuring signal is used to update the tendency vector of the cell. Dynamicity of *DICLA* is the result of modifications made to the tendency vectors of its constituting cells. gives a schematic of *DICLA*. A *DICLA* is formally defined in Definition 2.12. A DICLA is schematically depicted in Fig. 2.2.

Definition 2.12 Dynamic irregular cellular learning automaton is a structure $A = \left(G\langle V, E\rangle, \Psi, A, \Phi\langle \alpha, \underline{\psi}\rangle, \tau, F, Z \right)$ where

- *G* is an undirected graph, with *V* as the set of vertices (cells) and *E* as the set of edges (adjacency relations).
- Ψ is a finite set of interests. Cardinality of Ψ is denoted by $|\Psi|$.
- *A* is the set of learning automata each of which is assigned to one cell of *DICLA*.

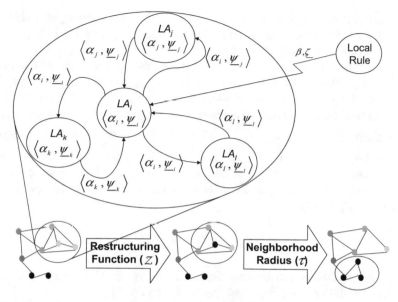

Fig. 2.2 Dynamic irregular cellular learning automaton

- $\Phi < \alpha, \underline{\psi} >$ is the cell state. State of a cell c_i (φ_i) consists of two parts; (1) α_i which is the action selected by the learning automaton of that cell, and (2) A vector $\underline{\psi}_i = \left(\psi_{i1}, \psi_{i2}, \ldots, \psi_{i|\Psi|} \right)^T$ called the tendency vector of the cell. Each element $\psi_{ik} \in [0, 1]$ in the tendency vector of the cell c_i shows the degree of tendency of c_i to the interest $\psi_k \in \Psi$.

- τ is the neighborhood radius. Two cells c_i and c_j of *DICLA* are neighbors if $\left\| \underline{\psi}_i - \underline{\psi}_j \right\| \leq \tau$. In other words, two cells of *DICLA* are neighbors if the distance between their tendency vectors is smaller than or equal to τ.

- $F : \underline{\varphi}_i \rightarrow \left\langle \beta, [0, 1]^{|\Psi|} \right\rangle$ is the local rule of *DICLA* in each cell c_i, where $\underline{\varphi}_i = \left\{ \varphi_j \middle| \left\| \underline{\psi}_i - \underline{\psi}_j \right\| < \tau \right\} \cup \{\varphi_i\}$ is the set of states of all neighbors of c_i, β is the set of values that the reinforcement signal can take, and $[0, 1]^{|\Psi|}$ is a $|\Psi|$-dimensional unit hypercube. From the current states of the neighboring cells of each cell c_i, local rule performs the followings: (1) gives the reinforcement signal to the learning automaton LA_i resides in c_i, and (2) produces a restructuring signal $\left(\underline{\zeta}_i = (\zeta_{i1}, \zeta_{i2}, \ldots, \zeta_{i|\Psi|})^T \right)$ which is used to change the tendency vector of c_i. Each element ζ_{ij} of the restructuring signal is a scalar value within the close interval $[-1, 1]$.

- $Z : [0, 1]^{|\Psi|} \times [-1, 1]^{|\Psi|} \rightarrow [0, 1]^{|\Psi|}$ is the restructuring function which modifies the tendency vector of the cell c_i using the restructuring signal produced by the local rule of the cell.

In what follows, we consider *DICLA* with *N* cells. The learning automaton LA_i which has a finite action set $\underline{\alpha}_i$ is associated to the cell c_i (for $i = 1, 2, \ldots, N$) of *DICLA*. Let the cardinality of $\underline{\alpha}_i$ be m_i.

The operation of *DICLA* takes place as the following iterations. At iteration k, each learning automaton chooses an action. Let $\alpha_i \in \underline{\alpha}_i$ be the action chosen by LA_i. Then, each *LA* receives a reinforcement signal. Let $\beta_i \in \underline{\beta}$ be the reinforcement signal received by LA_i. This reinforcement signal is produced by the application of the local rule $F : \underline{\varphi}_i \rightarrow \left\langle \underline{\beta}, [0, 1]^{|\Psi|} \right\rangle$. Higher values of β_i mean that the selected action of LA_i will receive higher penalties. Each *LA* updates its action probability vector on the basis of the supplied reinforcement signal and the action chosen by the cell. Next, each cell c_i updates its tendency vector using the restructuring function Z [Eq. (2.52)].

$$\underline{\psi}_i(k+1) = Z\left(\underline{\psi}_i(k), \underline{\zeta}_i(k)\right). \tag{2.52}$$

Like *ICLA*, *DICLA* can be either synchronous or asynchronous and an asynchronous *DICLA* can be either time-driven or step-driven.

2.3.1.1 Definitions and Notations

In this section, we give some definitions, which will be used later for the analysis of *DICLA*. Note that most of the definitions used for steady-state analysis of the behavior of the *ICLA* are redefined here for *DICLA* model.

Definition 2.13 A configuration of *DICLA* at step k is denoted by $\left\langle \underline{p}(k), \underline{\psi}(k) \right\rangle = \left\langle \left(\underline{p}_1, \ldots, \underline{p}_N\right)^T, \left(\underline{\psi}_1, \ldots, \underline{\psi}_N\right)^T \right\rangle$, where $\underline{p}_i$ is the action probability vector of the learning automaton LA_i, $\underline{\psi}_i$ is the tendency vector of the cell c_i, and T denotes the transpose operator.

Definition 2.14 A configuration $\left\langle \underline{p}, \underline{\psi} \right\rangle$ is called unit if the action probability vector of each learning automaton and the tendency vector of each cell are unit vectors, otherwise it is called general. Hence, the set of all unit configurations, $\langle K^*, Y^* \rangle$, and the set of all general configurations, $\langle K, Y \rangle$, in *DICLA* are

$$\langle K^*, Y^* \rangle = \left\{ \left\langle \underline{p}, \underline{\psi} \right\rangle \left| \begin{array}{l} \underline{p} = \left(\underline{p}_1, \ldots, \underline{p}_n\right)^T, \underline{p}_i = (p_{i1}, \ldots, p_{im_i})^T, p_{iy} = 0\, or\, 1\, \forall i, y, \sum\limits_y p_{iy} = 1\, \forall i, \\ \underline{\psi} = \left(\underline{\psi}_1, \ldots, \underline{\psi}_n\right)^T, \underline{\psi}_i = \left(\psi_{i1}, \ldots, \psi_{i|\Psi|}\right)^T, \psi_{ij} = 0\, or\, 1\, \forall i, j, \sum\limits_j \psi_{ij} = 1\, \forall i \end{array} \right. \right\},$$

$$\tag{2.53}$$

and

$$\langle K, Y \rangle = \left\{ \langle \underline{p}, \underline{\psi} \rangle \; \middle| \; \begin{array}{l} \underline{p} = \left(\underline{p}_1, \ldots, \underline{p}_n \right)^T, \underline{p}_i = \left(p_{i1}, \ldots, p_{im_i} \right)^T, 0 \le p_{iy} \le 1 \, \forall i, y, \sum_y p_{iy} = 1 \, \forall i, \\ \underline{\psi} = \left(\underline{\psi}_1, \ldots, \underline{\psi}_n \right)^T, \underline{\psi}_i = \left(\psi_{i1}, \ldots, \psi_{i|\Psi|} \right)^T, 0 \le \psi_{ij} \le 1 \, \forall i, j \end{array} \right\},$$

(2.54)

respectively. We refer to K as the probability space and to Y as the tendency space of *DICLA* hereafter.

Lemma 2.9 $\langle K, Y \rangle$ *is the convex hull of* $\langle K^*, Y^* \rangle$.

Proof Proof of this lemma is similar to the proof of lemma 1 given in (Beigy and Meybodi 2004). ∎

The application of the local rule to every cell allows transforming a configuration to a new one. The local rule of *DICLA* (F) consists of two parts; reinforcement signal generator $\left(F_\beta \right)$ and restructuring signal generator (F_ζ).

Definition 2.15 The global behavior of a *DICLA* is a mapping $G : <K, Y> \rightarrow <K, Y>$ that describes the dynamics of *DICLA*. The evolution of *DICLA* from a given initial configuration $\left\langle \underline{p}(0), \underline{\psi}(0) \right\rangle \in \langle K, Y \rangle$ is a sequence of configurations $\left\{ \left\langle \underline{p}(k), \underline{\psi}(k) \right\rangle \right\}_{k \ge 0}$, such that $\left\langle \underline{p}(k+1), \underline{\psi}(k+1) \right\rangle = G\left(\left\langle \underline{p}(k), \underline{\psi}(k) \right\rangle \right)$.

Definition 2.16 Neighborhood set of any particular LA_i in configuration $\left\langle \underline{p}, \underline{\psi} \right\rangle$, denoted by $N^{\underline{\psi}}(i)$, is defined as the set of all learning automata residing in the adjacent cells of the cell c_i, that is,

$$N^{\underline{\psi}}(i) = \left\{ LA_j \middle| \left\| \underline{\psi}_i - \underline{\psi}_j \right\| \le \tau \right\}.$$

(2.55)

Let $N_i^{\underline{\psi}}$ be the cardinality of $N^{\underline{\psi}}(i)$.

Definition 2.17 The average penalty for action r of the learning automaton LA_i for configuration $\left\langle \underline{p}, \underline{\psi} \right\rangle \in \langle K, Y \rangle$ is defined as

$$d_{ir}^\beta \left(\left\langle \underline{p}, \underline{\psi} \right\rangle \right) = E\left[\beta_i \middle| \left\langle \underline{p}, \underline{\psi} \right\rangle, \alpha_i = r \right]$$

$$= \sum_{y_{h_1}^{\underline{\psi}}, \ldots, y_{h_{N_i^{\underline{\psi}}}}^{\underline{\psi}}} \left(F_\beta^i \left(y_{h_1}^{\underline{\psi}}, y_{h_2}^{\underline{\psi}}, \ldots, y_{h_{N_i^{\underline{\psi}}}}^{\underline{\psi}}, r \right) \prod_{\substack{LA_l \in N^{\underline{\psi}}(i) \\ l \ne i}} p_{ly_{h_l}} \right),$$

(2.56)

and the average penalty for the learning automaton LA_i is defined as

$$
\begin{aligned}
D_i\left(\left\langle \underline{p}, \underline{\psi} \right\rangle\right) &= \mathrm{E}\left[\beta_i \middle| \left\langle \underline{p}, \underline{\psi} \right\rangle\right] \\
&= \sum_y d_{iy}^\beta\left(\left\langle \underline{p}, \underline{\psi} \right\rangle\right) p_{iy}.
\end{aligned}
\tag{2.57}
$$

The above definition implies that if the learning automaton LA_j is not a neighboring learning automaton for LA_i, then $d_{ir}^\beta\left(\left\langle \underline{p}, \underline{\psi} \right\rangle\right)$ does not depend on $\underline{p}_j$. We assume that $d_{ir}^\beta\left(\left\langle \underline{p}, \underline{\psi} \right\rangle\right) \neq 0$ for all i, r and $\left\langle \underline{p}, \underline{\psi} \right\rangle$, that is, in any configuration, any action has a non-zero chance of receiving penalty.

Definition 2.18 The total average penalty for *DICLA* at configuration $\left\langle \underline{p}, \underline{\psi} \right\rangle \in \langle \mathsf{K}, \mathsf{Y} \rangle$ is the sum of the average penalties for all learning automata in *DICLA*, that is,

$$
D\left(\left\langle \underline{p}, \underline{\psi} \right\rangle\right) = \sum_i D_i\left(\left\langle \underline{p}, \underline{\psi} \right\rangle\right).
\tag{2.58}
$$

Definition 2.19 The average restructuring signal for interest j of the cell c_i for configuration $\left\langle \underline{p}, \underline{\psi} \right\rangle \in \langle \mathsf{K}, \mathsf{Y} \rangle$ is defined as

$$
d_{ij}^\zeta\left(\left\langle \underline{p}, \underline{\psi} \right\rangle\right) = \sum_r \sum_{y_{h_1}^\psi, \dots, y_{h_{N_i^\psi}}^\psi} \left(\mathsf{F}_\zeta^i\left(y_{h_1}^\psi, y_{h_2}^\psi, \dots, y_{h_{N_i^\psi}}^\psi, r, \underline{\psi}_{*j} \right) \prod_{LA_l \in N_i^\psi(i)} p_{ly_{hl}} \right),
\tag{2.59}
$$

where $\underline{\psi}_{*j} = \left(\psi_{1j}, \psi_{1j}, \dots, \psi_{nj} \right)^T$.

The above definition implies that all interests are independent of each other, since the restructuring signal generator (F_ζ) for any interest of any cell depends only on the tendency values of the neighboring cells to that interest.

2.3.1.2 Dynamics of DICLA

Dynamics of *DICLA* within its environment is twofold; dynamics of the tendency vectors of its constituting cells and dynamics of the action probability vectors of its constituting learning automata. We refer to the former as the structural dynamics and to the latter as the behavioral dynamics of *DICLA*. In this subsection, we will first study the structural dynamics of *DICLA* and then analyze its behavioral dynamics.

Structural Dynamics of DICLA

Structural dynamics of *DICLA* can be described by the process $\left\{\underline{\psi}(k)\right\}_{k \geq 0}$ which evolves according to the following equation:

$$\underline{\psi}(k+1) = \underline{Z}\left(\underline{\psi}(k), \underline{\zeta}(k)\right),\tag{2.60}$$

where $\underline{Z}$ represents the restructuring function. In this chapter, we study the structural dynamics of *DICLA* for two different restructuring functions, namely simple restructuring function [Eq. (2.61)] and vanishing restructuring function [Eq. (2.62)].

$$\underline{\psi}(k+1) = \underline{\psi}(k) + \underline{b} \cdot \underline{\zeta}(k).\tag{2.61}$$

$$\underline{\psi}(k+1) = \underline{\psi}(k) + \underline{b} \cdot v(k) \cdot \underline{\zeta}(k).\tag{2.62}$$

In the above equations, $\underline{b}$ is an $n \times n$ diagonal matrix with $b_{ii} = b_i$ and $b_i > 0$ represents the restructuring rate for the cell c_i. $v(k)$ in Eq. (2.62), is a vanish at infinity function. A function v is said to vanish at infinity if $v(k) \to 0$ as $k \to \infty$. Note that the above restructuring functions can be used only if $-\psi_{ij} \leq b_i \cdot \zeta_{ij} \leq 1 - \psi_{ij}, \forall i,j$.

- **Structural Dynamics of DICLA using Simple Restructuring Function**

Components of $\underline{Z}$, using simple restructuring function, can be obtained as follows:

$$Z_{ij}\left(\psi_{ij}(k), \zeta_{ij}(k)\right) = \psi_{ij}(k) + b_i \cdot \zeta_{ij}(k).\tag{2.63}$$

From Eq. (2.60) it follows that $\left\{\underline{\psi}(k)\right\}_{k \geq 0}$ is a discrete-time Markov process (Isaacson and Madsen 1976) defined on the tendency space Y.

Now consider the following assumptions:

(A-1) There exists a set of fitness evaluation functions $\left(F^i, i = 1, 2, \ldots, n\right)$, each is able to evaluate the normalized fitness of the tendency vector of the cell c_i within the range [0, 1], where 0 is the fitness of the optimum tendency vector of the cell.

(A-2) The restructuring signal generator (F_ζ) can be described as the following equation:

$$F_\zeta^i\left(y_{h_1}^\psi, y_{h_2}^\psi, \ldots, y_{h_{N_i^\psi}}^\psi, r, \underline{\psi}_{*j}\right) = F^i\left(\underline{\psi}_i\right) \cdot G^i\left(y_{h_1}^\psi, y_{h_2}^\psi, \ldots, y_{h_{N_i^\psi}}^\psi, r, \underline{\psi}_{*j}\right),$$

$$\tag{2.64}$$

where $\mathbf{G}^i$ is a function within the range $[-1, 1]$.

(A-3) $\sum_j \left[\mathbf{G}^i \left(y_{h_1}^{\psi}, y_{h_2}^{\psi}, \ldots, y_{h_{N_i^{\psi}}}^{\psi}, r, \underline{\psi}_{*j} \right) \right]^2 \neq 0.$

Remark 2.1 Assumption (A-1) enables us to define a local optimum for the tendency vector of each cell of *DICLA*. The study of the structural dynamics of *DICLA* is then reduced to investigating whether the tendency vector of each cell converges to its local optimum or not. We refer to a point $\underline{\psi}^*$ within the tendency space of *DICLA* as the maximal tendency point, if for every cell c_i, $\underline{\psi}_i^*$ is the local optimum of the tendency vector of that cell.

Remark 2.2 Assumption (A-2) indicates that F_ζ^i is a multiplication of F^i and $\mathbf{G}^i$. Considering this assumption, $\mathbf{G}^i$ specifies the orientation and direction of the restructuring signal and F^i specifies its magnitude. When the tendency vector of a cell c_i is far from its optimal vector, F^i is large and as a result, the restructuring signal will be large. On the other hand, when the tendency vector of a cell c_i is near its optimal vector, F^i is small and hence, the restructuring signal will be small.

Remark 2.3 Assumption (A-3) indicates that the value of $\mathbf{G}^i$ must be non-zero for at least one of the interests. To explain the reason for this assumption, we have to note that $\mathbf{G}^i$ specifies the orientation and direction of the restructuring signal, and thus its value cannot be zero for all interests; otherwise it cannot specify any direction or orientation.

Using the above assumptions, the average restructuring signal for the interest j of the cell c_i for configuration $\langle \underline{p}, \underline{\psi} \rangle \in \langle \mathsf{K}, \mathsf{Y} \rangle$, defined previously through Eq. (2.59), can be rewritten as given below:

$$d_{ij}^r \left(\langle \underline{p}, \underline{\psi} \rangle \right) = F^i \left(\underline{\psi}_i \right) \cdot \sum_r \sum_{y_{h_1}^{\psi}, \ldots, y_{h_{N_i^{\psi}}}^{\psi}} \left(\mathbf{G}^i \left(y_{h_1}^{\psi}, y_{h_2}^{\psi}, \ldots, y_{h_{N_i^{\psi}}}^{\psi}, r, \underline{\psi}_{*j} \right) \prod_{LA_l \in N^{\psi}(i)} p_{lh_l} \right)$$

$$= F^i \left(\underline{\psi}_i \right) \cdot \overline{\mathbf{G}}^{ij} \left(\langle \underline{p}, \underline{\psi} \rangle \right).$$

(2.65)

Now define

$$\Delta \underline{\psi}(k) = \mathrm{E} \left[\underline{\psi}(k+1) \middle| \underline{\psi}(k) \right] - \underline{\psi}(k).$$ (2.66)

Since $\left\{ \underline{\psi}(k) \right\}_{k \geq 0}$ is Markovian and $\underline{\zeta}(k)$ depends only on $\langle \underline{p}(k), \underline{\psi}(k) \rangle$ and not on k explicitly, then $\Delta \underline{\psi}(k)$ can be expressed as a function of $\langle \underline{p}(k), \underline{\psi}(k) \rangle$. Hence we can write

$$\Delta\underline{\psi}(k) = \underline{bf}^{\zeta}\Big(\big\langle \underline{p}(k), \underline{\psi}(k)\big\rangle\Big). \tag{2.67}$$

The components of $\Delta\underline{\psi}(k)$ can be obtained as follows:

$$\begin{aligned}
\Delta\psi_{ij}(k) &= b_i \cdot f_{ij}^{\zeta}\Big(\big\langle \underline{p}(k), \underline{\psi}(k)\big\rangle\Big) \\
&= b_i \cdot d_{ij}^{\zeta}\Big(\big\langle \underline{p}(k), \underline{\psi}(k)\big\rangle\Big) \\
&= b_i \cdot \mathsf{F}^i\Big(\underline{\psi}_i(k)\Big) \cdot \overline{\mathsf{G}}^{ij}\Big(\big\langle \underline{p}(k), \underline{\psi}(k)\big\rangle\Big).
\end{aligned} \tag{2.68}$$

Lemma 2.10 *Function* $\underline{f}^{\zeta}\Big(\big\langle \underline{p}(k), \underline{\psi}(k)\big\rangle\Big)$ *whose components are given by Eq. (2.68) is Lipschitz continuous over the compact space* $\langle \mathsf{K}, \mathsf{Y}\rangle$.

Proof Function $\underline{f}^{\zeta}\Big(\big\langle \underline{p}(k), \underline{\psi}(k)\big\rangle\Big)$ has compact support (it is defined over $\langle \mathsf{K}, \mathsf{Y}\rangle$), is bounded because $-1 \le f_{ij}^{\zeta}\big(\langle \underline{p}, \underline{\psi}\rangle\big) \le 1, \forall \underline{p}, \underline{\psi}, i, j$, and is also continuously differentiable with respect to $\underline{\psi}$ over $\langle \mathsf{K}, \mathsf{Y}\rangle$. Therefore, its first derivative with respect to $\underline{\psi}$ is also bounded. Thus, using the Cauchy's mean value theorem, it can be concluded that $\underline{f}^{\zeta}\Big(\big\langle \underline{p}(k), \underline{\psi}(k)\big\rangle\Big)$ is Lipschitz continuous over the compact space $\langle \mathsf{K}, \mathsf{Y}\rangle$ with Lipschitz constant $\mathsf{K} = \sup_{\underline{\psi}}\Big\|\nabla_{\underline{\psi}}\underline{f}^{\zeta}\Big(\big\langle \underline{p}(k), \underline{\psi}(k)\big\rangle\Big)\Big\|$. ∎

For different values of $\underline{b}$, Eq. (2.61) generates a different process and we shall use $\underline{\psi}^{\underline{b}}(k)$ to denote this process whenever the value of $\underline{b}$ is to be specified explicitly. Define a sequence of continuous-time interpolation of (2.61), denoted by $\underline{\tilde{\psi}}^{\underline{b}}(t)$ and called an *interpolated process*, whose components are defined by:

$$\underline{\tilde{\psi}}_i^{\underline{b}}(t) = \underline{\psi}_i(k), \quad t \in [kb_i, (k+1)b_i), \tag{2.69}$$

where b_i is the restructuring rate of the simple restructuring function for the cell c_i. The interpolated process $\Big\{\underline{\tilde{\psi}}^{\underline{b}}(t)\Big\}_{t \ge 0}$ is a sequence of random variables that takes values from $[0,1]^{n \times |\Psi|}$, where $[0,1]^{n \times |\Psi|}$ is the space of all functions that, at each point, are continuous on the right and have a limit on the left over $[0, \infty)$ and take values in Y. The objective is to study the limit of the sequence $\Big\{\underline{\tilde{\psi}}^{\underline{b}}(t)\Big\}_{t \ge 0}$ as $\max\{\underline{b}\} \to 0$, which will be a good approximation to the asymptotic behavior of (2.69). Consider the following ordinary differential equation (ODE):

$$\dot{\underline{\psi}} = \underline{f}^{\zeta}\left(\left\langle \underline{p}, \underline{\psi} \right\rangle\right),$$ (2.70)

where $\dot{\underline{\psi}}$ is composed of the following components:

$$\frac{d\psi_{ij}}{dt} = \mathsf{F}^i\left(\underline{\psi}_i\right) \cdot \overline{\mathsf{G}}^{ij}\left(\left\langle \underline{p}(t), \underline{\psi}(t) \right\rangle\right).$$ (2.71)

Theorem 2.6 *Using the simple restructuring function* (2.61) *and considering* $\max\{\underline{b}\} \to 0, \psi(k)$ *is well approximated by the solution of the ODE* (2.70).

Proof Following conditions are satisfied by the restructuring function given by Eq. (2.61):

- $\left\{\underline{\psi}(k)\right\}_{k \geq 0}$ is a Markovian process.
- Given $\underline{\psi}(k), \underline{\zeta}(k)$ is independent of $\underline{\zeta}(k-1)$.
- $\underline{f}^{\zeta}\left(\left\langle \underline{p}(k), \underline{\psi}(k) \right\rangle\right)$ is independent of k.
- $\underline{f}^{\zeta}\left(\left\langle \underline{p}(k), \underline{\psi}(k) \right\rangle\right)$ is Lipschitz continuous over the compact space $\langle \mathsf{K}, \mathsf{Y} \rangle$ (Lemma 2.10) .
- $\mathrm{E}\left\|\underline{\zeta}(.) - \underline{f}^{\zeta}\left(\left\langle \underline{p}, \underline{\psi} \right\rangle\right)\right\|^2$ is bounded since $\underline{\zeta}(.) \in [-1, 1]^{N \times |\Psi|}$.
- Restructuring rates b_i, $i = 1, \ldots, n$ are sufficiently small since $\max\{\underline{b}\} \to 0$.

Therefore, using theorem (A.1) in Thathachar and Sastry (2004) we can conclude the theorem.   ∎

What Theorem 2.6 says is that $\underline{\psi}(k)$, given by Eq. (2.61), will closely follow the solution of the ODE (2.70), that is $\underline{\psi}(k)$ can be made to closely approximate the solution of its approximating ODE by taking $\max\{\underline{b}\}$ sufficiently small. Thus, if the ODE (2.70) has a locally asymptotically stable equilibrium point, then we can conclude that (by taking $\max\{\underline{b}\}$ sufficiently small), $\underline{\psi}(k)$, for large k, would be close to this equilibrium point if $\underline{\psi}(0)$ is within the region of attraction of that point.

To find the equilibrium points of the ODE (2.70), we have to solve equations of the following from:

$$\frac{d\psi_{ij}^*(t)}{dt} = 0 \quad \forall i, j.$$ (2.72)

Using Eqs. (2.71) and (2.72) can be rewritten as

$$\mathsf{F}^i\left(\underline{\psi}_i^*\right) \cdot \overline{\mathsf{G}}^{ij}\left(\left\langle \underline{p}(t), \underline{\psi}^*(t) \right\rangle\right) = 0, \quad \forall i, j.$$ (2.73)

Considering the assumption (A-3), the only solutions to the above equations are the set of tendency vectors $\underline{\psi}^*$ which satisfy the following equations:

$$\mathsf{F}^i\!\left(\underline{\psi}_i^*\right) = 0, \quad \forall i. \tag{2.74}$$

In other words, the equilibrium points of the ODE (2.70) coincide with the maximal tendency points of *DICLA*.

Since $\underline{\psi}^*$ is an absorbing state of the Markovian process $\left\{\underline{\psi}(k)\right\}_{k \geq 0}$, if $\underline{\psi}(0) = \underline{\psi}^*$, then *DICLA*, from the structural point of view, absorbed to $\underline{\psi}^*$. But if $\underline{\psi}(0) \neq \underline{\psi}^*$, then we cannot guarantee the convergence to a maximal tendency point. However, we give a sufficient condition under which convergence to a maximal tendency point can be assured.

Theorem 2.7 *Suppose there is a bounded differentiable function* $\mathsf{H} : \mathsf{R}^{m_1 + \cdots + m_n} \times \mathsf{R}^{N \times |\Psi|} \to \mathsf{R}$ *such that for some constants* $c_{ij} > 0$, *we have* $\dfrac{\partial \mathsf{H}(\langle \underline{p}, \underline{\psi} \rangle)}{\partial p_{iy}} = 0$ *and* $\dfrac{\partial \mathsf{H}(\langle \underline{p}, \underline{\psi} \rangle)}{\partial \psi_{ij}} = -c_{ij} \cdot \overline{\mathsf{G}}^{ij}\!\left(\langle \underline{p}, \underline{\psi} \rangle\right)$. *Then DICLA using the simple restructuring function and with sufficiently small value of the restructuring rate* $(\max\{\underline{b}\} \to 0)$, *converges, from the structural point of view, to a maximal tendency point, for any initial configuration in* $\langle \mathsf{K}, \mathsf{Y} \rangle$.

Proof We have

$$\begin{aligned}
\frac{d\mathsf{H}}{dt} &= \sum_i \sum_y \frac{\partial \mathsf{H}}{\partial p_{iy}} \cdot \frac{dp_{iy}}{dt} + \sum_i \sum_j \frac{\partial \mathsf{H}}{\partial \psi_{ij}} \cdot \frac{d\psi_{ij}}{dt} \\
&= -\sum_i \sum_j c_{ij} \cdot \mathsf{F}^i\!\left(\underline{\psi}_i\right) \cdot \left(\overline{\mathsf{G}}^{ij}\!\left(\langle \underline{p}, \underline{\psi} \rangle\right)\right)^2 \tag{2.75} \\
&\leq 0.
\end{aligned}$$

Thus, H is non-increasing along the trajectories of the ODE (2.70) and eventually converges to its minimum, where $\frac{d\mathsf{H}}{dt} = 0$. From (2.75), the derivative of H is zero if and only if $\mathsf{F}^i\!\left(\underline{\psi}_i\right) = 0$ for all i. This is an equilibrium point of the ODE (2.70). Thus, the ODE has to converge to some equilibrium point. Since an equilibrium point of the ODE (2.70) is also a maximal tendency point, the theorem follows. ∎

To see some samples of the H function with the properties specified in Theorem 2.7, please refer to Table 2.1.

- **Structural Dynamics of DICLA using Vanishing Restructuring Function**

Components of $\underline{Z}$, using vanishing restructuring function [Eq. (2.62)], can be obtained as given below:

Table 2.1 Some samples of the H function with the properties specified in Theorem 2.7

$\overline{G}^{ij}\big(\langle \underline{p}, \underline{\psi} \rangle\big)$	$H\big(\langle \underline{p}, \underline{\psi} \rangle\big)$	$\frac{\partial H(\langle \underline{p}, \underline{\psi} \rangle)}{\partial \psi_{ij}}$
k_{ij} (for some constants $k_{ij} > 0$)	$-\sum_i \sum_j k_{ij} \cdot \psi_{ij}$	$-k_{ij}$
$k_{ij} \cdot N_i^{\psi}$ (for some constants $k_{ij} > 0$)	$-\sum_i \sum_{l=1}^{N_i^{\psi}} \sum_j k_{ij} \cdot \psi_{ij}$	$-k_{ij} \cdot N_i^{\psi}$

$$Z_{ij}\big(\psi_{ij}(k), \zeta_{ij}(k)\big) = \psi_{ij}(k) + b_i \cdot v(k) \cdot \zeta_{ij}(k). \tag{2.76}$$

As it was mentioned earlier, $\left\{\underline{\psi}(k)\right\}_{k \geq 0}$ is a discrete-time Markov process (Isaacson and Madsen 1976) defined on the tendency space Y. Once again, consider the following difference equation:

$$\Delta\underline{\psi}(k) = \mathrm{E}\Big[\underline{\psi}(k+1)\big|\underline{\psi}(k)\Big] - \underline{\psi}(k). \tag{2.77}$$

The components of $\Delta\underline{\psi}(k)$ using vanishing restructuring function can be obtained as follows:

$$\Delta\psi_{ij}(k) = b_i \cdot v(k) \cdot d_{ij}^{\varsigma}\big(\langle \underline{p}(k), \underline{\psi}(k) \rangle\big). \tag{2.78}$$

Since $v(k) \to 0$ as $k \to \infty$, we can conclude that, for large k, $\Delta\psi_{ij}(k) \to 0$. This means that in the long run, the process $\left\{\underline{\psi}(k)\right\}_{k \geq 0}$ will eventually converges to a state $\underline{\psi}^*$, but no statement can be made about the properties of this state.

Remark 2.4 Note that the vanishing restructuring function must be used only if no fitness evaluation function with the properties specified in assumption (A-1) of the Section "Behavioral Dynamics of DICLA" can be found. In other words, when it is possible to find some fitness evaluation functions for evaluating the fitness of the tendency vectors of the cells of *DICLA*, it is rational to use such functions for guiding the exploration of *DICLA* through its tendency space until it reaches a well-fitted state. But, when it is hard or impossible to find such fitness evaluation functions, one can use the vanishing restructuring function to force *DICLA* to stop its exploration through its tendency space after a specific number of iterations. The rationale behind using a vanishing restructuring function is that explorations through tendency space, when thinking of the applications modeled by *DICLA* (such as mobile ad hoc and sensor networks), usually impose some costs (such as energy used for movements of mobile nodes), and such costs cannot be paid forever.

Behavioral Dynamics of DICLA

The behavioral dynamics of *DICLA* is affected by the structural dynamics of *DICLA*. This is due to the fact that the local rule of *DICLA*, which rules its behavioral and structural dynamics, in each cell is a function of the state of that cell and the states of its neighboring cells. But the neighboring cells of each cell change according to the structural dynamics of *DICLA* and thus, the behavioral dynamics of *DICLA* is affected accordingly. Considering this affection, we study the behavioral dynamics of *DICLA* from the time instant at which its evolution within the tendency space Y is stopped at the point $\underline{\psi}^*$. We refer to *DICLA*, whose evolution within the tendency space Y is stopped at the point $\underline{\psi}^*$, as $DICLA^{\underline{\psi}^*}$. A $DICLA^{\underline{\psi}^*}$ is indeed an *ICLA* and hence, the analysis given in Sect. 2.2.2 for the behavior of the *ICLA* are also valid for the behavioral dynamics of $DICLA^{\underline{\psi}^*}$. Thus, we can state the following three theorems which specify the behavioral dynamics of $DICLA^{\underline{\psi}^*}$.

Theorem 2.8 *A $DICLA^{\underline{\psi}^*}$ with SL_{RP} learning automata, regardless of its initial configuration, converges in distribution to a random configuration, in which the mean value of the action probability of any action of any LA is inversely proportional to the average penalty received by that action.*

Proof $DICLA^{\underline{\psi}^*}$ is indeed an *ICLA* and hence, the proof of this theorem is trivial considering the analysis given in Sect. 2.2.2 for the behavior of *ICLA*.   ∎

Theorem 2.9 *A $DICLA^{\underline{\psi}^*}$ with SL_{RP} learning automata, regardless of the local rule being used, is expedient.*

Proof $DICLA^{\underline{\psi}^*}$ is indeed an *ICLA* and hence, the proof of this theorem is trivial considering the analysis given in Sect. 2.2.2 for the behavior of *ICLA*.   ∎

Theorem 2.10 *Consider a $DICLA^{\underline{\psi}^*}$ with SL_{RP} learning automata. Assume that the number of actions of every learning automaton of DICLA is m and the reinforcement signal generator of every cell is both self-biased and biased-aggregated against the action r. Let $\rho > 1$ be the biasing factor. Then DICLA is totally expedient if the following inequality holds for every cell c_i:*

$$m^{N_i^{\underline{\psi}^*}+2}\rho\left(\rho^{N_i^{\underline{\psi}^*}}-1\right) - N_i^{\underline{\psi}^*}(\rho-1)(m-1+\rho)[(m-1)\rho+1]^{N_i^{\underline{\psi}^*}+1} < 0. \quad (2.79)$$

Proof $DICLA^{\underline{\psi}^*}$ is indeed an *ICLA* and hence, the proof of this theorem is trivial considering the analysis given in Sect. 2.2.2 for the behavior of *ICLA*.   ∎

2.3.2 Heterogeneous Dynamic Irregular Cellular Learning Automata

As it was indicated, *DICLA* is suitable for modeling problems in the areas with dynamic structure, such as mobile ad hoc and sensor networks. *DICLA* has been successfully used in deploying a wireless sensor network throughout the sensor field using mobile sensor nodes (Esnaashari and Meybodi 2011). In some applications, different cells need different learning parameters. For example, consider the *k*-coverage problem in *WSN*s, where different sub-regions of the sensor field require different degrees of coverage. Here, each sensor node has to consider a value for the parameter *k* which depends on the location of that node. *DICLA* cannot be used in such applications with heterogeneous learning parameters. Therefore, in this section, we propose heterogeneous dynamic irregular cellular learning automaton (*HDICLA*) as a generalization of *DICLA*, which can support heterogeneous learning parameters in different cells.

We define heterogeneous *DICLA* (*HDICLA*) as an undirected graph in which, each vertex represents a cell and a learning automaton is assigned to every cell (vertex). A finite set of interests and a finite set of attributes are defined for the *HDICLA*. In *HDICLA*, the state of each cell consists of the following three parts:

- *Selected action*: The action selected by the learning automaton residing in the cell.
- *Tendency vector*: Tendency vector of the cell is a vector whose *j*th element shows the degree of tendency of the cell to the *j*th interest.
- *Attribute vector*: Attribute vector of the cell is a vector whose *j*th element shows the value of the *j*th attribute within the cell's locality.

Two cells are neighbors in *HDICLA* if the distance between their tendency vectors is smaller than or equal to the neighborhood radius.

Like *DICLA*, there is a local rule that *HDICLA* operates under. The rule of *HDICLA*, the actions selected by the neighboring learning automata of any particular learning automaton LA_i, and the attribute vector of the cell in which LA_i resides (c_i) determine the followings: (1) the reinforcement signal to the learning automaton LA_i and (2) the restructuring signal to the cell c_i, which is used to update the tendency vector of the cell. Figure 2.3 gives a schematic of *HDICLA*. An *HDICLA* is formally defined below.

Definition 2.20 A heterogeneous dynamic irregular cellular learning automaton (*HDICLA*) is a structure $A = (G <E, V >, \Psi, \Lambda, A, \Phi <\alpha, \underline{\psi}, \underline{\lambda} >, \tau, F, Z)$ where

- G is an undirected graph, with V as the set of vertices and E as the set of edges. Each vertex represents a cell in *HDICLA*.
- Ψ is a finite set of interests. Cardinality of Ψ is denoted by $|\Psi|$.
- Λ is a finite set of attributes. Cardinality of Λ is denoted by $|\Lambda|$.
- A is the set of learning automata each of which is assigned to one cell of the *HDICLA*.

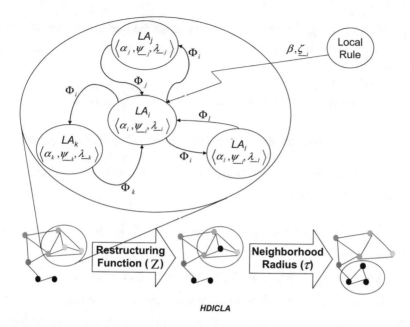

HDICLA

Fig. 2.3 Heterogeneous dynamic irregular cellular learning automaton (*HDICLA*)

- $\Phi < \alpha, \underline{\psi}, \underline{\lambda} >$ is the cell state. State of a cell c_i (φ_i) consists of the following three parts:

  - α_i which is the action selected by the learning automaton of that cell.
  - A vector $\underline{\psi}_i = \left(\psi_{i1}, \ldots, \psi_{i|\Psi|} \right)^T$ which is called tendency vector of the cell. Each element $\psi_{ik} \in [0,1]$ in the tendency vector of the cell c_i shows the degree of tendency of c_i to the interest $\psi_k \in \Psi$.
  - A vector $\underline{\lambda}_i = \left(\lambda_{i1}, \ldots, \lambda_{i|\Lambda|} \right)^T$ which is called attribute vector of the cell. Each element $\lambda_{ik} \in \mathbb{R}$ in the attribute vector of the cell c_i shows the value of the attribute $\lambda_k \subset \Lambda$ in the locality of the cell c_i.

- τ is the neighborhood radius. Two cells c_i and c_j of the *HDICLA* are neighbors if $\left\| \underline{\psi}_i - \underline{\psi}_j \right\| \leq \tau$. In other words, two cells of the *HDICLA* are neighbors if the distance between their tendency vectors is smaller than or equal to τ.

- $F : \underline{\varphi}_i \to \left\langle \underline{\beta}, [0,1]^{|\Psi|} \right\rangle$ is the local rule of *HDICLA* in each cell c_i, where $\underline{\varphi}_i = \left\{ \varphi_j \big| \left\| \underline{\psi}_i - \underline{\psi}_j \right\| \leq \tau \right\} \cup \{\varphi_i\}$ is the set of states of all neighbors of c_i, $\underline{\beta}$ is the set of values that the reinforcement signal can take, and $[0,1]^{|\Psi|}$ is a $|\Psi|$-dimensional unit hypercube. From the current states of the neighboring cells of each cell c_i, local rule performs the followings: 1. gives the reinforcement signal to the learning automaton LA_i resides in c_i and 2. produces a restructuring signal

$\left(\underline{\zeta_i} = \left(\zeta_{i1}, \ldots, \zeta_{i|\Psi|} \right)^T \right)$ which is used to change the tendency vector of c_i. Each element ζ_{ij} of the restructuring signal is a scalar value within the close interval $[0, 1]$.

- $Z : [0, 1]^{|\Psi|} \times [-1, 1]^{|\Psi|} \rightarrow [0, 1]^{|\Psi|}$ is the restructuring function which modifies the tendency vector of a cell using the restructuring signal produced by the local rule of the cell.

In what follows, we consider *HDICLA* with N cells. The learning automaton LA_i which has a finite action set $\underline{\alpha_i}$ is associated to cell c_i (for $i = 1, 2, \ldots, N$) of *HDICLA*. Let the cardinality of $\underline{\alpha_i}$ be m_i. The state of the *HDICLA* is represented by the triple $\langle \underline{p}, \underline{\psi}, \underline{\lambda} \rangle$, where

- $\underline{p} = \left(\underline{p_1}, \ldots, \underline{p_N} \right)^T$, where $\underline{p_i} = (p_{i1}, p_{i2}, \ldots, p_{im_i})^T$ is the action probability vector of LA_i.
- $\underline{\psi} = \left(\underline{\psi_1}, \ldots, \underline{\psi_N} \right)^T$, where ψ_i is the tendency vector of the cell c_i.
- $\underline{\lambda} = (\underline{\lambda_1}, \ldots, \underline{\lambda_N})^T$, where λ_i is the attribute vector of the cell c_i.

The operation of the *HDICLA* takes place as the following iterations. At iteration n, each learning automaton chooses an action. Let $\alpha_i \in \underline{\alpha_i}$ be the action chosen by LA_i. Then, each learning automaton receives a reinforcement signal. Let $\beta_i \in \underline{\beta}$ be the reinforcement signal received by LA_i. This reinforcement signal is produced by the application of the local rule $F\left(\underline{\varphi_i} \right) \rightarrow \left\langle \underline{\beta}, [0, 1]^{|\Psi|} \right\rangle$. Each LA updates its action probability vector on the basis of the supplied reinforcement signal and the action chosen by the cell. Next, each cell c_i updates its tendency vector using the restructuring function Z [Eq. (2.80)].

$$\underline{\psi_i}(k+1) = Z\left(\underline{\psi_i}(k), \underline{\zeta_i}(k) \right). \tag{2.80}$$

After the application of the restructuring function, the attribute vector of the cell c_i may change due to the modifications made to its local environment.

Like *DICLA*, *HDICLA* can be either synchronous or asynchronous and an asynchronous *DICLA* can be either time-driven or step-driven.

2.3.2.1   Definitions and Notations

In this section, we give some definitions, which will be used later for the analysis of *HDICLA*. Note that most of the definitions used for steady-state analysis of the behavior of the *DICLA* are redefined here for *HDICLA* model.

Definition 2.21 A configuration of *HDICLA* at step k is denoted by $\left\langle \underline{p}(k), \underline{\psi}(k), \underline{\lambda}(k) \right\rangle = \left\langle \left(\underline{p}_1, \ldots, \underline{p}_N\right)^T, \left(\underline{\psi}_1, \ldots, \underline{\psi}_N\right)^T, \left(\underline{\lambda}_1, \ldots, \underline{\lambda}_N\right)^T \right\rangle$, where $\underline{p}_i$ is the action probability vector of the learning automaton LA_i, $\underline{\psi}_i$ is the tendency vector of the cell c_i, $\underline{\lambda}_i$ is the attribute vector of the cell c_i, and T denotes the transpose operator.

Definition 2.22 A configuration $\left\langle \underline{p}, \underline{\psi}, \underline{\lambda} \right\rangle$ is called unit if the action probability vector of each learning automaton and the tendency vector of each cell are unit vectors, otherwise it is called general. Hence, the set of all unit configurations, $\langle K^*, Y^*, \Lambda \rangle$, and the set of all general configurations, $\langle K, Y, \Lambda \rangle$, in *HDICLA* are

$$\langle K^*, Y^*, \Lambda \rangle = \left\{ \left\langle \underline{p}, \underline{\psi}, \underline{\lambda} \right\rangle \middle| \begin{array}{l} \underline{p} = \left(\underline{p}_1, \ldots, \underline{p}_N\right)^T, \underline{p}_i = (p_{i1}, \ldots, p_{im_i})^T, p_{iy} = 0 \text{ or } 1\, \forall i, y, \sum_y p_{iy} = 1\, \forall i, \\ \underline{\psi} = \left(\underline{\psi}_1, \ldots, \underline{\psi}_N\right)^T, \underline{\psi}_i = \left(\psi_{i1}, \ldots, \psi_{i|\Psi|}\right)^T, \psi_{ij} = 0 \text{ or } 1\, \forall i, j, \sum_j \psi_{ij} = 1\, \forall i, \\ \underline{\lambda} = (\underline{\lambda}_1, \ldots, \underline{\lambda}_N)^T, \underline{\lambda}_i = (\lambda_{i1}, \ldots, \lambda_{i|\Lambda|})^T, \lambda_{ik} \in \Lambda \end{array} \right\},$$
(2.81)

and

$$\langle K, Y, \Lambda \rangle = \left\{ \left\langle \underline{p}, \underline{\psi}, \underline{\lambda} \right\rangle \middle| \begin{array}{l} \underline{p} = \left(\underline{p}_1, \ldots, \underline{p}_N\right)^T, \underline{p}_i = (p_{i1}, \ldots, p_{im_i})^T, 0 \leq p_{iy} \leq 1\, \forall i, y, \sum_y p_{iy} = 1\, \forall i, \\ \underline{\psi} = \left(\underline{\psi}_1, \ldots, \underline{\psi}_N\right)^T, \underline{\psi}_i = \left(\psi_{i1}, \ldots, \psi_{i|\Psi|}\right)^T, 0 \leq \psi_{ij} \leq 1\, \forall i, j, \\ \underline{\lambda} = (\underline{\lambda}_1, \ldots, \underline{\lambda}_N)^T, \underline{\lambda}_i = (\lambda_{i1}, \ldots, \lambda_{i|\Lambda|})^T, \lambda_{ik} \in \Lambda \end{array} \right\},$$
(2.82)

respectively. We refer to K as the probability space, to Y as the tendency space, and to Λ as the attribute space of *HDICLA* hereafter.

Lemma 2.11 $\langle K, Y \rangle$, *which represents the probability and tendency spaces of HDICLA, is the convex hull of* $\langle K^*, Y^* \rangle$.

Proof Proof of this lemma is similar to the proof of lemma 1 given in Beigy and Meybodi (2004). ∎

The application of the local rule to every cell allows transforming a configuration to a new one. The local rule of *HDICLA* (F) consists of two parts; reinforcement signal generator (F_β) and restructuring signal generator (F_ζ).

Definition 2.23 The global behavior of a *HDICLA* is a mapping $G : \langle K, Y, \Lambda \rangle \rightarrow \langle K, Y, \Lambda \rangle$ that describes the dynamics of *HDICLA*. The evolution of *HDICLA* from a given initial configuration $\left\langle \underline{p}(0), \underline{\psi}(0), \underline{\lambda}(0) \right\rangle \in \langle K, Y, \Lambda \rangle$ is a sequence of configurations $\left\{ \left\langle \underline{p}(k), \underline{\psi}(k), \underline{\lambda}(k) \right\rangle \right\}_{k \geq 0}$, such that $\left\langle \underline{p}(k+1), \underline{\psi}(k+1), \underline{\lambda}(k+1) \right\rangle = G\left(\left\langle \underline{p}(k), \underline{\psi}(k), \underline{\lambda}(k) \right\rangle \right)$.

Definition 2.24 Neighborhood set of any particular LA_i in configuration $\langle \underline{p}, \underline{\psi}, \underline{\lambda} \rangle$, denoted by $N^{\underline{\psi}}(i)$, is defined as the set of all learning automata residing in the adjacent cells of the cell c_i, that is,

$$N^{\underline{\psi}}(i) = \left\{ LA_j \Big| \left\| \underline{\psi}_i - \underline{\psi}_j \right\| \leq \tau \right\}. \tag{2.83}$$

Let $N^{\underline{\psi}}_i$ be the cardinality of $N^{\underline{\psi}}(i)$.

Definition 2.25 The average penalty for action r of the learning automaton LA_i for configuration $\langle \underline{p}, \underline{\psi}, \underline{\lambda} \rangle \in \langle \mathsf{K}, \mathsf{Y}, \Lambda \rangle$ is defined as

$$d_{ir}^{\beta}\left(\langle \underline{p}, \underline{\psi}, \underline{\lambda} \rangle\right) = \mathrm{E}\left[\beta_i \Big| \langle \underline{p}, \underline{\psi}, \underline{\lambda} \rangle, \alpha_i = r\right]$$

$$= \sum_{y_{h_1}^{\psi}, \ldots, y_{h_{N_i^{\psi}}}^{\psi}} \left(\mathsf{F}_{\beta}^{i}\left(y_{h_1}^{\psi}, y_{h_2}^{\psi}, \ldots, y_{h_{N_i^{\psi}}}^{\psi}, r, \underline{\lambda} \right) \prod_{\substack{LA_l \in N^{\underline{\psi}}(i) \\ l \neq i}} p_{ly_{h_l}} \right), \tag{2.84}$$

and the average penalty for the learning automaton LA_i is defined as

$$D_i\left(\langle \underline{p}, \underline{\psi}, \underline{\lambda} \rangle\right) = \mathrm{E}\left[\beta_i \Big| \langle \underline{p}, \underline{\psi}, \underline{\lambda} \rangle\right]$$

$$= \sum_y d_{iy}^{\beta}\left(\langle \underline{p}, \underline{\psi}, \underline{\lambda} \rangle\right) p_{iy}. \tag{2.85}$$

The above definition implies that if the learning automaton LA_j is not a neighboring learning automaton for LA_i, then $d_{ir}^{\beta}\left(\langle \underline{p}, \underline{\psi}, \underline{\lambda} \rangle\right)$ does not depend on $\underline{p}_j$. We assume that $d_{ir}^{\beta}\left(\langle \underline{p}, \underline{\psi}, \underline{\lambda} \rangle\right) \neq 0$ for all i, r and $\langle \underline{p}, \underline{\psi}, \underline{\lambda} \rangle$, that is, in any configuration, any action has a non-zero chance of receiving penalty.

Definition 2.26 The total average penalty for *HDICLA* at configuration $\langle \underline{p}, \underline{\psi}, \underline{\lambda} \rangle \in \langle \mathsf{K}, \mathsf{Y}, \Lambda \rangle$ is the sum of the average penalties for all learning automata in *HDICLA*, that is,

$$D\left(\langle \underline{p}, \underline{\psi}, \underline{\lambda} \rangle\right) = \sum_i D_i\left(\langle \underline{p}, \underline{\psi}, \underline{\lambda} \rangle\right). \tag{2.86}$$

Definition 2.27 The average restructuring signal for interest j of the cell c_i for configuration $\langle \underline{p}, \underline{\psi}, \underline{\lambda} \rangle \in \langle \mathsf{K}, \mathsf{Y}, \Lambda \rangle$ is defined as

$$d_{ij}^{\zeta}\left(\left\langle \underline{p}, \underline{\psi}, \underline{\lambda} \right\rangle\right) = \sum_{r} \sum_{\substack{y_{h_1}^{\psi}, \ldots, y_{N_i^{\psi}}^{\psi}}} \left(\mathsf{F}_{\zeta}^{i}\left(y_{h_1}^{\psi}, y_{h_2}^{\psi}, \ldots, y_{N_i^{\psi}}^{\psi}, r, \underline{\psi}_{*j}, \underline{\lambda} \right) \prod_{LA_l \in N_{-}^{\psi}(i)} p_{l y_{h_l}} \right),$$

$$(2.87)$$

where $\underline{\psi}_{*j} = \left(\psi_{1j}, \psi_{1j}, \ldots, \psi_{nj} \right)^{T}$.

The above definition implies that all interests are independent of each other, since the restructuring signal generator (F_{ζ}) for any interest of any cell depends only on the tendency values of the neighboring cells to that interest.

2.3.2.2 Dynamics of HDICLA

Like *DICLA*, dynamics of *HDICLA* within its environment is also twofold; dynamics of the tendency vectors of its constituting cells and dynamics of the action probability vectors of its constituting learning automata. We refer to the former as the structural dynamics and to the latter as the behavioral dynamics of *HDICLA*. In this subsection, we will first study the structural dynamics of *HDICLA* and then analyze its behavioral dynamics.

Structural Dynamics of HDICLA

Structural dynamics of *HDICLA* can be described by the process $\left\{ \underline{\psi}(k) \right\}_{k \geq 0}$ which evolves according to the following equation:

$$\underline{\psi}(k+1) = \mathsf{Z}\left(\underline{\psi}(k), \underline{\zeta}(k) \right), \tag{2.88}$$

where Z represents the restructuring function. In this cjapter, we study the structural dynamics of *HDICLA* for two different restructuring functions, namely simple restructuring function [Eq. (2.61)] and vanishing restructuring function [Eq. (2.62)].

- **Structural Dynamics of HDICLA using Simple Restructuring Function**

Consider the following assumptions:

(A-4) There exists a set of fitness evaluation functions $\left(F^i, i = 1, 2, \ldots, n \right)$, each is able to evaluate the normalized fitness of the tendency vector of the cell c_i within the range [0, 1], where 0 is the fitness of the optimum tendency vector of the cell.

(A-5) The restructuring signal generator (F_{ζ}) can be described as the following equation:

$$F_\zeta^i\left(y_{\overline{h_1}}^{\psi}, y_{\overline{h_2}}^{\psi}, \ldots, y_{\overline{h}_{\underset{N_i}{\psi}}}^{\psi}, r, \underline{\psi}_{*j}, \underline{\lambda}\right) = F^i\left(\underline{\psi}_i\right) \cdot$$

$$\tag{2.89}$$

$$G^i\left(y_{\overline{h_1}}^{\psi}, y_{\overline{h_2}}^{\psi}, \ldots, y_{\overline{h}_{\underset{N_i}{\psi}}}^{\psi}, r, \underline{\psi}_{*j}, \underline{\lambda}\right),$$

where G^i is a function within the range $[-1, 1]$.

$$(A\text{-}6) \sum_j \left[G^i\left(y_{\overline{h_1}}^{\psi}, y_{\overline{h_2}}^{\psi}, \ldots, y_{\overline{h}_{\underset{N_i}{\psi}}}^{\psi}, r, \underline{\psi}_{*j}, \underline{\lambda}\right)\right]^2 \neq 0.$$

Considering the above assumptions, studying the structural dynamics of *HDICLA* is reduced to investigating whether the tendency vector of each cell of *HDICLA* converges to its local optimum or not (We refer to a point $\underline{\psi}^*$ within the tendency space of *HDICLA* as the maximal tendency point, if for every cell c_i, $\underline{\psi}_i^*$ is the local optimum of the tendency vector of that cell). This study is similar to the study of the structural dynamics of *DICLA* using simple restructuring function given in Section "Behavioral Dynamics of DICLA". Thus, we can state the following results and theorems about the structural dynamics of *HDICLA* using simple restructuring function.

A point $\underline{\psi}^*$ within the tendency space of *HDICLA* which satisfy the set of Eqs. (2.90) is an absorbing state of the Markovian process $\left\{\underline{\psi}(k)\right\}_{k \geq 0}$.

$$F^i\left(\underline{\psi}_i^*\right) = 0, \quad \forall i. \tag{2.90}$$

Thus, if $\underline{\psi}(0) = \underline{\psi}^*$, then *HDICLA*, from the structural point of view, absorbed to $\underline{\psi}^*$. But if $\underline{\psi}(0) \neq \underline{\psi}^*$, then we cannot guarantee the convergence to a maximal tendency point. However, we can give a sufficient condition under which convergence to a maximal tendency point can be assured.

Theorem 2.11 *Suppose there is a bounded differentiable function* $H : R^{m_1 + \cdots + m_n} \times R^{N \times |\Psi|} \times R^{N \times |\Lambda|} \to R$ *such that for some constants* $c_{ij} > 0$, *we have* $\frac{\partial H(\langle \underline{p}, \underline{\psi}, \underline{\lambda}\rangle)}{\partial p_{iy}} = 0$, $\frac{\partial H(\langle \underline{p}, \underline{\psi}, \underline{\lambda}\rangle)}{\partial \lambda_{ik}} = 0$, *and* $\frac{\partial H(\langle \underline{p}, \underline{\psi}, \underline{\lambda}\rangle)}{\partial \psi_{ij}} = -c_{ij} \cdot \overline{G}^{ij}\left(\langle \underline{p}, \underline{\psi}, \underline{\lambda}\rangle\right)$.
Then HDICLA using the simple restructuring function and with sufficiently small value of the restructuring rate $(\max\{\underline{b}\} \to 0)$, *converges, from the structural point of view, to a maximal tendency point, for any initial configuration in* $\langle K, Y, \Lambda \rangle$.

Proof Proof is similar to the proof of Theorem 2.7 given in Section "Structural Dynamics of DICLA". ∎

- **Structural Dynamics of HDICLA using Vanishing Restructuring Function**

The structural dynamics of *HDICLA* using vanishing restructuring function is exactly similar to the structural dynamics of *DICLA* using vanishing restructuring function, and hence, we do not repeat it here.

Behavioral Dynamics of HDICLA

The behavioral dynamics of *HDICLA* is affected by the structural dynamics of *HDICLA*. This is due to the fact that the local rule of *HDICLA*, which rules its behavioral and structural dynamics, in each cell is a function of the state of that cell and the states of its neighboring cells. But the neighboring cells of each cell change according to the structural dynamics of *HDICLA* and thus, the behavioral dynamics of *HDICLA* is affected accordingly. Considering this affection, we study the behavioral dynamics of *HDICLA* from the time instant at which its evolution within the tendency space Y is stopped at the point $\underline{\psi}^*$. We refer to *HDICLA*, whose evolution within the tendency space Y is stopped at the point $\underline{\psi}^*$, as *HDICLA*$^{\underline{\psi}^*}$.

Now consider the following assumption:

(A-7) the attribute vector of each cell in *HDICLA* is independent of the probability vector of that cell and the probability vectors of its neighbors.

By this assumption, structural convergence of *HDICLA* to a point $\underline{\psi}^*$ within the tendency space results in convergence of *HDICLA* to a point $\underline{\lambda}^*$ within the attribute space. This is due to the fact that in an *HDICLA*$^{\underline{\psi}^*}$, the only source of dynamicity is the evolutions occur in the probability space. By assumption (A-7), such evolutions do not affect the attribute vectors.

We refer to an *HDICLA* whose evolutions within the tendency and attribute spaces are stopped at points $\underline{\psi}^*$ and $\underline{\lambda}^*$ respectively as *HDICLA*$^{\underline{\psi}^*, \underline{\lambda}^*}$. An *HDICLA*$^{\underline{\psi}^*, \underline{\lambda}^*}$ is indeed an *ICLA* and hence, the analysis given in Sect. 2.2.2 for the behavior of the *ICLA* are also valid for the behavioral dynamics of *HDICLA*$^{\underline{\psi}^*, \underline{\lambda}^*}$. Thus, we can state the following three theorems which specify the behavioral dynamics of *HDICLA*$^{\underline{\psi}^*, \underline{\lambda}^*}$.

Theorem 2.12 *An HDICLA*$^{\underline{\psi}^*, \underline{\lambda}^*}$ *with* SL_{RP} *learning automata, regardless of its initial configuration, converges in distribution to a random configuration, in which the mean value of the action probability of any action of any LA is inversely proportional to the average penalty received by that action.*

Proof HDICLA$^{\underline{\psi}^*, \underline{\lambda}^*}$ is indeed an *ICLA* and hence, the proof of this theorem is trivial considering the analysis given in Sect. 2.2.2 for the behavior of *ICLA*. ∎

Theorem 2.13 *An HDICLA$^{\underline{\psi}^*, \underline{\lambda}^*}$ with SL$_{RP}$ learning automata, regardless of the local rule being used, is expedient.*

Proof HDICLA$^{\underline{\psi}^*, \underline{\lambda}^*}$ is indeed an *ICLA* and hence, the proof of this theorem is trivial considering the analysis given in Sect. 2.2.2 for the behavior of *ICLA*. ∎

2.3.2.3 Numerical Examples

In this section, we give a number of numerical examples for better illustrating the analytical results specified in previous sections. The first two sets of examples are used to illustrate the analytical results given in Theorem 2.4 and Theorem 2.8. The next set of examples is used to study the behavior of the *ICLA* and *DICLA* in terms of expediency. The last example illustrates the theoretical results given in Theorem 2.7.

Numerical Example 1

This set of examples are given to illustrate that the action probability of any action of any learning automaton in an *ICLA* or *DICLA$^{\underline{\psi}^*}$* with *SL$_{RP}$* learning automata converges in distribution to a random variable, whose mean is inversely proportional to the average penalty received by that action. We consider 6 different models with different number of cells and different number of cell states. Table 2.2 gives the specifications used for this set of examples. In this table, *model$_{(n, m)}$* refers to an *ICLA* or a *DICLA$^{\underline{\psi}^*}$* with n cells and m states for each cell. Figures 2.4, 2.5, 2.6, 2.7, 2.8 and 2.9 compare the action probabilities of a randomly selected learning automaton in each model with their theoretical values obtained from Eq. (2.29) over the simulation time. As it can be seen from these figures, for large values of k, the action probability of all actions and their theoretical values approach each other. This is in coincidence with the results of the theoretical analysis given in Theorem 2.4 and Theorem 2.8, that is, an *ICLA* or a *DICLA$^{\underline{\psi}^*}$* with *SL$_{RP}$* learning automata converges in distribution to a random configuration, in which the mean value of the action probability of any action of any *LA* is inversely proportional to the average penalty received by that action.

Numerical Example 2

The goal of giving this set of numerical examples is to study the convergence behavior of an *ICLA* or a *DICLA$^{\underline{\psi}^*}$* when it starts to evolve within the environment from different initial configurations. For this study, we use a *model$_{5,5}$* with *SL$_{RP}$* learning automata. Table 2.3 gives the initial configurations of the model used in this study. In this table, the initial action probability vector for each learning

Table 2.2 Specifications used for numerical example 1

Model	$Model_{2,3}$	$Model_{2,5}$	$Model_{3,2}$	$Model_{3,5}$	$Model_{5,2}$	$Model_{5,5}$
Neighborhood	All cells are neighbors					
Learning Parameter	$a = 5 \times 10^{-5}$					
Initial Configuration	$p_{i1} = 1, p_{ir} = 0, \forall i, r, r \neq 1$					
Environment Response	$F^i(\alpha_1, \ldots, \alpha_n)$ is selected uniformly at random from the range [0, 1] at the beginning of the simulation					

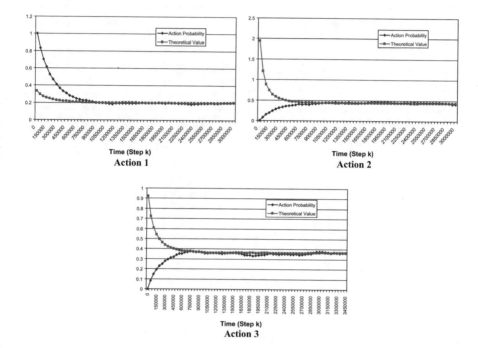

Fig. 2.4 Results of numerical example 1 for $model_{2,3}$

automaton in each configuration is given beneath that learning automaton. For example, in configuration 3, the initial action probability vector of the learning automaton LA_4 is $[.5, .1, .1, .2, .1]^T$. Figures 2.10, 2.11, 2.12, 2.13 and 2.14 plot the evolution of the action probabilities of two randomly selected actions of each learning automaton in different configurations. As it can be seen from these figures, no matter what the initial configuration of the model is, it converges to its equilibrium configuration. Thus, the results of this set of examples coincide with the results given in Theorem 2.4 and Theorem 2.8, that is, the convergence of an *ICLA* or a *DICLA*$^{\psi^*}$ to its equilibrium configuration is independent of its initial configuration.

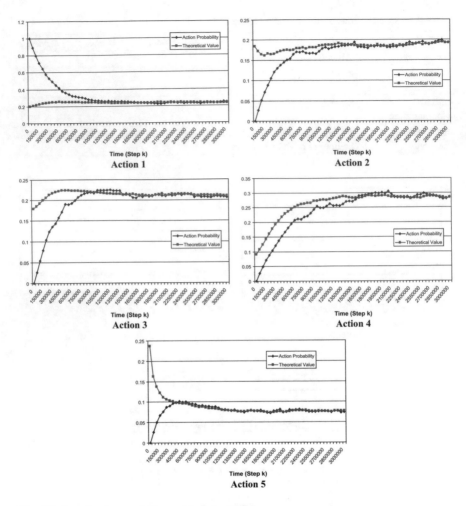

Fig. 2.5 Results of numerical example 1 for $model_{2,5}$

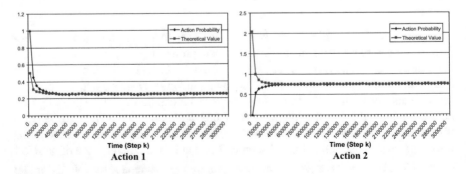

Fig. 2.6 Results of numerical example 1 for $model_{3,2}$

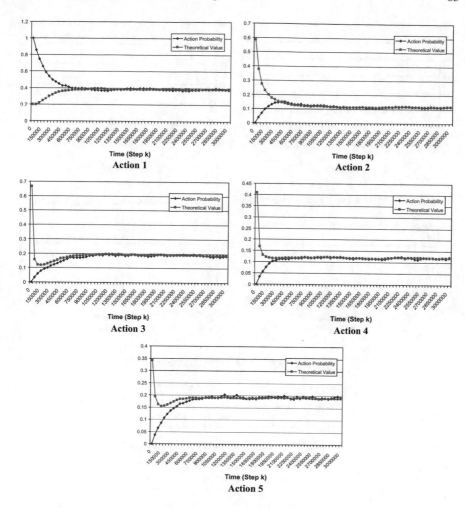

Fig. 2.7 Results of numerical example 1 for $model_{3,5}$

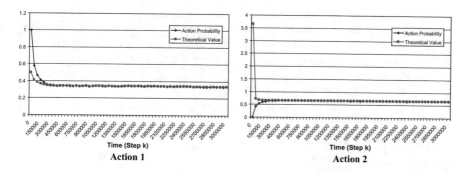

Fig. 2.8 Results of numerical example 1 for $model_{5,2}$

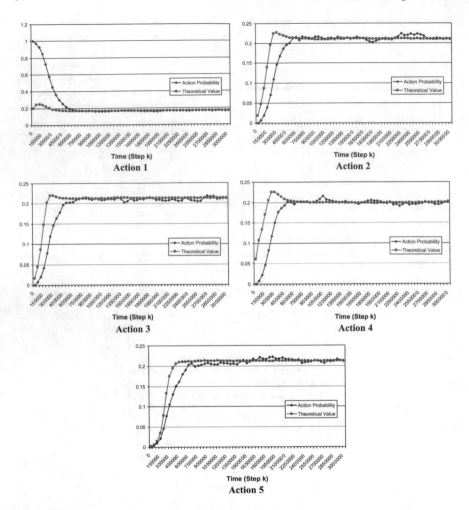

Fig. 2.9 Results of numerical example 1 for $model_{5,5}$

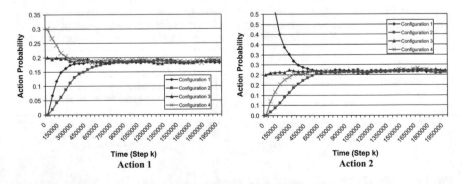

Fig. 2.10 Results of numerical example 2 for learning automaton LA_1

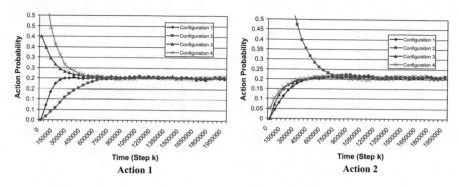

Fig. 2.11 Results of numerical example 2 for learning automaton LA_2

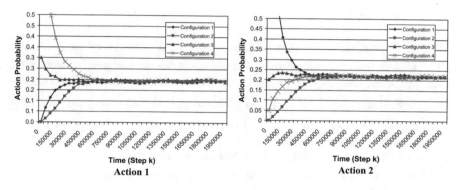

Fig. 2.12 Results of numerical example 2 for learning automaton LA_3

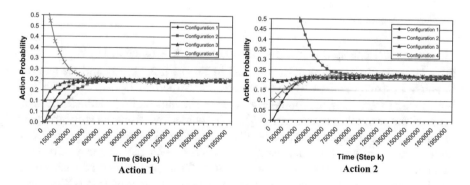

Fig. 2.13 Results of numerical example 2 for learning automaton LA_4

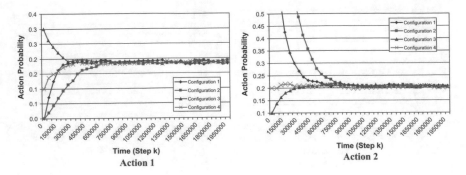

Fig. 2.14 Results of numerical example 2 for learning automaton LA_5

Numerical Example 3

This set of numerical examples is given to study the behavior of an *ICLA* or a *DICLA*$^{\psi^*}$ in terms of the expediency. For this study, we use the specifications of the first simulation study, given in Table 2.2. Figures 2.15, 2.16 and 2.17 compare $E\left[D_i\left(\underline{p}(k)\right)\right]$ and $\frac{1}{m_i}\sum_y d_{iy}\left(\underline{p}(k)\right)$ for all learning automata during the evolutions of all models given in Table 2.2. As it can be seen from these figures, the average penalty received by any learning automaton LA_i is less than that of a pure chance automaton. This is in coincidence with the theoretical results given in Theorems 2.5 and 2.9, that is, an *ICLA* or a *DICLA*$^{\psi^*}$ with SL_{RP} learning automata, regardless of the local rule being used, is expedient.

Numerical Example 4

This numerical example is given to illustrate that if a function $H : R^{m_1 + \cdots + m_n} \times R^{n \times |\Psi|} \rightarrow R$ with the properties specified in Theorem 2.7 exists and the assumptions (A-1), (A-2), and (A-3) given in Section "Structural Dynamics of DICLA" hold, then *DICLA* using the simple restructuring function and with sufficiently small value of the restructuring rate $(\max\{\underline{b}\} \rightarrow 0)$, converges, from the structural point of view, to a maximal tendency point. For this numerical example, we use a *DICLA* with three cells and three interests. This *DICLA* uses the simple restructuring function with the restructuring rate $b = 10^{-4}$. Tendency vector of each cell is specified randomly at the startup of the simulation. We assume that *DICLA* has a maximal tendency point (specified in Table 2.4) [Assumption (A-1)] and the restructuring signal generator (F_ζ) of every cell can be described as a multiplication of F^i and G^i [Assumptions (A-2) and (A-3)]. At each step k, $\overline{G}^{ij}\left(\left\langle\underline{p}(k),\underline{\psi}(k)\right\rangle\right)$ is selected randomly from the range $\left[-\psi_{ij}(k), 1 - \psi_{ij}(k)\right]$. Figures 2.18, 2.19 and 2.20 compare the tendencies of each cell to each of the interests of *DICLA* with

Table 2.3 Initial configurations for $model_{5,5}$ used in numerical example 2

Configuration 1					Configuration 2					Configuration 3					Configuration 4				
LA_1	LA_2	LA_3	LA_4	LA_5	LA_1	LA_2	LA_3	LA_4	LA_5	LA_1	LA_2	LA_3	LA_4	LA_5	LA_1	LA_2	LA_3	LA_4	LA_5
0	0	0	0	0	1	0	0	0	0	0.2	0.4	0.3	0.5	0.6	0.1	0.8	0.9	0.2	0.4
0	0	0	0	0	0	1	0	0	0	0.2	0	0.1	0.1	0.3	0.2	0.05	0.05	0.6	0.1
0	0	0	0	0	0	0	1	0	0	0.2	0.4	0.3	0.1	0	0.3	0	0	0.1	0.3
0	0	0	0	0	0	0	0	1	0	0.2	0	0.1	0.2	0	0.4	0.05	0	0.1	0
1	1	1	1	1	0	0	0	0	1	0.2	0.2	0.2	0.1	0.1	0	0.1	0.05	0	0.2

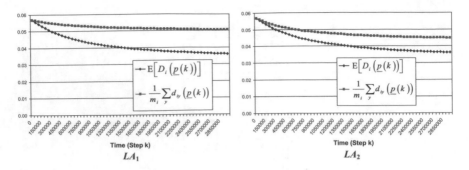

Fig. 2.15 Comparison of $E\left[D_i\left(\underline{p}(k)\right)\right]$ and $\frac{1}{m_i}\sum_y d_{iy}\left(\underline{p}(k)\right)$ for different learning automata during the evolution of $model_{2,3}$

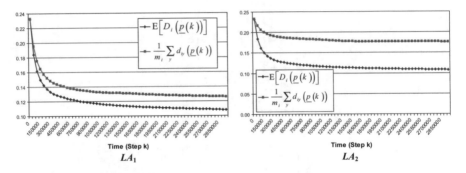

Fig. 2.16 Comparison of $E\left[D_i\left(\underline{p}(k)\right)\right]$ and $\frac{1}{m_i}\sum_y d_{iy}\left(\underline{p}(k)\right)$ for different learning automata during the evolution of $model_{2,5}$

their desired values (the values for tendencies at the maximal tendency point). As it can be seen from these figures, the tendencies of each cell to each of the interests gradually approaches their desired values. This is in coincidence with the theoretical analysis given in Theorem 2.7.

2.3.3   Closed Asynchronous Dynamic Cellular Learning Automata

In Closed Asynchronous Dynamic Cellular Learning Automata (*CADCLA*), we have a rule called structure updating rule which can be defined by the designer for controlling the evolution of the cellular structure (Saghiri and Meybodi 2017c).

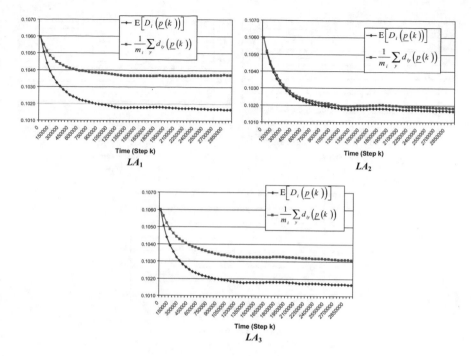

Fig. 2.17 Comparison of $\mathrm{E}\left[D_i\left(\underline{p}(k)\right)\right]$ and $\frac{1}{m_i}\sum_y d_{iy}\left(\underline{p}(k)\right)$ for different learning automata during the evolution of $model_{3,2}$

Definition 2.28 Closed Asynchronous Dynamic Cellular Learning Automaton (*CADCLA*) is an 7-tuple $CADCLA = (G, \mathrm{A}, \mathrm{N}, \Phi, \Psi, F_1, F_2)$, where:

- $G = (V, E)$ is an undirected graph which determines the structure of *CADCLA* where

  $V = \{cell_1, cell_2, \ldots, cell_n\}$ is the set of vertices and E is the set of edges.

- $A = \{LA_1, LA_2, \ldots, LA_n\}$ is a set of *LAs* each of which is assigned to one cell of *CADCLA*. The set of actions of automaton for a cell is the same as the set of states for that cell.

- $\mathrm{N} = \{\mathrm{N}_1, \mathrm{N}_2, \ldots, \mathrm{N}_n\}$ where $N_i = \{cell_j \in V | dist(cell_i, cell_j) < \theta_i\}$ where θ_i is the neighborhood radius of $cell_i$ and $dist(cell_i, cell_j)$ is the length of the shortest path between $cell_i$ and $cell_j$ in G. N_i^1 determines the immediate neighbors of $cell_i$.

- $\Phi = \{\Phi_1, \Phi_2, \ldots, \Phi_n\}$ where $\Phi_i = \{(j, \alpha_l) | cell_j \in N_i \ and \ action \ \alpha_l \ has \ been$ chosen by $LA_j\}$ denotes the state of $cell_i$. Φ_i^1 determines the state of $cell_i$ when $\theta_i = 1$.

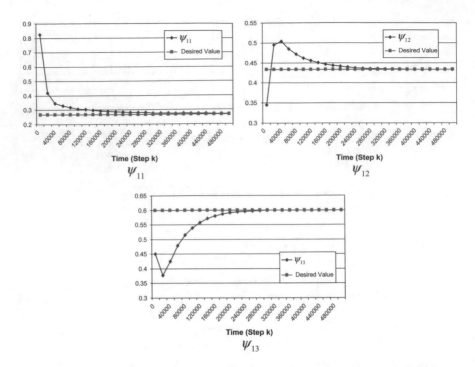

Fig. 2.18 Comparison of the tendencies of cell c_1 to each of the interests of *DICLA* with their desired values at the maximal tendency point

Table 2.4 Values of the tendencies of each cell to each of the interests of *DICLA* at the maximal tendency point

	Ψ_1	Ψ_2	Ψ_3
c_1	$\psi_{11} = 0.26$	$\psi_{12} = 0.43$	$\psi_{13} = 0.6$
c_2	$\psi_{21} = 0.91$	$\psi_{22} = 0.59$	$\psi_{23} = 0.09$
c_3	$\psi_{31} = 0.44$	$\psi_{32} = 0.73$	$\psi_{33} = 0.54$

- $\Psi = \{\Psi_1, \Psi_2, \ldots, \Psi_n\}$, $\Psi_i = \{(j, X_j) \mid cell_j \in N_i\}$ denotes the attribute of $cell_j$ where $X_j \subseteq \{x_1, x_2, \ldots, x_s\}\{x_1, x_2, \ldots, x_s\}$. is the set of allowable attributes. Ψ_i^1 determines the attribute of $cell_i$ when $\theta_i = 1$.

- $F_1 : (\underline{\Phi}, \underline{\Psi}) \rightarrow (\underline{\beta}, \underline{\zeta})$ is the local rule of *CADCLA*. In each cell, the local rule computes the reinforcement signal and the restructuring signal for the cell based on the states and attributes of that cell and its neighboring cells. The reinforcement signal is used to update the learning automaton of that cell.

- $F_2 : (\underline{N}, \underline{\Psi}, \underline{\zeta}) \rightarrow (\underline{N^1})$ is the structure updating rule. In each cell, the structure updating rule finds the immediate neighbors of the cell based on the restructuring signal computed by the cell, the attributes of the neighbors of the cell, and the neighbors of the cell. For example, in $cell_i$, structure updating rule takes $\langle N_i, \Psi_i, \zeta_i \rangle$ and returns N_i^1.

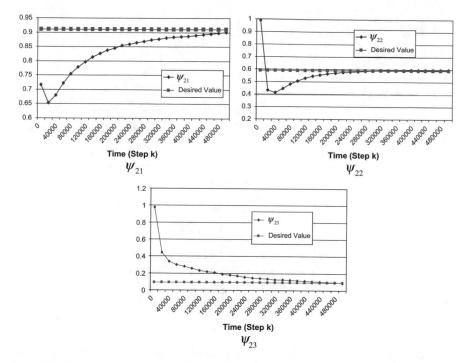

Fig. 2.19 Comparison of the tendencies of cell c_2 to each of the interests of *DICLA* with their desired values at the maximal tendency point

The internal structure of *cell$_i$* and its interaction with local environments is shown in Fig. 2.21.

2.3.3.1   Updating Process

The main loop for the operation of a *DCLA* be described in Fig. 2.22. The application determines which cell must be activated. Upon activation of a cell, the cell performs a process containing three phases: **preparation, structure updating and state updating**.

In what follows, we give a more detailed version of the operation of *DCLA* using the notations given for *DCLA*. Figure 2.23 shows the main loop describing the operation of *DCLA* and Fig. 2.24 shows the pseudo code of the process which is performed upon the activation of a cell of *DCLA*.

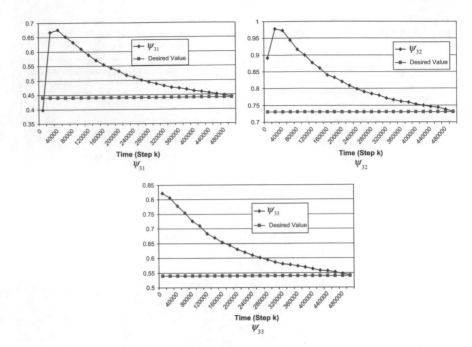

Fig. 2.20 Comparison of the tendencies of cell c_3 to each of the interests of *DICLA* with their desired values at the maximal tendency point

2.3.3.2   Analyzing the Expediency of *CADCLA*

In this section, required definitions are given and then the expediency of the *CADCLA* is analyzed.

Definition 2.29 The configuration of the *CADCLA* at iteration t, is defined as $Y(t) = \ <\underline{N}(t), \underline{P}(t), \underline{\Phi}(t)>\ $ where

- $\underline{N}(t) = (N_1(t), N_2(t), \ldots, N_n(t))^T$
- $\underline{P}(t) = (P_1(t), P_2(t), \ldots, P_n(t))^T$ where $P_i(t) = (p_{i1}(t), p_{i2}(t), \ldots, p_{ir}(t))^T$ in which $p_{ij}(t)$ is the probability of selecting action α_j of the learning automaton LA_i at iteration t. Each learning automaton has r actions.
- $\underline{\Phi}(t) = (\Phi_1(t), \Phi_2(t), \ldots, \Phi_n(t))^T$

The initial configuration of the *CADCLA* denoted by $Y(0) = \ <\underline{N}(0), \underline{P}(0), \underline{\Phi}(0)>$. As it was previously mentioned, upon activating the cells, a process takes the configuration, reinforcement signals, restructuring signals, and then updates the configuration of the *CADCLA*. The evolution of *CADCLA* can be described by sequence $\{Y(t)\}_{t \geq 0}$ which $\langle Y(t+1)\rangle = \underline{z}\Big(Y(t), \underline{\beta}(t), \underline{\zeta}(t)\Big)$.

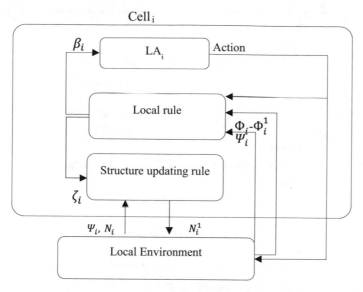

Fig. 2.21 Internal structure of *cell$_i$* and its interaction with the local environment

---

Algorithm mainloop()

Begin

   Repeat

   Choose a cell for activation;// the mechanism of choosing the cells is

   application dependent//

   Activate the selected cell and perform the following phases;

- **Preparation phase:** In this phase, the cell performs the following steps.
    - The cell set its attribute.
    - The cell and its neighboring cells computes its *restructuring signal* using the *local rule (F_1)*.
- **Structure updating phase:** In this phase, the cell performs the following steps.
    - The neighborhood structure of the cell is updated using the structure updating rule (F_2) if the value of the *restructuring signal* of that cell is 1.
- **State updating phase:** In this phase, the cell performs the following steps.
    - The *LA* of the cell select one of their actions. The action selected by the *LA* in the cell determines the new state for that cell.
    - The local rule (F_1) is applied and a *reinforcement signal* is generated
    - The probability vectors of the *LA* of the cell is updated.

   Until (there is no cell for activation)

End

Fig. 2.22 The main loop for the operation of *CADCLA*

Begin

  Repeat

    Choose a cell$_{z_t}$ for activation;   // The sequence

                                        according to which

  the

                                       cells are activated is

                                       application

  dependent//

    Call *Activate* (cell$_{z_t}$);   // Figure 2-25 is the pseudo

  code

                       for algorithm Activate//

  Until (there is no cell to be activated)

End

Fig. 2.23 The pseudo code for the operation of *DCLA*

Definition 2.30 A *CADCLA* is said to be expedient with respect to the cell *cell$_i$* if the following inequality holds:

$$\lim_{t \to \infty} E\left[D_i^{\beta}(Y(t))\right] > \lim_{t \to \infty} D_i^{pc}(Y^{pc}(t)) \tag{2.92}$$

Before we define the elements of the above equitation, we need to define the following items.

- *LG(i, j, k, t)* takes i (index of a cell), j (index of an action), k (index of an attribute), and t (iteration number) and then return $\left\langle \Phi_i^*, \Psi_i^* \right\rangle$ where Φ_i^* is a version of $\Phi_i(t)$ which its item (i, -) is replaced with (i, α_j) and Ψ_i^* is a version of $\Psi_i(t)$ which its item (i, -) is replaced with (i, b_k). Note that (i, -) refer to every item which its first element is equal to i.
- *LPG (i, j, k, t)* returns the probability of appearing a set which is returned by *LG* (i, j, k, t) in iteration t.
- $q_{ijk}^{\beta}(Y(t))$ denotes the reward probability of action α_j of learning automaton *LA$_i$* when the index of attribute of *cell$_i$* is k . $q_{ijk}^{\beta}(Y(t))$ is defined by Eq. (2.93) as given below

$$q_{ijk}^{\beta}(Y(t)) = E\left[\beta_i = 1 | (i, \alpha_j) \in \Phi_i(t), (i, b_k) \in \Psi_i(t)\right]$$
$$= \sum_{j}\sum_{k}\left(\text{LPG}(i, j, k, t) \times F_1^{\beta}(\text{LG}(i, j, k, t))\right) \tag{2.93}$$

Algorithm Activate ()

Input
> cell$_i$

Notations
> F$_1$ denotes the *local rule*
> F$_2$ denotes the *structure updating rule*
> Ψ_i denotes the attribute of *cell$_i$*
> Φ_i denotes the state of *cell$_i$*
> ζ_i denotes the restructuring signal of *cell$_i$*
> N_i denotes the set of neighbors of *cell$_i$*
> β_i denotes the reinforcement signal of the learning automata of *cell$_i$*

Begin

\\ **preparation phase**\\

Set the attribute of the cell$_i$.

Compute ζ_i using F$_1$;

Gather the *restructuring signals* of the Neighboring cells;

\\ **structure updating phase**\\

If (ζ_i is 1) Then
> Compute N_i and Ψ_i using F$_2$;

EndIf

\\state updating phase\\

Each *learning automaton* of *cell$_i$* chooses one of its actions;

Set Φ_i;// set Φ_i to be the set of actions chosen by the set of
> *learning automata* in cell$_i$;

Compute β_i using F_1;

Update the action probabilities of *learning automata* of *cell$_i$* using β_i;

End

Fig. 2.24 The pseudo code of the process which is executed upon the activation of a cell

- $d_{ij}^{\beta}(Y(t))$ denotes the reward probability of action α_j of learning automaton LA_i. $d_{ij}^{\beta}(Y(t))$ is defined by Eq. (2.94) as given below

$$d_{ij}^{\beta}(Y(t)) = \sum_k q_{ijk}^{\beta}(Y(t)) \tag{2.94}$$

Now we define the $D_i^{\beta}(Y(t))$ and $D_i^{pc}(Y^{pc}(t))$

$$D_i^{\beta}(Y(t)) = E[\beta_i = 1|\langle\Phi_i(t), \Psi_i(t)\rangle] = \sum_j \left(p_{ij}(t) \times d_{ij}^{\beta}(Y(t))\right) \tag{2.95}$$

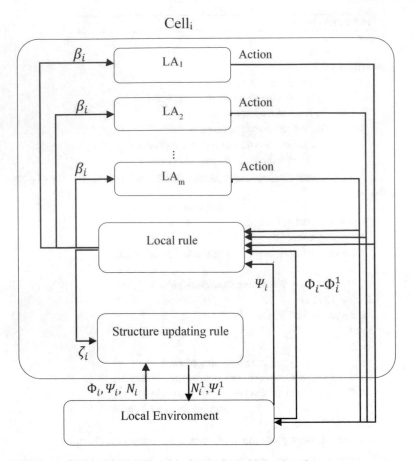

Fig. 2.25 Internal structure of celli and its interaction with local environment

$D_i^{pc}(Y^{pc}(t))$ where $Y^{pc}(t) = <\underline{N}^{pc}(t), \underline{P}^{pc}(t), \underline{\Phi}^{pc}(t)>$ denotes the average reward received by *cell$_i$* of in pure-chance *CADCLA*. A pure-chance *CADCLA* is a *CADCLA* in which, each *LA* is replaced with a pure-chance automaton. A pure-chance automaton is an automaton that always selects each of its actions with equal probabilities (Thathachar and Sastry 2004). The probability vectors of pure-chance automata of the *CADCLA* remains unchanged during the iterations of *CADCLA*. Note that, $LPG(i,j,k,t)^{pc}$ is the probability of appearing the set $LG(i,j,k,t)^{pc}$ in pure-chance *CADCLA*.

Definition 2.31 A *CADCLA* is said to be *expedient,* if it is *expedient* which respect to each cell.

Definition 2.32 A *CADCLA* is said to be ε-optimal with respect to cells, if for each *cell$_i$* in which $d_{il}^{\beta}(Y) > d_{ik}^{\beta}(Y)$ where $l \neq k$, we have the following inequality.

$$\lim_{t \to \infty} \inf p_{il}(t) > 1 - \varepsilon \quad \text{w.p.1} \tag{2.96}$$

Definition 2.33 The *structure updating rule* is called *potential-decreasing*, if applying this rule in each cell leading to decreasing the value of *restructuring signal* of that cell to zero.

Proposition 2.1 *if the structure updating rule is potential-decreasing and each cell is activated infinite times, then there is an iteration $t' < t$ which $T(t) = 0$.*

Proof In each iteration, utilizing *potential-decreasing structure updating rule* leading to decreasing the value of *restructuring signal* of a cell to zero. Since each cell is activated infinite times, the value of *restructuring signal* of that cell ultimately changes to zero. Note that, the *potential energy* is the summation of *restructuring signals* of all cells. Therefore, there is an iteration t' in which the *potential energy* is decreased to zero.

In the rest of this section, the environment under which LA_i in the CLA is operating can be modeled as follows,

1. There is function a $f_{ij}^{\beta}(\mathrm{P_i})$ which $f_{ij}^{\beta}(\mathrm{P_i}(t)) = d_{ij}^{\beta}(Y(t))$ and it is continuous in p_{ik} (j, k = 1, 2, ..., r).

2. $\dfrac{\partial f_{ij}^{\beta}(\mathrm{P_i})}{\partial p_{ij}} < 0$

3. $\dfrac{\partial f_{ik}^{\beta}(\mathrm{P_i})}{\partial p_{ij}} \gg \dfrac{\partial f_{ij}^{\beta}(\mathrm{P_i})}{\partial p_{ij}}$ for $k \neq j$

4. $f_{ij}^{\beta}(.)$ is continuously differentiable in all its arguments.

5. $f_{ij}^{\beta}(\mathrm{P_i})$ and $\left(\dfrac{\partial f_{ij}^{\beta}(\mathrm{P_i})}{\partial p_{ij}} \right)$ are Lipschitzian functions of all their arguments.

In this section, during evolution of the $CADCLA$, we assume that each cell is activated infinite times and $a \to 0$ where a is the reward parameter.

Expediency of $CADCLA$ with L_{RP} Learning Algorithm for the LAs

In this section, we suggest a set of conditions under which the $CADCLA$ with L_{RP} learning algorithm for the LAs is expedient.

Lemma 2.12 *If the non-stationary environment of a LA with L_{RP} Learning Algorithm satisfies the following conditions*

- $c_i (p_1, p_2, ..., p_r)$ is continuous in p_j (i, j = 1, 2, ..., r) if $p_j(t) = p_j$. $c_i (p_1, p_2, ..., p_r)$ denotes the penalty probability of choosing action α_i and p_i denotes the probability of choosing action α_i.
- $\dfrac{\partial c_i(p)}{\partial p_i} > 0$
- $\dfrac{\partial c_j}{\partial p_i} \ll \dfrac{\partial c_i}{\partial p_i}$ for $j \neq i$
- $c_i(.)$ is continuously differentiable in all its arguments.

and $a \to 0$ where a is the reward parameter, Then, the process $\{p_i(t)\}_{t \geq 0}$ is Markovian and ergodic, which satisfy the following equation.

$$\lim_{t \to \infty} E\left[p_i(t) - p_i^*\right]^2 = 0 \qquad (2.97)$$

Where p^* is the probability vector of the LA which satisfy the following equality.

$$p_1^* \times c_1(p^*) = (p_2^*) \times c_2(p^*) = \cdots = p_r^* \times c_r(p^*) \qquad (2.98)$$

Proof The proof is given in Section 7 of Narendra and Thathachar (1989).

Theorem 2.14 *In a CADCLA, if the learning algorithms of LAs are L_{RP}, then regardless to the initial configuration, restructuring function and structure updating rule, we have the following equation.*

$$\lim_{t \to \infty}(\underline{P}(t)) = \underline{P}^* \qquad (2.99)$$

Proof Since all conditions mentioned in Lemma 2.12 for the environment of the *LA* are also mentioned in the environment of the *LA* of each cell, the environment of a *LA* of the *CADCLA* is similar to the environment of the *LA* in the Lemma 2.12. As result of Lemma 2.12, the *LA* of $cell_i$ tries to find probability vector $P_i^* = (p_{i1}^*, p_{i2}^*, \cdots, p_{ir}^*)^T$ which satisfy the Eq. (2.99). This phenomenon occurs in all cells and therefore the probability vectors of all *LAs* approach to $\underline{P}^*$ and the proof is completed.

$$p_{i1}^* \times \left(1 - f_{i1}^\beta(P_i^*)\right) = p_{i2}^* \times \left(1 - f_{i2}^\beta(P_i^*)\right) = \cdots = p_{ir}^* \times \left(1 - f_{ir}^\beta(P_i^*)\right) \quad (2.100)$$

Lemma 2.13 *In a CADCLA, if the learning algorithms of LAs are L_{RP}, then regardless to the initial configuration, restructuring function and structure updating rule, we have* $1 - (r \times w_i) = \sum_{j=1}^{r} p_{ij}^* \times f_{ij}^\beta(P_i^*)$ *where* $w_i = p_{ij}^* \times (1 - f_{ij}^\beta(P_i^*))$ *which* $j \in \{1 \ldots r\}$.

Proof all conditions mentioned in Lemma 2.13 are satisfied for every cell of the *CADCLA*. According to the results of Lemma 2.13, Eq. (2.101) is correct for $cell_i$.

$$p_{i1}^* \times \left(1 - f_{i1}^\beta(P_i^*)\right) = p_{i2}^* \times \left(1 - f_{i2}^\beta(P_i^*)\right) = \cdots = p_{ir}^* \times \left(1 - f_{ir}^\beta(P_i^*)\right) = w_i$$
$$(2.101)$$

$$p_{i1}^* \times \left(1 - f_{i1}^\beta(P_i^*)\right) + p_{i2}^* \times \left(1 - f_{i2}^\beta(P_i^*)\right) + \cdots + p_{ir}^* \times \left(1 - f_{ir}^\beta(P_i^*)\right) = r \times w_i$$
$$(2.102)$$

Equation (2.102) is obtained by using Eq. (2.101).

$$\left(p_{i1}^{*} + p_{i2}^{*} + \cdots + p_{ir}^{*}\right) - p_{i1}^{*} \times f_{i1}^{\beta}\left(P_{i}^{*}\right) + p_{i2}^{*} \times f_{i2}^{\beta}\left(P_{i}^{*}\right) + \cdots + p_{ir}^{*} \times f_{ir}^{\beta}\left(P_{i}^{*}\right)$$
$$= r \times w_{i}$$

(2.103)

$$1 - \left(\sum_{j=1}^{r} \left(p_{ij}^{*} \times f_{ij}^{\beta}\left(P_{i}^{*}\right)\right)\right) = r \times w_{i} \tag{2.104}$$

$$1 - (r \times w_{i}) = \sum_{j=1}^{r} \left(p_{ij}^{*} \times f_{ij}^{\beta}\left(P_{i}^{*}\right)\right) \tag{2.105}$$

Equation (2.105) is obtained by simplifying Eq. (2.103) and the proof is complete.

Theorem 2.15 *In a CADCLA, if the learning algorithms of LAs are L_{RP}, and for every cell$_i$ $w_{i} < \frac{1}{r} - \frac{\sum_{j=1}^{r} f_{ij}^{\beta}(P_{i}^{pc})}{r^{2}}$, then the CADCLA is expedient regardless to its initial configuration, restructuring function and structure updating rule.*

Proof In order to prove that an *CADCLA* is expedient with respect to cells, we need to show that the inequality $\lim_{t \to \infty} E\left[D_{i}^{\beta}(Y(t))\right] > \lim_{t \to \infty} D_{i}^{pc}(Y^{pc}(t))$ is hold for every *cell$_i$*. By expanding both sides of this inequality, we have the following inequality.

$$\lim_{t \to \infty} E\left[\sum_{j}\left(p_{ij}(t) \times f_{ij}^{\beta}(P_{i}(t))\right)\right] > \lim_{t \to \infty} \sum_{j}\left(p_{ij}^{pc}(t) \times f_{ij}^{\beta}(P_{i}^{pc}(t))\right) \tag{2.106}$$

Again, by expanding both sides, we have the following inequality. Since the probability vector of a pure-chance automaton do not change over time, the index t of $P_{i}^{pc}(t)$ is omitted in Eq. (2.107).

$$\lim_{t \to \infty} E\left[p_{i1}(t) \times f_{i1}^{\beta}(P_{i}(t)) + p_{i2}(t) \times f_{i2}^{\beta}(P_{i}(t)) + \cdots + p_{ir}(t) \times f_{ir}^{\beta}(P_{i}(t))\right] > $$
$$\lim_{t \to \infty} \left(p_{i1}^{pc}(t) \times f_{i1}^{\beta}(P_{i}^{pc}) + p_{i2}^{pc}(t) \times f_{i2}^{\beta}(P_{i}^{pc}) + \cdots + p_{ir}^{pc}(t) \times f_{ir}^{\beta}(P_{i}^{pc})\right)$$

(2.107)

After applying expectation, we have the following inequality.

$$\lim_{t \to \infty} \left(E\left[p_{i1}(t) \times f_{i1}^{\beta}(P_{i}(t))\right] + E\left[p_{i2}(t) \times f_{i2}^{\beta}(P_{i}(t))\right] + \cdots + E\left[p_{ir}(t) \times f_{ir}^{\beta}(P_{i}(t))\right]\right) > $$
$$\lim_{t \to \infty} \left(p_{i1}^{pc}(t) \times f_{i1}^{\beta}(P_{i}^{pc}) + p_{i2}^{pc}(t) \times f_{i2}^{\beta}(P_{i}^{pc}) + \cdots + p_{ir}^{pc}(t) \times f_{ir}^{\beta}(P_{i}^{pc})\right)$$

(2.108)

After replacing $p_{ij}^{pc}(t)$ with $\frac{1}{r}$ in the right hand side, Eq. (2.108) changes to Eq. (2.109).

$$\lim_{t\to\infty}\left(E\left[p_{i1}(t)\times f_{i1}^{\beta}(P_i(t))\right]+E\left[p_{i2}(t)\times f_{i2}^{\beta}(P_i(t))\right]+\cdots+E\left[p_{ir}(t)\times f_{ir}^{\beta}(P_i(t))\right]\right)>\frac{\sum_{j=1}^{r}f_{ij}^{\beta}(P_i^{pc})}{r}$$

(2.109)

According to the result of Lemma 2.13, Eq. (2.109) changes to Eq. (2.110).

$$\sum_{j=1}^{r}\left(p_{ij}^{*}\times f_{ij}^{\beta}(P_i^{*})\right)>\frac{\sum_{j=1}^{r}f_{ij}^{\beta}(P_i^{pc})}{r}$$

(2.110)

By substituting Eq. (2.105) in Eq. (2.110) we have the following.

$$1-(r\times w_i)>\frac{\sum_{j=1}^{r}f_{ij}^{\beta}(P_i^{pc})}{r}$$

(2.111)

$$w_i<\frac{1}{r}-\frac{\sum_{j=1}^{r}f_{ij}^{\beta}(P_i^{pc})}{r^2}$$

(2.112)

And the proof is complete.

Proposition 2.2 *If the conditions mentioned in* Theorem 2.15 *are satisfied, then by proper choice of parameters of the LAs, the entropy of the CADCLA approaches to a constant value* h^*.

Proof Note that, H(t) refers to the entropy of the *CADCLA* in iteration t. By expanding $\lim_{t\to\infty}(H(t))$ we have $\lim_{t\to\infty}(-\sum_{k=1}^{n}\sum_{l=1}^{r_k}[p_{kl}(t)\times\ln(p_{kl}(t))])$. According to the result of Theorem 2.15, we have $\lim_{t\to\infty}(\underline{P}(t))=\underline{P}^*$. Therefore, H(t) approaches to $-\sum_{k=1}^{n}\sum_{l=1}^{r_k}[p_{kl}^*\times\ln(p_{kl}^*)]$ which is a constant value and the proof is complete.

Proposition 2.3 *If the conditions mentioned in* Theorem 2.15 *are satisfied, and* structure updating rule *is* potential-decreasing, *then regardless to its initial configuration, the configuration of the CADCLA approaches to configurations in which* $\lim_{t\to\infty}(T(t))=0$ *and* $\lim_{t\to\infty}(H(t))=h^*$ *where* h^* *is a constant value.*

Proof According to the results of Proposition 2.1 and Proposition 2.2, the proof is straightforward.

Expediency of *CADCLA* with $L_{R\varepsilon P}$ Learning Algorithm for the LAs

In this section, we suggest a set of conditions under which the *CADCLA* with $L_{R\varepsilon P}$ learning algorithm for the *LAs* is expedient. In this section, in every *cell$_i$*, we assume that there is an action α_l which $f_{il}^{\beta}(\mathrm{P}_i) > f_{ik}^{\beta}(\mathrm{P}_i)$ where $l \neq k$.

Lemma 2.14 *If the non-stationary environment of a LA with $L_{R\varepsilon P}$ learning algorithm satisfies the following conditions*

- $c_i(p_1, p_2, \ldots, p_r)$ is continuous in p_j $(i, j = 1, 2, \ldots, r)$.
- $\frac{\partial c_i(p)}{\partial p_i} > 0$
- $\frac{\partial c_j}{\partial p_i} \ll \frac{\partial c_i}{\partial p_i}$ for $j \neq i$
- $c_i(.)$ is continuously differentiable in all its arguments.
- $c_i(.)$ and $\left(\frac{\partial c_i}{\partial p_i}\right)$ are Lipschitzian functions of all their arguments.
- There exist an action α_l which $c_k(p) > c_l(p)$ for $k \neq l$

Then, by proper choice of parameters in the *LAs* and for any given $\varepsilon > 0$, the process $\{p(t)\}_{t \geq 0}$ is Markovian and ergodic, which satisfy the following inequality.

$$\lim_{t \to \infty} \inf p_l(t) > 1 - \varepsilon \quad \text{w.p.1} \tag{2.113}$$

Where $p_l(t)$ denotes the probability of selecting action α_l and $p(t)$ denotes the probability vector of the *LA* in iteration t.

Proof The proof is given in section 7 of Narendra and Thathachar (1989).

Theorem 2.16 *For any given $\varepsilon > 0$ and by proper choice of parameters in the LAs, if the learning algorithms of LAs are $L_{R\varepsilon P}$, then regardless to its initial configuration restructuring function and structure updating rule, the CADCLA is ε-optimal with respect to cells.*

Proof All conditions mentioned in Lemma 2.14 for the environment of the *LA* are also mentioned in the environment of the *LA* of each cell. Then the environment of a *LA* of the *CADCLA* is similar to the environment of the *LA* in the Lemma 2.14. According to the results of Lemma 2.14, the probability of selecting action α_l of LA_i (*cell$_i$*) which α_l has the highest reward probability among the actions in the action set of LA_i approaches to a value higher than $1 - \varepsilon$. This phenomenon occurs in all cells. In other word, the *CADCLA* is ε-optimal with respect to cells according to Definition 2.32 and the proof is complete.

Theorem 2.17 *If the learning algorithms of LAs are $L_{R\varepsilon P}$, and for every cell such as cell$_i$ we have $\lim_{t \to \infty} E\left[p_{il}(t) \times f_{il}^{\beta}(\mathrm{P}_i(t))\right] > \frac{\sum_{j=1}^{r} f_{ij}^{\beta}(\mathrm{P}_i^{pc})}{r} - \lim_{t \to \infty} E\left[\sum_{j \neq l} p_{ij}(t) \times f_{ij}^{\beta}(\mathrm{P}_i(t))\right]$, then the CADCLA is expedient with respect to cells regardless to its initial configuration, restructuring function and structure updating rule.*

Proof According to Definition 2.31, in order to prove this theorem, we need to show that the inequality $\lim_{t \to \infty} E\left[D_i^\beta(Y(t))\right] > \lim_{t \to \infty} D_i^{pc}(Y^{pc}(t))$ holds for every *cell$_i$*. By expanding both sides of this inequality, we have the following inequality.

$$\lim_{t \to \infty} E\left[D_i^\beta(Y(t))\right] > \lim_{t \to \infty} \sum_j \left(p_{ij}^{pc}(t) \times f_{ij}^\beta\left(P_i^{pc}(t)\right)\right) \tag{2.114}$$

Note that, the right hand side of (2.114) is not function of t because the probability vector of the pure-chance automata does not change over time. By expanding the left hand side and replacing $p_{ij}^{pc}(t)$ in right hand side of (2.114) with $\frac{1}{r}$, we have the following inequality.

$$\lim_{t \to \infty} E\left[p_{il}(t) \times f_{il}^\beta(P_i(t))\right] > \frac{\sum_{j=1}^r f_{ij}^\beta\left(P_i^{pc}\right)}{r} - \lim_{t \to \infty} E\left[\sum_{j \neq l} p_{ij}(t) \times f_{ij}^\beta(P_i(t))\right] \tag{2.115}$$

And the proof is complete.

Proposition 2.4 *if conditions mentioned in* Theorem 2.17 *are satisfied and the parameters of the LAs are chosen properly, then* $\lim_{t \to \infty}(H(t)) = h^\varepsilon$ *where* h^ε. *is a value which depends on* ε.

Proof by expanding $\lim_{t \to \infty}(H(t))$, we have the following equations.

$$\lim_{t \to \infty}\left(-\sum_{k=1}^n \sum_{l=1}^{r_k} [p_{kl}(t) \times \ln(p_{kl}(t))]\right) \tag{2.116}$$

$$\lim_{t \to \infty}\left(-\sum_{k=1}^n [p_{k1}(t) \times \ln(p_{k1}(t)) + p_{k2}(t) \times \ln(p_{k2}(t)) + \cdots + p_{kr_k}(t) \times \ln(p_{kr_k}(t))]\right) \tag{2.117}$$

As a result of Theorem 2.17, by proper choose of parameters in the *LAs*, we have $\lim_{t \to \infty} \inf p_{il}(t) > 1 - \varepsilon$ with probability 1 for action α_l of *LA$_i$*. Note that, if the value of $p_{il}(t)$ approaches to a value higher than $1 - \varepsilon$, the summation of probabilities of selecting other actions of *LA$_i$* approaches to a value lower than ε. This phenomenon occurs in all *LAs* of the *CADCLA* and therefore, $\lim_{t \to \infty}(H(t))$ approaches to h^ε which h^ε is a value which depends on ε.

Proposition 2.5 *If conditions mentioned in* Theorem 2.17 *are satisfied, and structure updating rule is potential-decreasing, then regardless to its initial configuration, the configuration of the CADCLA approaches to a configuration in which* $\lim_{t \to \infty}(H(t)) = h^\varepsilon$ *and* $\lim_{t \to \infty}(T(t)) = 0$.

Proof According to the results of Proposition 2.1 and Proposition 2.4, the proof is straightforward.

2.3.3.3   Notes for Designing the *DCLAs*

As it was previously mentioned, upon activation of a cell, the cell performs a process containing three phases: **preparation**, **structure updating** and **state updating**. In a cell, during the structure updating phase, the structure updating rule and local rule are conducted to find a neighborhood for the cell which receives low value for the restructuring signal and during the state updating phase, the *LAs* and local rule are conducted to find a state for the cell which receives high value for the reinforcement signal (reward) for the *LA* of that cell. the process of designing appropriate local rule and structure updating rule is application dependent and also very crucial. The following notes may be used to design appropriate rules.

(1) The *DCLAs* are able to find an appropriate cellular structure and also an appropriate set of states of the cells in a self-organized manner. Therefore, they can be used in the problems which can be mapped to a dynamic cellular structure and a set of states.

(2) *DCLAs* inherits the distributed and self-organized computational capabilities from cellular automata and learning capability in unknown environment form learning automata. Therefore, a designer may define the structure updating rule of the *DCLAs* based on self-organized models such as the Schelling segregation model (Schelling 1971) and the fungal growth model (Snyder et al. 2009). Note that, every algorithm capable with cellular automata may be used to design the structure updating rule and the local rule. For example, in (Saghiri and Meybodi 2017b), a distributed algorithm for constructing Voronoi diagrams is used to design a structure updating rule. Distributed construction algorithm for SOM[1] because of restructuring capability may be also used to design a new structure updating rule for the *DCLAs* (Bandeira et al. 1998; Kohonen 1998).

(3) As it was previously mentioned, the structure updating rule should be potential decreasing. it should be noted that the characteristics of the function of potential energy of a DCLA with potential decreasing structure updating rule is very similar to Lyapanov function. Therefore, in (Saghiri and Meybodi 2016a), Schelling segregation model is used to design an appropriate structure updating rule because this model has a Lyapanov function. In (Saghiri and Meybodi 2016a), swap operation is used to emulate the changes in the structure of the cells similar to Schelling segregation model.

(4) Each cell utilizes local information to change its neighborhood. The changes of neighborhood of the cells leading to gradually structuring of cellular structure which is appropriate with respect to criteria defined in the domain of the application. The structure updating rule utilizes cellular operations to change the structure of the cells and therefore the cellular operations must be implemented using local information about cells. In the literature, different types of cellular operators such as swap, join, shrink, and merge are reported (Saghiri and Meybodi 2016a, b, d).

---

[1]Self-organizing map.

2.3.4 Closed Asynchronous Dynamic Cellular Learning Automata with Varying Number of LAs in Each Cell

The Closed Asynchronous Dynamic Cellular Learning Automata with varying number of LAs in each cell (CADCLA-VL) is a version of *CADCLA* in which the number of LAs of the cells changes over time.

Definition 2.34 Closed Asynchronous Dynamic Cellular Learning Automaton with Varying number of Learning Automata in each cell (*CADCLA-VL*) is a network of cells whose structure changes with time and each cell contains a set of *LAs* and a set of attributes (Saghiri and Meybodi 2017b). In this model, the connections among cells, and the number of *LAs* of each cell changes during the course of evolution of *CLA*. This model can be formally defined by 7 tuples as follows

$$CADCLA - VL = (G, A, \Psi, \Phi, N, F_1, F_2),$$

where:

- $G = (V, E)$ is an undirected graph which determines the structure of *CADCLA-VL* where $V = \{cell_1, cell_2, \ldots, cell_n\}$ is the set of vertices and E is the set of edges.
- $A = \{LA_1, LA_2, \ldots, LA_v\}$ is a set of *LAs*. A subset of set A is assigned to a cell.
- $N = \{N_1, N_2, \ldots, N_n\}$ where $N_i = \{cell_j \in V | dist(cell_i, cell_j) < \theta_i\}$ where θ_i is the neighborhood radius of $cell_i$ and $dist(cell_i, cell_j)$ is the length of the shortest path between $cell_i$ and $cell_j$ in $G.N_i^1$ determines the immediate neighbors of $cell_i$.
- $\Psi = \{\Psi_1, \Psi_2, \ldots, \Psi_n\}$, $\Psi_i = \{(j, X_j, C_j) | cell_j \in N_i\}$ denotes the attribute of $cell_j$ where $X_j \subseteq \{x_1, x_2, \ldots, x_s\}$ and $C_j \subseteq A$. $\{x_1, x_2, \ldots, x_s\}$ is the set of allowable attributes. Ψ_i^1 determines the attribute of $cell_i$ when $\theta_i = 1$.
- $\Phi = \{\Phi_1, \Phi_2, \ldots, \Phi_n\}$ where $\Phi_i = \{(j, k, \alpha_l) | cell_j \in N_i \text{ and action } \alpha_l \text{ has been chosen by } LA_k \in C_i\}$ denotes the state of $cell_i$. Φ_i^1 determines the state of $cell_i$ when $\theta_i = 1$.
- $F_1 : (\underline{\Phi}, \underline{\Psi}) \rightarrow (\beta, \zeta)$ is the local rule of *CADCLA-VL*. In each cell, the local rule computes the reinforcement signal and the restructuring signal for the cell based on the states and attributes of that cell and its neighboring cells. for example, in $cell_i$, local rule takes $\langle \Phi_i, \Psi_i \rangle$ and returns $\langle \beta_i, \zeta_i \rangle$. The reinforcemenmt signal is used to update the learning automaton of that cell.
- $F_2 : (\underline{N}, \underline{\Psi}, \underline{\zeta}) \rightarrow (\underline{N}^1, \underline{\Psi}^1)$ is the structure updating rule. In each cell, the structure updating rule finds the immediate neighbors and attribute of that cell. For example, in $cell_i$, structure updating rule takes $\langle N_i, \Psi_i, \zeta_i \rangle$ and returns N_i^1 and Ψ_i^1.

Algorithm mainloop()

Begin

  Repeat

    Choose a cell for activation;// the mechanism of choosing the cells is

                                application dependent//

   Activate the selected cell and perform the following phases;

- **Preparation phase:** In this phase, the cell performs the following steps.
  - The cell set its attribute.
  - The cell and its neighboring cells compute their *restructuring signals* using the *local rule* (F_1).
- **Structure updating phase:** In this phase, the cell performs the following steps.
  - The neighborhood structure of the cell is updated using the structure updating rule (F_2) if the value of the *restructuring signal* of that cell is 1.
- **State updating phase:** In this phase, the cell performs the following steps.
  - Each *LAs* of the cell select one of their actions. The set of actions selected by the set of *LAs* in the cell determines the new state for that cell.
  - The local rule (F_1) is applied and a *reinforcement signal* is generated
  - The probability vectors of the *LAs* of the cell are updated.

  Until (there is no cell for activation)

End

Fig. 2.26 The main loop for the operation of *DCLA*

The internal structure of cell$_i$ and its interaction with local environments is shown in Fig. 2.25. In this model, each cell has a set of *LAs* which may vary during the operation of *CLA*.

The main loop for the operation of a *DCLA* be described in Fig. 2.26.

2.3.5   Norms of Behavior

To judge the behavior of the introduced learning models, it is necessary to set up quantitative norms of behavior. In this subsection, we define three such metrics, entropy, restructuring tendency, and potential energy. The first metric, namely entropy, can be used to study the behavior of models in terms of the actions selected by their constituting learning automata. The next two metrics, namely the restructuring tendency and potential energy, can be used to study the restructurings of the models with dynamic structure (*DICLA*, *HDICLA*, *CADCLA*, and *CADCLA-VL*).

2.3.5.1 Entropy

Entropy, as introduced in the context of information theory by Shannon (Shannon 1948), is a measure of uncertainty associated with a random variable and is defined according to Eq. (2.118),

$$H(X) = -\sum_{X \in \chi} P(X) \cdot \ln(P(X)), \tag{2.118}$$

where X represents a random variable with set of values χ and probability mass function $P(X)$. Considering the action chosen by a learning automaton LA_i as a random variable, the concept of entropy can be used to measure the uncertainty associated with this random variable at any given time instant k according to Eq. (2.119),

$$H_i(k) = -\sum_{j=1}^{m_i} p_{ij}(k) \cdot \ln(p_{ij}(k)), \tag{2.119}$$

where m_i is the cardinality of the action set of the learning automaton LA_i. In the learning process, $H_i(n)$ represents the uncertainty associated with the decision of LA_i at time instant k. Larger values of $H_i(n)$ mean more uncertainty in the decision of the learning automaton LA_i. H_i can only represent the uncertainty associated with the operation of a single learning automaton, but as the operation of an *ICLA*, a *DICLA*, or an *HDICLA* in the environment is a macroscopic view of the operation of its constituting learning automata, we extend the concept of entropy through Eq. (2.120) in order to provide a metric for evaluating the uncertainty associated with the operations of these models.

$$H(k) = \sum_{i=1}^{N} H_i(k). \tag{2.120}$$

In the above equation, N is the number of learning automata in the model. A value of zero for $H(k)$ means $p_{ij}(k) \in \{0, 1\}$ for every i and j. This means that no learning automaton in the model changes its selected action over time, or in other words, the behavior of the model remains unchanged over time. Higher values of $H(k)$ mean higher rates of changes in the behavior of the model. The entropy metric will be used in all models of the *CLAs*.

2.3.5.2 Restructuring Tendency

Restructuring tendency (υ), as defined by Eq. (2.121), is used to measure the dynamicity of the structure of a *DICLA* or an *HDICLA*.

$$v(k) = \sum_{i=1}^{N} |\zeta_i(k)|. \tag{2.121}$$

A value of zero for $v(k)$ means that the tendency vector of no cell of the model during the kth iteration has changed which means that no changes has occurred in the structure of the model during the kth iteration. Higher values of $v(k)$ mean higher changes in the structure of the model during the kth iteration.

2.3.5.3   Potential Energy

The potential energy of the *CLA* is defined by Eq. (2.122) given below

$$T(t) = \sum_{i=1}^{n} \zeta_i(t) \tag{2.122}$$

where $\zeta_i(t)$ is the restructuring signal of *cell*$_i$ at iteration t. **potential energy** can be used to study the changes in the structure of *CLA* as it interacts with the environment. If the value of $T(t)$ becomes zero then no further change needs to be made to the structure. Higher value of $T(t)$ indicates higher disorder in the structure of *CLA*. The potential energy metric will be used in *CADCLA* and *CADCLA-VL*.

2.4   Conclusion

In this chapter, we proposed five versions of the *CLAs*, namely irregular cellular learning automaton (*ICLA*), dynamic *ICLA* (*DICLA*), heterogeneous *DICLA* (*HDICLA*), closed asynchronous dynamic *CLA* (*CADCLA*), and *CADCLA* with varying number of LAs in each cell *(CADCLA-VL)*. In contrast to the CLA proposed by Beigy and Meybodi(Beigy and Meybodi 2004), the *ICLA* model have irregular structure which is more suitable for problem modeling in some areas such as ad hoc and sensor networks, web mining, and grid computing. Furthermore, in comparison to the *ICLA* model, *DICLA, HDICLA, CADCLA,* and *CADCLA-VL* models have dynamic structure, which makes them more suitable for problem solving and modeling of domains, which are dynamic in nature, such as mobile ad hoc and sensor networks. The steady-state behaviors of the introduced learning models were analytically studied. The concept of expediency was introduced for some of the proposed learning models. An *ICLA, DICLA, HDICLA, or CADCLA* is expedient with respect to one of its cells if, in the long run, the learning automaton resides in that cell performs better (receives less penalty) than a pure-chance automaton. An *ICLA, DICLA, HDICLA,* or *CADCLA* is expedient, if it is expedient with respect to all of its constituting cells. Accordingly, an *ICLA, DICLA, HDICLA,* or *CADCLA* is totally expedient if, in the long run, it performs better than its equivalent pure-chance model. Expediency is a notion of learning. Any model that is said to learn must then do at least better than its equivalent pure-chance model. The intended analytical studies showed that all of the proposed models, using L_{RP} learning algorithm, are expedient. Conditions under which these models become totally expedient were also studied. The proposed analytical results are valid only if the learning and restructuring rates are sufficiently small. The proposed model may be applied for designing self-organized mechanisms in the technology of kilobots reported in (Rubenstein et al. 2012, 2014).

Part II
Recent Applications

Chapter 3
Learning Automata for Wireless Sensor Networks

3.1 Introduction

In the recent few decades, substantial parts of our daily exchange of information have occurred in a virtual world, called the cyberspace, consists of several different networking technologies, such as telephone switching networks, television broadcasts, cellular mobile networks, and the internet. Prior to the genesis of the sensor networks, we, the human beings, have been the sole bridge between our physical world and the cyberspace. With the advent of the tiny and low-cost devices, capable of sensing the physical world and communicating over a wireless network, it becomes more and more evident that the status quo is about to change—sensor networks will soon close the gap between the cyberspace and the real world.

There are two main reasons why this vision is gaining momentum (Gama and Gaber 2007). The first reason is the inherent potential of sensor networks to improve our lives. With sensor networks we can expand our environmental monitoring, increase the safety of our buildings, improve the precision of military operations, provide better health care, and give well-targeted rescue aid, among many other applications. The second reason is the endless possibilities such networks offer for multi-disciplinary research, combining typical areas of electrical and computer engineering (sensor technology, integrated systems, signal processing, and wireless communications) with classical computer science subjects (routing, data processing, database management, machine learning, data mining, and artificial intelligence).

A wireless sensor network (*WSN*) is a collection of a large number of sensor nodes that may be randomly and densely deployed. A sensor node is a small electronic device with three capabilities: (1) sensing some types of information from the environment, (2) processing the gathered information, and (3) Communicating over a wireless channel (Ilyas and Mahgoub 2005). With the advent of the recent technologies, it is possible to make such devices small, powerful, and energy efficient. Sensor nodes can now be manufactured

© Springer International Publishing AG 2018
A. Rezvanian et al., *Recent Advances in Learning Automata*, Studies in
Computational Intelligence 754, https://doi.org/10.1007/978-3-319-72428-7_3

cost-effectively in quantity for specialized telecommunications applications. Considering the information processing capabilities and the compact size of the sensor nodes, sensor networks are often referred to as "smart dust." Nowadays, smart inexpensive sensors with several different sensing technologies are available. These include electric and magnetic field sensors; radio-wave frequency sensors; optical-, electro-optic-, and infrared sensors; radars; lasers; location/navigation sensors; seismic and pressure-wave sensors; environmental parameter sensors (e.g., wind, humidity, heat); and biochemical national security–oriented sensors (Sohrabi et al. 2007).

Sensors are internetworked via a series of multi-hop short-distance low-power wireless links. This wireless network is used for long-haul delivery of information to a central point of final data aggregation and analysis, called the sink. This usually employs contention-oriented random-access channel sharing and transmission techniques that are now incorporated in the IEEE 802.15.4 along with ZigBee (more specifically, ZigBee comprises the software layers above the newly adopted IEEE 802.15.4 standard and supports a plethora of applications) standards (Hatler 2004). ZigBee/IEEE 802.15.4 is designed to complement wireless technologies such as Bluetooth, Wi-Fi, and ultra-wideband (UWB), and is targeted at commercial point-to-point sensing applications where cabled connections are not possible and where ultra-low power and low cost are requirements.

Learning automaton and complex structures based on it have been proved to be efficient tools for problem solving within the area of *WSN*s. This chapter aims at providing a number of applications of learning automata in this area. Six different algorithms, which have been applied to four different *WSN* problems, will be proposed. Problems which will be reviewed and discussed are data aggregation, clustering, area coverage, and point coverage. For each problem, we will provide a short informal as well as formal definition, a number of learning automata-based algorithms, and the obtained results for each algorithm.

3.2 Data Aggregation

One way to reduce the energy consumption of the sensor nodes in a *WSN* is to reduce the number of data packets being transmitted. Since sensor networks are usually deployed with a number of redundant nodes, many nodes may have similar information which can be aggregated in intermediate nodes. Aggregation is thus a suitable way for reducing the number of transmitted data packets in the network. Aggregation ratio is maximized if all similar data packets are aggregated together. For this to occur, each node should forward its packets along a path on which maximum number of nodes with similar information exists. We refer to this path for each node as the *Aggregation Path* of that node hereafter. Finding *Aggregation Path* for each node is not so simple considering only local information. A simple approach to find an approximation of this path is to use a greedy algorithm. Each node tries to forward its packets to a neighbor which has similar information to that of the node as it is done in (Beaver et al. 2003). The problem of finding *Aggregation Path* for each node becomes even more challenging if the information possessed by each node differs during the lifetime of the network. In such a situation, *Aggregation Path* of each node changes from time to time and hence, nodes should adapt themselves to such changes.

In this section, a learning automata-based algorithm will be given which helps sensor nodes adapt themselves to the changes occur in their *Aggregation Paths* (Esnaashari and Meybodi 2010b). In the given algorithm, each node in the network is equipped with a learning automaton. These learning automata collectively learn the *Aggregation Paths* of all nodes of the network and help the nodes adapt themselves to the changes occur in the paths. To evaluate the performance of the algorithm, several experiments have been conducted and the results obtained for the given algorithm are compared with the results of three different algorithms: (i) LEACH (Heinzelman et al. 2000), (ii) the algorithm given in (Beaver et al. 2003) which performs data aggregation with no learning, and (iii) the algorithm given in (Beyens et al. 2005) which performs data aggregation using Q-learning. The results have shown that the proposed algorithm outperforms all these algorithms, especially when *Aggregation Paths* are highly dynamic.

3.2.1   Problem Statement

Consider N sensor nodes s_1, s_2, ..., s_N which are scattered randomly throughout a large $L \times L$ rectangular area (A) to monitor a certain phenomenon ζ such as temperature, humidity, etc. The location of each sensor node s_i is presented by (x_i, y_i). Every node s_i is activated periodically, measures ζ at the moment and reports the measured value of ζ $(\hat{\zeta}_i)$ to the sink node using a multi-hop routing scheme. We say $\hat{\zeta}_i$ and $\hat{\zeta}_j$ are *similar* and their values can be aggregated using MEAN operator if $\left| \hat{\zeta}_i - \hat{\zeta}_j \right| < \varepsilon$ for a small $\varepsilon > 0$. Considering a dense network, measured value of ζ is similar in some certain region $(\Omega \subset A)$ around each node i, and hence $\hat{\zeta}_i$ can be aggregated with $\hat{\zeta}_j$, $\forall s_j \in \Omega$. ζ partitions the A into M regions $\Omega_1, \Omega_2, ..., \Omega_M$ with the following properties:

- $\left| \hat{\zeta}_i - \hat{\zeta}_j \right| < \varepsilon;$ $\forall s_i, s_j \in \Omega_k,$ $k = 1, 2, ..., M$
- $\Omega_i \neq \varnothing;$ $for\ i = 1, 2, ..., M$
- $\Omega_i \cap \Omega_j = \varnothing;$ $for\ i \neq j$
- $\bigcup_{i=1}^{M} \Omega_i = A$

Assume that each Ω_k can be approximated with a circle centered at (X_k, Y_k) with radius R_k. Also assume that ζ changes from time to time in different parts of the network and hence, $\Omega_1, \Omega_2, ..., \Omega_M$ are not static regions, but their positions and radiuses vary during the network lifetime.

The aim of the network is to periodically collect for each region Ω_k, the estimate of its position, its radius and its estimated $\hat{\zeta}$ ($\hat{Z}_k$) at the sink node. For this to be achieved, each node s_i reports its position and estimated $\hat{\zeta}_i$ to the sink node periodically.

Define a data gathering round to be the time during which all sensor nodes of the network be activated once and report their information to the sink. Each data gathering round is identified by a unique number, called the data gathering round number, starts from 1 and increases by one at the end of each data gathering round. Using the process of data aggregation, one can reduce the number of packets received at the sink node in each data gathering round approximately to its ideal number, i.e. the number of regions in the environment at that round. Without data aggregation, N data packets, so many of which are redundant, are received by the sink node, one form each node. Performing data aggregation provides longer lifetime for the network. Network lifetime is defined to be the time elapsed from the network startup to the time at which a node in the network dies (Dasgupta and Namjoshi 2003). To measure the network lifetime, the *Round of Death* metric has been utilized, which is defined as the data gathering round number at which the first death in the network occurs.

To specify the problem more clearly, without loss of generality, we make use of a specific application in which nodes have the ability to sense the temperature, hence ζ is temperature in this application. We assume that several circular climates exist in the environment each having a different temperature, and moving along a specific direction with a specific velocity. The velocity and direction of the movement of each climate may be changed randomly during the operation of the network. These climates and their temperatures partition the area of the network into different parts (Ω) each having a different temperature. Figure 3.1 depicts such an environment with 18 sensor nodes and 5 climates. Dashed circles in this figure give different partitions of the network based on the estimate of the temperature in different nodes. Note that the dashed circles have some area in common, since each Ω is estimated by a circle. However, it's not a complicated task for the sink node to make a more accurate estimation of Ω s by removing the common areas.

Each sensor node reports the temperature of the climate in which it resides. If it happens for a node to reside in more than one climate region, its sensor will sense the temperature equal to the mean temperature of all climates it resides in. A large static climate is assumed to be in the environment which covers the whole area A of the network and its temperature is assumed to be 0. With the above assumptions, one can simply computes the temperature each node reports in Fig. 3.1. For example, s_2 and s_5 report 37, s_4 reports -6, and s_3 reports mean value of 37 and -6 which is 15.5. In addition s_1, s_7, s_8, s_9, s_{10}, s_{15} and s_{18} report 0 as their estimations.

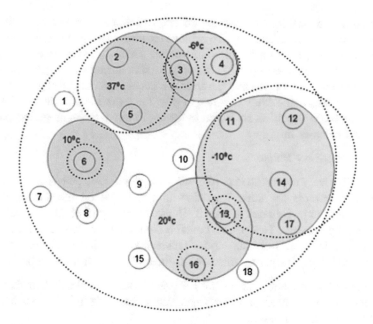

Fig. 3.1 A schematic of the simulated environment with 18 sensor nodes, and 5 different climates. Dashed circles give different partitions of the network based on the estimate of the temperature

3.2.2   The Algorithm

In the network, every node s_i has an internal clock which activates the node every t seconds to measure ζ and to report it to the sink using a multi-hop routing scheme. The routing scheme consists of two phases: route discovery phase and route selection phase. The route discovery phase of the routing scheme is first used to find, for each node, a number of alternative routes towards the sink. The route selection phase is then used to select a route among all the alternative routes using which more aggregation can be performed.

Data aggregation is performed along each route to the sink wherever it is possible. Data and location aggregation will be done at the same time; that is, wherever data aggregation is performed, location aggregation is also performed. In location aggregation, positions of the nodes whose data are aggregated are averaged to find an estimate of the center of the region to which they belong. The estimated radius of the region then becomes the distance between the estimated center and the farthest node. Without location aggregation, a single aggregated packet resulted from aggregating n different packets from n different nodes, must contain the positions of all n nodes. This makes the aggregated packet too long which consumes high energy for transmission.

In order to have more aggregation in the network, each node must try to route its packets toward the sink using a path along which more aggregation can be performed. Since positions of regions change dynamically, the routing algorithm must adapt itself to these changes. To achieve the above two goals, each node in the network is equipped with a learning automaton which is responsible for selecting the next best hop to forward packets towards the sink during a data gathering round. In the rest of this section, we first describe the first version of the algorithm, which we call it the basic algorithm in order to give the main idea. Then two improvements to the basic algorithm are given to further improve the aggregation ratio and prolong the network lifetime.

- **Route Discovery Phase**

The route discovery algorithm is used to discover all possible paths for each node of the network towards the sink. The algorithm is based on the simple flooding method. Sink node initiates a "Path Construction" packet and broadcasts it throughout the network. This packet contains a parameter called "DistanceToSink" which is set to 1 at the sink node. Other nodes wait to receive the "Path Construction". Upon receiving this packet, three different cases may be arisen:

- A "Path Construction" packet with lower "DistanceToSink" than that of the current packet has been previously received at this node. In this case, the newly received packet is ignored.
- The previously received "Path Construction" packets have equal "DistanceToSink" with that of the current packet. In this case, sender of the "Path Construction" is added to the routing table as a new alternative path to the sink, "DistanceToSink" is increased by 1, and the packet is broadcasted over the network again.
- The previously received "Path Construction" packets have greater "DistanceToSink" than that of the current packet. This means that a path with fewer number of hops to the sink is found. So, all the previously found paths are removed from the routing table. Then sender of the currently received "Path Construction" is added as a path to the sink, "DistanceToSink" is increased by 1, and finally the packet is broadcasted over the network again.

The algorithm continues until each node in the network discovers all its paths with the same number of hops toward the sink. For each path, only storing the next hop in the routing table of the node will be sufficient.

At the end of the route discovery phase, each node in the network has a list of all its neighbors using which it can forward its packets toward the sink. We use *RoutingList$_i$* (RL_i) to refer to this list for node s_i where $|RL_i|$ is the number of entries in this list.

- **Route Selection Phase**

After the route discovery phase is over, each node in the network is activated every t seconds and reports its information to the sink node.

In this phase, each node uses a learning automaton for selecting the next best hop to forward its packets towards the sink during each data gathering round. This way, each node gradually learns the best neighbor (the neighbor with the most related data) for forwarding its packets toward the sink.

The learning automaton associated to node s_i, referred to by LA_i, has $|RL_i|$ actions, probability of selecting each is initially set to $\frac{1}{|RL_i|}$. Each action of the LA_i corresponds to the selection of one of the neighboring nodes, listed in the RL_i, to be used for forwarding the data packets of s_i towards the sink. After *node* s_i is activated and has measured the temperature, it asks its learning automaton to select one of its neighboring nodes (actions) for forwarding the measured temperature toward the sink. Data are aggregated during the route selection phase. The action selected by the automaton will be rewarded or penalized based on the acknowledgment received from the selected neighboring node. How to penalize or reward the selected action will be described later.

Between two subsequent activations of the node s_i, the node may receive two different packet types from its neighbors; data packet and acknowledgement packet. A data packet contains measured temperature by the sender of the packet and its location whereas an acknowledgment packet is the response of a neighbor to a data packet sent by s_i to that neighbor. A data packet transmitted by node s_i contains three different parameters:

- K_i which specifies the number of packets aggregated into the packet,
- $\hat{\zeta}_{i,K_i}$ which is the aggregated data computed using Eq. (3.1).

$$\hat{\zeta}_{i,K_i} = \frac{\hat{\zeta}_i + \sum_{j=1}^{n} \hat{\zeta}_j}{n+1}. \tag{3.1}$$

In the above equation, n is the number of packets received at the node s_i.

- $\left(x_{i,K_i}, y_{i,K_i}\right)$ which is the aggregated location and computed using Eq. (3.2).

$$\begin{cases} x_{i,K_i} = \dfrac{x_i + \sum_{j=1}^{n} x_j}{n+1} \\ y_{i,K_i} = \dfrac{y_i + \sum_{j=1}^{n} y_j}{n+1} \end{cases}. \tag{3.2}$$

Received data packets are temporary stored and will be processed upon the next activation of s_i. In the following, we will discuss what a node performs when activated and what a node performs during the time between two subsequent activations.

Node s_i upon its nth activation performs the following operations:

1. Sensor node s_i sets $K_i = 1$.
2. Sensor node s_i senses its surrounding environment to measure $\hat{\zeta}_i$.
3. Sensor node s_i sets $\hat{\zeta}_{i,K_i} = \hat{\zeta}_i$, $x_{i,K_i} = x_i$, $y_{i,K_i} = y_i$.
4. **If** (no data packet is received during the $(n-1)$th activation and nth activation of s_i) **Then**

    4-1. Sensor node s_i creates a data packet containing $\hat{\zeta}_{i,K_i}$, (x_{i,K_i}, y_{i,K_i}) and K_i.

5. **Else if** (any data packet is received) **Then**

    5-1. **For** (each received packet from any neighbor s_j) **Do**

        5-1-1. **If** (the inequality Eq. (3.3) is satisfied) **Then**

            Sensor ode s_i performs data aggregation using a recursive version of the Eq. (3.1) given in Eq. (3.4).
            Sensor node s_i performs location aggregation using a recursive version of the Eq. (3.2) given in Eq. (3.5).
            Sensor node s_i sets $K_i = K_i + K_j$.
            Sensor node s_i uses Eq. (3.6) to compute the data aggregation ratio ($DAR_{i,j}$).
            $DAR_{i,j}$ is sent back to the node s_j via an acknowledgement packet as the feedback from the environment.

    5-2. Sensor node s_i creates a data packet containing $\hat{\zeta}_{i,K_i}$, (x_{i,K_i}, y_{i,K_i}), and K_i.

6. LA_i selects one of its actions (say action k) to determine the neighbor to which newly made data packet should be forwarded.
7. Newly created data packet and the data packets received from the neighbors of s_i, for which inequality Eq. (3.3) is not satisfied, are transmitted to the neighbor s_k.

$$\left| \hat{\zeta}_{i,K_i} - \hat{\zeta}_{j,K_j} \right| < \varepsilon. \tag{3.3}$$

ε is a threshold which specifies the maximum difference between measured temperatures below which aggregation can be performed.

$$\hat{\zeta}_{i,K_i} = \frac{K_i \cdot \hat{\zeta}_{i,K_i} + K_j \cdot \hat{\zeta}_{j,K_j}}{K_i + K_j}. \tag{3.4}$$

$$\begin{cases} x_{i,\mathrm{K}_i} = \dfrac{\mathrm{K}_i \cdot x_{i,\mathrm{K}_i} + \mathrm{K}_j \cdot x_{j,\mathrm{K}_j}}{\mathrm{K}_i + \mathrm{K}_j} \\[2mm] y_{i,\mathrm{K}_i} = \dfrac{\mathrm{K}_i \cdot y_{i,\mathrm{K}_i} + \mathrm{K}_j \cdot y_{j,\mathrm{K}_j}}{\mathrm{K}_i + \mathrm{K}_j} \end{cases} \tag{3.5}$$

$$DAR_{i,j} = U\left(\frac{Min(\hat{\zeta}_{i,\mathrm{K}_i} + \varepsilon, \ \hat{\zeta}_{j,\mathrm{K}_j} + \varepsilon) - Max(\hat{\zeta}_{i,\mathrm{K}_i} - \varepsilon, \ \hat{\zeta}_{j,\mathrm{K}_j} - \varepsilon)}{2\varepsilon} \right), \tag{3.6}$$

where

$$U(x) = \begin{cases} x; & x \geq 0 \\ 0; & x < 0 \end{cases}. \tag{3.7}$$

Upon receiving the acknowledgment packet, containing $DAR_{k,i}$, by $\underline{s_i}$, LA_i penalizes or rewards its selected action (action k). If $DAR_{k,i}$ is greater than a threshold $AcceptRate$, then action k is rewarded according to Eq. (3.8).

$$\begin{aligned} p_k(n+1) &= p_k(n) + \alpha \cdot (DAR_{k,i}) \cdot (1 - p_k(n)) \\ p_l(n+1) &= p_l(n) - \alpha \cdot (DAR_{k,i}) \cdot p_l(n) \qquad \forall l \, l \neq k, \end{aligned} \tag{3.8}$$

and otherwise, it is penalized according to Eq. (3.9).

$$\begin{aligned} p_k(n+1) &= (1 - \beta \cdot (1 - DAR_{k,i})) \cdot p_k(n) \\ p_l(n+1) &= \frac{\beta \cdot (1 - DAR_{k,i})}{r - 1} + (1 - \beta \cdot (1 - DAR_{k,i})) p_l(n) \quad \forall l \, l \neq k. \end{aligned} \tag{3.9}$$

In the above equations, α is the reward rate, and β is the punishment rate. If by any means, node s_i does not receive any acknowledgment from the node s_k, LA_i penalizes its selected action according to Eq. (3.9) considering $DAR_{k,i}$ as zero. Node s_i then retransmits its measured temperature via one of its neighboring nodes listed in RL_i other than the node s_k. If no other neighboring node exists in RL_i, node s_i transmits a *RouteRequest* packet to all of its neighboring nodes. If a neighboring node s_l receives the packet and s_i is not listed in its RL_l, it replies with a *RouteReply* message. Node s_i gathers the received *RouteReply* packets and adds them to its RL_i. The LA_i of s_i is reinitialized in such a way that all of its new actions have the same action probabilities. If the node s_i does not receive any *RouteReply* packet, then it can be regarded as a dead node. At this time, the network lifetime is considered to be over.

- **Improvements**

In this section we discuss two improvements to the algorithm given in the previous section. For the sake of clarity in presentation in the rest of this section, we

will call the previous algorithm as the "Basic Algorithm", and the basic algorithm
with the first improvement as the "Algorithm 3.1" and with the second improvement as the "Algorithm 3.2".

Algorithm 3.1 In Algorithm 3.1, unlike the basic algorithm, in which only the
aggregation ratio between a node and its next node in the routing path is considered
for choosing a neighboring node, the aggregation ratio between a node and its two
next nodes in the routing path is considered for choosing a neighboring node. For
the network of Fig. 3.2, if the basic algorithm is used, then node s_{10} chooses one of
its neighbors s_7, s_8 and s_9 at random for sending its packets towards the sink
whereas Algorithm 3.1 chooses node s_7. This is because the second node from s_{10}
in the routing path passed through s_7 is the node s_6 which is in the same region as
s_{10} and hence its information can be aggregated with the information of s_{10}.

Algorithm 3.1 performs the following. Each node s_i upon receiving a data packet
from its neighbor s_j in its nth activation time, computes $DAR_{i,j}$ using Eq. (3.6), and
then makes use of Eq. (3.10) for computing $Enhanced\text{-}DAR_{i,j}$. Once the $Enhanced\text{-}DAR_{i,j}$ is computed, this will be sent back to node s_j as the response of node s_i.

$$Enhanced - DAR_{i,j} = Max(DAR_{i,j}, MRR_i(n)). \qquad (3.10)$$

In Eq. (3.10), MRR_i stands for Maximum Received Reward and is defined
through Eq. (3.11).

$$MRR_i(n) = \max_{1 \le t \le n} \left(DAR_{k,i}; \quad for\ k \in RL_i \right). \qquad (3.11)$$

The max operation is performed over all responses received by node s_i from the
network startup until the time it is computed.

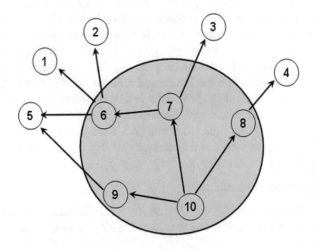

Fig. 3.2 Network of node s_{10}
and the environment around it

Using Algorithm 3.1, in the network of Fig. 3.2, the computed *Enhanced-DAR*$_{i,j}$ at the node s_7 is more than that of nodes s_8 and s_9. This is because *DAR*$_{i,j}$ in all three nodes are equal, but *MRR* at the node s_7 is more than that in the nodes s_8 and s_9. Therefore, the probability of selecting the node s_7 by node s_{10} will be higher than that of the nodes s_8 and s_9.

Algorithm 3.2 Algorithm 3.2 is Algorithm 3.1 in which when a node cannot find a neighboring node for data aggregation (a neighboring node in the same region) such as node s_6 in Fig. 3.2, it chooses a neighboring node with more residual energy for forwarding its packets. Using this algorithm, a node located at the boundary of a region, forward its packets to the outside of its region using a neighbor with highest residual energy. This means that routing inside a region is performed based on the data aggregation ratio between nodes whereas routing between different regions is performed based on the residual energy.

In Algorithm 3.2, upon receiving a packet in node s_i from the neighboring node s_j, *Enhanced-DAR*$_{i,j}$, is computed. If this ratio is more than *AcceptRate*, node s_i sends back a favorable response to LA_j. Otherwise, node s_i considers its residual energy to compute the response to action i of LA_j; if the normalized residual energy (*NRE*$_i$ is computed using Eq. (3.12)) is more than the *AcceptRate*, the response is favorable and is unfavorable otherwise.

$$NRE_i = \frac{EL_i}{MaxEnergyLevel}. \tag{3.12}$$

In the above equation, EL_i is the residual energy level of the sensor node s_i and *MaxEnergyLevel* is the residual energy level of a full battery charged node.

3.2.3 Experimental Results

To evaluate the performance of the given algorithms several experiments have been conducted. In the first two experiments, the basic algorithm is compared with three different methods: (i) LEACH algorithm proposed in (Heinzelman et al. 2000), (ii) method given in (Beaver et al. 2003) with no learning, and (iii) method given in (Beyens et al. 2005) which uses Q-learning. In the third experiment, the performance of the basic algorithm is studied by comparing the number of packets received at the sink node in each data gathering round by the number of packets which needed to be received at the sink at that round. The fourth experiment compares Algorithm 3.1 and Algorithm 3.2 with the basic algorithm.

All simulations have been implemented using NS2 simulator.[1] We have used IEEE 802.11 as the MAC layer protocol, two ray ground as the propagation model and omni-directional antenna. Nodes are placed randomly on a two-dimensional

---

[1]http://isi.edu/nsnam/ns/.

network area of size 100 m × 100 m. Energy model specified in Heinzelman et al. (2000) is used for estimating the amount of energy consumed for packet transmissions. All packets except for the data packets, which are 525 bytes long, are assumed to be 8 bytes long. Environment is modeled by 12 circular climates; each has its own temperature, direction and velocity of movement. Climates are moving using random way point movement strategy with a velocity which is randomly selected from a normal distribution with $\mu = 10^{-3}$ and $\sigma = 10^{-4}$. Movements of the climates are bounded to the boundaries of the network, and hence if a climate happens to get out of the environment boundaries, it changes its direction by 180° and comes back to the environment of the network. In all experiments, α and β for the proposed algorithm are set to 0.1, and ε (acceptable range of aggregation) is set to 5. Also α and β parameters in Q-learning method are set to 10^{-1} and 10^{-2} respectively. These values are selected experimentally. t which is the time interval between two consecutive sending of data to the sink in each node is equal to 10 s. *AcceptRate* is set to 0.85. Simulations are performed for 50, 100, 200, 300, 400, and 500 sensor nodes. The results are averaged over 50 runs.

- **Experiment 1**

In this experiment, the total number of packets received at the sink node is studied. Figure 3.3 depicts the results for the basic algorithm and the three existing algorithms mentioned earlier. This figure shows that the basic algorithm significantly outperforms algorithms *i*, *ii*, and *iii* by 66, 37, and 9%, respectively. Figure 3.4 compares the results obtained for the basic algorithm with the results obtained for the other three algorithms for $N = 500$ when μ varies in the range [0, 0.05]. This figure shows that the basic algorithm has no superiority over other algorithms when climates have no movement ($\mu = 0$), but it works better as μ increases.

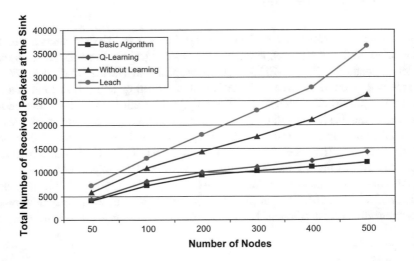

Fig. 3.3 Total number of the received packets at the sink

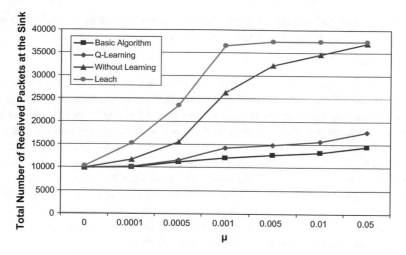

Fig. 3.4 Changes in the total number of the received packets at the sink node with respect to the changes in the value of μ

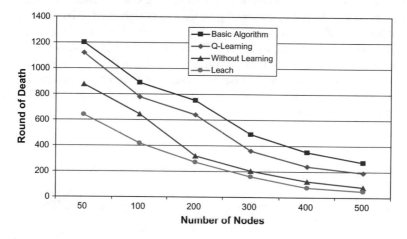

Fig. 3.5 Network lifetime based on the *Round of Death* metric

- **Experiment 2**

In this experiment, the lifetime of the network is studied. Figure 3.5 compares the lifetime of the network for the basic algorithm and the other three algorithms in terms of the *Round of Death* metric. Higher lifetime for the basic algorithm as shown in Fig. 3.5 is due to reduction in the number of transmitted packets in the network. As it can be seen from Fig. 3.5, when the number of nodes in the network increases, *Round of Death* decreases. This is mainly due to the fact that in a denser network, sensor nodes near the sink relay more packets and hence their energy is depleted earlier resulting in a lower *Round of Death*.

- **Experiment 3**

This experiment is conducted to study the efficiency of the basic algorithm with respect to the information received at the sink node during a data gathering round. For this purpose, we compare the number of packets actually received at the sink node using the proposed algorithm and the number of packets which needed to be received at the sink node in each data gathering round. Figure 3.6, which gives this comparison, shows that by applying the proposed algorithm, number of packets received at the sink node in each data gathering round gradually approaches its ideal amount, which is, the number of regions in the environment in that round. Peaks in this figure correspond to sudden changes in the number of regions due to the movements of the climates. It can be seen that the basic algorithm learns new routing paths and hence adapts the network to these changes.

- **Experiment 4**

In this experiment, the basic algorithm is compared with Algorithm 3.1 and Algorithm 3.2 with respect to the total number of packets received at the sink node. As it is shown in Fig. 3.7, comparing to the basic algorithm, Algorithm 3.1 and Algorithm 3.2 both results in lesser number of received packets at the sink. This indicates that higher data aggregation ratio can be obtained using Algorithms 3.1 or 3.2. Figure 3.8 compares the basic algorithm, Algorithm 3.1 and Algorithm 3.2 in terms of the network lifetime. As before, *Round of Death* metric is used for this purpose. It can be seen that Algorithm 3.1 and Algorithm 3.2 both result in longer lifetime for the network.

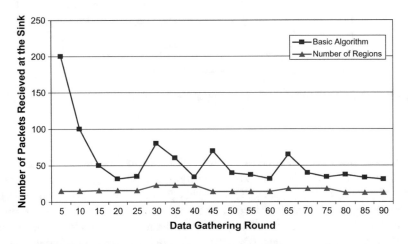

Fig. 3.6 Number of packets received at the sink in each data gathering round for Basic Algorithm versus the number of packets which must be received at the sink at that round

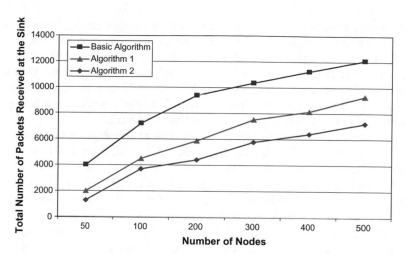

Fig. 3.7 Total number of packets received at the sink

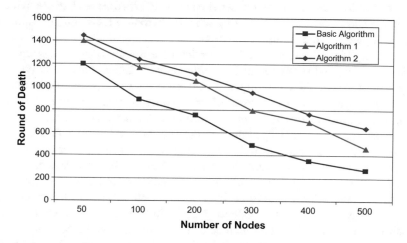

Fig. 3.8 Network lifetime based on the *Round of Death* metric

3.3   Clustering

The major design goal in a power-aware communication protocol for *WSN*s is to minimize the energy consumption and maximize the system lifetime. Clustering is one of the approaches for achieving such design goal. Clustering is the act of dynamically electing a cluster head among the eligible active nodes. In the clustering process, some of the sensor nodes with a defined radio range make a cluster together choosing a node as the cluster head (hereafter called the head), whose responsibility is to communicate with heads of the other clusters. It is possible that in the future, these clusters are combined to create a bigger cluster. Usually, at the end of a clustering algorithm, the sensor nodes are organized into disjoint sets (clusters). Each cluster consists of a head which manages the cluster and some followers (hereafter called the members) which are usually located within the communication radius of the head. Usually, each node is a member of only one cluster. The purpose of a clustering algorithm is to find a suitable clustering with proper number of clusters which cover the whole network.

Clustering has some advantages such as data aggregation in order to reduce energy consumption by the sensors (Heinzelman et al. 2000; Bandyopadhyay and Coyle 2003), facilitates queries on the sensor network (Demirbas et al. 2004), form an infrastructure for scalable routing (Lin and Gerla 1997; Banerjee and Khuller 2001), and efficient network-wide broadcast (Chen et al. 2004). Clustering algorithms for sensor networks fall into two categories; centralized clustering and distributed clustering (Liu 2007). Centralized clustering algorithms require the global network knowledge, introduce substantial storage, communication and computation overheads and thus are not desirable for resource-constrained sensor nodes. Distributed clustering algorithms, on the other hand, usually make decisions based on the localized information (Krishna et al. 1997; Amis et al. 2000; Wu et al. 2006). In general, distributed clustering schemes introduce less communication cost when compared with centralized schemes.

Clustering algorithms proposed in the literature usually consist of two phases; initial clustering and re-clustering (Heinzelman et al. 2000; Huang and Wu 2005; Ding et al. 2005). Initial clustering is performed at the network startup, resulting in an infrastructure in which each node of the network is either a head or a member. Using this infrastructure, the normal operation of the network is started. In many real scenarios, the normal operation of the network is a simple data gathering scheme which is performed as follows: members periodically gather data from the environment, process the data and then send it out to their heads. Heads periodically aggregate data received from their members and forward the aggregated data towards the sink using a multi-hop inter-cluster routing scheme. Heads consume lots of energy in such a data gathering scenario because of two main reasons; Receiving data from all of their members and inter-cluster communications which need higher transmission range than intra cluster communications. Thus, to prevent heads from rapid energy exhaustion, re-clustering is performed in the network from time to time to change the role of the nodes as heads or members.

In this section, we will give a clustering algorithm for *WSN*s which is fully distributed and does not require sensor nodes of the network to be fully synchronized with each other (Esnaashari and Meybodi 2008). This clustering algorithm consists of two phases; initial clustering and re-clustering. Unlike the static re-clustering schemes, the re-clustering scheme in this section is performed locally and adaptively whenever it is needed. A local re-clustering is only required if a head consumes so much energy that it cannot continue its role as a head. In such a situation, a local re-clustering is initiated in the cluster managed by that head. The role of the head is transferred to the newly elected head(s) at the end of the local re-clustering phase, thus normal operation of the network need not to be stopped during the re-clustering phase. Local re-clustering reduces the amount of energy consumed for changing the infrastructure of the network. The given clustering algorithm in comparison to the existing algorithms produces a clustering in which each cluster has higher number of nodes and higher residual energy for the cluster heads. Local re-clustering, higher residual energy in cluster heads and higher number of nodes in each cluster results in a network with longer lifetime. To evaluate the performance of the given algorithm several experiments have been conducted. The results of the experiments have shown that the given clustering algorithm outperforms the existing clustering algorithms in terms of the quality of clustering measured by the total number of clusters, the number of sparse clusters and the remaining energy levels of the cluster heads. Experiments have also shown that the given clustering algorithm in comparison to other existing algorithms better prolongs the network lifetime.

3.3.1   The Algorithm

This clustering algorithm consists of two phases: initial clustering and re-clustering. In the initial clustering phase, which is performed when the network starts operating, all nodes of the network participate. At the end of this phase, a fully clustered network is created. Re-clustering phase, on the other hand, is initiated whenever the energy level of a cluster head in the network degrades by a specified percent. In re-clustering phase, only those nodes which are members of the cluster whose head initiates re-clustering participate. The result of this phase is a new clustered infrastructure in that part of the network. Before the initial clustering phase starts, a time-driven asynchronous *ICLA* which is isomorphic to the sensor network topology is created. Each node s_i in the sensor network corresponds to the cell c_i in *ICLA*. Two cells c_i and c_j in *ICLA* are adjacent to each other if s_i and s_j in the sensor network are close enough to hear each other's signal. The learning automaton in each cell c_i of the *ICLA*, referred to as LA_i, has two actions a_0 and a_1. Action a_1 is "declaring the node as a head" and action a_0 is "declaring the node as a member". The probability of selecting each of these actions is initially set to .5. In the rest of this section algorithms for initial clustering and re-clustering phases are described in details.

- **Initial Clustering Phase**

During the initial clustering phase, the role of each sensor node, which is initially set to *unspecified*, is changed to either *head* or *member*. Each node in the network maintains an array called *NeighborsInfo* including several pieces of information about each of its neighbors: energy level, number of its neighbors and the selected action by the corresponding cell. *NeighborsInfo* array for a node is empty at the beginning of the *initial clustering* phase and will be updated every time the node receives a packet called *ClusterADV* from any of its neighbors. *ClusterADV* packet will be introduced later in this section.

Each cell of the *ICLA*, during the *initial clustering* phase, is activated asynchronously. The total number of times that a cell is activated is bounded by *MAX_ITERATION*. The nth activation of the cell c_i occurs at time $R_i + n \times t$ where R_i is a random number generated for cell c_i and t is a constant. We call n the local iteration number for the cell. Delays R_i are chosen randomly in order to reduce the probability of collisions between the neighboring nodes.

The clustering algorithm starts when a cell in *ICLA* is activated. If cell c_i is activated then the following operations are performed:

1. **If** (the activation of the cell c_i is its first activation (local iteration 1)) **Then**

    1-1. LA_i decides whether to declare node s_i in the sensor network as a head or as a member, that is, the cell chooses one of its actions using its action probability vector.

    1-2. Sensor node s_i creates a *ClusterADV* packet, containing the action chosen by cell c_i, its energy level, and the number of entries in *NeighborsInfo* array (which is equal to the number of neighbors of the node s_i from which *ClusterADV* packet is received so far) and transmits the packet to all of its neighbors.

2. **If** (the activation of the cell c_i is not its first activation (local iteration $n > 1$)) **Then**

    2-1. The node s_i in the sensor network interrupts its normal operation and computes the reinforcement signal β_i according to Eq. (3.13), using the information maintained in its *NeighborsInfo* array.

$$\beta_i(n) = \begin{cases} 0; & -\alpha_i(n) \cdot (\sum_{j=1}^{N_i(n)} \alpha_j(n)) + (1 - \alpha_i(n)) \cdot (\sum_{j=1}^{N_i(n)} \alpha_j(n) - 1) \geq 0 \\ 1; & Otherwise \end{cases} \quad (3.13)$$

In Eq. (3.13), $N_i(n)$ is the number of neighbors of the sensor node s_i at local iteration n and $\alpha_i(n)$ is the selected action of LA_i at local iteration n. $\alpha_j(n)$ is the last action selected by cell c_j observed by node s_i (that is the action included in the last *ClusterADV* packet sent by node s_j and received by node s_i).

    2-2. LA_i updates its action probability vector using the computed β_i according to the following learning algorithm with time varying parameters $a(n)$ and $b(n)$.

$$\begin{cases} p_i(n+1) = p_i(n) + a(n) \cdot (1 - p_i(n)) \\ p_j(n+1) = 1 - p_i(n+1) \end{cases}, \tag{3.14}$$

$$\begin{cases} p_i(n+1) = (1 - b(n)) \cdot p_i(n) \\ p_j(n+1) = 1 - p_i(n+1) \end{cases}. \tag{3.15}$$

Reward parameter $a(n)$ and penalty parameter $b(n)$ are time varying parameters which vary according to Eqs. (3.6) and (3.7). Values of these two parameters at time n at a given cell depend on the current selected action by the cell, current energy level and the number of the neighbors of the corresponding node.

$$a(n) = \alpha_i(n) \cdot (\phi \cdot \psi_i(n)) \\ + (1 - \alpha_i(n)) \cdot [\phi \cdot (1 - \psi_i(n))], \tag{3.16}$$

$$b(n) = \alpha_i(n) \cdot [\rho \cdot (1 - \psi_i(n))] \\ + (1 - \alpha_i(n)) \cdot (\rho \cdot \psi_i(n)), \tag{3.17}$$

where

$$\psi_i(n) = \frac{E_{s_i}(n)}{MaxEnergyLevel}. \tag{3.18}$$

In Eqs. (3.6) and (3.7) $\alpha_i(n)$ is the action selected by the cell at time n and ϕ and ρ are two constants. It can be shown that if the reward parameter $a(n)$ and the penalty parameter $b(n)$ vary according to Eqs. (3.16) and (3.17) and φ, $\rho < 1$, then we have $0 < a(n) < 1$, $0 < b(n) < 1$ (Lemma 3.1 given in Sect. 3.3.2). In Eq. (3.18), $E_{s_i}(n)$ is the residual energy level of the node s_i at time n and $MaxEnergyLevel$ is the energy level of a full battery charged node.

2-3. Cell c_i chooses one of its actions using its action probability vector; that is c_i declares the node s_i in the sensor network as a head or as a member.
2-4. Node s_i creates a *ClusterADV* packet and transmits it to its neighbors.

Initial clustering phase for a node stops if either its local iteration number exceeds *MAX_ITERATION* or if the role of the node changes from *unspecified* to head or member. A node becomes a head if the probability that the node declares itself as a head exceeds a predetermined threshold (*UP_THRESHOLD*) and becomes a member if it receives a signal (contained in *ClusterADV* packet) from a node which has already become a head.

If the clustering phase for a node is over because the local iteration number exceeds *MAX_ITERATION*, but its role is still *unspecified*, then it polls all of its neighbors whether or not any of them is a head. If there are such neighbors, then the polling node becomes a member of the cluster of one of those neighbors chosen at random. Otherwise, the polling node itself becomes a head.

- **Re-clustering Phase**

Re-clustering phase in many clustering algorithms, reported in the literature, is performed periodically for the entire network (Heinzelman et al. 2000; Ding et al. 2005; Huang and Wu 2005). In such algorithms, in predetermined time intervals, the normal operation of the network is interrupted, the clustering algorithm is performed on the entire network (producing a completely new clustered infrastructure) and then the normal operation of the network is resumed. Such periodical re-clustering schemes has a number of drawbacks. The very first problem with such schemes is that they consume too much energy because the re-clustering is performed on the entire network. Another problem is that the normal operation of the network is delayed until the re-clustering phase is over. In many critical applications, such as military domains, this delay is not acceptable. Finally, such re-clustering schemes are static and do not receive any feedback from the network and the state of the sensor nodes. Periodical re-clustering may change the role of a node from a head to a member though it still has enough residual energy to continue its role as a head. Unlike such static re-clustering schemes, the re-clustering scheme proposed in this section is performed locally and adaptively whenever it is needed. A local re-clustering is only required if a head consumes so much energy that cannot continue its role as a head. In such a situation, a local re-clustering is initiated in the cluster managed by that head. The role of the head is transferred to the newly elected head(s) at the end of the local re-clustering phase, thus normal operation of the network can be intermingled with the re-clustering phase.

Re-clustering phase is initiated by a head, whenever its energy level degrades to a specified percent (*RE-CLUSTERING_PERCENT*) below the energy level at the time it was selected as a head. Upon the initiation of a re-clustering phase by a node s_i, whose role is a head, the following operations are performed:

1. Sensor node s_i creates a packet called *Recluster* and transmits it to its neighbors. This packet is just for informing the neighbors of s_i the beginning of the re-clustering phase and contains no other information.
2. **If** (a sensor node s_j, which receives the *Recluster* packet, is a head) **Then**

   2-1. Sensor node s_j creates a *ClusterADV* packet and transmits it to its neighbors.

3. **Else if** (a sensor node s_j, which receives the *Recluster* packet, is a member) **Then**

3-1. **If** (sensor node s_j is a member of the cluster managed by s_i) **Then**

    3-1-1. The action probability vector $(p_1, p_2)^T$ of LA_j is set to $(.5, .5)^T$.

    3-1-2. Sensor node s_j changes its role from a member to *unspecified*.

    3-1-3. Sensor node s_j starts to execute the initial clustering phase algorithm with smaller value for parameter *MAX_ITERATION*.

3-2. **If** (sensor node s_j is not one of the members of the cluster managed by s_i) **Then**

    3-2-1. Sensor node s_j disregards the received *Recluster* packet and continues its normal operation.

Once the re-clustering phase is over, members of the cluster, whose energy was depleted, may have become a head or a member of an existing cluster or a member of a newly formed cluster. During the re-clustering phase, previous head (node s_i) maintains its role as a head to allow network performs its normal operation in parallel to re-clustering.

3.3.2   Analytical Study

In this section, we analytically proof that the proposed clustering algorithm produces a clustering infrastructure for the sensor network, in which the set of cluster heads form a connected graph asymptotically almost surely.

Definition 3.1 A simple monitoring application in a clustered sensor network is defined as follows: members of each cluster periodically obtain data from the environment and send it to the head of the cluster. The head of each cluster periodically aggregates data received from its members and then sends the aggregated data towards the sink using a multi-hop inter-cluster routing scheme. Inter-cluster communications are performed using transmission range of $3R_t$.

Definition 3.2 A data gathering round is the time during which the sink node receives one single packet from each head of the network.

Lemma 3.1 *If the reward parameter $a(n)$ and the penalty parameter $b(n)$ vary according to the Eqs. (3.16) and (3.17) and ϕ, $\rho < 1$, then we have $0 < a(n) < 1$ and $0 < b(n) < 1$.*

Proof Using Eq. (3.18) and knowing the fact that $E_{s_i}(n) \leq MaxEnergyLevel$ for every node s_i, then we have $\psi_i(n) \leq 1$. Using Eqs. (3.16) and (3.17) and knowing the facts that $\alpha_i(n)$ is either 0 or 1, $\psi_i(n) \leq 1$, and ϕ, $\rho < 1$, we can conclude that a $(n) < 1$ and $b(n) < 1$.

Lemma 3.2 *Assume that the normal operation of the network is a simple monitoring application which is performed using a multi-hop inter-cluster routing*

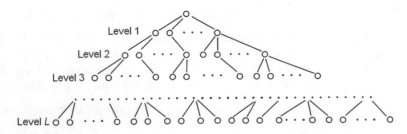

Fig. 3.9 Routing tree used for inter-cluster routing

scheme. Also assume that the routing scheme uses a routing tree rooted at the sink node, spanned throughout the network, and has a branching factor of d on average. Then, on average, each head in this routing tree relays $\eta = \frac{d^2 - \left(1 + (1-d) \cdot \left(\log_d^k - 2\right)\right) \cdot k}{k \cdot (1-d)^2}$ packets from other heads where k is the number of clusters in the network.

Proof Consider the routing tree depicted in Fig. 3.9. In this figure, the root of the tree is the sink and other nodes are head nodes. Any level of this tree relays the packets received from the levels beneath it. By the assumption of the lemma, each node has an average branching factor of d. Thus, the average number of packets relayed at level i of this tree (μ_i) and the total average number of packets relayed in the network (*ATPR*) are given by equations Eqs. (3.19) and (3.20), respectively.

$$\frac{d^{i+1} - d^{L+1}}{1 - d} = d^{i+1} + d^{i+2} + \cdots + d^L = \mu_i, \tag{3.19}$$

$$ATPR = \sum_{i=1}^{L} \mu_i = \frac{d^2 - (1 + (1 - d) \cdot (L - 1)) \cdot d^{L+1}}{(1 - d)^2}. \tag{3.20}$$

In the above equations, L is the depth of the routing tree. Using Eq. (3.20), the average number of packets, relayed by each node in the routing tree, is then computed using Eq. (3.21).

$$\eta = \frac{ATPR}{k} = \frac{d^2 - (1 + (1 - d) \cdot (L - 1)) \cdot d^{L+1}}{k \cdot (1 - d)^2}, \tag{3.21}$$

where k is the number of clusters in the network. Replacing L by $\log_d^k - 1$ in Eq. (3.21) we get

$$\eta = \frac{d^2 - \left(1 + (1 - d) \cdot \left(\log_d^k - 2\right)\right) \cdot k}{k \cdot (1 - d)^2}, \tag{3.22}$$

and hence the lemma.   ∎

Theorem 3.1 *Assume that the normal operation of a clustered network is a simple monitoring application. Furthermore, assume that the energy consumption for re-clustering is negligible. Then, if the transmission range for intra-cluster communication and inter-cluster communication are R_t and $3R_t$, respectively, then the optimum number of clusters (k) in order to minimize the total energy consumed by the network at each round of data gathering is given by*

$$k = d^{(1-d) \cdot \left(\frac{E_S^f - E_S^n - E_R}{E_S^f + E_R} \right) + \frac{1}{d-1} \frac{1}{\ln(d)} + 2},$$
(3.23)

where d is the average branching factor of the routing tree, E_S^f and E_S^n are the amount of energy consumed for sending a data packet using transmission ranges of $3R_t$ and R_t respectively, and E_R is the amount of energy consumed for receiving a data packet.

Proof Let E_{ch} be the amount of energy consumed by a head and E_{cm} be the amount of energy consumed by a member at each round of data gathering. Since the amount of energy consumed in the re-clustering phase is assumed to be negligible, E_{ch} can be expressed using Eq. (3.24).

$$E_{ch} = \left(\frac{N}{k} - 1 \right) \cdot E_R + E_S^f + \eta \cdot \left(E_R + E_S^f \right).$$
(3.24)

E_{ch}, as given in Eq. (3.24), consists of three terms; $\left(\frac{N}{k} - 1 \right) \cdot E_R$ is the amount of energy consumed for receiving data from the members of the head (number of members in a cluster is $\frac{N}{k} - 1$ on average), E_S^f is the amount of energy consumed for sending aggregated data towards the sink, and $\eta \cdot \left(E_R + E_S^f \right)$ is the amount of energy consumed for relaying data packets received from the other heads. η is the average number of packets relayed by each head (given in Lemma 3.2). The amount of energy consumed for aggregating data packets received from members is assumed to be neglected.

A member only sends its readings to the head of its cluster using R_t as its transmission range and hence, $E_{ch} = E_S^n$. Therefore the total energy dissipated in the network at each round of data gathering ($E(k)$) can be given by Eq. (3.25) where k is the number of clusters.

$$E(k) = k \cdot E_{ch} + (N - k) \cdot E_{cm}$$
$$= k \cdot \left[\left(\frac{N}{k} - 1 \right) \cdot E_R + E_S^f + \eta \cdot E_R + \eta \cdot E_S^f \right] + (N - k) \cdot \left[E_S^n \right].$$
(3.25)

To find the number of clusters which minimize $E(k)$, we set the derivative of $E(k)$ with respect to k to zero, that is,

$$\frac{dE(k)}{dk} = 0. \tag{3.26}$$

Since the transmission range for intra-cluster communication and inter-cluster communication are R_t and $3R_t$ respectively, then E_R, E_S^f, and E_S^n are constants and hence we have

$$\frac{dE(k)}{dk} = E_S^f - E_S^n - E_R + \left(E_R + E_S^f\right) \cdot \frac{d(k \cdot \eta)}{dk} = 0. \tag{3.27}$$

From Eq. (3.22), we have

$$\frac{d(k \cdot \eta)}{dk} = \frac{d}{dk} \left(\frac{d^2 - \left(1 + (1 - d) \cdot \left(\log_d^k - 2\right)\right) \cdot k}{(1 - d)^2} \right), \tag{3.28}$$

or

$$\frac{d(k \cdot \eta)}{dk} = \frac{1}{d - 1} \cdot \left(\frac{1}{1 - d} + \frac{1}{\ln(d)} + \left(\log_d^k - 2\right) \right). \tag{3.29}$$

After replacing $\frac{d(k \cdot \eta)}{dk}$ from Eq. (3.29) in Eq. (3.27) we have

$$\frac{dE(k)}{dk} = E_S^f - E_S^n - E_R + \frac{\left(E_R + E_S^f\right)}{d - 1} \cdot \left(\frac{1}{1 - d} + \frac{1}{\ln(d)} + \left(\log_d^k - 2\right) \right) \tag{3.30}$$
$$= 0.$$

Solving the above equation for k we obtain

$$k = d^{(1-d) \cdot \left(\frac{E_S^f - E_S^n - E_R}{E_S^f + E_R} \right) + \frac{1}{d-1} - \frac{1}{\ln(d)} + 2}, \tag{3.31}$$

and hence the theorem.   ■

Lemma 3.3 *At the termination of the initial clustering phase of the given algorithm, a node is either tagged as a head or as a member.*

Proof Initial clustering stops for a node either when the number of iterations reaches *MAX_ITERATION* or when the role of the node changes from *unspecified* to head or member. In the former case, the role of the node is still unknown. Such a node then polls all of its neighbors whether or not any of them is a head. If there are such neighbors, then the polling node becomes a member of the cluster of one of those neighbors chosen at random. Otherwise, the polling node itself becomes a head. So, at the end of the initial clustering phase of the proposed algorithm, a node is either tagged as a head or as a member.   ■

Lemma 3.4 *Assume that N nodes are uniformly and independently dispersed at random in an area* $R = [0, L]^d$ *for* $d = 1, 2, 3$ *and assume that* $R_t^d N = aL^d \ln L$ *for some constant* $a > 0$, *with* $R_t \ll L$ *and* $N \gg 1$. *If* $a > d \cdot a_d$, *or* $a = d \cdot a_d$ *and* $R_c \gg 1$, *then* $\lim_{L \to \infty} P_{conn}(L) = 1$, *where* $a_d = 2^d d^{d/2}$ *and* $P_{conn}(L)$ *denotes the probability that the communication graph is connected.*

Proof: This lemma is proved in Blough and Santi (2002). We rewrite the proof from Blough and Santi (2002) here.

Let $d = 1$, and subdivide $[0, L]$ into $M = \frac{2L}{R_c}$ non-overlapping cells of side $\frac{R_t}{2}$. It is immediate that if every segment contains at least one node, then the resulting communication graph is connected. Let $\mu_0(N, M)$ be the random variable denoting the number of empty cells. Since $\mu_0(N, M)$ is a non-negative integer random variable, then $P(\mu_0(N, M) > 0) \le E[\mu_0(N, M)]$, where $E[\mu_0(N, M)]$ is the expected value of $\mu_0(N, M)$ (Palmer 1985, pp. 10–11). We have (Kolchin et al. 1978):

$$E[\mu_0(N, M)] = M \left(1 - \frac{1}{M}\right)^N. \tag{3.32}$$

We want to investigate the asymptotic value of $E[\mu_0(N, M)]$ as $L \to \infty$, which, given the hypotheses $R_t \ll L$ and $N \gg 1$, is equivalent to the asymptotic for $M \to \infty$. Taking the logarithm, we obtain

$$\ln(E[\mu_0(N, M)]) = \ln(M) + N \ln\left(1 - \frac{1}{M}\right). \tag{3.33}$$

Replacing M with $\frac{2L}{R_t}$ in Eq. (3.33), we get

$$\ln(E[\mu_0(N, M)]) = \ln\left(\frac{2L}{R_t}\right) + N \ln\left(1 - \frac{R_t}{2L}\right). \tag{3.34}$$

Since $\frac{R_t}{2L} \to 0$ as $L \to \infty$, we can approximate the last term of Eq. (3.34) with the first term of its Taylor expansion, obtaining

$$\ln(E[\mu_0(N, M)]) = \ln\left(\frac{2L}{R_t}\right) - \frac{NR_t}{2L}. \tag{3.35}$$

Substituting the expression $R_t N = aL \ln L$ in Eq. (3.35), we obtain

$$\ln(E[\mu_0(N, M)]) = \ln\left(\frac{2}{R_t L^{\frac{a}{2}-1}}\right). \tag{3.36}$$

If $a > 2$, or if $a = 2$ and $R_c \gg 1$, then

$$\lim_{M\to\infty} \ln(E[\mu_0(N,M)]) = -\infty. \tag{3.37}$$

Hence, $\lim_{M\to\infty} \ln(E[\mu_0(N,M)]) = 0$ and $\lim_{L\to\infty} P(\mu_0(N,M) = 0) = 1$. It follows that each cell contains at least one node asymptotically almost surely, which implies $\lim_{L\to\infty} P_{conn}(L) = 1$.

The proof for the cases $d = 2$ and $d = 3$ are similar, and are obtained by subdividing R into non-overlapping d-dimensional cells of side $\frac{R_t}{\sqrt{2}}$ and $\frac{R_t}{\sqrt{3}}$ respectively. ∎

Corollary 3.1 *If the conditions of* Lemma 3.4 *hold for a two-dimensional network area, then each cell of the size* $\frac{R_t}{\sqrt{2}} \times \frac{R_t}{\sqrt{2}}$ *contains at least one node asymptotically almost surely.*

Proof The proof is trivial from Lemma 3.4.

Theorem 3.2 *Assume that the conditions of* Lemma 3.4 *hold. Then, if the transmission range of* $3R_t$ *is used for inter-cluster communications, the given clustering algorithm produces a clustered infrastructure in which the set of heads form a connected graph asymptotically almost surely.*

Proof Assume that the given clustering algorithm produces two connected sub-graphs of heads $G_1 = (V_1, E_1)$ and $G_2 = (V_2, E_2)$, such that any $v_1 \in V_1$ cannot communicate with any $v_2 \in V_2$ using transmission range of $3R_t$. We prove the theorem by contradiction. Assume that v_1 and v_2 are the two nearest heads in G_1 and G_2. Since v_1 and v_2 cannot communicate with each other, their distance is more than $3R_t$. Now consider a square S of the size $\frac{R_t}{\sqrt{2}} \times \frac{R_t}{\sqrt{2}}$ centered on the middle of the line passes through v_1 and v_2 and its edges are parallel to the edges of the area R (Fig. 3.10). It is clear that the nearest point on square S to v_1 is on the nearest edge of S to v_1 (edge AB in Fig. 3.10). Two cases can be considered:

- When the perpendicular line from v_1 to AB crosses AB in a point within AB (point E in Fig. 3.10a), then E is the nearest point to v_1 and therefore $d(v_1, E)$, the distance between points v_1 and E, is given by Eq. (3.38).

$$d(v_1, E) = \sqrt{d^2(v_1, O) - d^2(O, G)} - d(E, G). \tag{3.38}$$

Using Eq. (3.38) and the fact that $d(v_1, v_2) \geq 3R_t$ we get

$$d(v_1, E) = \sqrt{\frac{9R_t^2}{4} - d^2(O, G)} - \frac{R_t}{2\sqrt{2}}. \tag{3.39}$$

To obtain the minimum of $d(v_1, E)$, we let $d(O, G)$ to have its maximum value which is $\frac{R_t}{2\sqrt{2}}$. Therefore we get

Fig. 3.10 v_1 and v_2 are two heads in two disconnected graphs of heads and square S is centered on O which is in the middle of the line v_1v_2. **a** The nearest point on square S to v_1 is E, **b** the nearest point on square S to v_1 is B

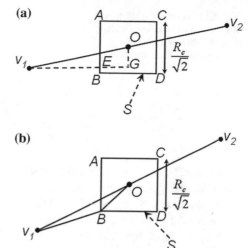

$$d(v_1, E) \geq \frac{\sqrt{17} - 1}{2\sqrt{2}} \cdot R_t \geq R_t. \tag{3.40}$$

- When the perpendicular line from v_1 to AB doesn't cross AB in a point within AB (Fig. 3.10b), then the nearest point to v_1 is either A or B (B in Fig. 3.10) and therefore $d(v_1, B)$, the distance between points v_1 and B, can be calculated using law of cosines as follows:

$$d(v_1, B) = \sqrt{d^2(v_1, O) + d^2(O, B) - 2 \cdot d(v_1, O) \cdot d(O, B) \cdot \cos\left(\widehat{BOv_1}\right)}. \tag{3.41}$$

To obtain the minimum of $d(v_1, B)$, we set angle $\widehat{BOv_1}$ to $0°$. Therefore we get

$$d(v_1, B) \geq R_t. \tag{3.42}$$

Since in the given clustering algorithm, members of a cluster can communicate with the head of the cluster using transmission range of R_t (i.e. members of a cluster are within the distance of at most R_t from the head of the cluster), the above two cases immediately imply that no point on square S can be a member of v_1. Using the same argument, it can be shown that no point on S can be a member of v_2 as well. This means that any sensor node within the square S cannot be a member of either v_1 or v_2.

In addition, assuming that the conditions of Lemma 3.4 hold, using Corollary 3.1 we conclude that at least one node, say node s_i, exists in square S asymptotically

Fig. 3.11 A sensor node s_i in square S must be either a head or a member of a cluster managed by a head say v_3

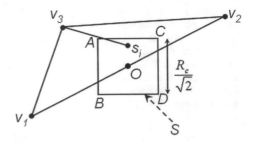

almost surely. As s_i located within square S, it cannot be a member of either v_1 or v_2. Thus according to Lemma 3.3, s_i is either a head or a member of another cluster managed by a head say v_3 (Fig. 3.11). We next show that s_i cannot be a head or a member of a cluster managed by v_3.

If s_i is a head, then either $s_i \in V_1$ or $s_i \in V_2$. Without loss of generality, assume $s_i \in V_1$. Since, s_i located on square S, it is clear that $d(s_i, v_2) < d(v_1, v_2)$. This contradicts with the assumption that v_1 and v_2 are the two nearest heads in G_1 and G_2, and hence s_i cannot be a head.

If s_i is a member of a cluster managed by v_3, then either $v_3 \in V_1$ or $v_3 \in V_2$. Without loss of generality, assume $v_3 \in V_1$. We can show that $d(v_3, v_2) < d(v_1, v_2)$ which contradicts with the assumption that v_1 and v_2 are the two nearest heads in G_1 and G_2. $d(v_3, v_2)$ can be calculated using the following equation:

$$d(v_3, v_2) = \sqrt{d^2(v_3, O) + d^2(O, v_2) - 2 \cdot d(v_3, O) \cdot d(O, v_2) \cdot \cos\left(\widehat{v_3 O v_2}\right)}.$$

$$(3.43)$$

Having the fact that $d(v_3, O) < \frac{3}{2} R_t$ and considering $d(v_1, v_2) = 3R_t + \varepsilon$ for some $\varepsilon > 0$, $d(v_3, v_2)$ can be computed as

$$d(v_3, v_2) \leq \sqrt{\frac{9}{4} R_t^2 + \frac{(3R_t + \varepsilon)^2}{4} - 3R_t \left(\frac{3R_t + \varepsilon}{2}\right) \cdot \cos\left(\widehat{v_3 O v_2}\right)}. \qquad (3.44)$$

To obtain the maximum of $d(v_3, v_2)$, we set the angle $\widehat{v_3 O v_2}$ to $180°$. Therefore we get

$$d(v_3, v_2) \leq 3R_t + \frac{\varepsilon}{2}. \qquad (3.45)$$

Inequality Eq. (3.45) shows that $d(v_3, v_2) < d(v_1, v_2)$. Thus, no head like v_3 can be found which manages a cluster to which the node s_i is assigned.

If s_i is neither a head, nor a member, then its status must be *unspecified*, which is in contradiction with the result of Lemma 3.3. Therefore G_1 and G_2 are connected asymptotically almost surely, and hence the theorem. ∎

3.3.3 Experimental Results

To evaluate the performance of the given clustering algorithm, a number of experiments have been conducted. In these experiments, the results obtained from the given algorithm are compared with the results obtained from the LEACH (Heinzelman et al. 2000), the basic HEED (Wu et al. 2006) and its extension (ExtendedHEED) (Huang and Wu 2005). We study the quality of the given clustering algorithm in terms of the *Number of clusters* in the clustering infrastructure produced at the end of the clustering, *Percentage of the sparse clusters* which is defined as the percentage of clusters having only one head and/or one member, *Mean energy level of heads* and *Ratio of the mean energy level of heads to the mean energy level of members* defined through Eq. (3.46), where n_h is the number of heads and n_m is the number of members.

$$\zeta = \frac{\sum_{i=1}^{n_h} E_{S_i}}{\sum_{j=1}^{n_m} E_{S_j}}. \tag{3.46}$$

In what follows, we first give the simulation scenario used in our experiments and then give the simulation results.

- **Simulation Scenario**

In the simulation scenario used in experiments, the clustering algorithm is performed first to determine the role of each node of the network as a head or as a member. In the clustering algorithm, nodes communicate with each other using R_t as their transmission range. Then, the normal operation of the network which is assumed to be a simple monitoring application (Definition 3.1) is started. As it was shown in Theorem 3.1, in such applications, the optimum number of clusters in order to minimize the total energy consumption of the network at each data gathering round is $k = d^{(1-d) \cdot \left(\frac{E_S^f - E_S^n - E_R}{E_S^f + E_R} \right) + \frac{1}{d-1} - \frac{1}{\ln(d)} + 2}$, where d is the averaged branching factor of the routing tree used for routing, E_S^f and E_S^n are the amount of energy consumed for sending a data packet using transmission range of $3R_t$ and R_t respectively, and E_R is the amount of energy consumed for receiving a data packet.

For the normal operation of the network to be performed correctly, heads must be able to communicate with each other. It was shown in Theorem 3.2 that using the transmission range of $3R_t$ for inter-cluster communications, a connected graph of all heads in the network is produced by the proposed clustering algorithm asymptotically almost surely.

The network operation continues until the time at which a node in the network dies. In other words, network lifetime is defined to be the time elapsed from the network startup to the time at which a node in the network dies (Mittal et al. 2003).

- **Simulation Results**

In the simple monitoring application considered in experiments, nodes periodically report 525 bytes of data to the sink node. Directed diffusion (Intanagonwiwat et al. 2003) is used as the multi-hop inter-cluster routing protocol. All simulations have been implemented using NS2 simulator. We use IEEE 802.11 as the MAC layer protocol. Nodes are placed randomly on a two-dimensional area of the size 100 m $\times$ 100 m. In all experiments, initial random delay R_i is selected uniformly and randomly in the range [0.06, 0.09](s), t is set to 10(s), *MAX_ITERATION* is set to 50, φ is set to 0.8, ρ is set to 0.2, R_t is set to 20 m, *MaxEnergyLevel* is set to 2.0 (J) and *RE-CLUSTERING_PERCENT* is set to 85%. *MAX_ITERATION* for the re-clustering phase is set to 10 and initial energy level of the nodes is selected uniformly and randomly from the range [1.5, 2.0](J). First order radio model specified in Heinzelman et al. (2000) is used for estimating the amount of energy consumed for packet transmissions. All packets except for data packets, which are 525 bytes long, are assumed to be 8 bytes. For the sake of simplicity, we assume that energy consumption in idle states is zero. Simulations are performed for 50, 100, 200, 300, 400, and 500 nodes. The results are averaged over 50 runs.

Experiment 1 In this experiment, the quality of the clustering infrastructure, produced using the given algorithm, LEACH, HEED, and ExtendedHEED is evaluated in terms of the number of clusters, percentage of the sparse clusters, and ratio of the mean energy level of the heads to the mean energy level of the members (ζ). Figure 3.12 compares the number of generated clusters using the proposed algorithm and the existing algorithms with the optimum number of clusters (given in Theorem 3.1). As it is shown, the number of clusters in the infrastructure resulted from the given clustering algorithm is very close to the optimum number of clusters. Figures 3.13 and 3.14 compares the given algorithm with other algorithms in terms

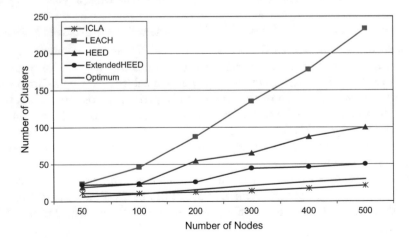

Fig. 3.12 Comparison of the number of clusters resulted from the clustering algorithms with the optimum number of clusters given by Theorem 3.1

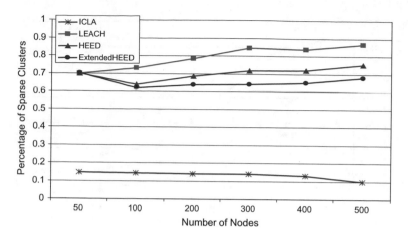

Fig. 3.13 Comparison of the clustering algorithms with respect to the Percentage *of sparse clusters*

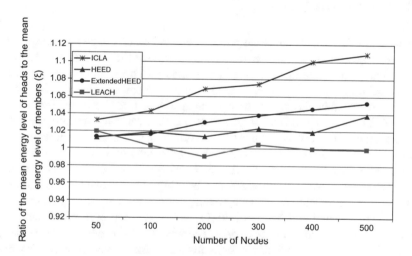

Fig. 3.14 Comparison of the clustering algorithms with respect the Ratio of the mean energy level of the heads to the mean energy level of the members (ζ)

of the percentage of sparse clusters and ζ, respectively. These figures show that the clustering infrastructure formed by the given algorithm is significantly better than the other three algorithms in terms of all of these parameters.

Experiment 2 This experiment, whose results are given in Fig. 3.15, compares the network lifetime for the given algorithm with LEACH, HEED and ExtendedHEED algorithms. Figure 3.15 shows that the network lifetime for the given algorithm is substantially higher than that for the other algorithms.

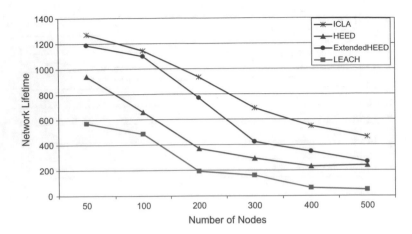

Fig. 3.15 Comparison of the clustering algorithms with respect to the network lifetime

3.4   Area Coverage

The most studied sensor coverage problem is the area coverage problem in which the main objective of the sensor network is to provide the desired coverage of the sensor field. Area coverage problem is classified into three sub-problems: deployment, k-coverage, and set cover. In deployment problem, the objective is to deploy sensor nodes within the sensor field in such a way that the desired coverage is achieved. The goal of the k-coverage problem is to cover every desired point within the area of the network with at least k sensor nodes. Set cover problem aims at reducing the number of active sensor nodes in the network to increase the network lifetime with the constraint that the desired coverage of the network area is fulfilled.

In this section, we will introduce two learning automata-based algorithms, for addressing sub-problems of the area coverage problem. The first algorithm, called CLA-DS (Esnaashari and Meybodi 2011), is a sensor deployment strategy which can be used for mobile sensor networks. The novelty of this deployment strategy is that it guides the movements of sensor nodes within the area of the network without any sensor to know its position or its relative distance to other sensors. Each sensor node is equipped with a learning automaton and the whole network is modeled by a dynamic irregular cellular learning automaton. The learning automaton in each sensor node in cooperation with the learning automata in the neighboring nodes controls the movements of the node in order to attain high coverage. The second algorithm, called CLA-EDS (Esnaashari and Meybodi 2013), is an extension of CLA-DS algorithm, which can provide k-coverage. The novelty of this algorithm is that it can address k-coverage requirement with different values of k in different sub-regions of the sensor field. In CLA-EDS, the sensor network is modeled by a heterogeneous dynamic irregular cellular learning automaton.

3.4.1   CLA-DS: A Cellular Learning Automata-Based Deployment Strategy

Sensor deployment strategies can be classified into the following four categories:

- **Predetermined**: This strategy is useful only if the network area is completely known (Musman et al. 1997; Petriu et al. 2000; Salhieh et al. 2001; Schwiebert et al. 2001; Chakrabarty et al. 2002; Dhillon et al. 2002).
- **Randomly undetermined**: In this strategy, sensor nodes are spread uniformly throughout the network area (Heinzelman 2000; Heinzelman et al. 2000, 2002; Lindsey et al. 2002; Lindsey and Raghavendra 2002; Tilak et al. 2002).
- **Biased distribution**: In some contexts, the uniform deployment of sensor nodes may not always satisfy the design requirements and biased deployment can then be a viable option (Willig et al. 2002).
- **Self-regulated**: In this strategy which is useful only in mobile sensor networks, sensor nodes are deployed randomly in some initial positions within the area of the network. After this initial placement, sensor nodes move around and find their best positions according to the positions of their neighboring nodes (Howard et al. 2002; Zou and Chakrabarty 2003, 2004a; Heo and Varshney 2005; Wang et al. 2006c).

One important problem which may arise in designing a deployment strategy for a *WSN* is how to deploy a specific number of sensor nodes so that the covered section of the area is maximized. Since this is an optimization problem, random based deployment strategies are not suitable approaches for solving it. Predetermined deployment strategies which are commonly centralized are not well suited to this problem as well. This is because a counterpart of this problem in computational geometry domain, which is to maximize the guarded interior of an art gallery by a specific number of vertex guards, is proved to be APX-hard (Emiris et al. 2006). This indicates that finding an approximation close enough to the optimal solution of the problem is impossible in polynomial time. Self-regulated deployment strategies on the other hand are well suited to this problem.

To the best of our knowledge, all of the self-regulated deployment strategies given in literature require for every node to be aware of either its geographical position within the area of the network or its relative distance to all of its neighbors. If none of this information is available to the nodes of a network, the problem becomes more complicated. In this section, we introduce a self-regulated deployment strategy based on cellular learning automata called CLA-DS which guides the movements of sensor nodes throughout the sensor field without any sensor to know its position or its relative distance to other sensors. Prior to give this algorithm, we first introduce a simplified algorithm called simple CLA-DS (SCLA-DS), in which sensor nodes are aware of their physical locations. Next, we modify SCLA-DS to reach the CLA-DS algorithm. In both SCLA-DS and CLA-DS deployment strategies, neighboring nodes apply forces to each other which make every node move according to the resultant force vector applied to it. Each node is equipped with a learning automaton. The Learning automaton of a node at any given time decides for the node whether to apply force to its neighbors or not. This way, each node in cooperation with its neighboring nodes gradually learns its best position within the area of the network so as to attain high coverage. Experimental results show that the SCLA-DS algorithm can compete with the existing deployment algorithms such as PF, DSSA, IDCA, and VEC in terms of the network coverage and distance moved by the sensor nodes. In terms of the node separation, which is a measure of the overlapping area between sensor nodes, SCLA-DS outperforms all of the existing algorithms. For CLA-DS algorithm, experimental results show that even though this algorithm does not use any information regarding the physical locations of the sensor nodes or their relative distances to each other, it still can compete with the existing deployment algorithms such as PF, DSSA, IDCA, and VEC when sensor nodes have precise location information and sensor movements are perfect. In addition, when the utilized location estimation techniques such as GPS-based devices and localization algorithms experience inaccuracies in their measurements, or the movements of the sensor nodes are probabilistic, the CLA-DS outperforms the mentioned algorithms in terms of the network coverage.

3.4.1.1 Problem Statement

Consider N mobile sensor nodes s_1, s_2, ..., s_N with equal sensing ranges of $R_s = r$ and transmission ranges of $R_t = 2 \cdot r$ which are initially deployed in some initial positions within an unknown 2 dimensional sensor field Ω. Assume that a rough estimate of the surface of Ω ($\hat{S}_\Omega$) is available (using Google maps for example). According to this estimate, expected number of neighbors for a sensor node (number of sensors residing within its transmission range) is equal to $E[N_{nei}] = N \cdot \frac{\pi \cdot R_t^2}{\hat{S}_\Omega} - 1$. $E[N_{nei}]$ is provided to all sensor nodes prior to their initial deployments. Sensor nodes are able to move in any desired direction within the area of the network at a constant speed, but they cannot cross the barrier of Ω.

Consider the following definitions:

Definition 3.3 Sensing region of a node s_i denoted by $C(s_i)$ is a circle with radius R_s centered on s_i.

Definition 3.4 Covered sub-area denoted by $C(\Omega)$ refers to any point within the network area which is in the sensing region of at least one of the sensor nodes. $C(\Omega)$ is stated using Eq. (3.47).

$$C(\Omega) = \bigcap \left(\Omega, \bigcup_{i=1}^{N} (C(s_i)) \right). \tag{3.47}$$

Definition 3.5 Covered section denoted by $S_{C(\Omega)}$ is the surface of the covered sub-area.

Definition 3.6 Deployment strategy is an algorithm which gives for any given sensor node s_i a certain position within the area of the network.

Definition 3.7 Deployed network refers to a sensor network which results from a deployment strategy.

Definition 3.8 Self-regulated deployment strategy is a deployment strategy in which each sensor node finds its proper position within the sensor field by exploring the area and cooperating with its neighboring sensor nodes.

Definition 3.9 A connected network is a network in which there is a route between any two sensor nodes.

Using the above definitions and assumptions, the problem considered in this section can be stated as follows: Propose a self-regulated deployment strategy which deploys N mobile sensor nodes throughout an unknown network area Ω with estimated surface $\hat{S}_\Omega$ so that the covered section of Ω ($S_{C(\Omega)}$) is maximized and the deployed network is connected.

3.4.1.2 SCLA-DS: Simplified CLA-Based Deployment Strategy

SCLA-DS deployment strategy consists of four major phases; Initial deployment, mapping, deployment, and maintenance. During the initialization phase, sensor nodes are initially deployed in some initial positions within the area of the network. Mapping the network topology into a *DICLA* is done in the mapping phase. Deploying sensor nodes throughout the area of the network is performed during the deployment phase. Finally, in the maintenance phase, deployed network is maintained in order to compensate the effect of possibly node failures on the covered sub-area ($C(\Omega)$). We explain these phases in more details in the subsequent sections.

Initial Deployment

Three main strategies exist for initially deploying sensor nodes in some initial positions within the area of the network (Ω):

- **Random deployment**: In random deployment strategy, a random based deployment strategy (Heinzelman 2000; Heinzelman et al. 2000, 2002; Lindsey et al. 2002; Lindsey and Raghavendra 2002; Tilak et al. 2002) is used to deploy sensor nodes uniformly at random throughout Ω. Using this strategy for initial deployment reduces the cost of the deployment phase due to the fact that after such initial deployment, sensor nodes need only to heel coverage holes within their vicinities rather than exploring Ω for finding their proper positions.

- **Sub-area deployment**: In this deployment strategy, sensor nodes are placed manually in some initial positions within a small accessible sub-area of the network.
- **Hybrid deployment**: In this approach, sensor nodes are deployed randomly within a small sub-area of the network.

Any of the random deployment, sub-area deployment, or hybrid deployment strategies can be used for the initial deployment phase of CLA-DS.

Mapping

In the mapping phase, a time-driven asynchronous *DICLA*, which is isomorphic to the sensor network topology, is created. Each sensor node s_i located at (x_i, y_i) in the sensor network corresponds to the cell c_i in *DICLA*. Interest set of *DICLA* consists of two members; X-axis and Y-axis of the network area. Initially (at time instant 0), tendency levels of each cell c_i to these interests are $\psi_{i1}(0) = \frac{x_i(0)}{\max(MaxX, MaxY)}$ and $\psi_{i2}(0) = \frac{y_i(0)}{\max(MaxX, MaxY)}$ respectively. (*MaxX, MaxY*) is the farthest location within the network area at which a sensor node can be located. Neighborhood radius (τ) of *DICLA* is equal to $\frac{R_t}{\max(MaxX, MaxY)}$, and hence two cells c_i and c_j in *DICLA* are adjacent to each other if s_i and s_j in the sensor network are close enough to hear each other's signals.

The learning automaton in each cell c_i of *DICLA*, referred to as LA_i, has three actions α_0, α_1, and α_2. Action α_0 is "attract the neighboring nodes", action α_1 is "repel the neighboring nodes", and action α_2 is "apply no force to the neighboring nodes". The probability of selecting each of these actions is initially set to 1/3.

Deployment

Each sensor node has a state which can be "*mobile*" or "*fixed*". Initially, each sensor node selects its state randomly with probability of selecting "*fixed*" state to be P_{fix}. P_{fix} is a constant which is known to all sensor nodes. A "*fixed*" sensor node leaves the deployment phase immediately and starts with the maintenance phase.

Deployment phase for each "*mobile*" sensor node s_i is divided into a number of rounds. Each round $R_i(n)$ is started by the asynchronous activation of the cell c_i of the *DICLA*. The nth activation of the cell c_i occurs at time $\delta_i + n \times ROUND\_DURATION$ where δ_i is a random number generated for cell c_i and *ROUND_DURATION* is an upper bound for the duration of a single round. We call n the local iteration number for the cell. Delays δ_i are chosen randomly in order to reduce the probability of collisions between neighboring nodes.

Upon the startup of a new round $R_i(n)$, LA_i selects one of its actions randomly according to its action probability vector. Selected action, which is one of "attract the neighboring nodes", "repel the neighboring nodes", or "apply no force to the neighboring nodes", specifies the force sensor node s_i will apply to its neighboring nodes during the current round. Sensor node s_i then creates a packet called *APPLIED_FORCE* containing its state and the selected action and broadcasts it in its neighborhood. After broadcasting the *APPLIED_FORCE* packet, sensor node s_i waits for certain duration (*RECIEVE_DURATION*) to receive *APPLIED_FORCE* packets from its neighboring nodes. Received packets are stored into a local database within the node.

When *RECIEVE_DURATION* is over, sensor node s_i collects following statistics from the stored information in its local database: number of received packets ($N_i^r(n)$), number of neighbors selecting "attract the neighboring nodes" action ($N_i^{att}(n)$), and number of neighbors selecting "repel the neighboring nodes" action ($N_i^{rep}(n)$). According to the collected statistics, local rule of the cell c_i computes the reinforcement signal $\beta_i(n)$ and the restructuring signal $\underline{\zeta}_i(n)$ which are used to update the action probability vector of LA_i and the tendency vector of c_i. Details on this will be given in Section "Applying Local Rule".

Next, sensor node s_i uses vector $\underline{\zeta}_i(n)$ as its movement path for the current round. In other words, if s_i is located at $(x_i(n),\ y_i(n))$, it moves to $(x_i(n+1),\ y_i(n+1)) = (x_i(n)+\zeta_{1,i},\ y_i(n)+\zeta_{2,i})$. Last step during the round $R_i(n)$ is the application of the restructuring function Z which updates the tendency levels of cell c_i using the restructuring signal $\underline{\zeta}_i(n)$ according to the Eq. (3.48).

$$\begin{cases} \psi_{i1}(n+1) = \psi_{i1}(n) + \dfrac{\zeta_{i1}(n)}{\max(MaxX,\ MaxY)} \\[2em] \psi_{i2}(n+1) = \psi_{i2}(n) + \dfrac{\zeta_{i2}(n)}{\max(MaxX,\ MaxY)} \end{cases}. \qquad (3.48)$$

When this step is done, sensor node s_i waits for the next activation time of its corresponding cell c_i in *DICLA* to start its next round ($R_i(n + 1)$).

Deployment phase for a sensor node s_i is completed upon the occurrence of one of the followings:

- Stability: Sensor node s_i moves less than a specified threshold (*LEAST_DISTANCE*) during its last N_r rounds.
- Oscillation: Sensor node s_i oscillates between almost the same positions for more than N_o rounds.

When a sensor node s_i completes the deployment phase of CLA-DS, its state is changed to "*fixed*" and it starts with the maintenance phase.

Applying Local Rule

Local rule of a cell c_i based on the states of the cell and its neighboring cells computes the reinforcement signal β_i and the restructuring signal $\underline{\zeta}_i$.

Reinforcement Signal: The reinforcement signal for the nth round in a node s_i is computed based on a comparison between the number of neighbors of s_i ($N_i^r(n)$) and the expected number of neighbors ($E[N_{nei}]$). According to this comparison, following three cases may occur:

- $N_i^r(n)$ is almost equal to $E[N_{nei}]$ or formally $\left|N_i^r(n) - E[N_{nei}]\right| < \varepsilon$ where ε is a specified constant: In this case, if the selected action of LA_i is "apply no force to the neighboring nodes", then the reinforcement signal is to reward the action. Otherwise, the reinforcement signal is to penalize the action. Both reward and penalty parameters (a and b) for this case are set to a constant called Mid_Value.

- $N_i^r(n)$ is greater than $E[N_{nei}]$ or formally $N_i^r(n) - E[N_{nei}] > \varepsilon$: In this case, the most suitable action is "repel the neighboring nodes". Therefore, if this action is selected by LA_i then the reinforcement signal is to reward the action. Otherwise, the reinforcement signal is to penalize the selected action. Reward and penalty parameters for this case are specified according to $N_i^{att}(n)$ and $N_i^{rep}(n)$ as follows:

  - If the selected action is "repel the neighboring nodes" and $\frac{N_i^{rep}(n)}{E[N_{nei}]} > \eta$, then a is set to a small constant ($Small\_Reward$). η is a predetermined constant.
  - If the selected action is "repel the neighboring nodes" and $\frac{N_i^{rep}(n)}{E[N_{nei}]} \leq \eta$, then a is set to a large constant ($Large\_Reward$).
  - If the selected action is "attract the neighboring nodes" and $\frac{N_i^{att}(n)}{E[N_{nei}]} > \eta$, then b is set to a large constant ($Large\_Penalty$).
  - If the selected action is "attract the neighboring nodes" and $\frac{N_i^{att}(n)}{E[N_{nei}]} > \eta$ then b is set to a small constant ($Small\_Penalty$).
  - If the selected action is "apply no force to the neighboring nodes" then b is set to 0.

- $N_i^r(n)$ is smaller than $E[N_{nei}]$ or formally $N_i^{rec}(n) - E[N_{nei}] < -\varepsilon$: In this case, the most suitable action is "attract the neighboring nodes". Therefore, if this action is selected by LA_i, then the reinforcement signal is to reward the action. Otherwise, the reinforcement signal is to penalize the selected action. Reward and penalty parameters for this case are specified according to $N_i^{att}(n)$ and $N_i^{rep}(n)$ as follows:

  - If the selected action is "attract the neighboring nodes" and $\frac{N_i^{att}(n)}{E[N_{nei}]} > \eta$, then a is set to $Small\_Reward$.
  - If the selected action is "attract the neighboring nodes" and $\frac{N_i^{att}(n)}{E[N_{nei}]} > \eta$, then a is set to $Large\_Reward$.
  - If the selected action is "repel the neighboring nodes" and $\frac{N_i^{rep}(n)}{E[N_{nei}]} > \eta$, then b is set to $Large\_Penalty$.

- If the selected action is "repel the neighboring nodes" and $\frac{N_i^{rep}(n)}{E[N_{nei}]} \leq \eta$, then b is set to *Small_Penalty*.
- If the selected action is "apply no force to neighboring nodes" then b is set to 0.

Equations (1.1) and (1.2) are used for rewarding or penalizing the selected action of LA_i. The idea behind the above method of computing the reinforcement signal is that to have a uniform deployment of sensor nodes, one way is to minimize the difference between the number of neighbors of each sensor node and the expected number of neighbors ($E[N_{nei}]$). When $N_i^r(n)$ is almost equal to $E[N_{nei}]$, it means that the number of neighbors of sensor node s_i is as it must be, and hence it is better for this node not to apply any forces to its neighbors. On the hand, when $\left| N_i^r(n) - E[N_{nei}] \right| > \varepsilon$, the number of neighbors of s_i differs from that expected, and hence it is better for it to apply forces to its neighbors.

Restructuring Signal: Selected action of each neighbor s_j of a sensor node s_i, applies a force vector to s_i referred to as $\overrightarrow{F}_{j,i}(n)$ for the nth round. The orientation of $\overrightarrow{F}_{j,i}(n)$ coincides with the angle of arrival ($A\hat{O}A$) of the *APPLIED_FORCE* packet received from s_j and its magnitude is unity. The direction of $\overrightarrow{F}_{j,i}(n)$ is towards s_j if the selected action of s_j is "attract the neighboring nodes" and in the opposite direction if the selected action of s_j is "repel the neighboring nodes". If the selected action of s_j is "apply no force to the neighboring nodes", then $\overrightarrow{F}_{j,i}(n) = \overrightarrow{0}$. The restructuring signal of the cell c_i for the nth round is the resultant force vector $\overrightarrow{F}$ which is computed according to Eq. (3.49).

$$\underline{\zeta}(n) = \overrightarrow{F}(n) = \sum_{j \in Nei(s_i)} \overrightarrow{F}_{j,i}(n). \tag{3.49}$$

In the above equation, $Nei(s_i)$ is the set of neighbors of sensor node s_i. This vector specifies the orientation, direction, and distance of movement for the node s_i during the current round.

Maintenance

Maintenance phase in a sensor node s_i is similar to the deployment phase except that s_i remains fixed during this phase and the force vectors applied by its neighbors have no effect on it. Additionally, during this phase, sensor node s_i collects a list of its "*fixed*" neighboring nodes (neighboring nodes which are in maintenance phase). If something happens to a member s_j of this list (its battery exhausted, it experiences some failures, it leaves the sensing region of s_i, and so on) then sensor node s_i does not receive *APPLIED_FORCE* packets from s_j anymore. This indicates that a hole may occur in the vicinity of s_i. As a result, s_i leaves the maintenance phase, set its state to "*mobile*" and starts over with the deployment phase in order to fill any probable holes. Since collisions may also result in not receiving *APPLIED_FORCE* packets from neighbors, a node s_i starts over with the deployment phase only if it does not receive *APPLIED_FORCE* packets from one of its "*fixed*" neighbors for more than l rounds.

3.4.1.3 CLA-DS Deployment Strategy

CLA-DS deployment strategy is a modification to the SCLA-DS deployment strategy in which, sensor nodes have no information regarding their physical locations or their relative distances to each other. Like SCLA-DS, this algorithm also consists of four major phases, namely initial deployment, mapping, deployment, and maintenance. The initial deployment and maintenance phases of CLA-DS are completely the same as that of SCLA-DS. Therefore, in what follows, we only concentrate on the mapping and deployment phases.

Mapping

In the mapping phase, a time-driven asynchronous *DICLA*, which is isomorphic to the sensor network topology, is created. Each sensor node s_i located at (x_i, y_i) in the sensor network corresponds to the cell c_i in *DICLA*. Interest set of *DICLA* consists of two members; X-axis and Y-axis of the network area. Initially (at time instant 0), tendency levels of each cell c_i to these interests are $\psi_{i1}(0) = \frac{x_i(0)}{\max(MaxX,\,MaxY)}$ and $\psi_{i2}(0) = \frac{y_i(0)}{\max(MaxX,\,MaxY)}$ respectively. (*MaxX, MaxY*) is the farthest location within the network area at which a sensor node can be located. Neighborhood radius (τ) of *DICLA* is equal to $\frac{R_t}{\max(MaxX,\,MaxY)}$, and hence two cells c_i and c_j in *DICLA* are adjacent to each other if s_i and s_j in the sensor network are close enough to hear each other's signals.

According to the definition of *DICLA*, values of tendency levels are used along with the neighborhood radius of *DICLA* to specify the neighboring cells of a cell. But here, values of $\psi_{i1}(0)$ and $\psi_{i2}(0)$ cannot be computed due to the fact that sensor nodes are not aware of their physical positions. This does not cause any problem due to the fact that in a *DICLA* which is mapped into a sensor network, neighboring cells of each cell are implicitly specified according to the topology of the network (two cells are adjacent to each other if their corresponding sensor nodes are within the transmission ranges of each other).

The learning automaton in each cell c_i of *DICLA*, referred to as LA_i, has two actions α_0, and α_1. Action α_0 is "apply force to neighboring nodes", and action α_1 is "do not apply force to neighboring nodes". The probability of selecting each of these actions is initially set to .5.

Deployment

Each sensor node has a state which can be "*mobile*" or "*fixed*". Initially, each sensor node selects its state randomly with probability of selecting "*fixed*" state to be P_{fix}. P_{fix} is a constant which is known to all sensor nodes. A "*fixed*" sensor node leaves the deployment phase immediately and starts with the maintenance phase.

Deployment phase for each "*mobile*" sensor node s_i is divided into a number of rounds. Each round $R_i(n)$ is started by the asynchronous activation of the cell c_i of the *DICLA*. The nth activation of the cell c_i occurs at time $\delta_i + n \times ROUND\_DURATION$ where δ_i is a random number generated for cell c_i and *ROUND_DURATION* is an upper bound for the duration of a single round. We call n the local iteration number for the cell. Delays δ_i are chosen randomly in order to reduce the probability of collisions between neighboring nodes.

Upon the startup of a new round $R_i(n)$, LA_i selects one of its actions randomly according to its action probability vector. Selected action, which is one of "apply force to neighboring nodes", or "do not apply force to neighboring nodes", specifies whether sensor node s_i will apply any forces to its neighboring nodes during the current round or not. Sensor node s_i then creates a packet called *APPLIED_FORCE* containing its state and the selected action and broadcasts it in its neighborhood. After broadcasting the *APPLIED_FORCE* packet, sensor node s_i waits for certain duration (*RECIEVE_DURATION*) to receive *APPLIED_FORCE* packets from its neighboring nodes. Received packets are stored into a local database within the node.

When *RECIEVE_DURATION* is over, sensor node s_i collects following statistics from the stored information in its local database: number of received packets $(N_i^r(n))$, and number of neighbors selecting "apply force to neighboring nodes" action $(N_i^f(n))$. According to the collected statistics, local rule of the cell c_i computes the reinforcement signal $\beta_i(n)$ and the restructuring signal $\underline{\zeta}_i(n)$ which are used to update the action probability vector of LA_i and the tendency vector of c_i. Details on this will be given in Section "Applying Local Rule".

Next, sensor node s_i uses vector $\underline{\zeta}_i(n)$ as its movement path for the current round. In other words, if s_i is located at $(x_i(n), y_i(n))$, it moves to $(x_i(n+1), y_i(n+1)) = (x_i(n) + \zeta_{1,i}, y_i(n) + \zeta_{2,i})$. Last step during round $R_i(n)$ is the application of the restructuring function Z which updates the tendency levels of cell c_i using the restructuring signal $\underline{\zeta}_i(n)$ according to the Eq. (3.48).

$$\begin{cases} \psi_{i1}(n+1) = \psi_{i1}(n) + \dfrac{\zeta_{i1}(n)}{\max(MaxX, \ MaxY)} \\ \psi_{i2}(n+1) = \psi_{i2}(n) + \dfrac{\zeta_{i2}(n)}{\max(MaxX, \ MaxY)} \end{cases} \quad (3.50)$$

When this step is done, sensor node s_i waits for the next activation time of its corresponding cell c_i in *DICLA* to start its next round ($R_i(n + 1)$).

Deployment phase for a sensor node s_i is completed upon the occurrence of one of the followings:

- Stability: Sensor node s_i moves less than a specified threshold (*LEAST_DISTANCE*) during its last N_r rounds.
- Oscillation: Sensor node s_i oscillates between almost the same positions for more than N_o rounds.

When a sensor node s_i completes the deployment phase of CLA-DS, its state is changed to "*fixed*" and it starts with the maintenance phase.

Applying Local Rule

Local rule of a cell c_i based on the states of the cell and its neighboring cells computes the reinforcement signal β_i and the restructuring signal $\underline{\zeta}_i$.

Reinforcement Signal: The reinforcement signal for the nth round in a node s_i is computed based on a comparison between the number of neighbors of s_i ($N_i^r(n)$) and the expected number of neighbors ($E[N_{nei}]$). According to this comparison, following two cases may occur:

- $N_i^r(n)$ is almost equal to $E[N_{nei}]$ or formally $|N_i^r(n) - E[N_{nei}]| < \varepsilon$ where ε is a specified constant: In this case, if the selected action of LA_i is "do not apply force to neighboring nodes", then the reinforcement signal is to reward the action. Otherwise, the reinforcement signal is to penalize the action.
- $N_i^r(n)$ is smaller or greater than $E[N_{nei}]$ or formally $|N_i^r(n) - E[N_{nei}]| > \varepsilon$: In this case, if the selected action of LA_i is "apply force to neighboring nodes", then the reinforcement signal is to reward the action. Otherwise, the reinforcement signal is to penalize the action.

Equations (1.1) and (1.2) are used for rewarding or penalizing the selected action of LA_i. The idea behind the above method of computing the reinforcement signal is that to have a uniform deployment of sensor nodes, one way is to minimize the difference between the number of neighbors of each sensor node and the expected number of neighbors ($E[N_{nei}]$). When $N_i^r(n)$ is almost equal to $E[N_{nei}]$, it means that the number of neighbors of sensor node s_i is as it must be, and hence it is better for this node not to apply any forces to its neighbors. On the other hand, when $|N_i^r(n) - E[N_{nei}]| > \varepsilon$, the number of neighbors of s_i differs from that expected, and hence it is better for it to apply forces to its neighbors.

Restructuring Signal: The restructuring signal $\underline{\zeta}_i(n) = (\zeta_{i1}(n), \zeta_{i2}(n))^T$ is a two- dimensional vector with a random orientation whose magnitude is $N_i^f(n)$ (number of neighbors selecting "apply force to neighboring nodes" action). The elements of this vector are computed according to Eq. (3.51) using an angle θ which is selected uniformly at random from the range $[0, 2\pi]$.

$$\begin{cases} \zeta_{i1}(n) = N_i^f(n) \cdot \cos(\theta) \\ \zeta_{i2}(n) = N_i^f(n) \cdot \sin(\theta) \end{cases}. \tag{3.51}$$

This vector specifies the orientation, direction, and distance of movement for the node s_i during the current round.

3.4.1.4 Experimental Results

To evaluate the performance of CLA-DS several experiments have been conducted and the results are compared with the results obtained for potential field-based algorithm given in (Howard et al. 2002) referred to as PF hereafter, DSSA and IDCA algorithms given in (Heo and Varshney 2005), and the basic VEC algorithm given in (Wang et al. 2006c). The Algorithms are compared with respect to three criteria: coverage, node separation, and distance.

- Coverage: Fraction of the area which is covered by the deployed network. Coverage is specified according to Eq. (3.52).

$$Coverage = \frac{S_{C(\Omega)}}{S_\Omega}. \tag{3.52}$$

- Node separation: Average distance from the nearest-neighbor in the deployed network. Node separation can be computed using Eq. (3.53). In this equation, $dist(s_i, s_j)$ is the Euclidean distance between sensor nodes s_i and s_j. Node separation is a measure of the overlapping area between the sensing regions of sensor nodes; smaller node separation means more overlapping.

$$Node\ separation = \frac{1}{N} \sum_{i=1}^{N} \min_{s_j \in Nei(s_i)} \left(dist\left(s_i, s_j\right) \right). \tag{3.53}$$

- Distance: The average distance traveled by each node. This criterion is directly related to the energy consumed by the sensor nodes. Energy of a sensor node in a virtual force-based deployment strategy is consumed in two ways: packet transmission and sensor movements. According to the work of Sibley et al. (2002), energy consumed for moving a sensor node 1 m is approximately equal to the energy consumed for 300 packet transmissions. Therefore, the energy

Fig. 3.16 Simulation area

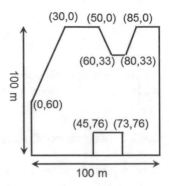

Table 3.1 Parameters of CLA-DS deployment strategy and their values

Parameter	Value
P_{fix}	0.1
ROUND_DURATION	11 (s)
RECEIVE_DURATION	10 (s)
LEAST_DISTANCE	1 (m)
N_o	6 rounds
ε	1
a (reward parameter)	0.25
b (penalty parameter)	0.25
Mid_Value	0.25
Small_Reward	0.1
Large_Reward	0.4
Small-Penalty	0.1
Large_Penalty	0.4
η	0.5
L	3 rounds

consumed for packet transmissions are assumed to be neglected in the proposed experiments.

Experiments are performed for the simulation area shown in Fig. 3.16. Networks of different sizes from $N = 50$ to $N = 500$ sensor nodes are considered for simulations. Sensing ranges ($R_s = r$) and transmission ranges ($R_t = 2 \cdot r$) of sensor nodes are assumed to be 5 and 10 m respectively. Energy consumption of nodes follows the energy model of the J-Sim simulator (Sobeih et al. 2006). Table 3.1 gives the values for different parameters of the algorithm.

All simulations have been implemented using J-Sim simulator. J-Sim is a java based simulator which is implemented on top of a component-based software architecture. Using this component-based architecture, new protocols and algorithms can be designed, implemented and tested in the simulator without any changes to the rest of the simulator's codes.

All reported results are averaged over 50 runs. We have used CSMA as the MAC layer protocol, free space model as the propagation model, binary sensing model and Omni-directional antenna.

Experiment 1

In this experiment the behavior of CLA-DS algorithm is compared with that of PS, DSSA, IDCA, and VEC algorithms in terms of the coverage as defined by Eq. (3.52). The experiment is performed for $N = 50$, 100, 200, 300, 400, and 500 sensor nodes which are initially deployed using a hybrid deployment method within a square with side length 10 m centered on the center of the network area. Figure 3.17 gives the results of this experiment. From the results we can conclude the following:

- Although CLA-DS algorithm does not use any information about sensor positions or their relative distances to each other, its performance in covering the network area is almost equal to that of PF algorithm in which sensor nodes have information about their relative distances to each other. This indicates the efficiency of the learning automata in guiding sensor nodes through the network area for finding their best positions.
- In sparse networks ($N < 400$), DSSA and IDCA algorithms better cover the network area than other algorithms. In dense networks, CLA-DS and PF algorithms outperform DSSA and IDCA algorithms in terms of coverage. This is due to the fact that the repulsive forces between neighboring nodes in DSSA and IDCA algorithms are stronger in sparse networks than in dense networks, and hence in sparse networks, sensor nodes can better spread through the network area.

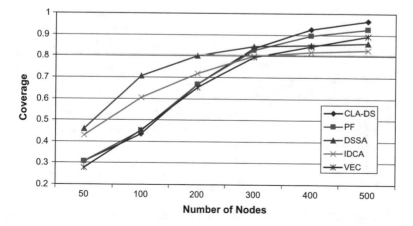

Fig. 3.17 Comparison of CLA-DS with existing deployment algorithms (coverage criterion(

- VEC algorithm performs the same as CLA-DS and PF algorithms in sparse networks, but in dense networks, its performance is degraded. The reason for low performance of VEC algorithm in a dense network is that in a dense network, Voronoi cells of sensor nodes are very small and are covered very quickly and therefore many of sensor nodes stop moving during initial rounds without enough exploration through the network area for finding better positions. Note that in VEC algorithm, a node moves only if its movement increases the coverage of its Voronoi cell.

Experiment 2

In this experiment, CLA-DS algorithm is compared with PS, DSSA, IDCA, and VEC algorithms in terms of the node separation criterion given by Eq. (3.53). The simulation settings of Experiment 1 are also used for this experiment. Figure 3.18 gives the results of this experiment. Node separation is a measure of the overlapping area between the sensing regions of sensor nodes; smaller node separation means more overlapping. Results of this experiment indicate that the overlapping area between sensor nodes in CLA-DS is more than in the PS, DSSA, IDCA, and VEC algorithms. Since the coverage of CLA-DS algorithm is better than or equal to the existing algorithms, having smaller node separation or more overlapped area makes CLA-DS superior to the existing algorithms due to the following reasons:

- The fraction of the network area, which is under the supervision of more than one sensor node, is higher in CLA-DS algorithm than the existing algorithms. This increases the tolerance of the network against node failures.

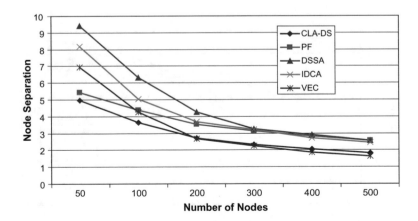

Fig. 3.18 Comparison of CLA-DS with existing deployment algorithms (node separation criterion)

- In occurrences of coverage holes (due to the node failures or deaths for example), neighboring nodes need fewer movements to heel the holes when node separation is smaller.

Experiment 3

In this experiment, CLA-DS algorithm is compared with PS, DSSA, IDCA, and VEC algorithms in terms of the distance criterion and its standard deviation. The simulation settings of Experiment 1 are also used for this experiment. Figures 3.19 and 3.20 give the results of this experiment. From the results obtained for this experiment one may conclude the following:

- In terms of the distance criterion, the most efficient algorithm among CLA-DS, PF, DSSA, IDCA, and VEC algorithms is PF algorithm. This is due to the fact that in PF algorithm, unlike other algorithms, movements of sensor nodes are directed by the application of Newton's second law of motion (Crowell 2004).
- In highly sparse networks ($N < 200$), CLA-DS outperforms DSSA and IDCA algorithms in terms of distance criterion, but in dense networks, DSSA and IDCA algorithms perform better than CLA-DS. This is again due to the fact that the repulsive forces between neighboring nodes in DSSA and IDCA algorithms are stronger in sparse networks than in dense networks, and hence in dense networks, movements of sensor nodes are more limited than in sparse networks. As it is shown in Experiment 1, this limited movement degrades the performance of DSSA and IDCA algorithms in covering the network area.
- For VEC algorithm, the average distance moved by sensor nodes for networks with different sizes is almost the same. This is due to the fact that in this algorithm, the number of sensor nodes which do not explore the network area sufficiently increases as the density of the network increases.

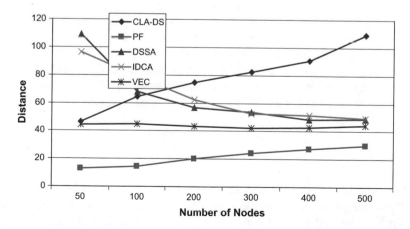

Fig. 3.19 Comparison of CLA-DS with existing deployment algorithms (distance criterion)

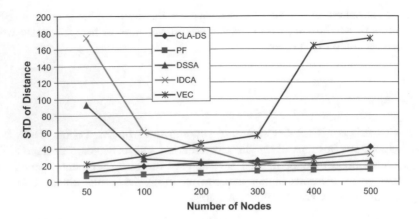

Fig. 3.20 Comparison of CLA-DS with existing deployment algorithms (standard deviation of the distance criterion)

- In terms of the standard deviation of the distance criterion, CLA-DS and PF algorithms are the best algorithms among the compared algorithms. This means that in these two algorithms, movements of sensor nodes, or equivalently energy consumptions of the sensor nodes, are approximately fair.

Experiment 4

In this experiment, CLA-DS algorithm is compared with PS, DSSA, IDCA, and VEC algorithms in terms of the coverage, node separation, and distance criteria when devices or algorithms used for location estimation in sensor nodes experience different levels of error. Such errors due to inaccuracies in measurements are common both in GPS-based location estimator devices (Zhou et al. 2009a) and localization techniques adopted to wireless sensor networks (Ramadurai and Sichitiu 2003; Lee et al. 2006). For simulating an error level of $0 < \lambda < 1$, for each sensor node s_i, two numbers $Rnd_i(x)$ and $Rnd_i(y)$ are selected uniformly at random from the ranges $[-MaxX, MaxX]$ and $[-MaxY, MaxY]$ respectively and are used for modifying the position (x_i, y_i) of the node according to Eq. (3.54). For this study, λ is assumed to be one of the following: 0.2, 0.25, 0.35, and 0.5.

$$\begin{cases} x_i^{inexact} = x_i + \lambda \cdot Rnd_i(x) \\ y_i^{inexact} = y_i + \lambda \cdot Rnd_i(y) \end{cases} .$$
(3.54)

The experiment is performed for $N = 500$ sensor nodes which are initially deployed using a hybrid deployment method within a square with side length 10 m centered on the center of the network area given in Fig. 3.16. Figures 3.21, 3.22

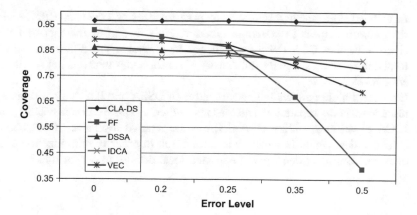

Fig. 3.21 Comparison of CLA-DS with existing deployment algorithms in terms of the coverage criterion in presence of inaccuracies in estimating the position of sensor nodes

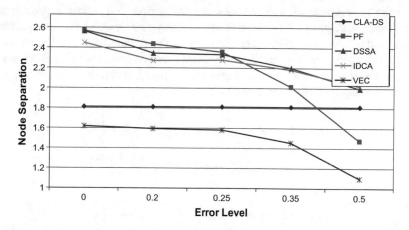

Fig. 3.22 Comparison of CLA-DS with existing deployment algorithms in terms of the node separation criterion in presence of inaccuracies in estimating the position of sensor nodes

and 3.23 give the results of this experiment for different criteria. These figures show that:

- Error level has no effect on the performance of CLA-DS algorithm with respect to the coverage, node separation and distance criteria. This is due to the fact that CLA-DS algorithm does not use any information about the position of the sensor nodes.
- PF algorithm is highly affected by increasing the error level. This is because in PF, movements of sensor nodes are directed by the application of Newton's second law of motion (Crowell 2004) which is highly sensitive to the exact locations of the sensor nodes.

- Error level does not highly affect the performances of DSSA and IDCA algorithms with respect to coverage criterion. This is due to the fact that these algorithms, like CLA-DS algorithm, try to minimize the difference between local density and expected local density of sensor nodes which is not sensitive to error level.
- Error level highly affects the performances of DSSA and IDCA algorithms with respect to node separation and distance criteria. This is because DSSA and IDCA algorithms, unlike CLA-DS, use the relative distances of neighboring sensor nodes, which is sensitive to error level, in order to minimize the difference between local density and expected local density of sensor nodes.

Experiment 5

In this experiment, behaviors of CLA-DS, PS, DSSA, IDCA, and VEC algorithms are compared with respect to the coverage, node separation and distance criteria when the movements of sensor nodes are not perfect and follow a probabilistic motion model. A probabilistic motion model can better describe the movements of sensor nodes in real world scenarios. We use the probabilistic motion model of sensor nodes given in Yap and Shelton (2008). In this probabilistic motion model, movements of a sensor node s_i for a given drive $(d_i(n))$ and turn $(r_i(n))$ command is described using the following equations:

$$
\begin{aligned}
x_i(n+1) = x_i(n) + D_i(n) \cdot \cos\left(\theta_i(n) + \frac{T_i(n)}{2}\right) \\
+ C_i(n) \cdot \cos\left(\theta_i(n) + \frac{T_i(n) + \pi}{2}\right),
\end{aligned}
\tag{3.55}
$$

$$
\begin{aligned}
y_i(n+1) = y_i(n) + D_i(n) \cdot \sin\left(\theta_i(n) + \frac{T_i(n)}{2}\right) \\
+ C_i(n) \cdot \sin\left(\theta_i(n) + \frac{T_i(n) + \pi}{2}\right),
\end{aligned}
\tag{3.56}
$$

$$
\theta_i(n+1) = (\theta_i(n) + T_i(n)) \mod (2\pi).
\tag{3.57}
$$

In the above equations, $\theta_i(n) + \frac{T_i(n)}{2}$ is referred to as the major axis of movement, $\theta_i(n) + \frac{T_i(n) + \pi}{2}$ is the minor axis of movement (orthogonal to the major axis), and $C_i(n)$ is an extra lateral translation term to account for the shift in the orthogonal direction to the major axis. $D_i(n)$, $T_i(n)$, and $C_i(n)$ are all independent and conditionally Gaussian given $d_i(n)$ and $r_i(n)$:

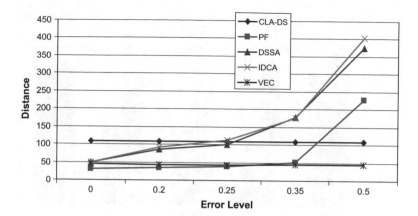

Fig. 3.23 Comparison of CLA-DS with existing deployment algorithms in terms of the distance criterion in presence of inaccuracies in estimating the position of sensor nodes

$$\begin{cases} D_i(n) \sim N\left(d_i(n), d_i^2(n) \cdot \sigma_{D_d}^2 + r_i^2(n) \cdot \sigma_{D_r}^2 + \sigma_{D_1}^2\right) \\ T_i(n) \sim N\left(r_i(n), d_i^2(n) \cdot \sigma_{T_d}^2 + r_i^2(n) \cdot \sigma_{T_r}^2 + \sigma_{T_1}^2\right) \\ C_i(n) \sim N\left(0, d_i^2(n) \cdot \sigma_{C_d}^2 + r_i^2(n) \cdot \sigma_{C_r}^2 + \sigma_{C_1}^2\right), \end{cases} \tag{3.58}$$

where $N(a, b)$ is a Gaussian distribution with mean a and variance b, and $\sigma_{D_d}^2$, $\sigma_{D_r}^2$, $\sigma_{D_1}^2$, $\sigma_{T_d}^2$, $\sigma_{T_r}^2$, $\sigma_{T_1}^2$, $\sigma_{C_d}^2$, $\sigma_{C_r}^2$, and $\sigma_{C_1}^2$ are all parameters of the specified motion model. Table 3.2 gives values of these parameters used in this experiment.

The experiment is performed for $N = 500$ sensor nodes which are initially deployed using a hybrid deployment method within a square with side length 10 m centered on the center of the network area given in Fig. 3.16. Figures 3.24, 3.25 and 3.26 show the results of this experiment. The results indicate the following facts:

- Using the probabilistic motion model instead of the perfect motion model degrades significantly the performances of PF, DSSA, and IDCA algorithms in terms all three criteria, but does not affect the performances of VEC and CLA-DS algorithms substantially.

Table 3.2 Parameters of the specified probabilistic motion model and their corresponding values

Parameter	Value	Parameter	Value	Parameter	Value
$\sigma_{D_d}^2$	2.1869×10^{-6}	$\sigma_{T_d}^2$	3.45×10^{-6}	$\sigma_{C_d}^2$	8.588×10^{-6}
$\sigma_{D_r}^2$	1.0731×10^{-6}	$\sigma_{T_r}^2$	3.38267×10^{-6}	$\sigma_{C_r}^2$	1.3427×10^{-6}
$\sigma_{D_1}^2$	10^{-6}	$\sigma_{T_1}^2$	6.66048×10^{-6}	$\sigma_{C_1}^2$	1.4×10^{-6}

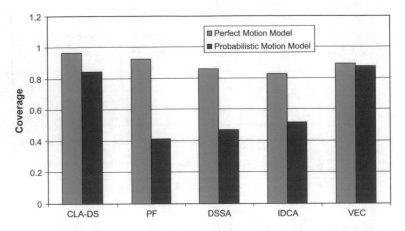

Fig. 3.24 Comparison of CLA-DS with existing deployment algorithms in terms of the coverage criterion when movements of sensor nodes follow a probabilistic motion model

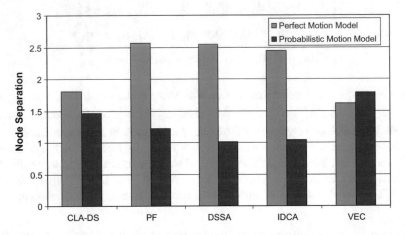

Fig. 3.25 Comparison of CLA-DS with existing deployment algorithms in terms of the node separation criterion when movements of sensor nodes follow a probabilistic motion model

- When the probabilistic motion model is used CLA-DS and VEC algorithms outperform the existing algorithms in terms of coverage criterion.
- Node separation of CLA-DS algorithm is smaller than that of VEC algorithm. This indicates the superiority of CLA-DS algorithm over VEC algorithm in terms of node separation criterion using a similar discussion to that given in Experiment 2.
- Node separation of CLA-DS algorithm is larger than that of PF, DSSA, and IDCA algorithms. This means that CLA-DS has more overlapping area between sensor nodes.

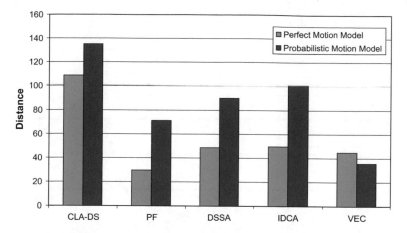

Fig. 3.26 Comparison of CLA-DS with existing deployment algorithms in terms of the distance criterion when movements of sensor nodes follow a probabilistic motion model

- For probabilistic motion model, the average distance moved by sensor nodes for all algorithms except for VEC algorithm is higher than when perfect motion model is used. When the probabilistic motion model is used the performances of CLA-DS, PF, DSSA, and IDCA algorithms are degraded by 24, 136, 81, and 100% respectively. This indicates that CLA-DS algorithm is more robust to deviations in the movements of sensor nodes than PF, DSSA, and IDCA algorithms.

- For VEC algorithm, the average distance moved by sensor nodes does not change significantly when the probabilistic motion model is used. This is due to following two reasons: (1) In VEC algorithm, many of the sensor nodes stop moving during initial rounds without enough exploration through the network area, and (2) Sensor nodes move within their Voronoi cells in VEC algorithm and hence their movements are very limited.

Experiment 6

This experiment is conducted to study the effect of parameter P_{fix} on the performance of CLA-DS algorithm. For this study, a network of $N = 500$ sensor nodes which are initially deployed using a hybrid deployment method within a square with side length 10 m centered on the center of the network area given in Fig. 3.16 is considered. Figures 3.27, 3.28 and 3.29 give the results of this experiment. These figures indicate that by increasing the value of parameter P_{fix} in CLA-DS algorithm, the covered section of the area, node separation and the average distance traveled by each sensor node decrease. In other words, higher values of P_{fix} results in more energy saving during the deployment process at the expense of poor coverage.

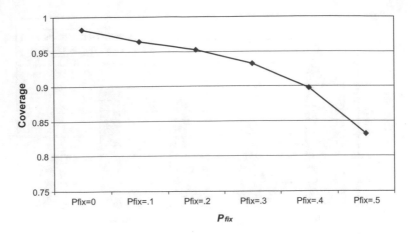

Fig. 3.27 Effect of parameter P_{fix} on the performance of CLA-DS in terms of the coverage criterion

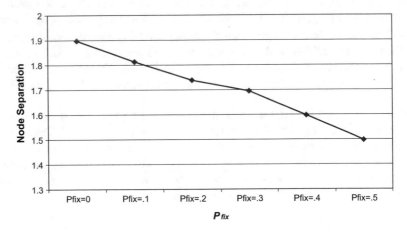

Fig. 3.28 Effect of parameter P_{fix} on the performance of CLA-DS in terms of the node separation criterion

Determination of P_{fix} for an application is very crucial and is a matter of cost versus precision. For better coverage, higher price must be paid.

Experiment 7

This experiment is conducted to study the behavior of *DICLA* as a learning model in CLA-DS algorithm. For this study, a network of $N = 500$ sensor nodes which are initially deployed using a hybrid deployment method within a square with side

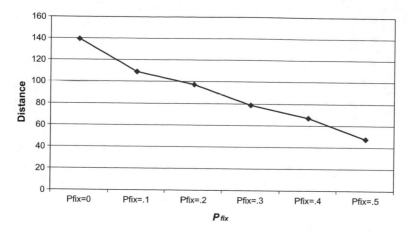

Fig. 3.29 Effect of parameter P_{fix} on the performance of CLA-DS in terms of the distance criterion

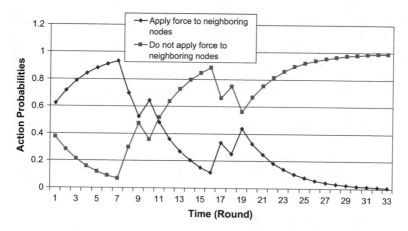

Fig. 3.30 Action probabilities of a randomly selected learning automaton from *DICLA*

length 10 m centered on the center of the network area given in Fig. 3.16 is considered. Figure 3.30 depicts the action probability vector of a randomly selected learning automaton from *DICLA*. As it can be seen in this figure, at the beginning of the deployment process, the action probability of "apply force to neighboring nodes" action increases. This is due to the fact that the density of sensor nodes in the initial hybrid deployment method is very high and hence, sensor nodes must apply force to each other to spread through the area. As time passes, the action probability of "do not apply force to neighboring nodes" action gradually increases and approaches unity. As a result, local node density gradually approaches its desired value ($E[N_{nei}]$).

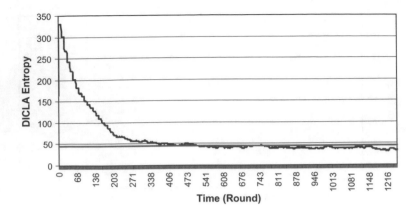

Fig. 3.31 Entropy of *DICLA*

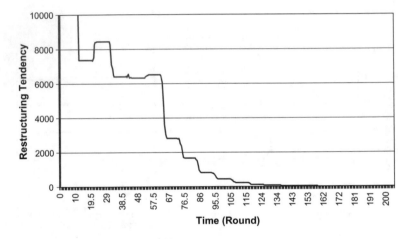

Fig. 3.32 Restructuring tendency of *DICLA*

Figure 3.31 shows the entropy of *DICLA* during the deployment process. This figure indicates that the entropy of *DICLA* is high at initial rounds of CLA-DS algorithm, but gradually decreases as time passes. It goes below 45 at about round number 750. This means that after this round, the entropy of each learning automaton LA_i in *DICLA* is on average below 0.09. If the entropy of a two-action learning automaton is below 0.09, then we can say that the action probability of one of its actions is higher than .982. This means that action switching in each learning automaton in *DICLA* rarely occurs after round number 750.

Figure 3.32 depicts changes in the restructuring tendency of *DICLA* during the deployment process. This figure shows that the restructuring tendency of *DICLA* is initially high and gradually approaches zero. It is initially high because during initial rounds, the magnitude of the force vector applied to each sensor node is large, and it gradually approaches zero because as time passes, local densities of sensor nodes approach its expected value which results in the magnitude of the force vector applied to each sensor node to approach zero.

3.4.1.5   Summary of Results

In this section, we compared the performance of CLA-DS deployment algorithm in terms of the coverage, node separation, and distance criteria with PF, DSSA, IDCA, and VEC deployment algorithms. Comparisons were made for different network sizes, and error free and with-error environments. From the results of this study we can conclude that:

- In error free environments, the given algorithm (CLA-DS) can compete with existing algorithms in terms of the coverage criterion, outperforms existing algorithms in terms of the node separation criterion, and performs worse than existing algorithms in terms of the distance criterion.
- CLA-DS algorithm, unlike existing algorithms, does not use any information regarding the position of the sensor nodes or their relative distances to each other and therefore, in environments, where utilized location estimation techniques such as GPS-based devices and localization algorithms experience inaccuracies in their measurements, CLA-DS algorithm outperforms existing algorithms in terms of all the criteria.
- The algorithms which are least affected by the selection of "node movement model" of the sensor nodes are CLA-DS and VEC.
- CLA-DS algorithm, unlike existing algorithms, has a parameter (P_{fix}) for controlling the tradeoff between the network coverage and the average distance traveled by sensor nodes.

3.4.2   CLA-EDS: An Extension to CLA-DS for Providing K-Coverage

CLA-DS algorithm, proposed in previous section, is robust against inaccuracies which may occur in the measurements of sensor positions or in the movements of sensor nodes. Its novelty is that it works without any sensor to know its position or its relative distances to other sensors. Despite its advantages, CLA-DS covers every point within the area of the network with only one sensor node, which is not enough for applications with k-coverage requirement such as intrusion detection (Mehta et al. 2003; Arora et al. 2004), data gathering (Zhao and Govindan 2003; Kumar et al. 2004), and object tracking (Llinas et al. 2001). In such applications, it is required for every point within the area of the network to be covered by at least k different sensor nodes. k is referred to as the degree of coverage. To make

CLA-DS capable of addressing the k-coverage requirement, in this section, we introduce an extension to this deployment strategy, called CLA-EDS, which is able to provide the k-coverage of the entire area of the network. This deployment strategy is also able to address k-coverage requirement with different values of k in different regions of the sensor field. Like CLA-DS, in CLA-EDS neighboring nodes apply forces to each other. Then, each node moves according to the resultant force vector applied to it from its neighbors. Each node is equipped with a learning automaton. The Learning automaton of a node at any given time decides for the node whether to apply force to its neighbors or not. This way, each node in cooperation with its neighboring nodes gradually learns its best position within the area of the network so as to fulfill the required degree of coverage. CLA-EDS uses *HDICLA* as its learning model.

To study the performance of this deployment strategy, several experiments have been conducted and the results obtained from CLA-EDS are compared with the results obtained from existing self-regulated deployment strategies, capable of providing k-coverage, such as DSSA, IDCA, and DSLE. Experimental results show that, in terms of the network coverage, the CLA-EDS strategy can compete with the existing deployment strategies in environments with exact location information and outperforms existing deployment strategies in environments with inexact location information. Results of the experiments also show that when the required degree of coverage differs in different regions of the network, CLA-EDS deployment strategy significantly outperforms DSSA, IDCA, DSLE, and CLA-DS algorithms with respect to the network coverage.

3.4.2.1 Problem Statement

Consider N mobile sensor nodes s_1, s_2, ..., s_N with equal sensing ($R_s = r$) and transmission ranges ($R_t = 2 \cdot r$) which are initially deployed in some initial region within an unknown two dimensional sensor field Ω. Assume that a rough estimate of the surface of Ω ($\hat{S}_\Omega$) is available (using Google maps for example). Sensor nodes are able to move along any desired direction within the area of the network at a constant speed, but they cannot cross the barrier of Ω. We assume that sensor nodes have no mechanism for estimating their physical positions or their relative distances to each other.

Definition 3.10 Coverage function $C_{x_l,y_l}(s_i)$ is defined according to the following equation:

$$C_{x_l,y_l}(s_i) = \begin{cases} 1; & (x_l, y_l) \in C(s_i) \\ 0; & otherwise \end{cases},$$
(3.59)

where $C(s_i)$ is as defined in Definition 3.3. In other words, $C_{x_l,y_l}(s_i) = 1$, if (x_l, y_l) is within the sensing region of the sensor node s_i.

We assume that different regions within the network area require different degrees of coverage. Let $\Omega_1, \Omega_2, \ldots, \Omega_M$, be M regions within the Ω with the following properties:

- $\Omega_i \cap \Omega_j = \varnothing$; *for* $i \neq j$
- $\cup_{i=1}^{M} \Omega_i = \Omega$

Let $\hat{S}_{\Omega_i}$ and $rdc(\Omega_i)$ be the estimated surface and the required degree of coverage of the ith region respectively. It is straight forward that the required degree of coverage of every point $(x_l, y_l) \in \Omega_i$ is equal to $rdc(\Omega_i)$, that is, $(x_l, y_l) \in \Omega_i \Rightarrow rdc(x_l, y_l) = rdc(\Omega_i)$. We assume that there exists a notification-based mechanism in the network (using local base stations for example) which notifies the values of $\hat{S}_{\Omega_i}$ and $rdc(\Omega_i)$ in each region Ω_i to the sensor nodes within that region.

Consider Definition 3.5 to Definition 3.9 from the previous section as well as the following definition:

Definition 3.11 Covered sub-area denoted by $C(\Omega)$ refers to the set of the points (x_l, y_l) within the network area, each is covered with at least $rdc(x_l, y_l)$ sensor nodes. $C(\Omega)$ is stated through Eq. (3.60).

$$C(\Omega) = \bigcup_{i=1}^{M} \left(\left\{ (x_l, y_l) \middle| (x_l, y_l) \in \Omega_i, \sum_{j=1}^{N} C_{x_l, y_l}(s_j) \geq rdc(\Omega_i) \right\} \right). \tag{3.60}$$

Using the above definitions and assumptions, the problem considered in this section can be stated as follows: Propose a self-regulated deployment strategy which deploys N mobile sensor nodes throughout an unknown network area Ω with estimated surface $\hat{S}_\Omega$ so that the covered section of Ω ($S_{C(\Omega)}$) is maximized and the deployed network is connected.

3.4.2.2 CLA-EDS Deployment Strategy

Like CLA-DS, CLA-EDS deployment strategy also consists of *initial deployment*, *mapping*, *deployment*, and *maintenance* phases. We explain these 4 phases in more details in the subsequent sections.

Initial Deployment

Like CLA-DS, initial deployment of the sensor nodes in CLA-EDS deployment strategy can be done using any of the random, sub-area, and hybrid deployment strategies.

Mapping

In the mapping phase, a time-driven asynchronous *HDICLA*, which is isomorphic to the sensor network topology, is created. Each sensor node s_i located at (x_i, y_i) in the sensor network corresponds to the cell c_i in the *HDICLA*.

Interest set of the *HDICLA* consists of two members; X-axis and Y-axis of the network area. Initially (at time instant 0), tendency levels of each cell c_i to these interests are $\psi_{i1}(0) = \frac{x_i(0)}{\max(MaxX, MaxY)}$ and $\psi_{i2}(0) = \frac{y_i(0)}{\max(MaxX, MaxY)}$ respectively. *(MaxX, MaxY)* is the farthest location within the network area at which a sensor node can be located.

Attribute set of the *HDICLA* has two member; *required degree of coverage* (*rdc*), and estimated surface ($\hat{S}$) of the region requires this *rdc*. As it was mentioned in the section of problem statement, the values of these attributes differ in different regions of the network and there exists a notification-based mechanism which notifies $rdc(\Omega_j)$ and $\hat{S}_{\Omega_j}$ of each region Ω_j to the sensor nodes within that region. According to these assumptions, the local values of the *rdc* (rdc_i) and $\hat{S}(\hat{S}_i)$ attributes are known to every cell c_i of the *HDICLA*. Initially, $rdc_i(0)$ and $\hat{S}_i(0)$ are set to 1 and $\hat{S}_\Omega$ (the rough estimate of the surface of the network area) respectively.

Neighborhood radius (τ) of the *HDICLA* is equal to $\frac{R_t}{\max(MaxX, MaxY)}$, and hence two cells c_i and c_j in the *HDICLA* are adjacent to each other if s_i and s_j in the sensor network are close enough to hear each other's signals.

The learning automaton in each cell c_i of the *HDICLA*, referred to as LA_i, has two actions; α_0 and α_1. Action α_0 is "apply force to the neighboring nodes" and action α_1 is "do not apply force to the neighboring nodes". The probability of selecting each of these actions is initially set to .5.

Deployment

Deployment phase of the CLA-EDS deployment strategy is mostly the same as that of the CLA-DS deployment strategy. The only differences are in the way, the reinforcement signal, $\beta_i(n)$, is computed and the addition of a new step, which is the reevaluation of the local attributes rdc_i and $\hat{S}_i$.

Computing the Reinforcement Signal

To compute the reinforcement signal, a sensor node s_i first computes the minimum number of sensor nodes ($N_i^{Min}(n)$) that if exist within its transmission range, the required degree of coverage of its local region is satisfied. Using the theorems and results given in (Yen et al. 2006) and by ignoring the border effect (Bettstetter and Krause 2001), $N_i^{Min}(n)$ can be computed using Eq. (3.61).

$$N_i^{Min}(n) = \underset{l \geq 1}{\text{Min}} \left\{ \mathrm{E}\left[C_l^{rdc_i(n)} \right] \geq 1 - \varepsilon \right\} \cdot \rho_i(n) - 1, \qquad (3.61)$$

where $\varepsilon \ll 1$ is a positive constant, $\rho_i(n) = \frac{\pi \cdot R_i^2}{S_i(n)}$, and $\mathrm{E}\left[C_l^{rdc_i(n)} \right]$ (given by the iterative Eq. 3.62) is the expected surface that is rdc_i-covered (covered with rdc_i sensor nodes) by randomly deploying l sensor nodes.

$$\mathrm{E}\left[C_l^{rdc_i(n)} \right] = (1 - \rho_i(n)) \cdot \mathrm{E}\left[C_{l-1}^{rdc_i(n)} \right] + \rho_i(n) \cdot \mathrm{E}\left[C_{l-1}^{rdc_i(n)-1} \right]. \qquad (3.62)$$

The reinforcement signal for the nth round in a node s_i is computed based on a comparison between the number of neighbors of s_i ($N_i^r(n)$) and $N_i^{Min}(n)$. According to this comparison, following cases may occur:

- $N_i^{Min}(n) \leq N_i^r(n) \leq 2 \cdot N_i^{Min}(n)$: In this case, if the selected action of LA_i is "do not apply force to the neighboring nodes", then the reinforcement signal is to reward the action. Otherwise, the reinforcement signal is to penalize the action.
- $N_i^r(n) < N_i^{Min}(n)$ or $N_i^r(n) > 2 \cdot N_i^{Min}(n)$: In this case, if the selected action of LA_i is "apply force to neighboring nodes", then the reinforcement signal is to reward the action. Otherwise, the reinforcement signal is to penalize the action.

In other words, if the number of neighboring nodes of a sensor node s_i is at least equal to the minimum required number of the sensor nodes ($N_i^{Min}(n)$), but not greater than the twice of it ($2 \cdot N_i^{Min}(n)$), then it is better for this node not to apply any forces to its neighbors. Otherwise, and if the number of neighbors of s_i is less than $N_i^{Min}(n)$ or more than $2 \cdot N_i^{Min}(n)$, then it is better for s_i to apply forces to its neighbors. Equations (1.1) and (1.2) are used for rewarding or penalizing the selected action of LA_i.

Reevaluation of the Local Attributes

After the application of the restructuring function Z, the values of the local attributes rdc_i and $\hat{S}_i$ in the new structure of the *HDICLA* must be reevaluated. To do this, sensor node s_i waits for certain duration (*WAIT_FOR_LOCAL_INFO_DURATION*) to receive a packet which notifies the local values of rdc_i and $\hat{S}_i$ in its new position. If such a notification packet is received within this duration, the local values of the rdc_i and $\hat{S}_i$ are updated accordingly.

Maintenance

Maintenance phase of the CLA-EDS deployment strategy is equivalent to the maintenance phase of the CLA-DS.

3.4.2.3 Experimental Results

To evaluate the performance of the CLA-EDS deployment strategy, several experiments have been conducted and the results are compared with the results obtained from DSSA and IDCA algorithms given in (Heo and Varshney 2005), DSLE algorithm given in (Li and Kao 2010), and CLA-DS algorithm. It should be mentioned here that DSSA and IDCA algorithms are not primarily designed for providing k-coverage. Instead, the goal of these two algorithms is to control the movements of sensor nodes in such a way that in the deployed network, the average distance between sensor nodes is almost equal to the desired distance of sensor nodes in a uniform distribution. This approach enables us to use these algorithms for providing k-coverage by simply changing the input parameter "desired distance of sensor nodes" from d_{avg} to $\frac{d_{avg}}{k}$. In the experiments given in this section, we use DSSA and IDCA algorithms with this modification.

For comparison of the mentioned algorithms, we use three criteria given in Section "Experimental Results", namely coverage, node separation, and distance. Sensor field is assumed to be a 100 m × 100 m rectangle. Networks of different sizes from $N = 500$ to $N = 1500$ sensor nodes are considered for simulations. Sensing ranges ($R_s = r$) and transmission ranges ($R_t = 2 \cdot r$) of sensor nodes are assumed to be 5 and 10 m respectively. Energy consumption of the sensor nodes follows the energy model of the J-Sim simulator (Sobeih et al. 2006). Table 3.3 gives the values for different parameters of the algorithm.

All simulations have been implemented using J-Sim simulator. All reported results are averaged over 50 runs. We have used CSMA as the MAC layer protocol, free space model as the propagation model, binary sensing model and Omni-directional antenna.

Table 3.3 Parameters of the CLA-EDS algorithm and their values

Parameter	Value
P_{fix}	0
ROUND_DURATION	11 (s)
RECEIVE_DURATION	10 (s)
WAIT_FOR_LOCAL_INFO_DURATION	2 (s)
LEAST_DISTANCE	1 (m)
N_r	3 rounds
N_o	6 rounds
ε	0.01
a (reward parameter)	0.25
b (penalty parameter)	0.25
l	3 rounds

Experiment 1

In this experiment the behavior of CLA-EDS algorithm is compared with that of DSSA, IDCA, DSLE, and CLA-DS algorithms in terms of coverage as defined by Eq. (3.52). For this experiment, sensor nodes are initially deployed using a hybrid deployment method within a square with side length 10 m centered on the center of the network area. We assume that $M = 1$, that is, the required degree of coverage (k) throughout the network area is uniform. The experiment is performed for $k = 1$, 2, 3, 4, and 5. Figures 3.33, 3.34 and 3.35 give the results of this experiment for networks of different sizes ($N = 500$, 1000, and 1500). From the results we can conclude the following:

- The performance of CLA-EDS algorithm in covering the network area, for different values of k and networks of different sizes, can compete with that of DSSA and IDCA algorithms. Such performance, when considering the fact that CLA-EDS algorithm does not use any information regarding sensor positions or their relative distances to each other, indicates the efficiency of the learning automata in guiding the sensor nodes throughout the sensor field for finding their best positions.
- Performances of CLA-EDS and DSLE algorithms in covering the network area are almost the same in networks of small sizes ($N < 1000$), but in networks of large sizes ($N > 1000$), CLA-EDS outperforms DSLE in terms of the coverage criterion. The reason behind this phenomenon is that DSLE algorithm makes use of the Voronoi diagram for guiding the movements of sensor nodes throughout the network area, but the information coded in the Voronoi diagram of a highly

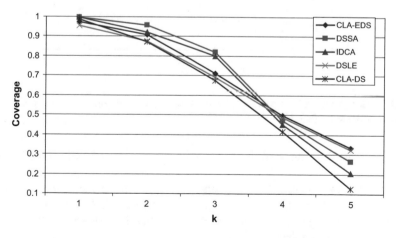

Fig. 3.33 Comparison of CLA-EDS with existing deployment algorithms in terms of the coverage criterion for $N = 500$ sensor nodes

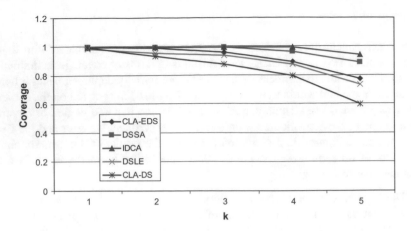

Fig. 3.34 Comparison of CLA-EDS with existing deployment algorithms in terms of the coverage criterion for $N = 1000$ sensor nodes

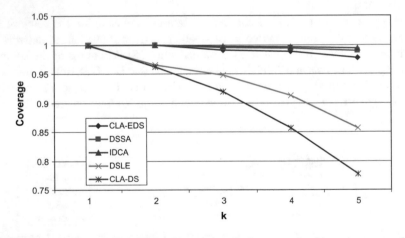

Fig. 3.35 Comparison of CLA-EDS with existing deployment algorithms in terms of the coverage criterion for $N = 1500$ sensor nodes

dense network is not as useful as that coded in the Voronoi diagram of a sparse network.

• CLA-EDS outperforms CLA-DS algorithm in terms of the coverage criterion when $k > 1$. This superiority is expected since CLA-DS is designed for providing 1-coverage of the network area whereas its extension, CLA-EDS, is designed for providing k-coverage.

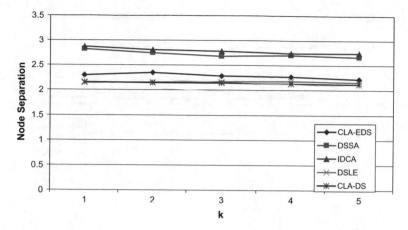

Fig. 3.36 Comparison of CLA-EDS with existing deployment algorithms in terms of the node separation criterion for $N = 500$ sensor nodes

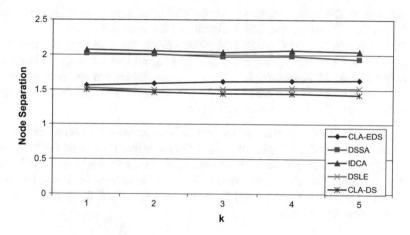

Fig. 3.37 Comparison of CLA-EDS with existing deployment algorithms in terms of the node separation criterion for $N = 1000$ sensor nodes

Experiment 2

In this experiment, CLA-EDS algorithm is compared with DSSA, IDCA, DSLE, and CLA-DS algorithms in terms of the node separation criterion given by Eq. (3.53). The simulation settings of Experiment 1 are also used for this experiment. Figures 3.36, 3.37 and 3.38 give the results of this experiment for networks of different sizes. Results of this experiment indicate that:

- Performances of CLA-EDS, CLA-DS, and DSLE algorithms in terms of the node separation criterion for different values of k and networks of different sizes

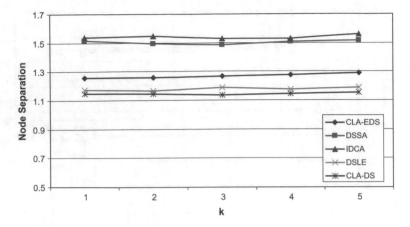

Fig. 3.38 Comparison of CLA-EDS with existing deployment algorithms in terms of the node separation criterion for $N = 1500$ sensor nodes

are almost the same. In other words, the overlapping areas resulted from CLA-EDS, CLA-DS, and DSLE algorithms are almost the same.

- Node separation of the CLA-EDS algorithm is less than that of DSSA and IDCA algorithms. Since the coverage of CLA-EDS algorithm is almost equal to that of DSSA and IDCA algorithms in almost all cases, having smaller node separation or more overlapped area makes CLA-EDS superior to DSSA and IDCA algorithms due to the following two reasons:

  - The fraction of the network area, which is under the supervision of more than one sensor node, is higher in CLA-EDS algorithm than in DSSA and IDCA algorithms. This increases the tolerance of the network against node failures.
  - In occurrences of coverage holes (due to node failures or deaths for example), neighboring nodes need fewer movements to heal the holes when node separation is smaller.

Experiment 3

In this experiment, CLA-EDS algorithm is compared with DSSA, IDCA, DSLE, and CLA-DS algorithms in terms of the distance criterion and its standard deviation. The simulation settings of Experiment 1 are also used for this experiment. Figures 3.39, 3.40, 3.41, 3.42, 3.43 and 3.44 give the results of this experiment for networks of different sizes. These figures show that in terms of the distance criterion, CLA-EDS is the worst algorithm among the compared algorithms. This is due to the fact that CLA-EDS algorithm does not use any information regarding the position of the sensor nodes or their relative distances to each other. To compensate this weakness, CLA-EDS algorithm has a parameter p_{fix} which can be used to make

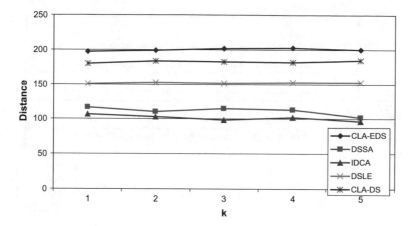

Fig. 3.39 Comparison of the CLA-EDS with existing deployment algorithms in terms of the distance criterion for $N = 500$ sensor nodes

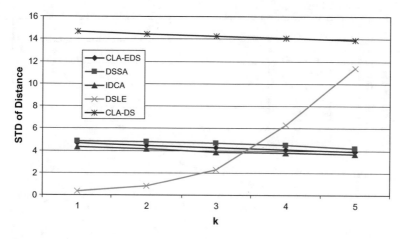

Fig. 3.40 Comparison of the CLA-EDS with existing deployment algorithms in terms of the standard deviation of the distance criterion for $N = 500$ sensor nodes

a tradeoff between the distance and coverage criteria. By increasing the value of p_{fix}, one can decrease the average distance moved by sensor nodes at the expense of less network coverage (refer to Experiment 6). In addition, the standard deviations of the distance, given in these figures, show that using the CLA-EDS, DSSA, and IDCA algorithms, the distance moved by each sensor node, or equivalently the energy consumed by each sensor node, is almost the same for all of the nodes. In other words, these algorithms can fairly balance the energy consumed by the sensor nodes in the network during the deployment process.

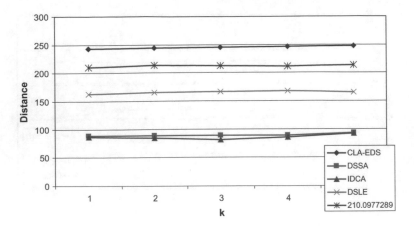

Fig. 3.41 Comparison of the CLA-EDS with existing deployment algorithms in terms of the distance criterion for $N = 1000$ sensor nodes

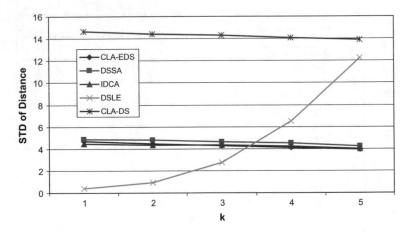

Fig. 3.42 Comparison of the CLA-EDS with existing deployment algorithms in terms of the standard deviation of the distance criterion for $N = 1000$ sensor nodes

Experiment 4

In this experiment, CLA-EDS algorithm is compared with DSSA, IDCA, DSLE, and CLA-DS algorithms in terms of the coverage, node separation, and distance criteria when devices or algorithms used for location estimation in sensor nodes experience different levels of error. Simulating the error is done in a similar way to that given in Section "Experiment 4". For this study, error level, λ, is assumed to be one of the followings: 0.1, 0.2, 0.3, 0.4, and 0.5.

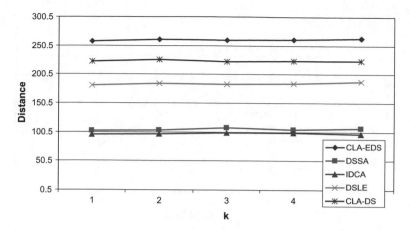

Fig. 3.43 Comparison of the CLA-EDS with existing deployment algorithms in terms of the distance criterion for $N = 1500$ sensor nodes

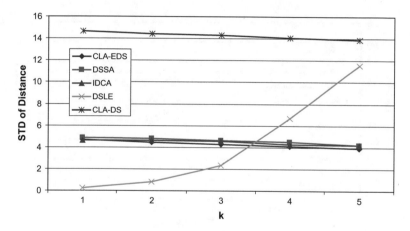

Fig. 3.44 Comparison of the CLA-EDS with existing deployment algorithms in terms of the standard deviation of the distance criterion for $N = 1500$ sensor nodes

We assume that $M = 1$, $N = 1500$, and $k = 5$. Sensor nodes are initially deployed using a hybrid deployment method within a square with side length 10 m centered on the center of the network area. Figures 3.45, 3.46 and 3.47 give the results of this experiment for different criteria. These figures show that:

- Error level has no effect on the performance of CLA-EDS and CLA-DS algorithms with respect to the coverage, node separation and distance criteria. This is due to the fact that CLA-EDS and CLA-DS algorithms do not use any information about the position of the sensor nodes.

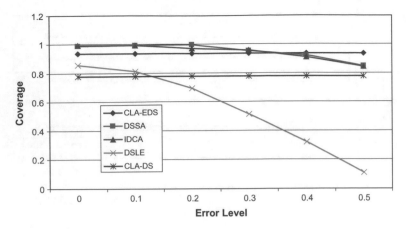

Fig. 3.45 Comparison of CLA-EDS with existing deployment algorithms in terms of the coverage criterion in the presence of inaccuracies in estimating the position of sensor nodes

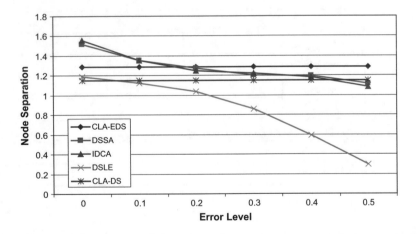

Fig. 3.46 Comparison of CLA-EDS with existing deployment algorithms in terms of the node separation criterion in the presence of inaccuracies in estimating the position of sensor nodes

- DSLE algorithm is highly affected by increasing the error level. This is because the deployment strategy used in this algorithm is highly dependent on the exact positions of sensor nodes.
- The impact of the error level on the performances of DSSA and IDCA algorithms with respect to the coverage and node separation criteria is not too significant. This is due to the fact that these algorithms, like CLA-EDS algorithm, try to minimize the difference between local density and expected local density of sensor nodes which is not sensitive to the error level.

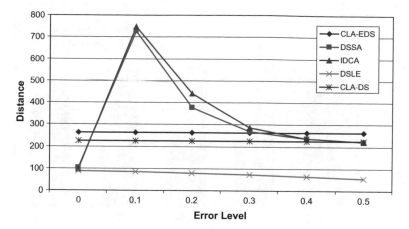

Fig. 3.47 Comparison of CLA-EDS with existing deployment algorithms in terms of the distance criterion in the presence of inaccuracies in estimating the position of sensor nodes

- Error level highly affects the performances of DSSA and IDCA algorithms with respect to the distance criterion. This is because DSSA and IDCA algorithms, unlike CLA-EDS, use the relative distances of neighboring sensor nodes, which is sensitive to error level, in order to minimize the difference between local density and expected local density of sensor nodes.

Experiment 5

In this experiment, behaviors of CLA-EDS, DSSA, IDCA, DSLE, and CLA-DS algorithms are compared with respect to the coverage, node separation and distance criteria when the movements of sensor nodes are not perfect and follow a probabilistic motion model. We use the probabilistic motion model, specified in Section "Experiment 5". We assume that $M = 1$, $k = 5$, and $N = 1500$. Sensor nodes are initially deployed using a hybrid deployment method within a square with side length 10 m centered on the center of the network area. Figures 3.48, 3.49 and 3.50 show the results of this experiment. The results indicate the following facts:

- Using the probabilistic motion model instead of the perfect motion model significantly degrades the performances of DSSA, IDCA, and DSLE algorithms in terms all three criteria, but does not affect the performances of CLA-EDS and CLA-DS algorithms substantially.
- When the probabilistic motion model is used, CLA-EDS algorithm outperforms the existing algorithms in terms of the coverage and node separation criteria.
- For probabilistic motion model, the average distance moved by sensor nodes for all algorithms is higher than when perfect motion model is used. When the probabilistic motion model is used, the performances of CLA-EDS, DSSA,

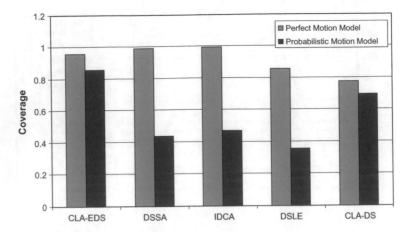

Fig. 3.48 Comparison of CLA-EDS with existing deployment algorithms in terms of the coverage criterion when movements of sensor nodes follow a probabilistic motion model

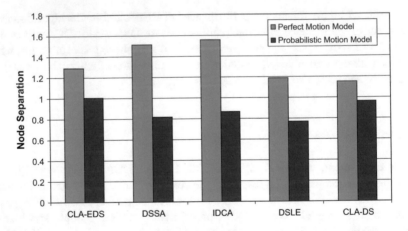

Fig. 3.49 Comparison of CLA-EDS with existing deployment algorithms in terms of the node separation criterion when movements of sensor nodes follow a probabilistic motion model

IDCA, DSLE, and CLA-DS algorithms in terms of the distance criterion are degraded by 6, 200, 201, 153, and 4% respectively. This indicates that CLA-EDS and CLA-DS algorithms are more robust to the deviations in the movements of sensor nodes than DSSA, IDCA, and DSLE algorithms.

Experiment 6

This experiment is conducted to study the effect of the parameter P_{fix} on the performance of CLA-EDS algorithm. For this study, we let $M = 1$ and $k = 5$.

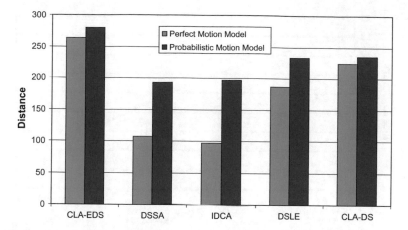

Fig. 3.50 Comparison of CLA-EDS with existing deployment algorithms in terms of the distance criterion when movements of sensor nodes follow a probabilistic motion model

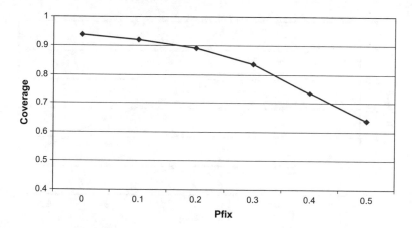

Fig. 3.51 Effect of parameter P_{fix} on the performance of CLA-EDS in terms of coverage criterion

We consider a network of $N = 1500$ sensor nodes which are initially deployed using a hybrid deployment method within a square with side length 10 m centered on the center of the network area. Figures 3.51, 3.52 and 3.53 give the results of this experiment. These figures indicate that by increasing the value of parameter P_{fix} in CLA-EDS algorithm, the covered section of the area, node separation and the average distance traveled by each sensor node decrease. In other words, higher values of P_{fix} results in more energy saving during the deployment process at the expense of poor coverage. Determination of P_{fix} for an application is very crucial and is a matter of cost versus precision. For better coverage, higher price must be paid.

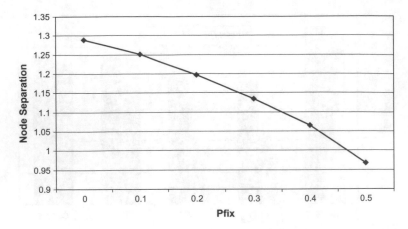

Fig. 3.52 Effect of parameter P_{fix} on the performance of CLA-EDS in terms of the node separation criterion

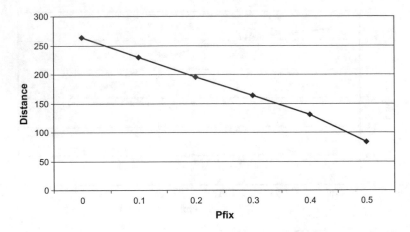

Fig. 3.53 Effect of parameter P_{fix} on the performance of CLA-EDS in terms of the distance criterion

Experiment 7

In this experiment, behaviors of CLA-EDS, DSSA, IDCA, DSLE, and CLA-DS algorithms are compared with respect to the coverage, node separation, and distance criteria when the required degree of coverage differs in different regions of the network. For this experiment, we consider $M = 10$ different regions. Each region is a circular sub-area within the network area with a radius, which is randomly selected from the range [5, 40] meters, and a required degree of coverage, which is selected randomly from the range [1, 5]. The experiment is performed for $N = 500$,

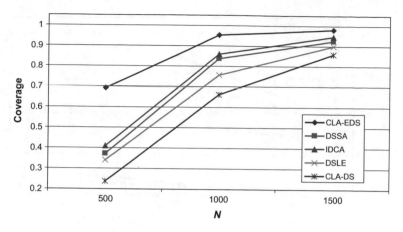

Fig. 3.54 Comparison of CLA-EDS with existing deployment algorithms in terms of the coverage criterion when the required degree of coverage differs in different regions of the network

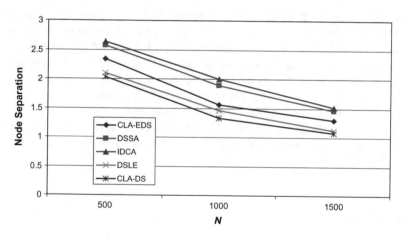

Fig. 3.55 Comparison of CLA-EDS with existing deployment algorithms in terms of the node separation criterion when the required degree of coverage differs in different regions of the network

1000, and 1500 sensor nodes which are initially deployed using a hybrid deployment method within a square with side length 10 m centered on the center of the network area. Figures 3.54, 3.55 and 3.56 give the results of this experiment. From these figures, one may conclude the following facts:

- CLA-EDS algorithm significantly outperforms DSSA, IDCA, DSLE, and CLA-DS with respect to the coverage criterion in networks of different sizes, especially when $N < 1000$. This indicates that the learning mechanism utilized in CLA-EDS algorithm is able to adapt itself to the different degrees of coverage needed in different regions of the network.

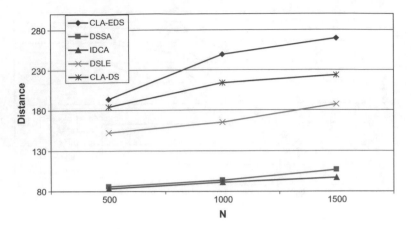

Fig. 3.56 Comparison of CLA-EDS with existing deployment algorithms in terms of the distance criterion when the required degree of coverage differs in different regions of the network

- According to Fig. 3.55, CLA-EDS is superior to CLA-DS and DSLE algorithms in terms of the node separation criterion, that is CLA-EDS better spreads sensor nodes throughout the network area than CLA-DS and DSLE algorithms. This figure also shows that the node separation of DSSA and IDCA algorithms are higher than that of CLA-EDS algorithm. But with a similar discussion given in Experiment 2, this also indicates that CLA-EDS algorithm is superior to IDCA and DSSA algorithms with respect to the node separation criterion.
- In terms of the distance criterion, CLA-EDS has the worst performance among the compared algorithms. This inferiority is expected as it was mentioned before in Experiment 3.

Experiment 8

The aim of conducting this experiment is to study the behavior of CLA-EDS deployment strategy in controlling the local density of sensor nodes within different regions, each having a different requirement of degree of coverage, during the deployment process. The simulation settings of the Experiment 8 are also used for this experiment. The experiment is performed for $N = 1500$ sensor nodes which are initially deployed using a hybrid deployment method within a square with side length 10 m centered on the center of the network area. Figures 3.57, 3.58 and 3.59 give the results of this experiment for three randomly selected regions with the following properties:

- Region 1: Radius = 24 m, $k = 5$
- Region 2: Radius = 16 m, $k = 3$
- Region 3: Radius = 36 m, $k = 4$

The figures show that CLA-EDS algorithm, without any node knowing its physical position or its relative distances to its neighbors, controls the local density of sensor nodes in such a way that the local density in each region approaches its expected value in that region. The figures also show that DSSA, IDCA, DSLE, and CLA-DS algorithms are not able to perfectly control the local density of sensor nodes when the expected local density differs in different regions of the network.

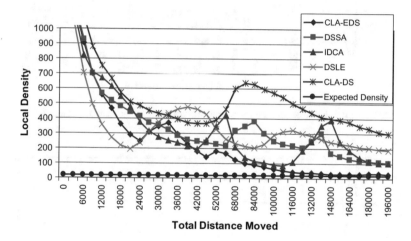

Fig. 3.57 Local density of sensor nodes during the deployment process in region 1

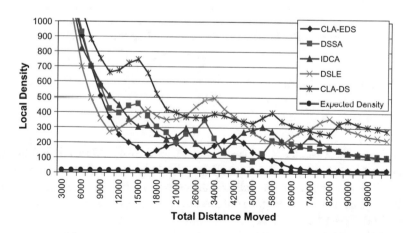

Fig. 3.58 Local density of sensor nodes during the deployment process in region 2

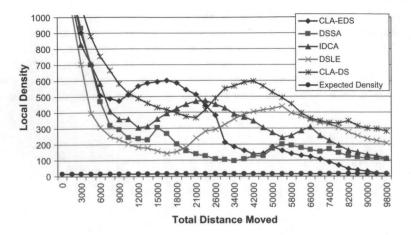

Fig. 3.59 Local density of sensor nodes during the deployment process in region 3

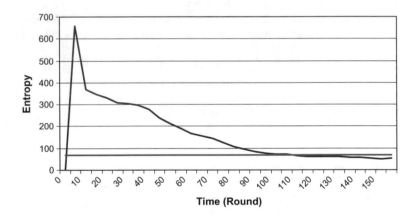

Fig. 3.60 Entropy of *HDICLA*

Experiment 9

This experiment is conducted to study the behavior of the *HDICLA* as a learning model used in CLA-EDS algorithm. For this study, we set $M = 1$, $k = 5$, and $N = 1500$. Sensor nodes are initially deployed using a hybrid deployment method within a square with side length 10 m centered on the center of the network area. Figure 3.60 shows the *HDICLA* entropy during the deployment process. This figure indicates that the entropy of the *HDICLA* is high at initial rounds of CLA-EDS algorithm, but gradually decreases as time passes. It goes below 70 at about round number 100. This means that after this round, the entropy of each learning automaton LA_i in the H*DICLA* is on average below 0.047. If the entropy of a two-action learning automaton is below 0.047, then it can be concluded that the

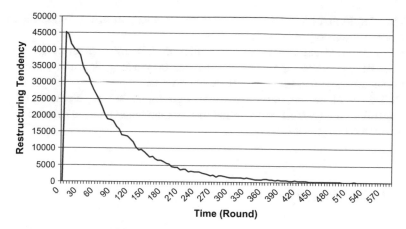

Fig. 3.61 Restructuring tendency of *HDICLA*

action probability of one of its actions is higher than 0.992. This means that action switching in each learning automaton in the *HDICLA* rarely occurs after round number 100.

Figure 3.61 depicts changes in the *HDICLA* restructuring tendency during the deployment process. This figure shows that the restructuring tendency of the *HDICLA* is initially high and gradually approaches zero. It is initially high because during initial rounds, the magnitude of the force vector applied to each sensor node is large, and it gradually approaches zero because as time passes, the local density of the sensor nodes approaching its expected value which results in the magnitude of the force vector applied to each sensor node to approach zero.

3.4.2.4 Summary of Results

In this section, performance of CLA-EDS deployment algorithm was compared with respect to the coverage, node separation, and distance criteria with DSSA, IDCA, DSLE, and CLA-DS deployment algorithms. Comparisons were made for different network sizes, different degrees of coverage, and error-free and with-error environments. From the results of this study we can conclude that:

- In error-free environments, the proposed algorithm (CLA-EDS) can compete with existing algorithms in terms of the coverage criterion, outperforms existing algorithms in terms of the node separation criterion, and performs worse than the existing algorithms in terms of the distance criterion.
- CLA-EDS, unlike DSSA, IDCA, and DSLE algorithms, does not use any information regarding the position of the sensor nodes or their relative distances to each other and therefore, in environments with inaccurate location information, outperforms existing algorithms in terms of the coverage and node separation criteria.

- The algorithm which is least affected by the selection of the "node movement model" of sensor nodes is CLA-EDS.
- When required degree of coverage differs in different regions of the network, CLA-EDS performs significantly better than the existing algorithms with respect to the coverage and the node separation criteria.
- CLA-EDS algorithm, unlike existing algorithms, has a parameter (P_{fix}) for controlling the tradeoff between the network coverage and the average distance traveled by sensor nodes.

3.5   Point Coverage

Point coverage problem in WSNs is the problem of covering a set of stationary or non-stationary target points residing within the sensor field. When the target points are stationary, point coverage problem becomes a special case of the area coverage problem, in which the aim is to cover only a set of specific target points, instead of covering every point within the sensor field. On the other hand, when the target points are non-stationary, the point coverage problem becomes very similar to the target tracking problem in WSNs. The difference is that in non-stationary point coverage problem, the aim is to cover the target points whereas in the target tracking problem, the main objective is to track and report the almost exact positions of the targets.

This section is devoted to the dynamic point coverage problem which is the point coverage problem with non-stationary target points. This problem can be addressed in different ways. One way is to design a deployment strategy which can best address the criterion of minimum number of required nodes (Dhillon et al. 2002; Chen et al. 2004; Zou and Chakrabarty 2004b, 2005; Dingxing et al. 2006; Pedraza et al. 2006). This is an inflexible solution which cannot deal with the topology changes, commonly occur in sensor networks. Another more flexible solution to the dynamic point coverage problem is to dynamically rearrange sensor nodes, assuming that they are mobile (Shucker and Bennett 2005; Schwagre et al. 2006; Saber 2007). In this approach, nodes are first deployed randomly around the target points. After this initial random positioning, each node tries to find its best position using the position information of the target points and its surrounding nodes. Using this approach, dynamic changes in the topology of the network or positions of the target points can be adaptively addressed. One major problem with this solution is the high overhead of controlling packets which are required to control the position of each sensor node according to the position of its neighboring nodes. This can lead to fast energy exhaustion of the sensor nodes, and hence shortening the lifetime of the network. Many researchers try to overcome the dynamic point coverage problem by designing a suitable sleep-scheduling (or simply scheduling) mechanism which selects a subset of sensor nodes in the network as active nodes for sake of monitoring the target points (Gupta and Das 2003;

Ye et al. 2003, 2006; Chen et al. 2004; Gui and Mohapatra 2004; Sanli et al. 2004; Bhattacharya et al. 2005; Zhang and Hou 2005; Zheng et al. 2005; Zou and Chakrabarty 2005; Dingxing et al. 2006; He et al. 2006; Jeong et al. 2006, 2007; Wang et al. 2006a, b, 2007a, b; Watfa 2006; Wu et al. 2006; Bagheri and Hefeeda 2007; Cai et al. 2007; Giusti et al. 2007; Khalifa et al. 2007; Jiang et al. 2008; Soe 2008). In terms of dynamicity, this solution can deal with changes that occur in the topology of the network and position of target points using a dynamic scheduling mechanism. In terms of energy consumption, this method best fits the sensor networks, since in each period of time, only nodes which can sense the target points in that period are awakened, and rest of the nodes are asleep, saving their energies.

Active node scheduling can be done through a fixed or a dynamic mechanism. In fixed scheduling mechanisms (Ye et al. 2003; Chen et al. 2004; Sanli et al. 2004; Zhang and Hou 2005; Zheng et al. 2005; Zou and Chakrabarty 2005; Dingxing et al. 2006; Wang et al. 2006a, b; Wu et al. 2006; Bagheri and Hefeeda 2007; Cai et al. 2007; Giusti et al. 2007), the set of sensors in the network is divided into disjoint sets so that every set completely covers the entire area of the network. These disjoint sets are activated successively in such a way that at, any given time, only one set is active. Since all target points are monitored by every sensor set, the goal of this approach is to determine a maximum number of disjoint sets in such a way that the time interval between two successive activations of any given sensor is maximized. Besides its complexity, this approach needs the information of the network topology to be available in a central node which is not always possible. In dynamic scheduling mechanisms, sensor nodes are locally scheduled to be active or inactive based on the movement paths of the target points (Gupta and Das 2003; Gui and Mohapatra 2004; Zhang and Cao 2004; Bhattacharya et al. 2005; Yang and Sidkar 2005; He et al. 2006; Jeong et al. 2006; Watfa 2006; Watfa and Commuri 2006; Jeong et al. 2007; Khalifa et al. 2007; Wang et al. 2007a, b; Jiang et al. 2008; Soe 2008). In such schemes, usually some of the sensor nodes, which have higher residual energies than other sensor nodes, are active all of the times and monitor the whole sensor field for detecting target points. Whenever a target point is detected, such active nodes track the target and estimate its movement path. This estimation leads to a prediction about the location of the target point in the near future. Sleeping nodes in the vicinity of the predicted location are then activated by currently active nodes. This activation is performed using some sort of notification messages. This dynamic scheduling approach has two major drawbacks; one is the overhead of the notification messages required and the other is that sleeping nodes must have the ability to receive messages, and hence they cannot power off their receiving antennas. This means that a sleeping node can only switch off its processing unit, but its communicating unit must be in the idle mode, waiting to receive activation messages. According to Raghunathan et al. (2002), energy consumption of a sensor node in receiving and idle states is nearly equal to the energy consumed during the transmission mode. As the energy consumed by a processing unit is in the order of 10^{-3} of the energy consumed by a communicating unit, it is concluded that using these dynamic scheduling mechanisms, not so much energy saving can be gained in the sleeping nodes.

To overcome the above drawbacks, in this section, two learning algorithms that deal with the problem of dynamic point coverage and have no such drawbacks will be introduced. The first algorithm is a learning automata-based algorithm, called SALA (Esnaashari and Meybodi 2010c). In SALA, no notification message is required to be exchanged between sensor nodes and as a consequence, sleeping nodes can switch off their communicating units as well as their processing units in order to save energy. Thus, SALA can better prolong the network lifetime than the existing dynamic scheduling algorithms. In this algorithm each node in the network is equipped with a set of learning automata which try to learn the proper times for sleeping and awakening of that node (schedule) based on the movement patterns of the target points passing through its sensing region. In the second algorithm, called SACLA (Esnaashari and Meybodi 2010a), the sensor network is mapped into an irregular cellular learning automaton (*ICLA*) proposed in this chapter. In this mapping, each sensor node in the network is mapped into a cell in the *ICLA*. Each cell is equipped with a learning automaton. The learning automaton residing in each cell, in cooperation with the learning automata residing in the neighboring cells, dynamically learns (predicts) the existence of any target points in the vicinity of the corresponding node in the network in near future. This prediction is then used to schedule the active times of that node. Instead of notification messages which are exchanged between neighboring nodes in the existing dynamic scheduling schemes, in the SACLA, a local base station in each neighborhood is always active and is responsible for queuing and relaying control packets between the neighboring nodes during their active times. As a consequence, sleeping nodes in SACLA can switch off both their communicating and processing units, just like the case in SALA algorithm. Experimental results show that the proposed scheduling algorithms outperform the existing methods such as LEACH (Heinzelman et al. 2000), GAF (Xu et al. 2001), PEAS (Ye et al. 2003, 2006) and PW (Gui and Mohapatra 2004) in terms of the energy consumption.

The rest of this section is organized as follows. Section 3.5.1 will be devoted to the SALA algorithm which is a scheduling algorithm based on learning automata model. Irregular cellular learning automata-based scheduling algorithm, SACLA, will be proposed in Sect. 3.5.2.

3.5.1 SALA: A Learning Automata-Based Scheduling Solution to the Dynamic Point Coverage Problem

In this section, we will introduce the scheduling algorithm based on learning automata (SALA) for solving the dynamic point coverage problem.

3.5.1.1 Problem Statement

In this section, we first give two sample applications of the problem and then give its formal definition.

Sample Applications

A trial wireless sensor network in San Francisco, is a sensor network that announces which of the parking spaces of the city is free at any moment.[2] This network uses a wireless sensor embedded in a 4-inch-by-4-inch piece of plastic, fastened to the pavement adjacent to each parking space. In this application, each sensor node has to monitor its parking space and reports the free times of the parking space to a local base station. The local base station then prepares the information of free parking spaces for drivers passing the area and requesting such information. From the viewpoint of the sensor network, cars coming into the parking spaces and getting out of them are moving target points which must be monitored. If a sensor node is able to switch periodically between active and sleep operation modes, it will have a longer lifetime than if it always remains in the active operation mode. Therefore, each node can utilize a local scheduling scheme for switching its operation mode between sleep and active modes. However this makes it possible for a driver to be mistakenly guided to a node's parking space, while it is occupied by another car. Such incorrect guides are acceptable while their rate is below an acceptable level.

As an alternative application, one may consider the habitat monitoring (Polastre 2003) such as the one reported by the College of Atlantic (COA) in the Great Duck Island for habitat monitoring of 5000 pairs of petrels nesting on the island (Ambagis 2002). To analyze the patterns of the organisms, the sensors are deployed both in the organisms' burrows as well as on the surface surrounding the burrows to monitor the differences. In order to determine why petrels nest in specific patches, data are gathered from both populated and unpopulated petrel patches. In order to monitor an entire field season (corresponding to a single petrel reproductive cycle), the sensors manage their power in such a way that they provide at least 7 months of continuous operation. In such application, it is rational to sacrifice precision to some degree in order to save more energy and prolong the network lifetime.

---

[2]http://www.nytimes.com/2008/07/12/business/12newpark.html.

Formal Definition of the Problem

Consider a sensor network with N sensor nodes $s_1, s_2, \ldots, s_N$ which are scattered randomly throughout a large $L \times L$ rectangular field (Ω) in such a way that Ω is completely covered. All sensor nodes have the same sensing ranges (r). Each sensor node s_k has 4 different modes of operation (Wang and Xiao 2006) as given below.

- **On-duty** $(S_A C_A)$: Both sensing and communicating units are switched on referred to as active mode.
- **Sensing Unit On-duty** $(S_A C_S)$: The sensing unit is switched on, but the communicating unit is switched off.
- **Communicating Unit On-duty** $(S_S C_A)$: The communicating unit is switched on, but the sensing unit is switched off.
- **Off-duty** $(S_S C_S)$: Both sensing and communicating units are switched off referred to as sleep mode.

Note that in $S_A C_A$, $S_A C_S$, $S_S C_A$ and $S_S C_S$, index A stands for active and index S stands for sleep. For further simplicity in notation, we use index x in the above notations to refer to more than one operation mode, i.e. $S_A C_x$ refers to both $S_A C_A$ and $S_A C_S$ modes, and $S_x C_x$ refers to all 4 modes of operation. At any instance of time, a sensor node can be only in one of the above 4 operation modes. The operation mode of a sensor s_k at time instant t is denoted by $O_{s_k}(t)$.

Let TP be a finite set of target points residing in Ω. TP is divided into two disjoint sets; *Moving objects* (TP^M) and *Events* (TP^E). A target point $tp_i^M \in TP^M$ is a moving object and has a continuous movement trajectory. We assume that the movement path of a target point tp_i^M is fixed throughout the lifetime of the network and repeated every T_e seconds, but the velocity of tp_i^M may change over time. On the other hand, a target point $tp_i^E \in TP^E$ is an event that occurs somewhere in Ω repeatedly every T_e seconds or randomly following a Poisson distribution and lasts for a short static or random duration.

Every T_e seconds is called an epoch (Ep). Each Ep^n starts at $\tau_{Ep^n}^S = (n-1) \cdot T_e$, $n = 1, 2, \ldots$. Each epoch Ep^n is further divided into N^R rounds (R) all having equal durations (τ_R). We refer to the mth round of epoch Ep^n by $R^{m,n}$ and to the mth round of all epochs by R^m. $R^{m,n}$ starts at $\tau_{R^{m,n}}^S = \tau_{Ep^n}^S + (m-1) \cdot \tau_R$ and lasts for τ_R. Since usually people have a repetitive movement paths in their day to day lives (from home to work in the morning and from work to home in the evening), in the first sample application, every 24 h can be considered as one epoch. Furthermore, as the traffic pattern of a city changes in different hours of a day (and hence, occupancy rates of parking spaces differ in different hours), each hour of a day can be considered as one round.

We denote the Euclidean distance between a sensor node s_k located at $(x(s_k), y(s_k))$ and a target point tp_i located at $(x(tp_k), y(tp_k))$ as $d(s_k, tp_i)$, i.e.

$$d(s_k, tp_i) = d(s_k, tp_i) = \sqrt{(x(s_k) - x(tp_i))^2 + (y(s_k) - y(tp_i))^2}.$$

Assuming the binary sensing model (Chakrabarty et al. 2002) and sensing range of r for all sensor nodes in the network, we say a target point $tp_i \in TP$ is sensed, detected or monitored by a sensor node s_k at time t if and only if $d(s_k, tp_i) < r$ and $O_{s_k}(t) = S_A C_x$. The network detects a target point $tp_i \in TP$ at time t if and only if at least one of the sensor nodes of the network detects tp_i at time t.

Definition 3.12 Network lifetime (T) is the duration between the startup of the network and the first point in time when Ω is not further completely covered by the network.

It should be noted that this definition of the network lifetime coincides with the *total network lifetime* metric given in (Dietrich and Dressler 2009) where the utilized criterion function (ψ_{**}) for the network liveliness is *area coverage* and the length of the time interval during which the criterion must be satisfied (Δt^y_{**}) is set to zero.

Definition 3.13 Activation time of a target point $tp_i \in TP$ denoted by τ_{tp_i} is the summation of all the times during which tp_i can be detected by the network.

Definition 3.14 Activation time of a sensor node s_k denoted by τ_{s_k} is the summation of all the times during which the sensor is in $S_A C_x$ operation mode; that is $O_{s_k}(t) = S_A C_x$.

Definition 3.15 Detection time of a target point $tp_i \in TP$ denoted by $\tau^d_{tp_i}$ is the summation of all the times during which tp_i is detected by the network.

Definition 3.16 Detection time of a target point $tp_i \in TP$ by a sensor node s_k denoted by τ_{s_k, tp_i} is the summation of all the times during which tp_i is detected by s_k.

Definition 3.17 Network detection rate denoted by η_D is the rate of the target detection in the network and is defined according to Eq. (3.63). Network detection rate in a single round $R^{m,n}$ is referred to as $\eta_D^{m,n}$.

$$\eta_D = \frac{\sum_{tp_i \in TP} \tau^d_{tp_i}}{\sum_{tp_i \in TP} \tau_{tp_i}}. \tag{3.63}$$

Definition 3.18 Network sleep rate denoted by η_S is defined as the ratio of the times during which nodes of the network are in the sleep mode to the times during which they can be in the sleep mode. η_S is defined according to Eq. (3.64) given below. Network sleep rate in a single round $R^{m,n}$ is referred to as $\eta_S^{m,n}$.

$$\eta_S = \frac{\sum_{s_k} \left(\frac{T - \tau_{s_k}}{T - \sum_{tp_i \in TP} \tau_{s_k, tp_i}} \right)}{N}. \tag{3.64}$$

The objective of the network is to detect target points and report their locations to a central node called the sink. We assume that the network is clustered. Cluster heads are powerful and rechargeable nodes which are always in the active mode.

Each node can directly communicate with its cluster head and cluster heads form an infrastructure through which network data can be sent to the sink.

To prolong the network lifetime, a node will switch to the $S_x C_A$ operation mode only if it wants to communicate with its cluster head; otherwise, the communication unit of the node will be switched off. The sensing unit of a sensor node has to be switched on only if a target point is in its sensing range, but since nodes have no knowledge about the movement patterns of the target points, they cannot calculate the exact times for switching their sensing units on or off. Deciding for switching from $S_A C_x$ to $S_S C_x$ is quite simple; if no target point can be detected by the node, it turns off its sensing unit. For turning the sensing unit on, each node needs to know the time at which a target point will enter its sensing region. Since this time is not known by a node a priori, it instead uses a local scheduler which determines a duration after which the node must wake up. The node is informed of this duration before it switches to the sleep mode. The local scheduling or simply scheduling of a sensor node $s_k(\Delta_{s_k})$ is defined as follows:

Definition 3.19 The scheduling of a sensor node s_k in epoch Ep^n is denoted by the ordered list $\Delta_{s_k}^n = \left\langle T_{Sleep}^{1,n}(s_k), T_{Sleep}^{2,n}(s_k), \ldots, T_{Sleep}^{N^R,n}(s_k) \right\rangle$ where $T_{Sleep}^{m,n}(s_k)$ specifies the sleep duration (i.e. the duration of $S_S C_x$ operation mode) for node s_k in the round $R^{m,n}$.

Having the above definitions and assumptions, the problem is to locally learn the scheduling of each sensor node s_k using the movement patterns of the target points during the lifetime of the network such that the network sleep rate (η_S) is maximized while the network detection rate (η_D) does not drop below an acceptable level.

3.5.1.2 Required Analysis

In this section, we first give some definitions and then prove a theorem whose result will be used later by the SALA algorithm.

Definition 3.20 $TP^{m,n}(s_k) = \left\langle tp_1^{m,n}(s_k), tp_2^{m,n}(s_k), \ldots, tp_{last}^{m,n}(s_k) \right\rangle$ is defined as the list of target points which pass through the sensing region of a sensor node s_k during the round $R^{m,n}$ ordered according to their arrival times.

Definition 3.21 $\tau_{tp_i^{m,n}}^E(s_k)$ is defined as the time at which the target point $tp_i^{m,n}(s_k)$ enters into the sensing region of node s_k.

Definition 3.22 $\tau_{tp_i^{m,n}}^D(s_k)$ is defined as the time at which the target point $tp_i^{m,n}(s_k)$ exits from the sensing region of node s_k.

Definition 3.23 $E\left(T_{sleep}^{m,n}(s_k)\right)$ is defined as the amount of energy consumed by a node s_k during the round $R^{m,n}$ using the sleeping duration of $T_{sleep}^{m,n}(s_k)$.

Theorem 3.3 *If for a sensor network with the specifications given in* Section "Formal Definition of the Problem" *we have that*

- *every node is activated at the beginning of any round,*
- *each node in the network immediately switches to the sleep operation mode if it cannot sense any target point,*
- *the network detection rate is one ($\eta_D = 1$),*

then the energy consumption of the network is minimized if and only if the sleeping duration of any node s_k in any round $R^{m,n}$ ($\mathrm{T}_{sleep}^{m,n}(s_k)$) is set to $\underline{\mathrm{T}}_{sleep}^{m,n}(s_k)$ where

$$
\underline{\mathrm{T}}_{sleep}^{m,n}(s_k) = \gcd \left(\begin{array}{c} \left(\tau_{tp_1^{m,n}}^{E}(s_k) - \tau_{R^{m,n}}^{S}(s_k) \right), \\ \left(\tau_{tp_2^{m,n}}^{E}(s_k) - \tau_{tp_1^{m,n}}^{D}(s_k) \right), \ldots, \\ \left(\tau_{tp_{last}^{m,n}}^{E}(s_k) - \tau_{tp_{last-1}^{m,n}}^{D}(s_k) \right) \end{array} \right) ; \quad \forall s_k, m, n. \tag{3.65}
$$

Here, gcd is the greatest common divisor function.

Before we prove Theorem 3.3, we state and prove several lemmas.

Lemma 3.5 *If in a sensor network with the specifications given in* Section "Formal Definition of the Problem" *we assume that*

- *a node s_k be activated at the beginning of a round $R^{m,n}$,*
- *node s_k immediately switches to the sleep operation mode if it cannot sense any target point,*

then node s_k can fully detects all the target points passing through its sensing region during the round $R^{m,n}$ if and only if $\mathrm{T}_{sleep}^{m,n}(s_k)$ is a divisor of $\underline{\mathrm{T}}_{sleep}^{m,n}(s_k)$ given by Eq. (3.65).

Proof **The proof of necessity condition**: We will prove the necessity condition of the lemma by induction on the order of arrival of target points into the sensing region of the node s_k.

Basis of induction: We will show that the first target point can be detected. Let $tp_1^{m,n}(s_k)$ be the first target point passing through the sensing region of the node s_k during the round $R^{m,n}$. Target point $tp_1^{m,n}(s_k)$ enters the sensing region of the node s_k either before ($\tau_{tp_1^{m,n}}^{E}(s_k) \le \tau_{R^{m,n}}^{S}(s_k)$) or after ($\tau_{tp_1^{m,n}}^{E}(s_k) > \tau_{R^{m,n}}^{S}(s_k)$) the beginning of the round. Considering $\tau_{tp_1^{m,n}}^{E}(s_k) \le \tau_{R^{m,n}}^{S}(s_k)$, s_k is able to fully detect target point $tp_1^{m,n}(s_k)$ due to the fact that node s_k is activated at the beginning of the round $R^{m,n}$ (the first assumption of the lemma). Considering $\tau_{tp_1^{m,n}}^{E}(s_k) > \tau_{R^{m,n}}^{S}(s_k)$ and the second assumption of the lemma, we can conclude that node s_k switches to the sleep operation mode at $\tau_{R^{m,n}}^{S}(s_k)$. Now let $t_{d-e,0}^{m,n}(s_k) = \tau_{tp_1^{m,n}}^{E}(s_k) - \tau_{R^{m,n}}^{S}(s_k)$ be the difference between $\tau_{R^{m,n}}^{S}(s_k)$ and the time that target point $tp_1^{m,n}(s_k)$ enters the sensing region of node s_k ($\tau^E_{tp^{m,n}_1}(s_k)$). If $\mathrm{T}^{m,n}_{sleep}(s_k)$ is a divisor of $\underline{\mathrm{T}}^{m,n}_{sleep}(s_k)$, then $t^{m,n}_{d-e,0}(s_k)$ is divisible by $\mathrm{T}^{m,n}_{sleep}(s_k)$. Since node s_k switches to the sleep operation mode at the time $\tau^S_{R^{m,n}}(s_k)$ and $t^{m,n}_{d-e,0}(s_k)$ is divisible by $\mathrm{T}^{m,n}_{sleep}(s_k)$, then it can be concluded that it certainly wakes up at $\tau^E_{tp^{m,n}_1}(s_k)$ and fully detects $tp^{m,n}_1(s_k)$.

Induction hypothesis: The ith target point that enters the sensing region of the node s_k ($tp^{m,n}_i(s_k)$) can be fully detected.

Induction proof: We will show that if the node s_k fully detects the ith target point $tp^{m,n}_i(s_k)$, then it can fully detect the next target point that enters into its sensing region ($tp^{m,n}_{i+1}(s_k)$) as well. Assume that the target point $tp^{m,n}_i(s_k)$ has been fully detected by the node s_k. By the second assumption of the lemma, node s_k switches to the sleep operation mode at the departure time of $tp^{m,n}_i(s_k)$, that is, it switches to the sleep operation mode at $\tau^D_{tp^{m,n}_i}(s_k)$. Let $t^{m,n}_{d-e,i}(s_k) = \tau^E_{tp^{m,n}_{i+1}}(s_k) - \tau^D_{tp^{m,n}_i}(s_k)$ be the difference between the time that the ith target point $tp^{m,n}_i(s_k)$ departs the sensing region of the node s_k and the time that the next target point $tp^{m,n}_{i+1}(s_k)$ enters the sensing region of the node s_k. If $\mathrm{T}^{m,n}_{sleep}(s_k)$ is a divisor of $\underline{\mathrm{T}}^{m,n}_{sleep}(s_k)$, then $t^{m,n}_{d-e,i}(s_k)$ is divisible by $\mathrm{T}^{m,n}_{sleep}(s_k)$. Since the node s_k switches to the sleep operation mode at $\tau^D_{tp^{m,n}_i}(s_k)$ and $t^{m,n}_{d-e,i}(s_k)$ is divisible by $\mathrm{T}^{m,n}_{sleep}(s_k)$, then it can be concluded that it certainly wakes up at $\tau^E_{tp^{m,n}_{i+1}}(s_k)$ and fully detects $tp^{m,n}_{i+1}(s_k)$.

The proof of sufficiency condition: We will show that if the node s_k fully detects every target point which passes through its sensing region during the round $R^{m,n}$, then $\mathrm{T}^{m,n}_{sleep}(s_k)$ is a divisor of $\underline{\mathrm{T}}^{m,n}_{sleep}(s_k)$. To show this, we use contradiction. Assume that $\mathrm{T}^{m,n}_{sleep}(s_k)$ is not a divisor of $\underline{\mathrm{T}}^{m,n}_{sleep}(s_k)$. This implies that there exists $t^{m,n}_{d-e,i}(s_k) = \tau^E_{tp^{m,n}_{i+1}}(s_k) - \tau^D_{tp^{m,n}_i}(s_k)$ which is not divisible by $\mathrm{T}^{m,n}_{sleep}(s_k)$. Now we consider the following two cases:

- $tp^{m,n}_i(s_k)$ is not detected by the sensor node s_k: This case can not occur because it is in contradiction with the assumption that s_k fully detects every target point which passes through its sensing region during the round $R^{m,n}$.
- $tp^{m,n}_i(s_k)$ is detected by sensor node s_k: If $tp^{m,n}_i(s_k)$ is detected by the sensor node s_k, then by the second assumption of the lemma, node s_k switches to the sleep operation mode at $\tau^D_{tp^{m,n}_i}(s_k)$. Since node s_k switches to the sleep operation mode at $\tau^D_{tp^{m,n}_i}(s_k)$ and $t^{m,n}_{d-e,i}(s_k)$ is not divisible by $\mathrm{T}^{m,n}_{sleep}(s_k)$, it can be concluded that it will not wake up at $\tau^E_{tp^{m,n}_{i+1}}(s_k)$. This implies that $tp^{m,n}_{i+1}(s_k)$ cannot be fully detected. This is again in contradiction with the assumption that s_k fully detects every target point which passes through its sensing region during round $R^{m,n}$. ∎

Lemma 3.6 *If for a sensor network with the specifications given in Section "Formal Definition of the Problem" we have that*

- *all the nodes are activated at the beginning of every round,*
- *any node in the network immediately switches to the sleep operation mode if it cannot sense any target point,*

then the network fully detects all the target points ($\eta_D = 1$) if and only if $T_{sleep}^{m,n}(s_k)$ is a divisor of $\underline{T}_{sleep}^{m,n}(s_k)$ given by Eq. (3.65).

Proof By Lemma 3.5 and the fact that all nodes are activated at the beginning of every round (assumption 1) we can conclude that the network can fully detects all the target points, that is $\eta_D = 1$ if and only if $T_{sleep}^{m,n}(s_k)$ is a divisor of $\underline{T}_{sleep}^{m,n}(s_k)$. ■

Lemma 3.7 *The energy consumed by a node, working in a network with the specifications given in Section "Formal Definition of the Problem", during a round is a decreasing function of the sleeping duration of that node.*

Proof Let $T_{1,sleep}^{m,n}(s_k)$ and $T_{2,sleep}^{m,n}(s_k)$ be two different sleeping durations for a node s_k during a round $R^{m,n}$. If $T_{1,sleep}^{m,n}(s_k) < T_{2,sleep}^{m,n}(s_k)$, then we can conclude that the node s_k switches between the active operation mode and the sleep operation mode more frequently when it uses $T_{1,sleep}^{m,n}(s_k)$ as the sleeping duration. This implies that

$$E\left(T_{1,sleep}^{m,n}(s_k)\right) > E\left(T_{2,sleep}^{m,n}(s_k)\right)$$ and hence the lemma. ■

Lemma 3.8 *If the sleeping duration of any node, which works in a network with the specifications given in Section "Formal Definition of the Problem", during any round is set to its maximum possible value, then the energy consumed by the network will be minimized.*

Proof The proof of this lemma is immediate from Lemma 3.7. ■

Now we are ready to give the proof of Theorem 3.3.

Proof of Theorem 5.1: Lemma 3.6 states that when $\eta_D = 1$, $T_{sleep}^{m,n}(s_k)$ is a divisor of $\underline{T}_{sleep}^{m,n}(s_k)$. This implies that if $\eta_D = 1$, then the maximum possible value for the sleeping duration of a node s_k during a round $R^{m,n}$ is $\underline{T}_{sleep}^{m,n}(s_k)$. Lemma 3.8 states that if the sleeping duration of any node during any round is set to its maximum possible value, then the energy consumption of the network will be minimized. Therefore, if $T_{sleep}^{m,n}(s_k)$ is set to $\underline{T}_{sleep}^{m,n}(s_k)$ for any node s_k in any round $R^{m,n}$, then the energy consumption of the network will be minimized.

Conversely, assume that the energy consumed by the network is minimum. By Lemma 3.6, $T_{sleep}^{m,n}(s_k)$ is a divisor of $\underline{T}_{sleep}^{m,n}(s_k)$. By Lemma 3.7, if $T_{sleep}^{m,n}(s_k) < \underline{T}_{sleep}^{m,n}(s_k)$, then $E\left(T_{sleep}^{m,n}(s_k)\right) > E\left(\underline{T}_{sleep}^{m,n}(s_k)\right)$ which is in contradiction with the assumption that the energy consumed by the network is minimum. This implies that $T_{sleep}^{m,n}(s_k)$ must be equal to $\underline{T}_{sleep}^{m,n}(s_k)$. ■

3.5.1.3 SALA Scheduling Algorithm

Algorithm Outline

The purpose of the SALA is to find N^R sleeping durations, each for one of the R^m rounds. To do this, each node s_k in the network is equipped with N^R learning automata, $LA_k^1, LA_k^2, \ldots, LA_k^{N^R}$. Learning automaton LA_k^m gets activated during R^m rounds to help node s_k learn its proper sleeping duration during these rounds $(T_{sleep}^m(s_k))$.

At the beginning of each round $R^{m,n}$ (mth round of the nth epoch), all nodes are in the active operation mode; that is $O_{s_k}(t) = S_A C_A$. The round is started asynchronously in each node s_k by activating its learning automaton LA_k^m. Learning automaton LA_k^m upon its activation, selects a sleeping duration for the node s_k $(T_{sleep}^m(s_k))$ to be used during that round. During the round $R^{m,n}$, if the node s_k senses any target point, it collects information about that target and sends collected data to the local base station. In addition, node s_k logs the entrance and exit times of the target point in its local database. During the round $R^{m,n}$, whenever the node s_k cannot sense any target point, it switches immediately to the sleep operation mode $(S_S C_S)$ and stays in this mode for the duration of $T_{sleep}^m(s_k)$. At the end of the round $R^{m,n}$, node s_k uses its local database to evaluate the suitability of the selected sleeping duration $(T_{sleep}^m(s_k))$. The result of this evaluation is then given to the learning automaton LA_k^m as the reinforcement signal of the environment. This way learning automaton LA_k^m gradually learns the best sleeping duration for the node s_k during R^m rounds.

Detailed Description

Each learning automaton has three actions; *Extending*, *Shrinking* and *No change* which we call α_1, α_2 and α_3 respectively. Let the probabilities of selecting each of these actions to be denoted by $p_1, p_2,$ and p_3. Node s_k extends (shrinks) its sleeping duration $T_{sleep}^m(s_k)$ if the learning automaton LA_k^m selects *Extending* (*Shrinking*). It does not change its sleeping duration $T_{sleep}^m(s_k)$ if the learning automaton LA_k^m selects *No change*.

At the beginning of the algorithm, we have $T_{sleep}^{m,0}(s_k) = MinSleepDuration$, $\forall R^m$, $\forall s_k$, $p_1 = .8$ and $p_2 = p_3 = .1$. *MinSleepDuration* is the minimum allowable sleep time for the sensing unit of a node. Since $T_{sleep}^{m,0}(s_k)$ is set to the *MinSleepDuration* at the beginning of the algorithm, we set p_1 to a higher value than p_2 and p_3 in order to allow $T_{sleep}^m(s_k)$ getting close to its proper value more quickly.

Each round $R^{m,n}$ is started asynchronously in each node s_k by the activation of its learning automaton LA_k^m. At the startup of the round, the node s_k uses its learning automaton LA_k^m to select its sleeping duration (more details on this step will be given

in Section "Selecting the Sleeping Duration") for the current round ($\mathrm{T}_{sleep}^{m,n}(s_k)$).
Node s_k then enters a loop during which it checks if it can detect any target, and if so,
starts reporting its information to the local base station. Whenever the node s_k finds
out that no target exists in its sensing range, it goes to the sleep operation mode for
the duration of $\mathrm{T}_{sleep}^{m,n}(s_k)$ seconds. After the sleep duration, s_k continues the loop from
its beginning, i.e. checking for the presence of any target. The loop continues for the
whole duration of the current round at the end of which s_k computes and sends the
reinforcement signal (more details on this step will be given in Section "Computing
the Reinforcement Signal") to the LA_k^m. The above procedure then repeats for the
next round. In a step by step manner, the operation of the SALA algorithm in a
sensor node s_k during a round $R^{m,n}$ can be described as given below.

1. Sensor node s_k sets $t = \tau_{R^{m,n}}^S$.
2. Sensor node s_k sets $O_{s_k}(t) = S_A C_x$.
3. Sensor node s_k uses LA_k^m to select $\mathrm{T}_{sleep}^{m,n}(s_k)$.
4. Sensor node s_k checks if any target can be detected.
5. **If** (sensor node s_k detects any target point) **Then**

   5-1. Sensor node s_k stores t as the entrance time of the target.
   5-2. Sensor node s_k sets $O_{s_k}(t) = S_A C_A$.
   5-3. **While** (Sensor node s_k detects the target point) **Do**

      5-3-1. Sensor node s_k reports data to the local base station.
      5-3-2. Sensor node s_k sets $t = t + 1$.

   5-4. Sensor node s_k sets $O_{s_k}(t) = S_A C_S$.
   5-5. Sensor node s_k stores t as the exit time of the target.

6. Sensor node s_k sets $O_{s_k}(t) = S_S C_x$.
7. Sensor node s_k sleeps for $\mathrm{T}_{sleep}^{m,n}(s_k)$ seconds.
8. **If** $\mathrm{T}_{sleep}^{m,n}(s_k) > 0$ **Then**

   8-1. Sensor node s_k sets $t = t + \mathrm{T}_{sleep}^{m,n}(s_k)$.

9. **Else**

   9-1. Sensor node s_k sets $t = t + 1$.

10. **If** ($m = N^R$) **Then**

    10-1. **If** ($t \geq \tau_{R^{1,n+1}}^S$) **Then**

       10-1-1. Sensor node s_k sends reinforcement signal to LA_k^m.
       10-1-2. Sensor node s_k starts the procedure for round $R^{1,n+1}$ from step 1.

    10-2. **Else**

       10-2-1. Sensor node s_k continues from step 4.

11. **Else**

  11-1. **If** $(t \geq \tau^{S}_{R^{m+1,n}})$ **Then**

      11-1-1. Sensor node s_k sends reinforcement signal to LA^m_k.
      11-1-2. Sensor node s_k starts the procedure for round $R^{m+1,n}$ from step 1.

  11-2. **Else**

      11-2-1. Sensor node s_k continues from step 4.

The following subsections provide details for how to use LA^m_k to select $T^{m,n}_{sleep}(s_k)$ at the startup of the $R^{m,n}$ round and how to compute the reinforcement signal for LA^m_k at the end of the $R^{m,n}$ round.

Selecting the Sleeping Duration

As it was stated before, at the beginning of each round $R^{m,n}$, each node s_k uses its learning automaton LA^m_k to select its sleeping duration for that round ($T^{m,n}_{sleep}(s_k)$). This is done as follows. Learning automaton LA^m_k selects one of its actions randomly based on its action probability vector. We refer to the selected action as $\hat{\alpha}$. If the selected action ($\hat{\alpha}$) is *Extending*, then the current sleeping duration is extended by a predetermined constant value (*ExtendValue*). If $\hat{\alpha}$ is *Shrinking*, then the current sleeping duration is shrunk by a predetermined constant value (*ShrinkValue*). Otherwise, the sleeping duration remains unchanged. More attention must also be paid not to extend or shrink the sleeping duration beyond the predetermined range [*MinSleepDuration*, *MaxSleepDuration*]. This procedure can be described in a step by step manner as follows:

1. LA^m_k selects one of its actions randomly based on its action probability vector. Let $\hat{\alpha}$ be the selected action.
2. **If** $(\hat{\alpha} = \alpha_1)$ **Then**

   2-1. $T^{m,n}_{sleep}(s_k) = T^{m,n-1}_{sleep}(s_k) + ExtendValue$.
   2-2. **If** $(T^{m,n}_{sleep}(s_k) > MaxSleepDuration)$ **Then**

      2-2-1. $T^{m,n}_{sleep}(s_k) = MaxSleepDuration$.

3. **Else If** $(\hat{\alpha} = \alpha_2)$ **Then**

   3-1. $T^{m,n}_{sleep}(s_k) = T^{m,n-1}_{sleep}(s_k) - ShrinkValue$.
   3-2. **If** $(T^{m,n}_{sleep}(s_k) < MinSleepDuration)$ **Then**

      3-2-1. $T^{m,n}_{sleep}(s_k) = MinSleepDuration$.

4 **Else If** $(\hat{\alpha} = \alpha_3)$ **Then**

   4-1. $T^{m,n}_{sleep}(s_k) = T^{m,n-1}_{sleep}(s_k)$.

The above procedure is used at the beginning of the round $R^{m,n}$ to select the sleeping duration of the node s_k for the round $R^{m,n}$ ($T_{sleep}^{m,n}(s_k)$). $T_{sleep}^{m,n}(s_k)$ will also be changed at the end of the round $R^{m,n}$ according to the movement patterns of the target points passed through the sensing region of s_k in that round. This is described in the next section.

Computing the Reinforcement Signal

Theorem 3.3 states that the best sleeping duration in terms of the energy consumption and network detection rate for node s_k during the round $R^{m,n}$ is $\underline{T}_{sleep}^{m,n}(s_k)$. This implies that the reinforcement signal to the learning automaton of the node s_k better be computed based on $\underline{T}_{sleep}^{m,n}(s_k)$. But since the exact value of $\underline{T}_{sleep}^{m,n}(s_k)$ cannot be computed during the operation of the network (because $\tau_{tp_i^{m,n}}^{E}(s_k)$ is not known to s_k), we instead use an estimate of $\underline{T}_{sleep}^{m,n}(s_k)$ ($\hat{\underline{T}}_{sleep}^{m,n}(s_k)$) by recording the times at which target points enter and exit the sensing region of the node s_k (Steps 5-1 and 5-5 of the SALA step by step procedure). $\hat{\underline{T}}_{sleep}^{m,n}(s_k)$ is computed at the end of the round $R^{m,n}$. Using $\hat{\underline{T}}_{sleep}^{m,n}(s_k)$, the reinforcement signal for the learning automaton of the node s_k is specified as follows:

- If the selected sleeping duration ($T_{sleep}^{m,n}(s_k)$) is below $\hat{\underline{T}}_{sleep}^{m,n}(s_k)$ and the selected action of the learning automaton was *Extending*, then the reinforcement signal is to reward the selected action.
- If the selected sleeping duration ($T_{sleep}^{m,n}(s_k)$) is below $\hat{\underline{T}}_{sleep}^{m,n}(s_k)$ and the selected action of the learning automaton was *Shrinking* or *No change*, then the reinforcement signal is to penalize the selected action.
- If the selected sleeping duration ($T_{sleep}^{m,n}(s_k)$) is above $\hat{\underline{T}}_{sleep}^{m,n}(s_k)$ and the selected action of the learning automaton is *Shrinking*, then the reinforcement signal is to reward the selected action.
- If the selected sleeping duration ($T_{sleep}^{m,n}(s_k)$) is above $\hat{\underline{T}}_{sleep}^{m,n}(s_k)$ and the selected action of the learning automaton is *Extending* or *No change*, then the reinforcement signal is to penalize the selected action.
- If the selected sleeping duration ($T_{sleep}^{m,n}(s_k)$) is equal to $\underline{T}_{sleep}^{m,n}(s_k)$ and the selected action of the learning automaton is *No change*, then the reinforcement signal is to reward the selected action.
- If the selected sleeping duration ($T_{sleep}^{m,n}(s_k)$) is equal to $\hat{\underline{T}}_{sleep}^{m,n}(s_k)$ and the selected action of the learning automaton is *Extending* or *Shrinking*, then the reinforcement signal is to penalize the selected action.

At the end of the round $R^{m,n}$, node s_k modifies the sleeping duration ($T_{sleep}^{m,n}(s_k)$) according to the movement patterns of the target points passed through its sensing region during the round. This is done as follows:

1. **If** $(\hat{\underline{T}}_{sleep}^{m,n}(s_k) < MaxSleepDuration)$ **Then**

   1-1. $T_{sleep}^{m,n}(s_k) = \hat{\underline{T}}_{sleep}^{m,n}(s_k)$.

2. **Else If** $(\gamma \cdot \hat{\underline{T}}_{sleep}^{m,n}(s_k) < MinSleepDuration)$ **Then**

   2-1. $T_{sleep}^{m,n}(s_k) = MinSleepDuration$.

3. **Else**

   3-1. $T_{sleep}^{m,n}(s_k) = \gamma \cdot \hat{\underline{T}}_{sleep}^{m,n}(s_k)$.

In the above procedure, parameter $\gamma \geq 0$ is a constant which controls $T_{sleep}^{m,n}(s_k)$. As γ increases, more energy saving and lower detection rate are obtained (see Experiment 5).

Figure 3.62 gives the complete flowchart of the SALA algorithm.

3.5.1.4   Experimental Results

To evaluate the performance of the SALA algorithm, several experiments have been conducted and the results are compared with the results obtained for LEACH (Heinzelman et al. 2000) from the routing layer algorithms, GAF (Xu et al. 2001) from the grid-based algorithms, PEAS (Ye et al. 2003, 2006) from the coverage-related algorithms, Proactive Wakeup based on PEAS (PW) (Gui and Mohapatra 2004) from the tracking-based algorithms and SALA in which $T_{sleep}^{m}(s_k) = 60(s)$, $m = 1, 2, \ldots, N^R$ which we call hereafter Periodic Sleep (PS). PS is an example of a static scheduling algorithm. The reason for comparison of LEACH with SALA is to demonstrate that for the problem of dynamic point coverage, using a simple algorithm like LEACH, in which sensing devices of all sensor nodes are always active, is not efficient due to high energy overhead. We have also compared GAF with SALA as an example of a static scheduling algorithm to show that they are not suitable for dynamic environments such as the one in the dynamic point coverage problem. GAF algorithm has no mechanism for predicting the movement paths of the target points to be used for scheduling the sensors.

Unless otherwise stated, the simulation environment is a 100 m × 100 m area through which 100 sensor nodes are scattered randomly. Sensing ranges (r) of sensor nodes are assumed to be 15 m. For energy consumption of the nodes, we use the specifications of MEDUSA II sensor node given in (Raghunathan et al. 2002). Based on this specification, the power consumption of a node during the $S_A C_A$, $S_A C_S$, $S_S C_A$ and $S_S C_S$ operation modes are 24.58(mW), 9.72(mW), 14.86(mW), 0.02(mW) respectively. Energy required to switch a node from one operation mode to another operation mode is assumed to be negligible. *MinSleepDuration MaxSleepDuration, ExtendValue, ShrinkValue*, and γ are set to 0(s), 600(s), 2(s), 2 (s), and 0.01 respectively. Learning parameters α and β are both set to 0.01.

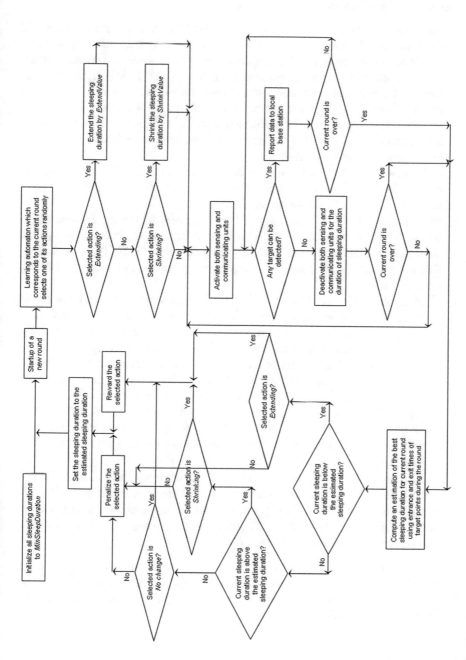

Fig. 3.62 Flowchart of the SALA algorithm

In all experiments, following different types of moving target points are considered to exist in the area of the network:

- **Constant Events**: Constant events have constant start times and durations. At the startup of the simulation, number of events is selected uniformly and randomly from the range [1, 200]. The start time of each event is a constant which is uniformly and randomly selected from the range [1, 86400] (number of seconds per day) and the duration of each event is a constant which is uniformly and randomly selected from the range [60, 900] (at least 1 min and at most 15 min). Each event has a static occurrence position which is selected uniformly and randomly in Ω. Events are repeated with the same start times and durations every epoch of the simulation.
- **Noisy Events**: Noisy events are like constant events except that the number of events, their start times and their durations are affected in every epoch by a normally distributed random noise.
- **Poisson Events**: The occurrences of Poisson events follow the Poisson distribution. To generate such events, a number of Poisson random number generators (selected uniformly and randomly from the range [1, 100]) are used. Each Poisson random number generator separately generates events in a randomly selected location in Ω.
- **Straight Path Moving Objects**: Each straight path moving object has a start position, stop position and a velocity. A straight path moving object starts from its start position and moves directly towards its stop position using its velocity. The start and stop positions are selected uniformly and randomly in Ω and the velocity is selected uniformly and randomly from the range [0, *MaxVelocity*]. There are 24 straight path moving objects each moves only in one of the rounds R^m, $m = 1, 2, \ldots, N^R$ and repeats its movement path in every epoch.
- **Complex Path Moving Objects**: These target points are like straight path moving objects except that they are not moving from their start positions directly towards their stop positions. Instead, each complex path moving object has an ordered list of points $\langle (x^{start}, y^{start}), (x^1, y^1), (x^2, y^2), \ldots, (x^{Stop}, y^{Stop}) \rangle$. A complex path moving object starts from its start point and moves directly towards (x^1, y^1) using its velocity. Whenever the target point reaches (x^1, y^1), it changes its direction towards (x^2, y^2). This movement continues until the target point reaches its stop position. The number of intermediate points for each moving object is selected randomly from [1, 5].

All simulations have been implemented using J-Sim simulator (Sobeih et al. 2006). All reported results are averaged over 50 runs. We have used CSMA as the MAC layer protocol, free space model as the propagation model, binary sensing model and Omni-directional antenna.

Experiment 1

In this experiment, SALA is compared with GAF, LEACH, PEAS, PW and PS algorithms in terms of the network detection rate (η_D) and mean energy consumption of the nodes.

Table 3.4 Parameters used for simulating different target points in Experiment 1

Target point	Parameter	Distribution	Value
Constant events	Number of events	Uniform	[1, 200]
	Event start time	Uniform	[1, 86400]
	Event duration	Uniform	[60, 900]
Noisy events	Number of events	Normal	μ = Randomly selected from the range [1, 200] $\sigma = 10$
	Event start time	Normal	μ = Randomly selected from the range [1, 86400] $\sigma = 120$
	Event duration	Normal	μ = Randomly selected from the range [60, 900] $\sigma = 120$
Poisson events	λ	Poisson	1.5
	Event duration	Constant	600
Straight path moving objects	MaxVelocity	Uniform	[0, 0.5]
Complex path moving objects	MaxVelocity	Uniform	[0, 0.5]
	Number of intermediate points	Uniform	[1, 5]

Table 3.4 gives the parameters used for simulating the moving target points.

Figures 3.63 and 3.64 give the results of comparison in terms of η_D and mean consumed energy of all nodes of the network respectively. As it can be seen from Fig. 3.63, network detection rate in SALA outperforms PEAS, GAF and PS, and about 5% worse than the LEACH for which the network detection rate is equal to 1. From this figure, it can also be concluded that it takes about 20 epochs for the SALA algorithm to stabilize its detection rate and learn the movement patterns of the target points. Note that in every epoch of the SALA algorithm, each learning automaton performs one action and receives one feedback from the environment.

Figure 3.64 shows that for SALA, the mean energy consumption is lower than the mean energy consumption of all the existing algorithms except for PS algorithm.

From the results given in Figs. 3.63 and 3.64 we can conclude that the LEACH algorithm has the highest network detection rate and energy consumption and PS algorithm has the lowest network detection rate and energy consumption among the above mentioned algorithms. High network detection rate and energy consumption of LEACH is due to the fact that in this algorithm, all sensor nodes are always in the active operation mode. Similarly, low network detection rate and energy consumption of PS algorithm is due to the fact that in this algorithm, sensor nodes stay in the sleep operation mode more often than the other mentioned algorithms. From this we may say that in order to obtain a higher network detection rate, more energy must be paid. GAF, PEAS and PW algorithms are not always able to fully cover the entire area and hence a network detection rate less than 1 may be experienced.

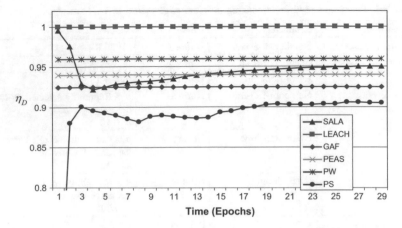

Fig. 3.63 Comparison of the scheduling algorithms in terms of the network detection rate (η_D)

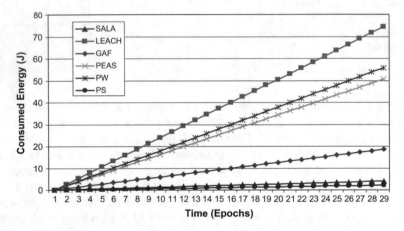

Fig. 3.64 Comparison of the scheduling algorithms in terms of the mean consumed energy of a node

Experiment 2

In this experiment we compare SALA with LEACH, GAF, PEAS, PW and PS in terms of η_D and mean consumed energy for dense networks. The simulation settings of Experiment 1 are also used for this experiment except for the number of sensor nodes which is set to 1000. Figures 3.65 and 3.66 show the results of this experiment. The results indicate that GAF, PEAS and PW outperform SALA. This is because these algorithms unlike SALA use the operation modes of neighbors of a node to schedule that node. In dense networks, the average number of neighbors for a node is higher and hence such algorithms schedule more sensor nodes to be in the sleep operation mode because of the availability of more information about the environment of the sensor nodes.

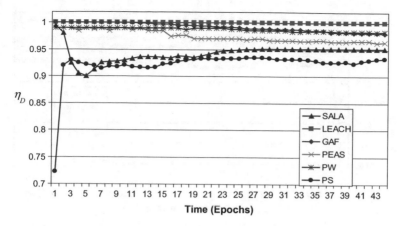

Fig. 3.65 Network detection rate (η_D) in different algorithms for dense networks

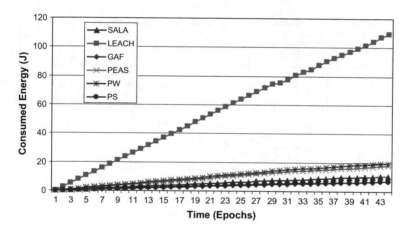

Fig. 3.66 Mean consumed energy of a node in different algorithms for dense networks

Experiment 3

This experiment is conducted to study the behavior of SALA in comparison to LEACH, PEAS, PW and PS algorithms in terms of η_D and mean consumed energy for large networks. The simulation settings of Experiment 1 are also used for this experiment except for the number of sensor nodes which is set to 900 and the network area which is a 300 m × 300 m rectangular area. Figures 3.67 and 3.68 give the simulation results of this experiment. From the results of this experiment we can say that the number of nodes and dimension of the network do not affect the performance of the SALA significantly. The reasons that the performance does not vary significantly as these two parameters change are (1) the network considered in this section is a clustered network with cluster heads as rechargeable nodes, (2) the

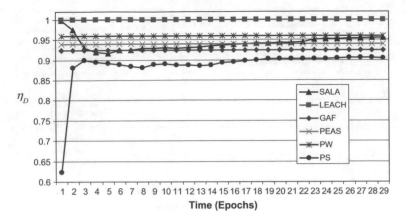

Fig. 3.67 Network detection rate (η_D) in different algorithms for large networks

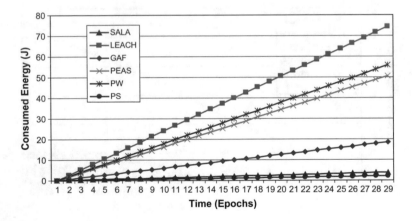

Fig. 3.68 Mean consumed energy of a node in different algorithms for large networks

algorithm executed in each sensor node uses local information within the node's cluster.

Experiment 4

In this experiment, we study the effect of the noise level in noisy events on the performance of the SALA algorithm. For this purpose, we change the standard deviation of the normally distributed noise from 120 to 800. Figures 3.69 and 3.70 show the results of this study. As it is shown, increasing the noise level does not affect the performance of SALA substantially. The robustness of SALA is because of two reasons: (1) utilization of a learning mechanism for making SALA adaptable

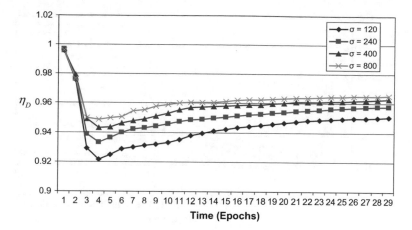

Fig. 3.69 Effect of the STD of the noise on the network detection rate (η_D)

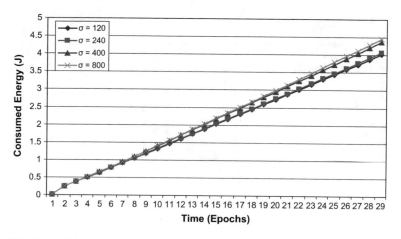

Fig. 3.70 Effect of the STD of the noise on the mean consumed energy of a node

to the environmental changes and (2) utilization of a controlling mechanism by the learning mechanism in order to have a controlled adaptation of the environmental changes. Controlled adaptation is achieved by proper choices of *MinSleepDuration* and *MaxSleepDuration* parameters.

Experiment 5

In this experiment, we evaluate the performance of the SALA algorithm when the mean number of Poisson events per round (λ) varies. Figures 3.71 and 3.72 show

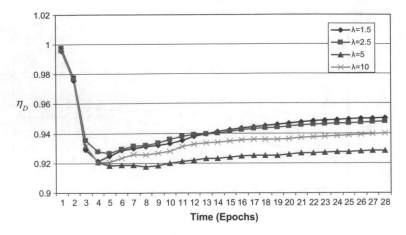

Fig. 3.71 Effect of the mean number of Poisson events on the network detection rate (η_D)

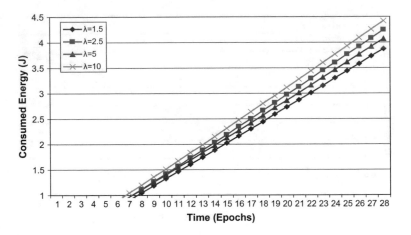

Fig. 3.72 Effect of the mean number of Poisson events on the mean consumed energy of a node

the results of this experiment when λ varies from 1.5 to 10. It can be seen from these figures that the performance of the SALA algorithm does not depend significantly on the value of λ.

Experiment 6

This experiment is designed to evaluate the performance of the SALA algorithm when the *MaxVelocity* of the moving objects varies. For this purpose, we change *MaxVelocity* from 0.1 to 2. Figures 3.73 and 3.74 show the results in terms of the

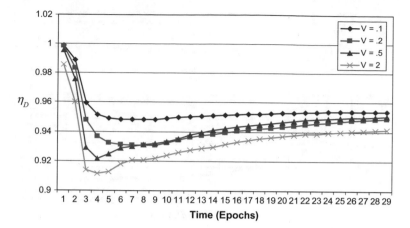

Fig. 3.73 Effect of the *MaxVelocity* of the moving objects on the network detection rate (η_D)

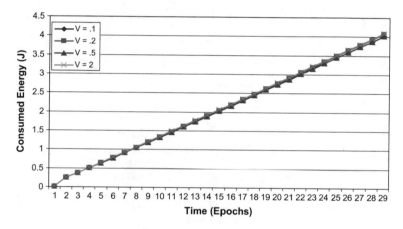

Fig. 3.74 Effect of the *MaxVelocity* of the moving objects on the mean consumed energy of a node

network detection rate (η_D) and the mean energy consumption of the nodes. As indicated by these figures, the changes in the *MaxVelocity* do not affect the performance of the SALA algorithm significantly.

Experiment 7

This experiment is conducted to study the effect of the parameter γ on the performance of the SALA. In this experiment we use the target points of Experiment 1. We also let γ varies in the range [0, 0.2]. Figures 3.75 and 3.76 show the results of this experiment in terms of the network detection rate (η_D) and the mean energy

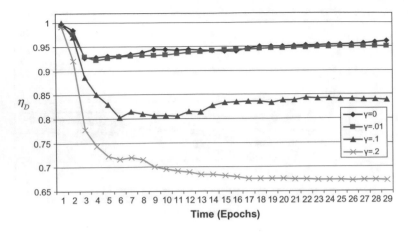

Fig. 3.75 Effect of the parameter λ on the network detection rate (η_D)

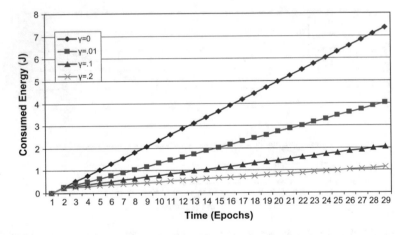

Fig. 3.76 Effect of the parameter λ on the mean consumed energy of a node

consumption of the nodes. From these figures we can conclude that (1) Lower values of γ results in more network detection rate at the expense of more energy consumption, more redundant active times and less sleep times. (2) Higher values of γ results in more energy saving in the network at the expense of poor network detection rate.

Figure 3.77 gives the network detection rate versus mean energy consumption of the nodes as γ varies. As it can be seen from this figure, changing γ in the range [0.01, 0.5] affects both the network detection rate and the mean energy consumption of the nodes. Changing γ outside of this range does not affect these two parameters significantly. To explain this phenomenon, we use the concept of sensitivity of a node to γ. A node is sensitive to γ if changing in the value of γ changes its detection

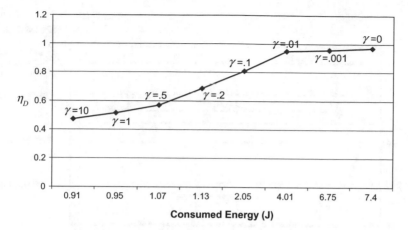

Fig. 3.77 Network detection rate versus Mean energy consumption of the nods as γ varies

rate and energy consumption. From the SALA algorithm, it is evident that a node s_k is sensitive to γ if $\hat{T}_{sleep}^{m,n}(s_k) > MaxSleepDuration$ and $\gamma \cdot \hat{T}_{sleep}^{m,n}(s_k) > MinSleepDuration$. Changing γ out of the range [0.01, 0.5], decreases the number of sensitive nodes. This is because choosing large values for γ results in $\hat{T}_{sleep}^{m,n}$ to be selected equal to *MaxSleepDuration* and choosing small values for γ results in $\hat{T}_{sleep}^{m,n}$ to be selected equal to *MinSleepDuration*. Decreasing the number of sensitive nodes causes the rate of the changes in the network detection rate and the mean energy consumption of nodes to decrease. For $\gamma = 0$ or $\gamma \geq \frac{MaxSleepDuration}{MinSleepDuration}$, all nodes become insensitive to γ which causes the mean energy consumption of the nodes and the detection rate to remain fixed during the lifetime of the network.

Figure 3.78 compares SALA with LEACH, GAF, PEAS and PW in terms of the network detection rate and the mean energy consumption of the nodes. This figure shows that there exists a value for γ ($\gamma = 0.001$) below which SALA outperforms GAF, PEAS and PW in terms of the network detection rate and mean energy consumption of the nodes. Comparing SALA with LEACH in which nodes are always active, the proposed algorithm reaches a network detection rate very near to that of LEACH by consuming only about 9% of the energy consumed by LEACH.

Determination of γ for an application is very crucial and is a matter of cost versus precision. For higher network detection rate, higher price must be paid. For example, as it is indicated in Fig. 3.78, for network detection rate to be above 95%, each node must consume about 6.503(J) on average during the lifetime of the network whereas to obtain a network detection rate of about 85%, each node must consume only about 1.981(J) on average. Reaching highest precision which can be obtained by setting γ to zero results in maximum energy consumption by the network.

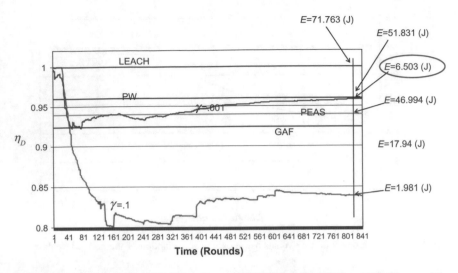

Fig. 3.78 Comparison of the SALA algorithm with LEACH, GAF, PEAS and PW in terms of the network detection rate and the mean energy consumption of the nodes

3.5.1.5 Summary of Results

In this section, we compared the performance of the SALA algorithm in terms of the network detection rate and the mean consumed energy with LEACH, GAF, PEAS, PW and PS algorithms. Comparisons were made for small, large, dense and non-dense networks and showed that:

- In terms of the mean consumed energy, SALA and PS algorithms outperform almost all of the other algorithms in any of the small, large, dense and non-dense networks. Of course, superiority of these two algorithms over other algorithms in non-dense networks is significantly higher than in dense networks. The only exception to this superiority is the mean consumed energy of GAF algorithm in dense networks which is almost equal to that of SALA and PS algorithms. Note that PS algorithm is a simple case of SALA algorithm in which sleeping duration of any node during any round is equal to *MinSleepDuration*.
- In terms of the network detection rate, SALA and LEACH algorithms are the only two algorithms that have robust performance over networks of different sizes and densities. Network detection rates of GAF, PEAS, PW and PS algorithms in non-dense networks are significantly lower than in dense networks.
- There is a tradeoff between the network detection rate and the mean consumed energy; for a higher network detection rate, more energy must be consumed. SALA algorithm, unlike other algorithms, has a parameter (γ) for controlling this tradeoff. There exists a value for parameter γ ($\gamma = 0.001$) below which SALA reaches a network detection rate very near to that of LEACH and outperforms GAF, PEAS, PW and PS. On the other hand, there exists a value for

parameter γ ($\gamma = 0.1$) above which SALA outperforms LEACH, GAF, PEAS, PW and PS in terms of the mean consumed energy. In fact, determination of γ for an application is very crucial and is a matter of cost versus precision.

Furthermore, a number of experiments conducted to study the behavior of the SALA algorithm when parameters of the movement patterns of the target points, such as the velocity of the moving targets and the frequency of occurrences of the events, change. These experiments showed that such changes do not affect the performance of the SALA algorithm in terms of the network detection rate and the mean consumed energy substantially. In other words, SALA algorithm is robust against changes in the parameters of the target points' movement patterns.

3.5.2 SACLA: A Cellular Learning Automata-Based Algorithm for Dynamic Point Coverage Problem

In this section, we will introduce the scheduling algorithm based on the irregular cellular learning automata (SACLA) for solving the dynamic point coverage problem.

3.5.2.1 Problem Statement

Consider a sensor network consists of N sensor nodes s_1, s_2, ..., s_N, M reporter nodes r_1, r_2, ..., r_M and one sink within a $L \times L$ rectangular, completely known and accessible field (Ω). Sensor nodes, which are responsible for sensing and monitoring the field, are scattered randomly throughout the area of the network so that Ω is completely covered. Reporter nodes are placed manually in specified positions within Ω and form an infrastructure through which packets received from the sensor nodes are relayed towards the sink. This manual placement of reporter nodes is feasible since we assume that Ω is completely known and accessible. Reporter nodes are powerful and rechargeable nodes which are always active. All sensor nodes have the same sensing ranges (R_s) and transmission ranges (R_t). Each sensor node s_k has 4 different modes of operation (Wang and Xiao 2006) as follows:

- **On-duty ($CPU_AS_AC_A$)**: CPU, sensing and communicating units are switched on referred to as active mode.
- **Sensing Unit On-duty ($CPU_AS_AC_S$)**: The CPU and the sensing units are switched on, but the communicating unit is switched off.
- **Communicating Unit On-duty ($CPU_AS_SC_A$)**: The CPU and the communicating units are switched on, but the sensing unit is switched off.
- **Off-duty ($CPU_SS_SC_S$)**: CPU, sensing and communicating units are switched off referred to as sleep mode.

Note that in $CPU_AS_AC_A$, $CPU_AS_AC_S$, $CPU_AS_SC_A$ and $CPU_SS_SC_S$, index A stands for active and index S stands for sleep. For further simplicity in notation, we use index x in the above notations to refer to more than one operation mode, i.e. $CPU_AS_AC_x$ refers to both $CPU_AS_AC_A$ and $CPU_AS_AC_S$ modes, and $CPU_xS_xC_x$ refers to all 4 modes of operation. At any instance of time, a sensor node can be only in one of the above 4 operation modes. The operation mode of a sensor s_k at time instant t is denoted by $O_{s_k}(t)$.

Let TP be a finite set of target points residing in Ω. TP is divided into two disjoint sets; *Moving objects* (TP^M) and *Events* (TP^E). A target point $tp_i^M \in TP^M$ is a moving object and has a continuous movement trajectory. On the other hand, a target point $tp_i^E \in TP^E$ is an event which occurs somewhere in Ω repeatedly or randomly following a Poisson distribution and lasts for a short static or random duration.

We denote the Euclidean distance between a sensor node s_k located at $(x(s_k), y(s_k))$ and a target point tp_i located at $(x(tp_k), y(tp_k))$ as $d(s_k, p_i)$, i.e.

$$d(s_k, p_i) = \sqrt{(x(s_k) - x(tp_i))^2 + (y(s_k) - y(tp_i))^2}.$$

Assuming the binary sensing model (Chakrabarty et al. 2002) and sensing range R_s for all sensor nodes in the network, we say a target point $tp_i \in TP$ is sensed, detected or monitored by a sensor node s_k at time t if and only if $d(s_k, tp_i) < R_s$ and $O_{s_k}(t) = CPU_AS_AC_x$. The network detects a target point $tp_i \in TP$ at time t if and only if at least one of the sensor nodes of the network detects tp_i at time t.

Consider Definition 3.12 to Definition 3.18 from Section "Formal Definition of the Problem".

The objective of the network is to detect the target points and report their locations to the sink. We assume that the reporter nodes are placed so that each sensor node can directly communicate with at least one r_i. Sensor nodes send their packets directly to reporter nodes and reporter nodes forward received packets towards the sink.

To prolong the network lifetime, a sensor node will switch to the $CPU_AS_xC_A$ operation mode only if it wants to communicate with a reporter node; otherwise, the communicating unit of the sensor node will be switched off. The sensing unit of a sensor node has to be switched on only if a target point is in its sensing range, but since sensor nodes have no knowledge about the movement paths of the target points, they cannot calculate the times for switching their sensing units on or off. As an alternative way, we assume the time to be divided into a number of very short epochs (Ep) having equal durations (τ_{Ep}). The sensing units of the sensor nodes can be switched on or off only at the startup of each epoch. Therefore, at the startup of each epoch Ep_k^n, a sensor node s_k has to predict if any target points will pass through its sensing region during Ep_k^n. Based on this prediction, s_k switches its sensing unit on or off at the startup of the epoch Ep_k^n. Index k in Ep_k^n states that we assume no synchronization between sensor nodes, and hence, each sensor node has its own timing and epochs. We refer to the operation mode of a sensor node s_k during the epoch Ep_k^n as $O_{s_k}(Ep_k^n)$.

Having the above definitions and assumptions, the problem is to locally predict the status of the sensing unit of each node at the startup of each epoch such that the network sleep rate (η_S) is maximized while the network detection rate (η_D) does not drop below an acceptable level.

3.5.2.2 SACLA Scheduling Algorithm

SACLA scheduling algorithm consists of three major phases; Initialization, mapping and operation. During the initialization phase, each sensor node in the network finds a reporter node in its neighborhood through which it can sends its packets towards the sink. Mapping the network topology to an irregular cellular learning automaton (*ICLA*) is done in the mapping phase. Finally, scheduling the active times of the sensor nodes for different epochs is performed during the operation phase. We explain these three phases in more details in the subsequent sections.

Initialization

Each sensor node has to find a reporter node through which it can send its' packets towards the sink. We refer to the reporter node of a sensor node s_k as $r(s_k)$. During the initialization phase, all sensor nodes are in $CPU_A S_C C_A$ operation mode; that is both CPU and communicating units of all sensor nodes are switched on. Each reporter node r_i periodically broadcasts *ReporterADV* packets which contain the id and location of r_i. A reporter node r_i broadcasts *ReporterADV* packets at times $Rnd_i + \tau_{Adv}$. τ_{Adv} is the period of transmitting *ReporterADV* packets and Rnd_i is a random delay which is used to reduce the probability of collisions between neighboring reporter nodes. A sensor node s_k, upon receiving a *ReporterADV* packet from a reporter node r_i, performs one of the followings:

- If s_k has not seen any *ReporterADV* packet before, then it sets $r(s_k)$ to r_i.
- If $r(s_k) \neq r_i$ and the distance between r_i and s_k is less than the distance between $r(s_k)$ and s_k, then s_k sets $r(s_k)$ to r_i.
- Otherwise, s_k ignores the received packet.

Since we assume that the reporter nodes are placed so that each sensor node can directly communicate with at least one reporter node, using the above procedure, each sensor node s_k is able to find its $r(s_k)$ during the initialization phase.

Mapping

In the mapping phase, a time-driven asynchronous *ICLA* which is isomorphic to the sensor network topology is created. Each sensor node s_k in the sensor network corresponds to the cell c_k in *ICLA*. Two cells c_k and c_m in *ICLA* are adjacent to each other if $r(s_k)$ is equal to $r(s_m)$; that is, the reporter nodes of s_k and s_m are identical. The learning automaton in each cell c_k of the *ICLA*, referred to as LA_k, has two actions α_0 and α_1. Action α_0 is "switch off the sensing unit" and action α_1 is "switch on the sensing unit". The probability of selecting each of these actions is initially set to .5.

Operation

During the operation mode, normal operation of the network is performed; that is sensor nodes sense the environment for the existence of target points and if any target is detected, send information about the detected target to the sink through the infrastructure of reporter nodes. The operation phase is divided into a number of epochs. Each epoch starts with an *evaluation* phase, followed by a *status selection* phase and ends with a *monitoring* phase. The first epoch is an exception, since it starts with the *status selection* phase and it has no *evaluation* phase.

Each cell of the *ICLA* during the operation phase is activated asynchronously. The nth activation of the cell c_k occurs at the startup of the epoch Ep_k^n. We call n the local iteration number for the cell. The operation phase starts when a cell in *ICLA* is activated.

1. **If** (the activation of the cell c_k of the *ICLA* is its first activation (local iteration $n = 1$)) **Then**

    1-1. *Status Selection* phase: sensor node s_k makes decision for the status of its sensing unit for epoch Ep_k^n. The operation mode of s_k during the *status selection* phase is $CPU_A S_A C_S$.

    1-2. *Monitoring* phase: sensor node s_k switches its sensing unit on or off based on the decision made in the previous step. If the sensing unit is active, then s_k monitors its sensing region and sends any required information about detected target points in its sensing region to $r(s_k)$. Required information about a detected target is application specific, and hence we do not specify any details for it. The operation mode of s_k during the *monitoring* phase is $CPU_A S_A C_S$ (if the sensing unit is selected to be active) or $CPU_S S_S C_S$ (if the sensing unit is selected to be sleep).

2. **If** (the activation of cell c_k is not its first activation (local iteration $n > 1$)) **Then**

    2-1. *Evaluation* phase: sensor node s_k evaluates its selected status and sends it to $r(s_k)$. The operation mode of s_k during the *evaluation* phase is $CPU_A S_A C_A$.

    2-2. *Status Selection* phase: This step is similar to the step 1-1.

    2-3. *Monitoring* phase: This step is similar to the step 1-2.

Figure 3.79 gives the transition diagram of the operation modes of a sensor node in different phases of the proposed scheduling algorithm.

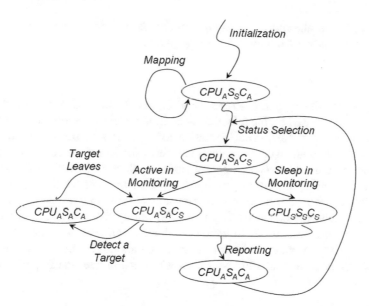

Fig. 3.79 Transition diagram of the operation modes of a sensor node in different phases of the proposed scheduling algorithm

Status Selection Phase

As we stated before, at the startup of each epoch, every sensor node has to select the status of its sensing unit for the monitoring phase of that epoch. This selection must be done based on a short-term prediction about the movement paths of the target points in the vicinity of each sensor node. Learning automaton resides in the cell c_k of the *ICLA* helps the sensor node s_k to select the status of its sensing unit for the monitoring phase of each epoch Ep_k^n. This is done through the following algorithm.

1. Sensor node s_k checks its sensing region for at most *CheckingDuration* to see if any target points can be detected.
2. **If** (any target points can be detected) **Then**

    2-1. s_k sets its operation mode for the monitoring phase of the epoch Ep_k^n to $CPU_A S_A C_S$.

3. **Else**

    3-1. LA_k decides whether to switch the sensing unit of s_k on or off for the monitoring phase of the epoch Ep_k^n; that is, the cell c_k chooses one of its actions using its action probability vector. We refer to this action by $\alpha_{i,k}^n$.

4. Sensor node s_k enters the monitoring phase of the epoch Ep_k^n with the specified operation mode.

In the above algorithm, *CheckingDuration* is a constant which specifies the maximum duration of the *selection status* phase.

Monitoring Phase

During the *monitoring* phase, a sensor node s_k may be in $CPU_A S_A C_S$ or $CPU_S S_S C_S$ operation mode based on the selection done in the *status selection* phase. A sensor node s_k which is in $CPU_A S_A C_S$ operation mode, senses its sensing region and if detects a target point, switches its communicating device on and sends any required information about it to $r(s_k)$. A sensor node s_k which is in $CPU_S S_S C_S$ operation mode, does nothing during the *monitoring* phase. In other words, the short-term prediction which is performed by each sensor node s_k in the *status selection* phase is applied to the status of the sensing unit of s_k during the *monitoring* phase. Each *monitoring* phase lasts for the duration of *MonitoringDuration*. Since a short-term prediction can be valid only for a short duration of time, long values of *MonitoringDuration* results in lesser network detection rate (see Experiment 5).

Evaluation Phase

At the startup of each epoch Ep_k^n, each sensor node s_k has to evaluate its prediction for the status of its sensing unit during the epoch Ep_k^{n-1} (which is used as a reinforcement signal to LA_k). The evaluation phase of the epoch Ep_k^n will be performed only if in the status selection phase of epoch Ep_k^{n-1}, the status of the sensing device of the sensor node s_k is selected by LA_k (step 3-1 in the status selection phase).

The evaluation process consists of two parts; computing the internal feedback (denoted by $\underline{\delta}_k^{n-1}$) and computing the external feedback (denoted by $\overline{\delta}_k^{n-1}$). For computing the internal feedback, each sensor node s_k evaluates its predicted status based on its own information about the target points passed through its sensing region whereas for computing the external feedback, the evaluation is performed using the information of the neighboring sensor nodes. The external feedback can enhance the internal feedback due to the fact that a target point which passes through the sensing region of a sensor node s_k may also pass through the sensing region of its neighboring sensor nodes with a high probability.

Depending on the action selected by LA_k in epoch Ep_k^{n-1}, the internal feedback is computed differently as follows:

- **The action selected by LA_k was "switch on the sensing device"**: Sensor node s_k uses Eq. (3.66) for computing the internal feedback ($\underline{\delta}_k^{n-1}$).

$$
\underline{\delta}_k^{n-1} = \begin{cases} 1 - \dfrac{\tau_{M,s_k}^{n-1}}{MonitoringDuration}; & \dfrac{\tau_{M,s_k}^{n-1}}{MonitoringDuration} < NDR_M \\[2ex] -\dfrac{\tau_{M,s_k}^{n-1}}{MonitoringDuration}; & Otherwise \end{cases} . \tag{3.66}
$$

In the above equation, τ_{M,s_k}^{n-1} is a fraction of the *monitoring* phase of the epoch Ep_k^{n-1} during which no target point is detected by s_k. If τ_{M,s_k}^{n-1} is a substantially small fraction of the *MonitoringDuration*, then it can be concluded that the selected action of LA_k (switch on the sensing device) was a proper selection for the epoch Ep_k^{n-1}. In other words, if $\frac{\tau_{M,s_k}^{n-1}}{MonitoringDuration}$ is lower than a specified threshold (NDR_M), then the internal feedback of s_k for epoch $Ep_k^{n-1}(\underline{\delta}_k^{n-1})$ is positive. Otherwise, $\underline{\delta}_k^{n-1}$ is negative.

- **The action selected by LA_k was "switch off the sensing device"**: In this case, sensor node s_k has no information about the passage of target points through its sensing region during the *monitoring* phase of the epoch Ep_k^{n-1}. Therefore, it cannot evaluate the action selected by its learning automaton. To compensate this, sensor node s_k checks its sensing region for a very short duration called *EvaluationDuration* (*EvaluationDuration* $\ll$ *MonitoringDuration*) to see if any target point can be detected. This short term checking is used by the sensor node s_k for computing the internal feedback. In this case, the internal feedback of the sensor node s_k is computed using Eq. (3.67).

$$\underline{\delta}_k^{n-1} = \begin{cases} 1 - \dfrac{\tau_{E,s_k}^n}{EvaluatingDuration}; & \dfrac{\tau_{E,s_k}^n}{EvaluatingDuration} > NDR_E \\[2ex] -\dfrac{\tau_{E,s_k}^n}{EvaluatingDuration}; & Otherwise \end{cases} \qquad (3.67)$$

In the above equation, τ_{E,s_k}^n is a fraction of the *EvaluationDuration* during which no target point is detected by s_k. If τ_{E,s_k}^n is a substantially small fraction of the *EvaluationDuration*, then we conclude that the action selected by LA_k (switch off the sensing device) was not a proper selection for epoch Ep_k^{n-1}. In other words, if $\frac{\tau_{E,s_k}^n}{MonitoringDuration}$ is higher than a specified threshold (NDR_E), the internal feedback of s_k for the epoch $Ep_k^{n-1}(\underline{\delta}_k^{n-1})$ is positive. Otherwise, $\underline{\delta}_k^{n-1}$ is negative.

Each node s_k computes its internal feedback using Eqs. (3.66) or (3.67) as described above and then creates an *InternalEvaluation* packet which contains $\alpha_{i,k}^{n-1}$ and $\underline{\delta}_k^{n-1}$. This packet is then sent to $r(s_k)$. $r(s_k)$ upon the reception of an *InternalEvaluation* packet from a sensor node s_k, replies to this packet with a *NeighborEvaluations* packet which contains for each neighbor s_l of s_k, its $\alpha_{i,l}^{n-1}$ and $\underline{\delta}_l^{n-1}$. Using the information in the *NeighborEvaluations* packet, each sensor node s_k computes the external feedback ($\overline{\delta}_k^{n-1}$) according to the Eq. (3.68). In this equation, $N(s_k)$ is the set of neighboring nodes of s_k, i.e. the set of sensor nodes s_l for which $r(s_l) = r(s_k)$ and $|S|$ represents the cardinality of the set S. N_A is a subset of $N(s_k)$ and is defined according to Eq. (3.69).

$$
\bar{\delta}_k^{n-1} = \begin{cases} 0; & |N(s_k)| = 0 \\ \frac{|N_A|}{|N(s_k)|}; & \frac{|N_A|}{|N(s_k)|} > LNSA \text{ and } \alpha_{i,k}^{n-1} = \alpha_1 \\ -\frac{|N_A|}{|N(s_k)|}; & \frac{|N_A|}{|N(s_k)|} > LNSA \text{ and } \alpha_{i,k}^{n-1} = \alpha_0 \\ \frac{|N_A|}{|N(s_k)|} - 1; & \frac{|N_A|}{|N(s_k)|} \leq LNSA \text{ and } \alpha_{i,k}^{n-1} = \alpha_1 \\ 1 - \frac{|N_A|}{|N(s_k)|}; & \frac{|N_A|}{|N(s_k)|} \leq LNSA \text{ and } \alpha_{i,k}^{n-1} = \alpha_0 \end{cases} \tag{3.68}
$$

$$
N_A = \left\{ s_l \in N(s_k) \,\middle|\, \left(\alpha_{i,l}^{n-1} = \alpha_1\right) \wedge \left(\underline{\delta}_l^{n-1} > 0\right) \right\}. \tag{3.69}
$$

N_A is the subset of $N(s_k)$ whose sensing devices were active in their previous *monitoring* phases and their internal evaluations are positive. $\frac{|N_A|}{|N(s_k)|}$ gives the fraction of such neighbors. If this fraction falls below a certain threshold ($LNSA$), it can be concluded that α_1 was not a proper selection for the sensor node s_k in the epoch Ep_k^{n-1}. According to the Eq. (3.68), only a proper selection will result in a positive external feedback.

Using the computed internal and external feedbacks, each sensor node s_k computes its feedback for the previous epoch (δ_k^{n-1}) using the Eq. (3.70).

$$
\delta_k^{n-1} = \begin{cases} \psi_k^{n-1} \cdot \underline{\delta}_k^{n-1} + \left(1 - \psi_k^{n-1}\right) \cdot \bar{\delta}_k^{n-1}; & \bar{\delta}_k^{n-1} > 0 \\ \underline{\delta}_k^{n-1}; & otherwise \end{cases} \tag{3.70}
$$

In the above equation, ψ_k^{n-1} is a coefficient which specifies the impact of the internal and external feedbacks on the value of δ_k^{n-1}. It can be a time varying coefficient (as it is indicated in Eq. 3.70) or a constant coefficient (See Experiments 6 and 7). δ_k^{n-1} is then used for the computation of the reinforcement signal β_k^{n-1} as follows:

$$
\beta_k^{n-1} = \begin{cases} 0; & \delta_k^{n-1} \geq 0 \\ 1; & Otherwise \end{cases}. \tag{3.71}
$$

This reinforcement signal is given to LA_k. LA_k updates its action probability vector based on the selected action $\alpha_{i,k}^{n-1}$ and the given β_k^{n-1} according to the following learning algorithm:

$$
\begin{cases} p_{i,k}^{n+1} = p_{i,k}^n + a_k^n \cdot \left(1 - p_{i,k}^n\right) \\ p_{j,k}^{n+1} = 1 - p_{i,k}^{n+1} \end{cases}, \tag{3.72}
$$

$$
\begin{cases} p_{i,k}^{n+1} = \left(1 - b_k^n\right) \cdot p_{i,k}^n \\ p_{j,k}^{n+1} = 1 - p_{i,k}^{n+1} \end{cases}. \tag{3.73}
$$

Reward parameter a_k^n and penalty parameter b_k^n are time varying parameters which vary according to the Eqs. (3.74) and (3.75).

$$a_k^n = a \cdot \delta_k^n,$$ (3.74)

$$b_k^n = -b \cdot \delta_k^n.$$ (3.75)

In Eqs. (3.74) and (3.75), a and b are two constants which control the rate of learning. Higher values of a and b results in faster learning and lesser precision.

In short, on the nth activation of each cell c_k of the *ICLA*, if LA_k selected any action on the $(n-1)$th activation of the *ICLA*, the reinforcement signal β_k^n is computed and the selected action is rewarded or penalized accordingly. Then, the node s_k checks if any target points exists in its sensing region and if so, it selects its operation mode as $CPU_A S_A C_S$ for the upcoming *monitoring* phase. Otherwise, LA_k decides whether to switch the sensing device of s_k on or off for the upcoming *monitoring* phase.

3.5.2.3 Experimental Results

To evaluate the performance of the SACLA, several experiments have been conducted and the results are compared with the results obtained for LEACH (Heinzelman et al. 2000) from the routing layer algorithms, GAF (Xu et al. 2001) from the grid-based algorithms, PEAS (Ye et al. 2003, 2006) from the coverage-related algorithms, Proactive Wakeup based on PEAS (PW) (Gui and Mohapatra 2004) from the tracking-based algorithms.

The simulation environment is a 100 m × 100 m area through which 100 sensor nodes are scattered randomly. Sensing range (r) of sensor nodes is assumed to be 10 m. For placing the reporter nodes, the simulation environment is divided into a number of square cells with dimensions 14 m × 14 m. One reporter node is placed on each boundary point of the cells.

Energy consumption of the sensor nodes follows the energy model of the J-sim simulator (Sobeih et al. 2006). Based on this model, the power consumption of a node during the $CPU_A S_A C_A$, $CPU_A S_A C_S$, $CPU_A S_S C_A$ and $CPU_S S_S C_S$ operation modes are 18.9(mW), 2.901(mW), 18.9(mW), 0.001(mW) respectively. Energy required to switch a node from one operation mode to another operation mode is assumed to be negligible. *CheckingDuration*, *MonitoringDuration*, *EvaluatingDuration*, τ_{Adv}, NDR_M, NDR_E and *LNSA* are set to 5(s), 100(s), 20(s), 1 (s), 0.6, 0.6 and 0.2 respectively. ψ_k^n is assumed to be constant and is set to 0.4 for all sensor nodes. Learning parameters a and b are both set to 0.1.

In all experiments, following different types of moving target points are considered to exist in the area of the network:

- **Constant Events**: Constant events have constant start times and durations. At the startup of the simulation, number of constant events is selected uniformly at random from the range [1, 10]. The start time of each event is a constant which is uniformly and randomly selected from the range [1, 1000] and the duration of each event is a constant which is uniformly and randomly selected from the range [1, 10]. Each event has a static occurrence position which is selected uniformly at random in Ω. Events are repeated with the same start times and durations every 1000 s of the simulation.

- **Noisy Events**: Noisy events are like constant events except that the start times and durations of events are affected every 1000 s by a normally distributed random noise.
- **Poisson Events**: The occurrences of Poisson events follow the Poisson distribution. To generate such events, a number of Poisson random number generators (selected uniformly and randomly from the range [1, 10]) are used. Each Poisson random number generator separately generates events in a randomly selected location in Ω.
- **Straight Path Moving Objects**: Each straight path moving object has a start position, stop position and a velocity. A straight path moving object starts from its start position and moves directly towards its stop position using its velocity. The start and stop positions are selected uniformly at random in Ω and the velocity is selected uniformly and randomly from the range [0, *MaxVelocity*]. The number of straight path moving objects is selected randomly from the range [1, 10]. A straight path moving object restarts its movement from its start position with a newly random velocity when it reaches its stop position.
- **Complex Path Moving Objects**: These target points are like straight path moving objects except that they are not moving from their start positions directly towards their stop positions. Instead, each complex path moving object has a sorted list of points $\langle (x^{start}, y^{start}), (x^1, y^1), (x^2, y^2), \ldots, (x^{Stop}, y^{Stop}) \rangle$. A complex path moving object starts from its start point and moves directly towards (x^1, y^1) using its velocity. Whenever the target point reaches (x^1, y^1), it changes its direction towards (x^2, y^2) with a newly random velocity. This movement continues until the target point reaches its stop position. The number of intermediate points for each moving object is selected randomly from the range [1, 5]. A complex path moving object restarts its movement from its start position with a newly random velocity when it reaches its stop position.
- **Random Waypoint Moving Objects**: A random waypoint moving object follows the random waypoint movement model (Bettstetter et al. 2004).

All simulations have been implemented using J-Sim simulator (Sobeih et al. 2006) and the results are averaged over 25 runs.

Experiment 1

In this experiment, the SACLA algorithm is compared with GAF, LEACH, PEAS and PW algorithms in terms of the network detection rate (η_D) and the mean energy consumption of the nodes. Table 3.5 gives the parameters used for simulating the moving target points.

Table 3.5 Parameters used for simulating different target points in Experiment 1

Target point	Parameter	Distribution	Value
Constant events	Number of events	Uniform	[1, 10]
	Event start time	Uniform	[1, 1000]
	Event duration	Uniform	[1, 10]
Noisy events	Number of events	Normal	μ = Randomly selected from the range [1, 10] $\sigma = 20$
	Event start time	Uniform	μ = Randomly selected from the range [1, 1000] $\sigma = 20$
	Event duration	Uniform	μ = Randomly selected from the range [1, 10] $\sigma = 20$
Poisson events	λ	Poisson	2.0
	Event duration	Constant	2
Straight path moving objects	*MaxVelocity*	Uniform	1
Complex path moving objects	*MaxVelocity*	Uniform	1
	Number of intermediate points	Uniform	[1, 5]

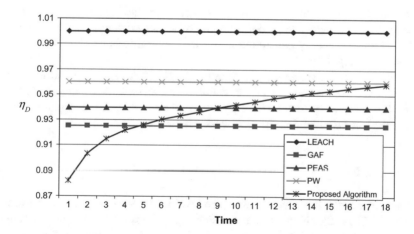

Fig. 3.80 Comparison of the scheduling algorithms in terms of the network detection rate (η_D)

Figures 3.80 and 3.81 give the results of the comparison in terms of η_D and the mean consumed energy of all nodes of the network respectively. As it can be seen from Fig. 3.80, network detection rate in the proposed algorithm outperforms PEAS and GAF, approaches PW, and about 4% worse than the LEACH for which the network detection rate is equal to 1.

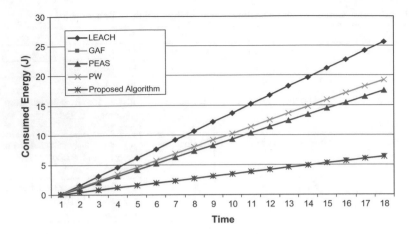

Fig. 3.81 Comparison of scheduling algorithms in terms of the mean consumed energy of the nodes

Table 3.6 Network lifetime

Algorithm	LEACH	PEAS	PW	GAF	Proposed method
Network lifetime	198	269	238	317	319

Figure 3.81 shows that for the SACLA algorithm, the mean energy consumption is lower than the mean energy consumption of all the existing algorithms except for GAF. The mean energy consumption of GAF algorithm is nearly equal to that of the proposed algorithm.

Table 3.6 gives the network lifetime for the SACLA algorithm and the existing methods. We use the number of simulation round at which the network area (Ω) is no further completely covered by the network. As it is shown, the proposed SACLA algorithm can better prolong the network lifetime.

Experiment 2

In this experiment, the effect of the noise level in noisy events on the performance of the SACLA algorithm is studied. For this purpose, we change the standard deviation of the normally distributed noise from 20 to 400. Figures 3.82 and 3.83 show the results of this study. As it is shown, increasing the noise level does not affect the performance of the proposed method substantially.

Experiment 3

In this experiment, the performance of the SACLA algorithm is evaluated when the mean number of Poisson events per round (λ) varies. Figures 3.84 and 3.85 show

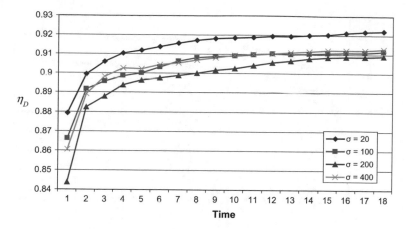

Fig. 3.82 Effect of the STD of the noise on the network detection rate (η_D)

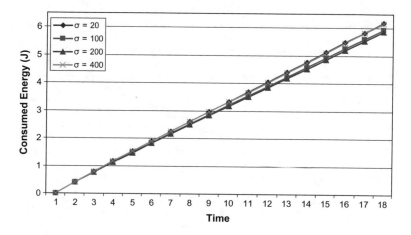

Fig. 3.83 Effect of the STD of the noise on the mean consumed energy of the nodes

the results of this experiment when λ varies from 1 to 8. It can be seen from these figures that the performance of the SACLA algorithm does not depend significantly on the value of λ.

Experiment 4

This experiment is designed to evaluate the performance of the SACLA algorithm when the *MaxVelocity* of the moving objects varies. For this purpose, we change *MaxVelocity* from 0 to 1. Figures 3.86 and 3.87 show the results in terms of the network detection rate (η_D) and the mean energy consumption of the nodes. As

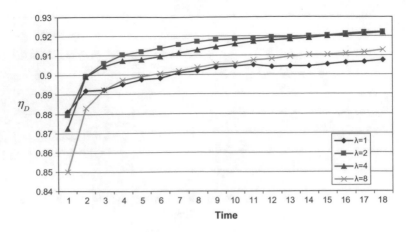

Fig. 3.84 Effect of the mean number of Poisson events on the network detection rate (η_D)

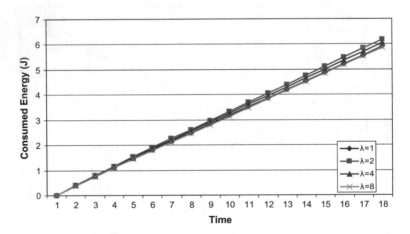

Fig. 3.85 Effect of the mean number of Poisson events on the mean consumed energy of the nodes

indicated by these figures, when moving objects have no movement at all (*MaxVelocity* = 0), the performance of the proposed method in terms of η_D and mean energy consumption of the nodes is very high. Increasing the value of *MaxVelocity* from 0 to 0.5, degrades the performance of the SACLA method, but further increasing in the value of *MaxVelocity* does not affect it significantly. In other words, performance of the method is highly affected by the movement of the target points (in comparison to the case when target points have no movement), but it is not affected too much by increasing the movement velocity.

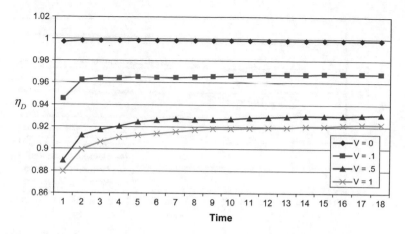

Fig. 3.86 Effect of the *MaxVelocity* of moving objects on the network detection rate (η_D)

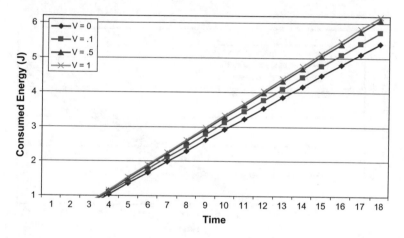

Fig. 3.87 Effect of the *MaxVelocity* of moving objects on the mean consumed energy of the nodes

Experiment 5

This experiment is conducted to study the effect of the *MonitoringDuration* on the performance of the SACLA method. For this experiment we use the target points of experiment 1. We also let *MonitoringDuration* varies in the range [25, 200]. Figures 3.88 and 3.89 show the results of this experiment in terms of the network detection rate (η_D) and the mean energy consumption of the nodes. From these figures we can conclude that: (1) Lower values of *MonitoringDuration* results in more network detection rate at the expense of more energy consumption, more redundant active times and less sleep times and (2) Higher values of *MonitoringDuration* results in more energy saving in the network at the expense of poor network detection rate.

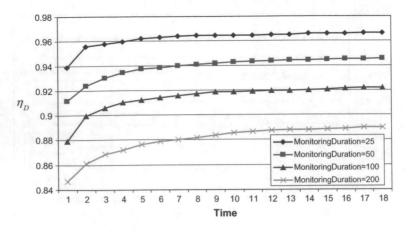

Fig. 3.88 Effect of the *MonitoringDuration* parameter on the network detection rate (η_D)

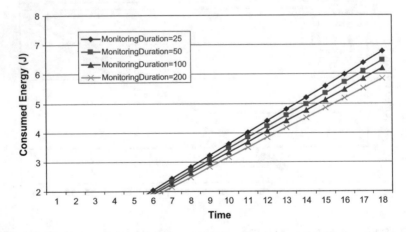

Fig. 3.89 Effect of the *MonitoringDuration* parameter on the mean consumed energy of the nodes

Figure 3.90 gives the mean energy consumption of the nodes versus network detection rate as *MonitoringDuration* varies. As it can be seen from this figure, increasing the value of *MonitoringDuration* results in the network detection rate to decrease. This can be explained as follows: By increasing the value of *MonitoringDuration*, the length of the *monitoring* phase in the SACLA algorithm increases. As a result, the short-term prediction of the *status selection* phase of the algorithm must apply to a longer period of time which results in the precision of the algorithm to decrease.

Figure 3.90 also indicates that as *MonitoringDuration* increases, the mean energy consumption of the nodes decreases to a minimum level (*MonitoringDuration* = 200) and then remains almost fixed. This can also be

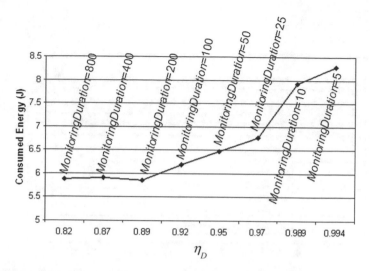

Fig. 3.90 Mean energy consumption of the nods versus Network detection rate as *MonitoringDuration* parameter varies

explained by the effect of a long *MonitoringDuration* on the precision of the short-term prediction performed in the *status selection* phase. A low precise prediction results in a pure chance selection, i.e. the probability of selecting for the sensing unit of the node to be off or on in each epoch becomes equal. Therefore, number of epochs in which the sensor node is in $CPU_A S_A C_S$ or $CPU_S S_S C_S$ operation mode is equal in expected sense. In addition, considering a large value for *MonitoringDuration*, the times a sensor node spends in *CheckingDuration* and *EvaluatingDuration* would be negligible. This indicates that no matter how long is the *MonitoringDuration*, each sensor node spends half of its lifetime in the $CPU_A S_A C_S$ operation mode and the other half in $CPU_S S_S C_S$ operation mode in expected sense, and hence the mean energy consumption of the nodes remains almost fixed for large values of *MonitoringDuration*.

Figure 3.91 compares the SACLA algorithm with LEACH, GAF, PEAS and PW in terms of the network detection rate and the mean energy consumption of the nodes. This figure shows that there exists a value for *MonitoringDuration* (*MonitoringDuration* = 10) below which the proposed algorithm outperforms GAF, PEAS and PW and approaches LEACH in terms of the network detection rate and there exists another value for *MonitoringDuration* (*MonitoringDuration* = 200) above which SACLA algorithm outperforms GAF, PEAS, PW and LEACH in terms of the mean energy consumption of the nodes.

Determination of the *MonitoringDuration* for an application is very crucial and is a matter of cost versus precision. For higher network detection rate, higher price must be paid. For example, as it is indicated in Fig. 3.91, for network detection rate to be above 98%, each node must consume about 7.34(J) on average during the lifetime of the network whereas to obtain a network detection rate of about 92%,

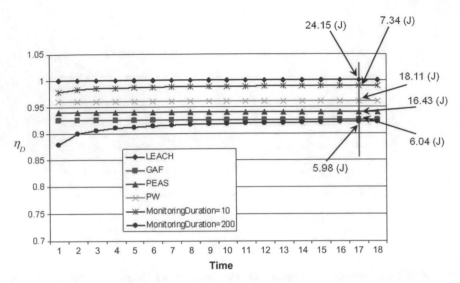

Fig. 3.91 Comparison of the SACLA algorithm with LEACH, GAF, PEAS and PW in terms of the network detection rate and the mean energy consumption of the nodes

each node must consume only about 5.98(J) on average. Reaching highest precision which can be obtained by setting *MonitoringDuration* to zero results in maximum energy consumption by the network.

Experiment 6

In this experiment, the effect of the internal and external feedbacks on the performance of the SACLA method is evaluated. To do this, we let ψ varies in the range [0, 1] ($\psi = 0$ means ignoring the effect of the internal feedback and $\psi = 1$ means ignoring the effect of the external feedback). Figures 3.92 and 3.93 show the results of this experiment in terms of the network detection rate (η_D) and the mean energy consumption of the nodes. From these figures we can conclude that: (1) decreasing the effect of the internal feedback in Eq. (3.70) by decreasing parameter ψ results in more network detection rate at the expense of more energy consumption and (2) increasing the effect of the internal feedback in Eq. (3.70) by increasing parameter ψ results in more energy saving in the network at the expense of lower network detection rate.

Experiment 7

Experiment 6 has shown that the internal and external feedbacks have different effects on the performance of the SACLA algorithm in terms of the network

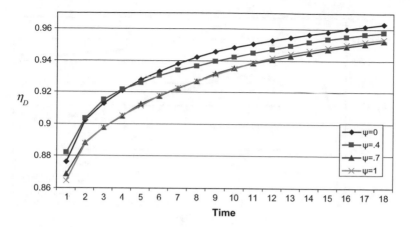

Fig. 3.92 Effect of the parameter ψ on the network detection rate (η_D)

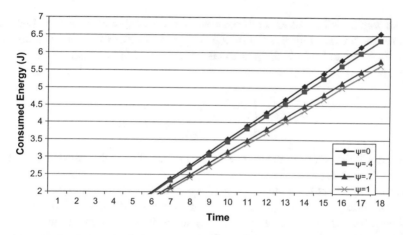

Fig. 3.93 Effect of the parameter ψ on the mean consumed energy of the nodes

detection rate (η_D) and mean energy consumption of the nodes. The results reported in Experiment 6 hold only if the number of sensors in the network remains unchanged. This is because of the fact that the accuracy of the external feedback of the sensor node s_k starts degrading as the number of its neighboring sensors decreases due to malfunctioning or energy depletion. Lesser number of neighboring sensors for a sensor means that the probability of accuracy of information obtained about the surrounding environment decreases. When the accuracy of the external feedback decreases, a node has to depend on its own information about the movement patterns of the target points obtained using its sensing unit. Therefore, if a node wants its detection rate not to fall too much, then it has to let its sensing unit be in the active state more often, which means more energy consumption.

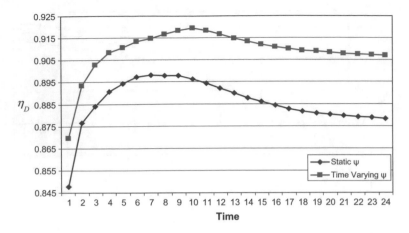

Fig. 3.94 Comparison of static and time varying ψ in terms of the network detection rate (η_D)

One way to reduce the effect of decrease in the accuracy of the external feedback is to use a time varying ψ such as the one given in Eq. (3.76). In this equation, ψ for each sensor node is defined to be a function of the number of neighbors of the node.

$$\psi_k^n = \begin{cases} \frac{N(s_k)-N^n(s_k)}{N(s_k)}; & N(s_k) \neq 0 \\ \qquad 1; & N(s_k) = 0 \end{cases}. \tag{3.76}$$

In Eq. (3.76), $N^n(s_k)$ is the number of neighbors of s_k in the nth activation of the cell c_k of the *ICLA*.

In this experiment, we study the performance of the SACLA algorithm when ψ varies according to the Eq. (3.76). We compare the results obtained for the SACLA algorithm for a fixed value of ψ ($\psi = 0.4$) and a time varying ψ. Figures 3.94 and 3.95 show the results of this experiment in terms of the network detection rate (η_D) and the mean energy consumption of the nodes. These figures show that using a time varying ψ enhances the network detection rate at the expense of more energy consumption, more redundant active times and less sleep times.

3.5.2.4   Summary of Results

In this section, we evaluate the performance of the SACLA algorithm in terms of the network detection rate and the mean consumed energy. The evaluation provides the following results:

- In terms of the network detection rate, SACLA algorithm outperforms PEAS and GAF, approaches PW, and about 4% worse than the LEACH for which the network detection rate is equal to 1.

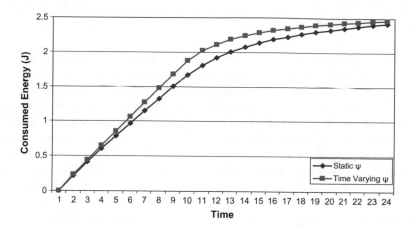

Fig. 3.95 Comparison of static and time varying ψ in terms of mean consumed energy of the nodes

- In terms of the mean energy consumption, SACLA algorithm outperforms all the existing algorithms except for GAF. The mean energy consumption of the GAF algorithm is nearly equal to that of the SACLA algorithm.
- The performance of the SACLA algorithm in terms of the network detection rate and the mean consumed energy is not highly affected by changing the parameters of the movement patterns of the target points, such as the velocity of the moving targets and the frequency of occurrences of the events.
- There is a tradeoff between the network detection rate and the mean consumed energy; for a higher network detection rate, more energy must be consumed. SACLA algorithm has a parameter (*MonitoringDuration*) for controlling this tradeoff. There exists a value for *MonitoringDuration* (*MonitoringDuration* = 10) below which SACLA algorithm outperforms GAF, PEAS and PW and approaches LEACH in terms of the network detection rate and there exists another value for *MonitoringDuration* (*MonitoringDuration* = 200) above which, SACLA algorithm outperforms GAF, PEAS, PW and LEACH in terms of the mean energy consumption of the nodes. In fact, determination of *MonitoringDuration* for an application is very crucial and is a matter of cost versus precision.
- The tradeoff between the network detection rate and the mean consumed energy in SACLA algorithm can also be controlled through adjusting another parameter, namely ψ. Lower values for parameter ψ results in more network detection rate at the expense of more energy consumption and higher values of this parameter results in more energy saving in the network at the expense of lower network detection rate.

3.6 Conclusion

Wireless sensor network is a rapidly growing research area in wireless communications. A key feature which separates *WSNs* from traditional wireless networks such as cellular networks is the primary importance of the energy consumption in these networks. This importance is due to the fact that the energy sources in *WSNs* can be hardly replaced or recharged. As a consequence, the algorithms and protocols, designed for the sensor networks, must be both energy-aware and energy-efficient.

On the other hand, the environment of a *WSN* is highly dynamic due to a number of reasons. Topological changes, which occasionally occur in the network as a consequence of the addition of new nodes, node failures, or energy exhaustion of sensor nodes is the main source of dynamism. Wireless errors, caused by unreliable communication links, packet collisions, congestion, presence of the transitional regions (Zuniga and Krishnamachari 2004), and the funneling effect (Wan et al. 2003) is another source of dynamism in *WSNs*. Furthermore, *WSNs* have a distributed environment, within which, sensor nodes can often be regarded as autonomous agents with self-interested motivations, working together to fulfill the overall purpose of the network. An autonomous agent may change its operation mode, its role in the network, its manner of cooperation with its neighbors, and so on. As a result of such changes, the dynamism of the network's environment increases. Yet another source of dynamism in the area of the *WSNs* is the resource limitations. Handling such limitations affects the operation of the sensor nodes and compels them to change their strategies from time to time during the operation of the network. Considering such dynamism, it is evident that the algorithms and protocols designed for *WSNs* should be made adaptive so that they can cope with the environmental changes occur in such networks.

Considering the above mentioned characteristics of *WSNs*, this chapter aimed at justifying the suitability of the learning automaton model in designing algorithms and protocols for *WSNs*. There are strong evidences which support this idea. First, *LA* is proved to perform well in distributed environments like the environments of *WSNs*, where the number of distributed elements is very large and the overhead of using centralized algorithms is very high (Lakshmivarahan and Narendra 1982; Mason and Gu 1986; Economids and Silvester 1988; Economides 1997; Atlasis et al. 1998). Second, *LA* has a very low computational and communicational overhead which makes it an outstanding model to be used in resource limited environments as of *WSNs* (Esnaashari and Meybodi 2008, 2010a, b, c, 2011). Third, *LA* model is highly adaptive to the environmental changes, and hence, is well-suited to highly dynamic environments like the environments of *WSNs* (Glockner and Pasquale 1993; Howell et al. 1997; Lima and Saridis 1999). Finally, the reinforcement signal used by the *LA* is considered as a random variable and hence, its instant values do not affect the performance of the *LA* in the long run.

To justify the idea, a number of learning automata-based algorithms proposed for addressing five different problems within the area of *WSNs*; data aggregation, clustering, deployment, *k*-coverage, and dynamic point coverage.

Two learning automata-based algorithms, one for increasing the aggregation ratio of the data packets in the network and the other for producing a clustering infrastructure of the sensor network were proposed. Both algorithms aimed at reducing the amount of energy dissipated by the network, and in consequence, extending the network lifetime. It was shown that the proposed algorithms outperform similar algorithms in terms of the energy dissipated in the network and the network lifetime.

CLA-DS algorithm, proposed to address the deployment problem, was a deployment strategy which aimed at guiding the movements of sensor nodes throughout the sensor field in order to attain high coverage. Unlike existing algorithms, CLA-DS did not use any information regarding the positions of the sensor nodes or their relative distances to each other. This made the algorithm capable of working even in environments, where utilized location estimation techniques such as GPS-based devices and localization algorithms experience inaccuracies in their measurements, or the movements of sensor nodes are not perfect and follow a probabilistic motion model.

CLA-EDS, was an extension of the CLA-DS algorithm for providing k-coverage. The algorithm was designed to guide the movements of the sensor nodes throughout the sensor field in such a way that in the resultant deployed network, any point within the area of the network is k-covered. The novelty of the CLA-EDS deployment strategy was that it can address the k-coverage requirement even when k differs in different sub-regions of the sensor field.

Finally, we proposed two algorithms for addressing the dynamic point coverage problem, which is the point coverage problem with non-stationary target points. One algorithm was based on the learning automaton model, called SALA, and the other one was based on the *ICLA* model, called SACLA. Both algorithms were dynamic scheduling strategies which activated and deactivated the sensor nodes according to the movement patterns of the target points throughout the sensor field. Unlike other solutions to the dynamic point coverage problem in which some sort of notification messages were used to activate sleeping nodes, in SALA, no notification messages needed to be exchanged between sensor nodes. Instead, each node separately learnt the best scheduling strategy using its set of learning automata. Therefore, sensor nodes did not need to wait for receiving notification messages from their neighboring nodes. This resulted in a substantially high energy saving in sensor nodes in contrast to the existing algorithms. In SACLA algorithm, each sensor node was mapped into one cell of an *ICLA*. The learning automaton residing in each cell of the *ICLA* in cooperation with the learning automata residing in the neighboring cells dynamically predicted the existence of any target point in the vicinity of its corresponding node in the network in near future. This prediction was then used to schedule the active times of that node. It was illustrated through simulations that the usage of the learning automata for learning and predicting movement patterns of target points can significantly reduce the amount of energy consumed in the network, without substantially affecting the network detection rate.

Chapter 4
Learning Automata for Cognitive Peer-to-Peer Networks

4.1 Introduction

This section gives an approach for designing *DCLA* based cognitive engines in cognitive peer-to-peer networks (Saghiri and Meybodi 2016a). A cognitive peer-to-peer network learns from the past experiences and improves its decisions about its configuration. Cognitive peer-to-peer networks are defined as networks with cognitive processes capable of learning from the results of their actions. A cognitive process recognizes current network situations (plans, conditions, etc.) and acts based on them (Mahmoud 2007). The cognitive peer-to-peer network consists of a set of cognitive peers. Each cognitive peer in the cognitive peer-to-peer network corresponds to a peer in the peer-to-peer overlay network. The topological structure of the cognitive peer-to-peer network is isomorphic to the topological structure of the peer-to-peer overlay network (Fig. 4.1). The structure of the framework that is used in a cognitive peer is shown in Fig. 4.2. Each cognitive peer uses a framework which consists of three layers: *Requirement Layer, Cognitive Process Layer* and *SAN Layer*. The definitions of the layers in the presented framework are given in the next three subsections.

Requirement Layer: In the *Requirement Layer*, the goals and the behavior of the network are described by *Cognitive Specification Language*. Each cognitive peer finds a file celled *configuration* file that is shared in the peer-to-peer overlay network. In the *configuration* file, the *Cognitive Specification Language* is used to determine the goals of the cognitive network. Several approach considering distributed nature of peer-to-peer network for sharing the *configuration* file are suggested in (Saghiri and Meybodi 2016a).

SAN Layer: In the *SAN Layer*, the network status sensors and modifiable elements are designated based on local configurations of the peers in the peer-to-peer overlay network. In other words, each cognitive peer uses its network status sensors for gathering local information about its corresponding peer in the overlay network

© Springer International Publishing AG 2018
A. Rezvanian et al., *Recent Advances in Learning Automata*, Studies in
Computational Intelligence 754, https://doi.org/10.1007/978-3-319-72428-7_4

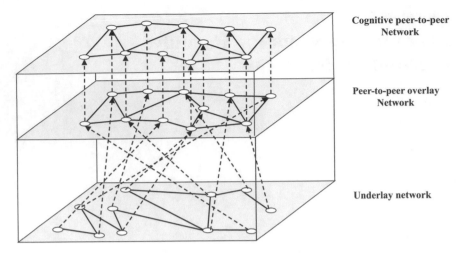

Fig. 4.1 Cognitive peer-to-peer network (Saghiri and Meybodi 2016a)

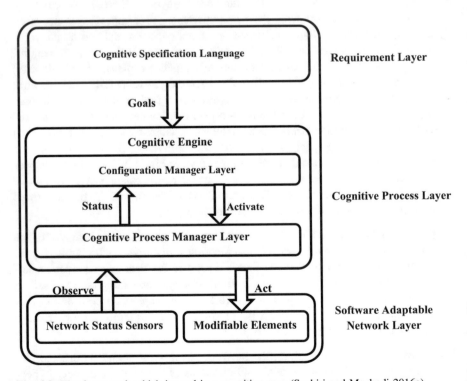

Fig. 4.2 The framework which is used in a cognitive peer (Saghiri and Meybodi 2016a)

to observe its environment. Each cognitive peer also acts on its modifiable elements to change the local configuration of its corresponding peer in the overlay network.

Cognitive Process Layer: The *Cognitive Process Layer* is implemented using a set of cognitive engines resided in the cognitive peers. The cognitive engine consists of two layers (Fig. 4.2): *Configuration Manager Layer* and *Cognitive Process Manager Layer*. The descriptions of the layers of the *Cognitive Process Layer* are given bellows (Saghiri and Meybodi 2016a).

- *Cognitive Process Manager Layer*: This layer manages the operations of the cognitive peers using a network of cognitive processes (Fig. 4.3). In this layer, a network of cognitive processes observes the status of the peer-to-peer overlay network using network status sensors of the cognitive peers. The network of cognitive processes of this layer also cooperatively changes the configuration of the cognitive peers using the modifiable elements of the cognitive peers considering the goals of the cognitive network. It should be note that, the *Cognitive Process Manager Layer* is aware of the goals of the cognitive network and therefore the network of cognitive processes of the cognitive peer-to-peer network update the configuration of the cognitive peers according to the goals of the cognitive network.
- *Configuration Manager Layer*: In this layer, each cognitive peer manages the configuration of its corresponding peer of the peer-to-peer overlay network. This layer has two major responsibilities described as follow. The first responsibility is to provide required information for the network of cognitive processes of the *Cognitive Process Manager Layer*. The second responsibility is to activate the network of cognitive processes and execute appropriate management algorithm determined by the cognitive processes of the *Cognitive Process Manager Layer* using SAN layer. This layer also executes other required management algorithms of the peer-to-peer overlay network which do not use the *Cognitive Process Manager Layer*.

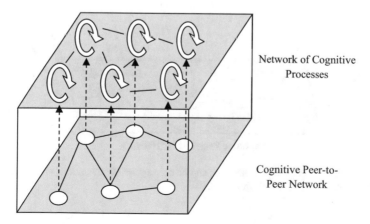

Network of Cognitive
Processes

Cognitive Peer-to-
Peer Network

Fig. 4.3 Network of cognitive processes (Saghiri and Meybodi 2016a)

Now, we propose an approach based on *CADCLA* for designing cognitive engines for solving network management problems in peer-to-peer overlay networks. In this approach, the *Cognitive Process Manager Layer* of the cognitive engine of the presented framework uses an *CADCLA* whose structure is isomorphic to the cognitive peer-to-peer network (Fig. 4.4). Each cognitive peer of the cognitive peer-to-peer network corresponds to a cell of *CADCLA*. In the *Cognitive Process Manager Layer*, executing the activation function of the *CADCLA* which results in updating the structure and the states of the cells of the *CADCLA* leads to changing the structure and the parameters of the peers of the peer-to-peer overlay network in order to solve a network management problem of the peer-to-peer overlay network. In each cell, the activation function is in charge of implementing the cognitive process of its corresponding cognitive peer. In other word, a network of cognitive processes based on *CADCLA* is conducted to solve the network management problem. Note that, in each cognitive peer, several modifiable elements are defined for changing the configurations of the corresponding peer of that cognitive peer in the peer-to-peer overlay network.

In the rest of this section, three cognitive engines based on *CADCLA* for solving network management problems are designated in peer-to-peer networks.

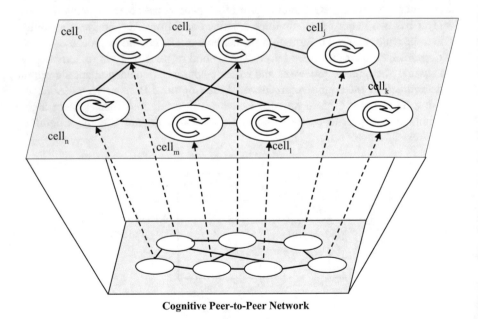

Cognitive Peer-to-Peer Network

Fig. 4.4 A cognitive peer-to-peer network and its corresponding network of cognitive processes based on *CADCLA* (Saghiri and Meybodi 2016a)

4.2   A Cognitive Engine for Solving Topology Mismatch Problem Based on Schelling Segregation Model

In this section, the restructuring rule of *Scehlling segregation model* is fused with *DCLA* for designing the cognitive engine (Saghiri and Meybodi 2016a). *Scehlling segregation model* is a well-known multi agent model (Schelling 1971; Domic et al. 2011). The remainder of this section is organized as follows. Section 4.2.1 is dedicated to some preliminaries used in this chapter. In Sect. 4.2.2, an adaptive algorithm for super-peer selection has been proposed. The results of simulations are reported in Sect. 4.2.3. In order to study the performance of *DCLA*, two metrics: entropy and *potential energy* will be used in Sect. 4.2.3.

4.2.1   Preliminaries

In this section, in order to provide basic information for the remainder of the chapter, we present a brief overview of topology mismatch problem, PROP-O algorithm, Schelling segregation model, and a framework introduced by Thomas et al. for cognitive networks.

Topology mismatch problem: Peer-to-Peer networks are overlay networks that are constructed over underlay networks. In these networks, peers choose their neighbors without considering underlay positions, and therefore the resultant overlay network may have large number of mismatched paths. In a mismatched path a message may meet an underlay position several times which causes redundant network traffic and end-to-end delay (Moustakas et al. 2016). In some of the topology matching algorithms such as *PROP-O* (Qiu et al. 2009), *THANCS* (Liu 2008), *X-BOT* (Leitão et al. 2012), and one reported in (Rostami and Habibi 2007), each peer uses a local search operator for gathering information about the neighbors of that peer located in its neighborhood radius. In these algorithms, each peer also uses a local operator for changing the connections among the peers. These matching algorithms reconfigure the overlay structure (using the local operators) in an online fashion in order to solve the topology mismatch problem. These algorithms reconfigure a given overlay graph $G = (V, E)$ to another graph $G^o = (V, E^o)$ by solving the following problem.

$$\min z = \sum_{v \in V} \sum_{u \in V - \{v\}} x_{uv} d_{uv} \tag{4.1}$$

$$\text{s.t.} \sum_{v \in V} \sum_{u \in V - \{v\}} x_{uv} = 2 \times |E| \tag{4.2}$$

$$\forall_{U \subset V, \ U \neq \emptyset} \sum_{uv \in l(U, \ V)} x_{vu} \geq 1 \tag{4.3}$$

$$x_{uv} \in \{0, 1\} \tag{4.4}$$

In (4.1), x_{uv} indicates the existence or absence of a connection between $peer_u$ and $peer_v$. If $x_{uv} = 1$ then there is a connection between $peer_u$ and $peer_v$. d_{uv} is the end-to-end delay from $peer_u$ and $peer_v$. Constraint (4.2) means that the matching algorithm must not change the number of links in the overlay. In Constraint (4.3), $l(U, V) = \{vu | v \in U, u \in V - U\}$ is the set of overlay links incident to the peers in the subset $U \subset V$. Constraint (4.3) indicates that for any subset $U \subset V$ there is at least one overlay link connecting the two components U and $V - U$. The objective of matching algorithms is to minimize z subject to constraint (4.2), (4.3) and (4.4).

PROP-O algorithm: *PROP-O* algorithm is a topology matching algorithm which is able to solve topology mismatch problem in both structured and unstructured peer-to-peer networks (Qiu et al. 2009). In this algorithm, to solve the topology mismatch problem, a local operator called exchange operator is presented. Figure 4.5 shows how the exchange operator exchange equal number of neighbors between $peer_i$ and $peer_j$ if $d_{ki} + d_{lj} > d_{kj} + d_{li}$. In this operator, the edges $(peer_i, peer_k)$ and $(peer_j, peer_l)$ will be replaced by $(peer_j, peer_k)$ and $(peer_i, peer_l)$ only if $peer_i$ and $peer_j$ are adjacent to each other. With variable neighborhood radius, the exchange operator can be extended if there is a path between $peer_i$ and $peer_j$. In this chapter, $\{peer_i, peer_j\}$ and $\{peer_k, peer_l\}$ are called *corresponding peers* and *candidate peers,* respectively. The execution of exchange operator for decreasing the delays of paths of the overlay network leads to decreasing the mismatched paths of the overlay network. The detailed descriptions about this algorithm are given in the rest of this part.

In *PROP-O* algorithm, the process executed by each peer when joining to the network has two major phases: (1) warm-up phase, and (2) maintenance phase. Each phase of this algorithm has two main sub-phases: local search, and exchange. When a new peer $peer_i$ joins to the network starts the warm-up phase.

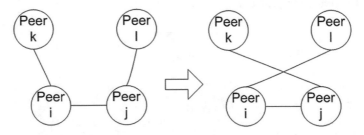

Fig. 4.5 Exchange operator

During warm-up phase, $peer_i$ gathers information about delays to its neighbors and then starts its local search. The information about neighbors is stored in a priority queue called *neighborQ*. During the local search, $peer_i$ selects one of its neighbors such as $peer_s$ which is used to find *corresponding* peers then contact to a random $peer_v$ as *corresponding* peer which is in a neighborhood radius (variable r_i determines the neighborhood radius) to $peer_i$. Then $peer_i$ selects some of its neighbors as *candidate* peers and calculates the difference between end-to-end delays before and after exchange operation with $peer_v$. The difference between end-to-end delays is stored in a variable called *Var*. After computing the value of variable *Var*, if the exchange operation can decrease the end-to-end delays of links of $peer_i$ and $peer_v$, then $peer_i$ starts its exchange phase. During the exchange phase, $peer_i$ utilizes the exchange operator to exchange its *candidate* peers with the *candidate* peers of $peer_v$. At the end of warm-up phase, $peer_i$ decides to repeat the warm-up phase or goes to maintenance phase. The warm-up phase will be repeated for *MAX_INIT_TRIAL* times which the value of parameter *MAX_INIT_TRIAL* was set by a designer. After warm-up phase, $peer_i$ goes to the maintenance phase.

Maintenance phase starts with local search. The local search of the maintenance phase is similar to the local search of warm-up phase. During local search, $peer_i$ gathers some information about its neighbors and then contact to a random $peer_v$ which is in a neighborhood radius (determined by variable r_i). After computing the value of variable *Var* the peer decides to apply exchange operator or not. If the value of variable *Var* is greater than threshold *MIN_VAR* then $peer_i$ change its neighbors using exchange operator. Otherwise the priority of the selected neighbor $peer_s$ will be decreased. If an exchange occurs, $peer_i$ will decrease the priority number of $peer_s$ by a small number like 1 so that $peer_s$ could be chosen in near future. Otherwise, $peer_s$ will be replaced at the tail of *neighborQ*. After updating the *neighborQ*, $peer_i$ decides to repeat the maintenance phase or not using a Timer. In the maintenance phase, the optimization process (consist of local search and exchange) will be repeated till the end of a duration computed based on the result of exchange operation and two parameters *MAX_TIMER* and *INIT_TIMER*. In the maintenance phase, each peer modifies its *neighborQ* if it receives join or leave messages or when its connections changes. In *PROP-O*, a threshold called m is defined to determine the numbers of *candidate* peers which can be used in each exchange operation. For more information about *PROP-O* algorithm we refer to (Qiu et al. 2009).

The *PROP-O* algorithm suffers from two problems which are described as follows. The first problem is that there is no adaptive mechanism for setting the neighborhood radius parameter. Finding an appropriate value for this parameter manually is a time consuming process and also error prone. Large neighborhood radius speeds up the convergence of the matching algorithm (because the number of *candidate* peers at each step increases) and it decreases the number of exchanges that must be endured until the convergence of the algorithm. Also, large neighborhood radius causes higher traffic and computational overhead of the network. Small neighborhood radius decreases the number of *candidate* peers at each step of the algorithm which causes the total number of exchanges to be increased. Small neighborhood radius results in lower traffic and computational overhead. Because of the dynamicity of peer-to-peer networks, the operational environment and the neighbors of each peer may change over time (peers continually join and leave the network) and for this reason using a fixed neighborhood radius may not be appropriate. The second problem is the lack of an adaptive mechanism for managing the execution of the exchange operator. Non-adaptive mechanism for managing the execution of the exchange operator leads to performing unnecessary exchange operations which results in increasing the overhead of the matching algorithm (higher number of peers to be reconfigured and extra control messages).

Schelling segregation model: *Schelling segregation* model (Schelling 1971; Domic et al. 2011) is composed of independent and identical agents. Each agent cares only about the composition of its own local neighborhood. Each agent using a function (called similarity function) calculates the portion of its neighbors which have similar attribute with it. According to a rule called happiness rule each agent decides whether or not to change its neighbors. If the value of similarity function is lower than a parameter z, the agent is unhappy and prefers to change its neighbors in order to increase the number similar neighbors. This process continues until no agent wants to change its neighbors any longer. The happiness rule controls the process of changing the neighborhood in the model.

4.2.2 An Approach Based on CADCLA for Designing Cognitive Engines and Its Application to Solve Topology Mismatch Problem

In this section, a cognitive engine called *PROP-OL* for solving topology mismatch problem in unstructured peer-to-peer networks is presented.

4.2.2.1 Cognitive Process Manager Layer

This layer manages a network of cognitive processes based on *CADCLA*. In each cognitive peer, a cognitive process is executed by the activation function of its corresponding cell. In the rest of this section, at first, the required variables are defined, then the design of components of the *CADCLA* are given, and finally the activation function of the *CADCLA* is described.

Required notations: In the proposed cognitive engine, each cognitive peer contains the data structure and algorithms of a cell of *CADCLA*. For example, $peer_i$ contains the data structure and algorithms of $cell_i$. Therefore, each cognitive peer has variables for saving the attribute, state, neighbors, and the probability vector of learning automaton of its corresponding cell. In addition, The following items are defined for designing the components of the *CADCLA*.

- Neighborhood radius of $peer_i$ is denoted by r_i.
- Neighborhood set of $peer_i$ denoted by NP_i^r contains all peers residing in the neighborhood radius of $peer_i$, that is $NP_i^r = \{peer_j \in V | dist(peer_i,\ peer_j) \le r_i\}$ where $dist(peer_i,\ peer_j)$ is the length of the shortest path (with minimum number of hops) between $peer_i$ and $peer_j$ in the overlay network. The immediate neighbors of $peer_i$ is denoted by NP_i^1.
- Candidate peer set of $peer_i$ for $peer_j$ denoted by C_{ij}. This set contains some of the neighbors of $peer_i$ which change their connections from $peer_i$ to $peer_j$ during the exchange operation.
- Corresponding peer set of $peer_i$ denoted by MP_i. This set contains some of neighbors of $peer_i$ such as $peer_j \in NP_i^r$ which $peer_j$ has a candidate peer set $C_{ji} \subseteq NP_j^1$ and $peer_i$ has a candidate peer set $C_{ij} \subseteq NP_i^1$ such that $|C_{ji}| = |C_{ij}|$ and

$$\left[\sum\nolimits_{peer_k \in C_{ij}} d_{ki} + \sum\nolimits_{peer_l \in C_{ji}} d_{lj} \right] > \left[\sum\nolimits_{peer_l \in C_{ji}} d_{li} + \sum\nolimits_{peer_k \in C_{ji}} d_{kj} \right].$$

In other words, MP_i contains some of the neighbors of $peer_i$ which can participate in the exchange operation to decrease the delays of paths and also mismatched paths of the overlay network.

- $\lambda_i = \dfrac{|NP_i^1| - |MP_i|}{|NP_i^1|}$ is the portion of neighboring peers of $peer_i$ which do not have mismatched path with $peer_i$ (the portion of the neighbors of $peer_i$ which cannot participate in the exchange operation with $peer_i$). λ_i is called similarity function. The value of λ_i increases during the exchange operation and reaches 1 when $peer_i$ has no mismatch path ($|MP_i| = 0$). Function λ_i gives valuable information about the neighbors of $peer_i$ and it is used to tune some components of $ADCLA$ which implement the process of changing neighbors of cells.

In the $CADCLA$, each cell has two states: "Increase parameter" and "Decrease parameter". The initial state of all cells is set to "Increase parameter". Each cell is equipped with a learning automaton which has two actions: "Increase parameter" and "Decrease parameter" to determines the state of that cell. In $peer_i$, the state of $cell_i$ will be used to increase (or decrease) the value of variable r_i. According to the definition of $CADCLA$, we need to determine attributes of cells to use them in restructuring function, structure updating rule, and local rule. But here, attributes cannot be computed due to the fact that peers are not aware of their underlay positions. This does not cause any problems due to the fact that the proposed model of $CADCLA$ uses the similarity function (λ_i) which does not need to use attributes of cells. The similarity function gives required information about the neighbors of the peer.

Design of components of the $CADCLA$: The components of the $CADCLA$ are described as below.

- **Local rule** takes information of immediate neighbors of a cell $cell_i$ as input and then returns reinforcement signal (β_i) and restructuring signal (ζ_i) of $cell_i$.

- In $cell_i$, the restructuring signal ζ_i^1 is equal to 1 if $1 - (\lambda_i - \frac{1}{|NP_i^1|}) > z$ and 0 otherwise.
- In $cell_i$, the reinforcement signal is equal to 1 in only two cases described as follows. In the first case, the value of λ_i is higher than a parameter t (parameter t is initially set to a given value) and the action selected by the learning automaton of $cell_i$ is equal to "Decrease parameter". In the second case, the value of λ_i is lower than or equal the parameter t and the action selected by the learning automaton of $cell_i$ and the majority of states of immediate neighboring cells of $cell_i$ are equal to "Increase parameter". In other cases, the reinforcement signal is 0.

- **Structure updating rule** is implemented using an operator called swap operator. Figure 4.6 shows an example of usage of swap operator. In this figure, if $\zeta_i = 1$ then the structure updating rule selects $cell_j$ using a function called *prop-selector()* and then exchanges equal number (determined by parameter m which is initially set to a given value) of neighbors between $cell_i$ and $cell_j$ in order to decrease the value of ζ_i^1. In $cell_i$, function *prop-selector()* takes information about $cell_i$, information about a neighboring cell called *s-cell*, and the neighbors of $cell_i$ determined by the neighborhood radius r_i and then returns one cell as output. The detailed descriptions about the structure updating rule and also function *prop-selector()* are given later.

Remark 4.1 The ideas according to which the structure updating rule is obtained are borrowed from *PROP-O* algorithm and *Schelling segregation* model which are described as bellows.

- The swap operator of *CADCLA* implements the exchange operator of *PROP-O* algorithm.
- The definition of restructuring function is borrowed from the similarity function of *Schelling segregation* model. The goal of increasing the value of similarity function in the *Schelling segregation* model is the same as the goal of increasing the value of restructuring signal in the *CADCLA*.
- In the *CADCLA*, similar to the *Schelling segregation* model, $cell_i$ changes its neighbors using the structure updating rule in order to increase the portion of

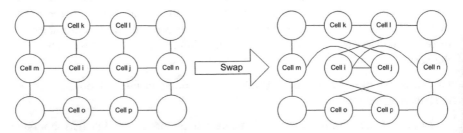

Fig. 4.6 An example of swap operator of the *CADCLA*

neighboring cells that they have similar attribute with $cell_i$ which leads to increasing the value of restructuring signal ζ_i.

- Function *prop-selector()* is designated based on the local search of *PROP-O* algorithm. The process of finding appropriate cell for executing swap operation in the *CADCLA* plays the same role as the process of finding a *corresponding* peer for executing exchange operation in the *PROP-O* algorithm. For example, in $cell_i$, *prop-selector()* returns $cell_j$ which $peer_j$ considering the local search of *PROP-O* algorithm is an appropriate *corresponding* peer for executing exchange operation with $peer_i$. Cell *s-cell* which is used as input in *prop-selector()* denotes the cell of the peer which is used to start the *local search* of *PROP-O* algorithm.

Design of Activation function of CADCLA: Figure 4.7 shows the pseudo code of the function which each cell of *CLA* executes after activation. It should be noted that, the *Configuration Manager Layer* which will be described in the next part uses function *activate_cell()* of the *Cognitive Process Manager Layer* to activate its corresponding cell.

4.2.2.2   Configuration Manager Layer

This layer utilizes the *Cognitive Process Manager Layer* to change the connections of the peers of the peer-to-peer overlay network in order to solve the topology mismatch problem. In each peer, the algorithm which is used in this layer starts with *extended-warm-up* phase and ends with *extended-maintenance* phase. In the rest of this part, the phases of this algorithm are described.

In the *extended-warp-up* phase, $peer_i$ gathers information about its neighboring peers, selects one of its neighbors called $peer_s$, and calls *activate_cell()* function from *Cognitive Process Manager Layer*. Then $peer_i$ changes the value of its neighborhood radius r_i according to the state of $cell_i$ (the cell of $peer_i$). Note that when the state of $cell_i$ is equal to "Increase parameter" ("Decrease parameter") we increase (decrease) the value of variable r_i one unit. If the value of neighborhood radius is lower than 1, $peer_i$ set the value of neighborhood radius to 1. Finally, $peer_i$ refines the list of its neighbors considering the management operations executed by the *activate_cell()* function of the *Cognitive Process Manager Layer*, and then decides to repeat *extended-warm-up* phase or not using a counter called *ntrial*. Counter *ntrial* is use to control the iterations of the extended-warm-up phase. The pseudo code of the *extended-warp-up* phase is given in Fig. 4.8.

In the *extended-maintenance* phase, $peer_i$ gathers information about its neighboring peers, selects one of its neighbors called $peer_s$, and calls *activate_cell()* function from *Cognitive Process Manager Layer*. Then $peer_i$ changes the value of its neighborhood radius r_i according to the state of $cell_i$ (the cell of $peer_i$). Note that when the state of $cell_i$ is equal to "Increase parameter" ("Decrease parameter") we

Algorithm *activate_cell*()

    Inputs:

      z// the parameter for the portion of the neighbors of
         the *cell* that have similar attribute with the *cell*.
         parameter z is used by the automaton trigger
         function and the structure updating rule of the
         cell.

      t// the parameter for the portion of the neighbors of
         the *cell* that have similar attribute with the *cell*.
         parameter *t* is used by the local rule of the *cell*.

      r// the parameter neighborhood radius of the *cell*.

      m// the number of cells which can be exchanged
         among cells.

      s-cell //the cell which is used to find other
         neighboring cells.

01	**Begin**
02	- gather information of the neighboring cells;
03	- compute the *restructuring signal* of the *cell* using the *local rule* (F_1);
04	- find the immediate neighbors of the cell using the *structure updating rule* (F_2);
05	- ask from *learning automaton* of the *cell* to chose an action;
06	- set the state of the *cell* to the action chosen by the *learning automaton*;
07	- compute the *reinforcement signal* using the *local rule* (F_1); //the *local rule* uses parameter *t*
08	- update the *learning automaton* of the *cell* using *reinforcement signal*;
09	**End**

Fig. 4.7 The pseudo code of the procedure which each cell executes after activation

---

Algorithm extended-warm-up()

    Input:

        r // the initial value of parameter neighborhood radius

        INIT_TIMER// the initial value for variable timer in the algorithm

        MAX_INIT_TRIAL // the maximum number of iteration in the

                algorithm

        z,t, and m // the parameters used by the cell of the peer

    Notations:

        Let **timer** denotes a waiting time which is used after each exchange operation.

        Let **neighborQ** denotes a priority queue which saves the information about neighbors of the peer.

        Let **ntrial** denotes the iteration number at each iteration.

01	**Begin**
02	- timer ← INIT_TIMER;
03	- add all neighbors into neghborQ;
04	**While** ntrial < *MAX_INIT_TRIAL* **do**
05	- s ← neighborQ. pop;
06	- neighborQ. addTail(s);
07	- make *peer$_s$* as destination of the first hop;
	// *s-cell* denotes the cell of *peer$_s$*
08	- gather required information for the *cell* of the *peer;*// prepare the *cell* for
	executing function *activate_cell*
09	- call *activate_cell(z, t, r, m, s-cell)*;
	//the pseudo code of this function is given in Figure 4-7
10	**If** (the state of the *cell* of the *peer* is equal to "Increase parameter") **Then**
11	- $r \leftarrow r + 1$;
12	**Else**
13	- $r \leftarrow r - 1$;
14	**EndIf**
15	**If** (r<1) **Then**
16	- $r \leftarrow 1$;
17	**EndIf**
18	- refine the list of neighbors of the peer;
	// considering the operation executed by the *cell* of the *peer*
19	- ntrial ← ntrial + 1
20	- wait timer before next ntrial;
21	**EndWhile**
22	**End**

Fig. 4.8 Pseudo code of extended-warm-up phase of PROP-OL algorithm

increase (decrease) the value of variable r_i one unit. If the value of neighborhood radius is lower than 1, $peer_i$ set the value of neighborhood radius to 1. If the neighbors of the $peer_i$ has been changed, $peer_i$ refines the list of its neighbors, decreases the priority of $peer_s$. Otherwise, the priority of the selected neighbor $peer_s$ will be decreased and the timer of the $peer_i$ will be changed. After updating the $neighborQ$, $peer_i$ decides to repeat the *extended-maintenance* phase or not using a Timer. In the *extended-maintenance* phase, the matching process will be repeated till the end of a duration computed based on two parameters *MAX_TIMER* and *INIT_TIMER*. In the *extended-maintenance* phase, each peer resets its neighborhood radius and the variables of its corresponding cell if it receives join or leave messages or when its connections changes. The pseudo code of the *extended-maintenance* phase is given in

Remark 4.2 The algorithm used in the *Configuration Manager Layer* is a modified version of *PROP-O* algorithm. The phases of this algorithm are modified version of *warm-up* phase and *maintenance* phase of *PROP-O* algorithm respectively. Some parts of this algorithm such as the algorithms of selecting neighboring peers, selecting a peer for starting the matching procedure (called $peer_s$), setting the timers and changing the priority of peers in $neighborQ$ are similar to *PROP-O* algorithm (Fig. 4.9).

Remark 4.3 The rationale behind of designing the network of cognitive processes is described as follows. The process of changing neighbors in the network of cognitive processes which is based on *PROP-O* algorithm plays the same role as the process of changing neighbors of agents in the *Schelling segregation* model. Figure 4.10 shows the process of changing neighbors in both *Schelling segregation* model and *CADCLA*. Figure 4.11 shows the network of cognitive processes and its corresponding Schelling segregation model and also illustrates the neighbors of $cell_i$ when the neighborhood radius is equal to 1.

As it was previously mentioned, *PROP-O* algorithm suffer from two problems; neither the neighborhood radius nor the exchange operator can adapt themselves to the dynamicity of the network. In the network of cognitive processes of the proposed cognitive engine, these problems were solved with the following solutions.

- The similarity function of *Schelling segregation* model was used to manage the execution of exchange operation. In the network of cognitive processes, similar to the *Schelling segregation* model, $peer_i$ changes its neighbors in order to increase the portion of neighboring peers (λ_i) which do not have mismatched path with $peer_i$. Since the value of similarity function gives valuable information about the position of $peer_i$, it is used in the stopping condition of the structure updating rule of the *CADCLA* which manages the execution of exchange operation. In other word, the cognitive processes conducted by the *CADCLA* observes the information about the positions of the peers in order to adaptively manage the exchange operator considering the current status of the network.
- The learning automata of the *CADCLA* are used to manage the neighborhood radiuses of the peers. In the network of cognitive processes, learning automata

---

Algorithm extended-maintain()

Input:

   r// the initial value of parameter neighborhood radius

   INIT_TIMER// the initial value for variable timer in the algorithm

   MAX_TIMER // the maximum value for waiting time

   t,z, and m // the parameters used by the cell of the peer

Notations:

   Let **timer** denotes a waiting time which is used after each exchange

   Let **neighborQ** denotes a priority queue which saves the information about neighbors of the *peer*.

01	**Begin**
02	**While** time expires **do**
03	- s ← neighbor. pop;
04	- make *peer$_s$* as destination of the first hop;
05	- gather required information for the *cell* of the *peer*;
06	- call *activate_cell(z, t, r, m, s-cell)*;
	//the pseudo code of this function is given in Figure 4-7
07	**If** (the state of the *cell* of the *peer* is equal to "Increase parameter") **Then**
08	- $r \leftarrow r + 1$;
09	**Else**
10	- $r \leftarrow r - 1$;
11	**EndIf**
12	**If** (r<1) **Then**
13	- $r \leftarrow 1$;
14	**EndIf**
15	**If** (the neighbors of the *peer* has been changed) **Then**
16	- refine the list of neighbors of the *peer*;
	// according to the changes made by the *cell* of the *peer*
17	- s. priority ← s. priority − 1;
18	**Else**
19	-s. priority ← neighborQ. minPriority − 1;
20	timer ← min(timer * 2, MAX_TIMER);
21	**EndIf**
22	- refresh neighborQ;
23	**EndWhile**
24	**If** (receive join/leave message or detect failure entries) **Then**
25	- timer ← INIT_TIMER;
26	- add new entries to the front of neighborQ;
27	- reset variables of the *cell* of the *peer*;
28	**EndIF**
29	**End**

Fig. 4.9 Pseudo code of extended-maintenance phase of PROP-OL algorithm

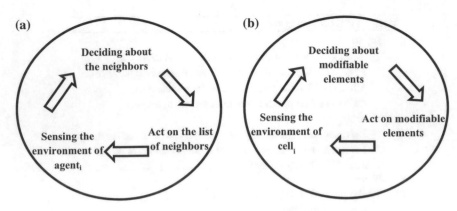

(a)
Deciding about
the neighbors

Sensing the
environment of
agent_i

Act on the list
of neighbors

(b)
Deciding about
modifiable
elements

Sensing the
environment of
cell_i

Act on modifiable
elements

Fig. 4.10 a The process of changing neighbor in agent_i of schelling segregation model. **b** The process of changing neighbor in cell_i of *CADCLA* (Saghiri and Meybodi 2016a)

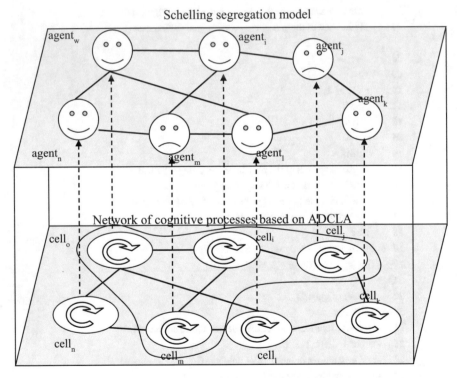

Schelling segregation model

agent_w   agent_i   agent_j

agent_n   agent_m   agent_l   agent_k

Network of cognitive processes based on ADCLA

cell_o   cell_i   cell_j

cell_k

cell_n   cell_m   cell_l

Fig. 4.11 Schelling segregation model and its corresponding network of cognitive processes (Saghiri and Meybodi 2016a)

are used to find appropriate values for the neighborhood radiuses of the peers. The current status of the network determines appropriate values for the neighborhood radiuses. Therefore, in each peer, the local rule of the *CADCLA* which is in charge of generating reinforcement signal of the learning automaton of that peer utilizes the value of similarity function and also information about the neighboring peers of that peer.

The detailed descriptions of the proposed algorithm is given in (Saghiri and Meybodi 2016a).

4.2.3   Experimental Results

All simulations have been implemented using *OverSim* simulator (Baumgart et al. 2009). This simulator supports different kinds of underlay network models: *Simple*, *SingleHost*, *INET*, and *ReaSE*. *Simple model* is the most scalable one. Experiments reported in this section are conducted on one underlying networks called *Topology.1* as an example of *Simple model* as given in Table 4.1. For overlay network *Gnutella* (Chawathe et al. 2003) which is an unstructured peer-to-peer network is used. We used the Gnutella dataset of (2014) to generate the overlay topology. This data set contains a graph which its nodes, edges, and diameter are 10,876, 39,994, and 9 respectively.

The parameters of *PROP-OL* and *PROP-O* algorithms are set based on Table 4.2. For all experiments, each peer is equipped with a variable structure learning automaton of type L_{RP} with parameters reported in Table 4.3. Based on our experimental study using different parameter settings, we have chosen the best parameters values for *PROP-O* algorithm. Simulations are performed for 100 rounds. Results reported are averages over 50 different runs. The algorithms are

Table 4.1 Underlay topologies

Underlay topology	Descriptions
Topology. 1	In this underlay topology, peers are placed on a N-dimensional euclidean space and the internet latencies are based on CAIDA/Skitter (Huffaker et al. 2002; Mahadevan et al. 2005) data

Table 4.2 The parameters of PROP-OL and PROP-O algorithms

Parameters	Value
M	3
MIN_VAR	0
MAX_INIT_TRIAL	10
INIT_TIMER	10 min
MAX_TIMER	1000 min

Table 4.3 The parameters of
the learning automata

Parameters	Value
Reward parameter α	0.25
Penalty parameter b	0.25

compared with respect to five metrics: Overlay Communication Delay (OCD), and
Control Message Overhead (CMO). These metrics are briefly explained below:

- **Overlay Communication Delay (OCD)** is the sum of end-to-end delays of
  links in the overlay network [using (4.1)].
- **Control Message Overhead (CMO)** is the number of extra control messages
  generated by the matching algorithm of the overlay network.

Experiment 1:

This experiment is conducted to study the effect of parameter t on the performance
of the proposed cognitive engine when parameter z is set to 1 and parameter r is set
to 2. For this study, the proposed cognitive engine is tested for three initial values
for parameter $t = 0.3$, 0.6 and 0.9, and *Topology.1* is used as the underlay network.
The results are compared with respect to five above mentioned criteria. According
to the results of this experiment that are shown in Figs. 4.12 and 4.13 and Table 4.4
one may conclude the following:

- In terms of *OCD*, the proposed cognitive engine with $t = 0.6$ and $t = 0.9$ performs better than the proposed cognitive engine with $t = 0.3$. This is because
  low value for parameter t leads to low sensitivity of the cognitive engine of a
  peer to the information about other peers of the network. It should be noted that,
  in each peer, parameter t implicitly determines the information about neighboring peers can be used for calculating the reinforcement signal of the learning
  automaton in the cognitive engine or not. The information about the neighbors
  of peers gives valuable information about the state of the network and therefore
  low sensitivity of the cognitive engines conduct the learning automata of cognitive engine to converge to inaccurate actions which lead to low performance of
  the proposed cognitive engine (with respect to high *OCD*).

Fig. 4.12 The impact of
parameter t on the
performance of *PROP-OL*
with respect to *OCD* when
parameter z is1

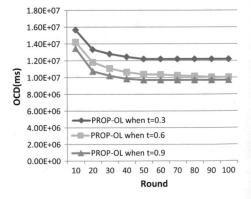

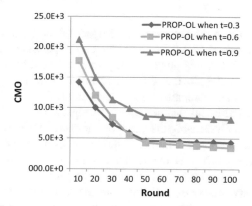

Fig. 4.13 The impact of parameter t on the performance of *PROP-OL* with respect to *CMO* when parameter z is1

Table 4.4 The required number of rounds to the lowest values of OCD for PROP-OL when $z = 1$

	PROP-OL when $t = 0.3$	PROP-OL when $t = 0.6$	PROP-OL when $t = 0.9$
The required number of rounds	34	73	42
The lowest value of OCD	1.22E + 07 ms	1.02E + 07 ms	9.70E + 06 ms

- Increasing the value of parameter t leading to increasing *CMO* but decreasing *OCD* at early rounds of the simulation. In other word, by increasing the value of parameter t we can find an appropriate overlay with low *OCD* but we will have high overhead (with respect to high *CMO*) at early rounds of the simulation.
- In terms of *CMO*, the proposed cognitive engine with $t = 0.6$ performs better than the proposed cognitive engine with $t = 0.3$ and $t = 0.9$ except for the early rounds of the simulation.
- The proposed cognitive engine when $t = 0.9$ performs well with respect to *OCD*, but does not perform well with respect to *CMO*. This means that, for high value for parameter t, when the overlay topology is transformed to an appropriate overlay (with respect to low *OCD*) the overhead of the proposed cognitive engine remains high (with respect to high *CMO*). High value for parameter t leads to high sensitivity of the learning automata of cognitive engines to information about other peers of the network. Since the information about the neighbors of some peers may not reflect the whole state of the network high sensitivity of the cognitive engine may conduct the learning automata of cognitive engines to set the neighborhood radius of the peers to large values. It is obvious that large values for parameter neighborhood radius lead to creating many control messages and performing many unnecessary changes in the network which leads to high *CMO*.

Experiment 2:

This experiment is conducted to show the impact of the parameter z on the performance of the proposed cognitive engine when the parameter t is set to 0.6 and parameter r is set to 2. For this purpose, the proposed cognitive engine is tested for three values for parameter $z = 0.5$, 0.8 and 1, and *Topology.1* is used as underlay topology. The results are compared with respect to *OCD,* and *CMO.* According to the results of this experiment that are shown in Figs. 4.14 and 4.15 and Table 4.5 one may conclude that increasing the value of parameter z leading to improving the performance of the proposed cognitive engine in terms of *OCD.* In each peer, parameter z implicitly activates the learning automaton of the cognitive engine. High value of parameter z leading to increasing the activity of learning automata to adjust with their environments, and therefore the learning automata of cognitive engines are able to gradually find appropriate actions which results in high accuracy of the proposed cognitive engine (with respect to low *OCD*). It should be noted that, increasing the value of parameter z leading to increasing the required rounds to achieve the lowest value of OCD.

Fig. 4.14 The impact of parameter z on the performance of *PROP-OL* with respect to *OCD* when parameter t is 0.6

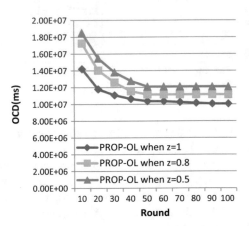

Fig. 4.15 The impact of parameter z on the performance of *PROP-OL* with respect to *CMO* when parameter t is 0.6

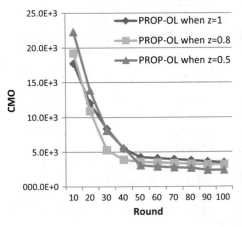

Table 4.5 The required number of rounds to the lowest values of OCD for PROP-OL when $t = 0.6$

	PROP-OL when $z = 0.5$	PROP-OL when $z = 0.8$	PROP-OL when $z = 1$
The required number of rounds	44	54	73
The lowest value of OCD	1.21E + 07 ms	1.12E + 07 ms	1.02E + 07 ms

Experiment 3:
This experiment is designated to support the theoretical results given in Sect. 3. For this purpose, the proposed algorithm is compared with the following algorithms.

- *PROP-OP* algorithm in which each *learning automaton* is replaced with a *pure chance automaton*. In a *pure chance automaton* the actions of automaton are always selected with equal probabilities (Narendra and Thathachar 1989).
- *PROP-OX* algorithm is a version of the proposed algorithm. This algorithm is obtained by applying the following local rule in the CLA of the *PROP-OL* algorithm.

  - **Local rule** takes information of immediate neighbors of a cell $cell_i$ as input and then returns a reinforcement signal (β_i) and a restructuring signal for $cell_i$. The reinforcement signal of $cell_i$ is equal to 1 in only two cases described as follows. In the first case, the value of λ_i is higher than a parameter t (parameter t is initially set to a given value), the action selected by the learning automaton of $cell_i$ is equal to "Decrease parameter" and the value of r is higher than one. In the second case, the value of λ_i is lower than or equal the parameter t, the value of r is lower than 2, and the action selected by the learning automaton of $cell_i$ and the majority of states of immediate neighboring cells of $cell_i$ are equal to "Increase parameter". In other cases, the local rule returns 0.

For all version of the proposed algorithm, the values of parameters z and t are set to 1 and 0.6 respectively. the initial value for the parameter r is set to 2 and *Topology.1* is used as underlay topology. The reward and penalty parameters are set to 0.001. The results of this experiment are organized into two parts described as bellow.

Part A: Analyzing the expediency of the *CADCLA*

The results of this part are shown in Figs. 4.16, 4.17, 4.18, 4.19 and 4.20. In order to use the results given in Sect. 2, some conditions mentioned for Theorem 2–17 were checked. Noted that, in the simulation of the proposed algorithm, many parameters affects on the evolution of the *CADCLA* and therefore finding a closed form for a function which generates the reward probabilities of the LAs is not possible. In this experiment, we studied the reward probabilities for a random learning automaton. The Figs. 4.16 and 4.17 show that the reward probabilities are

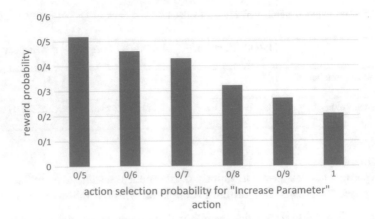

Fig. 4.16 The reward probability for selecting "increase parameter" action

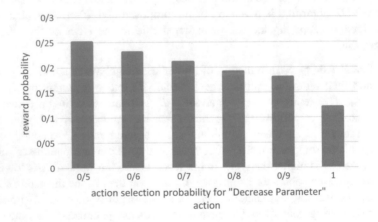

Fig. 4.17 The reward probability for selecting "increase parameter" action

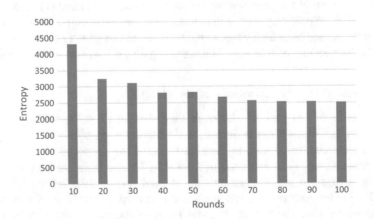

Fig. 4.18 Entropy

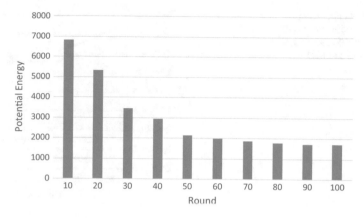

Fig. 4.19 Potential energy

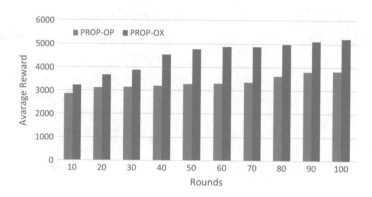

Fig. 4.20 Average reward

monotonically decreasing function of p_{ij}. In other word, the condition $\frac{\partial f_{ij}^{\beta}(\mathrm{P_i})}{\partial p_{ij}} < 0$ can be satisfied by the environment of the LA. Figures 4.18 and 4.19 show that the entropy and *potential energy* decreases over time which means that the changes in actions and cellular topology decreases over time. From Fig. 4.20 we may conclude that in terms of *average reward*, *PROP-OX* algorithms perform better than *PROP-OP* algorithm which indicates that the *CADCLA* used by *PROP-OX* algorithm is expedient.

Part B: Analyzing the performance of CADCLA based matching algorithm

The results of this part are shown in Figs. 4.21 and 4.22. From the results one may conclude that PROP-OX performs better than PROP-OP and PROP-OL algorithms with respect to OCD, and CMO.

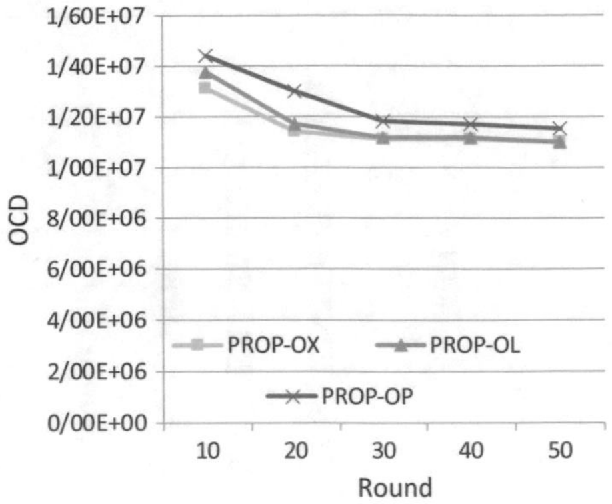

Fig. 4.21 Comparison of PROP-OL with PROP-OP and PROP-OX algorithms with respect to OCD

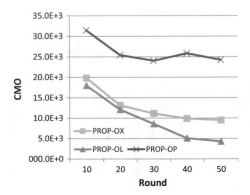

Fig. 4.22 Comparison of PROP-OL with PROP-OP and PROP-OX algorithms with respect to CMO

4.3   A Cognitive Engine for Solving Super-Peer Selection Problem Based on Fungal Growth Model

In this section, a cognitive engine for adaptive super-peer selection considering peers capacity based on *CADCLA* will be proposed (Saghiri and Meybodi 2017d). The *CADCLA* of the proposed engine uses the model of fungal growth to adjust the attributes of the cells in order to take into consideration the dynamicity that exists in peer-to-peer networks in the process of super-peers selection. The difference between the super-peer selection algorithm proposed in this chapter and *Myconet* algorithm (Snyder et al. 2009) which also uses a model of fungal growth is that (1) The fungal growth model is fused with a dynamic *CLA* used for the purpose of super peer selection and (2) The fungal growth model fused with *CLA* is different from the one used in *Myconet* algorithm. In order to study the performance of *DCLA*, two metrics: entropy and *potential energy* will be used. Computer experimentations have been conducted to study the performance of the proposed super-peer selection algorithm. The results of experiments show that the proposed

CLA based super selection algorithm outperforms the existing algorithms with respect to the number of super-peers, and capacity utilization. The remainder of this section is organized as follows. Section 4.3.1 is dedicated to some preliminaries used in this chapter. In Sect. 4.3.2, an adaptive algorithm for super-peer selection has been proposed. The results of simulations are reported in Sect. 4.3.3.

4.3.1 Preliminaries

In this section, in order to provide basic information for the remainder of the chapter, we present a brief overview of super-peer selection problem, and growth pattern of fungi.

Super-peer selection problems: Consider n peers which are connected to each other through a peer-to-peer network. The topology of the peer-to-peer network can be represented by each graph such as $G = (V, E)$ in which $V = \{peer_1, peer_2, \ldots, peer_n\}$ is a set of peers and $E \subseteq V \times V$ is a set of links connecting the peers in the network. In super-peer networks, some peers must be selected as super-peers. The peers which are not selected as super-peers are called as ordinary-peers. The topology of the super-peers network can be represented by a graph $G^s = (V^s, E^s)$ in which $V^s \subseteq V$ is a set of super-peers and $E^s \subseteq V \times V$ is a set of links connecting the super-peers in the super-peer network. In a super-peer network, each super-peer in V^s is mapped to several ordinary-peers in V according to an one-to-many function $H : V^s \to V$. In the super-peer networks, the network management responsibilities are handled by the super-peers. In these networks, the network communications are done among super-peers. A super-peer network with a large number of super-peers imposes a large overhead to the network with respect to control messages generated by the management algorithms of the super-peers. Therefore the existing algorithms such as those reported (Montresor 2004; Snyder et al. 2009; Liu et al. 2013; Gholami et al. 2014) try to adaptively select a small set of super-peers considering some metrics such as peers capacity in distributed fashion.

   In (Lo et al. 2005), different types of the super-peer selection problems has been compared with classic problems such as dominating set, p-centers, and leader election. In super-peer networks such as those reported in (Montresor 2004; Snyder et al. 2009; Liu et al. 2013; Gholami et al. 2014), variable c_i (called capacity) to save the number of peers that can be handled by $peer_i$ if $peer_i$ is selected as super-peer in the network is defined. The value of variable c_i is determined when $peer_i$ joins the network and it remains constant throughout the operation of the network. Since, the goal of super-peer selection considering peers capacity sounds a lot like capacitated minimum vertex cover algorithms, some required definitions about capacitated minimum vertex cover are given as below.

Definition 4.1 A vertex cover V^c of graph is a subset of V such that $(u, v) \in E \to u \in V^c$ or $v \in V^c$. Such a set is said to vertex cover of G (Irit and Safra 2005).

Definition 4.2 A minimum vertex cover is a vertex cover which it has the smallest possible size. The problem of finding a minimum vertex cover is an NP-hard problem (Irit and Safra 2005).

Definition 4.3 A capacitated minimum vertex cover is a minimum vertex cover in which there is a limit to the number of edges that a vertex can cover (Rajiv et al. 2006).

According to the mathematical formulation of capacitated vertex cover problem reported in (Rajiv et al. 2006), In order to solve the super-peer selection problem considering peers capacity the following problem should be solved.

$$\min z = \sum_v x_v \tag{4.5}$$

$$y_{eu} + y_{ev} = 1 \quad e = \{u, v\} \in E \tag{4.6}$$

$$c_v x_v - \sum_{e \in E(v)} y_{ev} \geq 0 \quad v \in V \tag{4.7}$$

$$x_v \geq y_{ev} \quad v \in e \in E \tag{4.8}$$

$$y_{ev} \in \{0, 1\} \quad v \in e \in E \tag{4.9}$$

$$x_v \in \{0, 1\} \quad v \in V \tag{4.10}$$

In (4.5), the values of x_v correspond to cost of selecting a peer $peer_v$ as a super-peer for a super-peer network. In (4.6), $y_{ev} = 1$ if and only if the link corresponding to edge $e \in E$ is connected to peer $peer_v$. Constraint (4.6) says that every peer must be connected to one super-peer. For any peer $peer_v$, let E(v) denote the set of links $peer_v$. In (4.7), c_v denotes the capacity of peer $peer_v$ (we assume that c_v is an integer). Constraint (4.7) guarantees that the number of links of a peer $peer_v$ cannot be more than its capacity. Constraint (4.8) guarantees that ordinary-peers cannot connect to another ordinary-peer.

Since no global knowledge about the network exists and the conditions of the network are highly dynamic, minimizing (4.5) subject to (4.6), (4.7), (4.8), (4.9) and (4.10) by the super-peer selection algorithm leading to a challenging problem. Therefore the existing algorithms such as those reported (Montresor 2004; Snyder et al. 2009; Liu et al. 2013; Gholami et al. 2014) try to adaptively select a small set of super-peers considering capacity of the peers.

A brief description about the growth pattern of fungi: In the nature, fungi reproduce itself by extending filamentous strands through a growth medium such as the soil (Fig. 4.23). The filamentous strands are called *Hyphae*. A *Hypha* (plural *Hyphae*) is a long, branching filamentous structure of a fungus. Fungi follow an interesting pattern of growth. They do not follow a fixed evolutionary pathway. The growth pattern of a fungus is very flexible because all cells of a *Hyphae* may initiate a colony. The mechanism used for formation of colonies of fungi is determined by

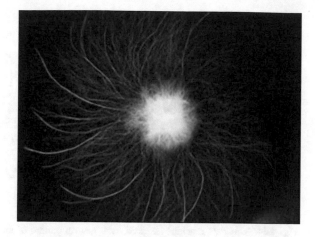

Fig. 4.23 An example for growth pattern of fungi (Meškauskas et al. 2004)

the water and nutrients of the soil of the environment. *Hypha* cells are able to sense reproductive cells from distance, and grow towards them. In order to find new resources, *Hypha* cells are also able to penetrate to the permeable surfaces during reproduction. Different classifications considering the cell, structure, and growth pattern of fungi are reported in the literature (Meškauskas et al. 2004; Robson et al. 2007).

4.3.2   Proposed Algorithm: X-NET

In this section, we first present a state machine inspired from growth pattern of fungi, then outline the proposed super-peer selection algorithm.

4.3.2.1   A State Machine Inspired from Growth Pattern of Fungi

The state machine is described as follows (Fig. 4.24). Each cell takes one of three types: *Unattached-Cell*, *Attached-Cell*, and *Colony-Manager*. The initial type of all cells is set to *Unattached-Cell*. Each *Unattached-Cell* cell tries to find a *Colony-Manager* cell from its neighbors. In *Unattached-Cell $cell_i$*, after finding a *Colony-Manager $cell_j$*, $cell_i$ changes its type to *Attached-Cell* (Transition 1). If $cell_i$ couldn't find any *Colony-Manager* cell then it changes its type to *Colony-Manager* (Transition 2). If a *Colony-Manager $cell_i$* is connected to a *Colony-Manager $cell_j$* which the capacity of $cell_j$ is greater than the capacity of $cell_i$ then $cell_i$ changes its type to *Attached-Cell* (Transition 3). If an *Attached-Cell $cell_i$* is connected to *Colony-Manager $cell_j$* and the capacity of $cell_i$ is greater than the capacity of $cell_j$ and all *Attached-Cell* cells that are connected to the $cell_j$ then $cell_i$ changes its type to *Colony-Manager* (Transition 4).

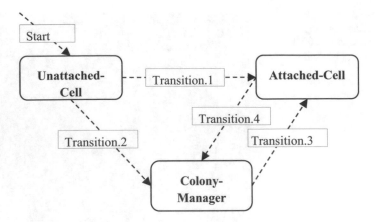

Fig. 4.24 The state machine inspired from growth pattern of fungi

Each *Colony-Manager* cell takes one of two states: *Colony-Extender*, and *Colony-Immobilizer*. If the state of a cell is equal to *Colony-Extender*, the cell can be connected to *Colony-Extender*, and *Colony-Immobilizer* cells. If the state of a cell is equal to *Colony-Immobilizer*, the cell can be connected to *Colony-Immobilizer* cells. A *Colony-Immobilizer* cell is able to absorb the *Attached-Cell* cells of other *Colony-Immobilizer* and *Colony Extender* cells. A *Colony-Extender* cell is able to absorb the *Attached-Cell* cells of other *Colony-Extender* cells.

Since, this state machine is merged into the *CADCLA*, the mechanisms used for selecting the state of a cell, and changing the connections among the cell will be described in more details later in the proposed algorithm.

4.3.2.2 Proposed Algorithm

Initially, an *CADCLA* isomorphic to the peer-to-peer network is created which involves defining the initial structure, local rule, structure updating rule, automata trigger function, restructuring function, and local environments (Fig. 4.25). each peer *peer$_i$* corresponds to the cell *cell$_i$* in *CADCLA*. Each peer may play one of three roles: unattached, ordinary or super. Each peer uses its corresponding cell to set its role and execute appropriate management operation. Each cell may have one of two states: *Colony-Extender*, and *Colony-Immobilizer*. Each cell is equipped with a *LA* which has two actions: *Colony-Extender*, and *Colony-Immobilizer* to determine the state of the cell. The attribute of *cell$_i$* consists of two parts: capacity c_i and type t_i. For a cell, capacity is defined as maximum number of cells which can connect to the cell simultaneously. A cell may take one of three types: *Unattached-Cell, Attached-Cell,* and *Colony-Manager.* In each peer, the role of the peer is determined by the type of its corresponding cell (will be described in more details later). The remaining parts of the *CADCLA* are described later in the rest of this section.

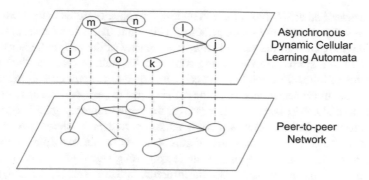

Fig. 4.25 Asynchronous dynamic cellular learning automata and peer-to-peer network

Once the *CADCLA* is created, the proposed algorithm utilizes it to manage the roles of the peers. The process executed by each peer *peer_i* when joining to the network consists of three phases: *initialization* Phase, *construction* phase, and *maintenance* phase. These phases are briefly described as below.

- *Initialization* **phase**: During this phase performed by a peer, the peer establishes its connections to other peers of the network, and initializes its corresponding cell. Then the cell goes to *construction* phase. During the initialization of the cell, the following settings are used.

  - The initial state of the cell is set to *Colony-Extender*.
  - The initial type of the cell is set to *Unattached-Cell*.
  - The neighborhood radius of the cell is set to 2.
  - The value of capacity is determined.

- *Construction* **phase**: During this phase performed by a peer, the peer determines its role using its corresponding cell. In this phase, the peer activates its corresponding cell. After executing the activation procedure of *cell_i*, *peer_i* sets its role using the type of *cell_i* and goes to *maintenance* phase. *peer_i* sets its role to *super* if the type of *cell_i* is equal to *Colony-Manager*. *peer_i* sets its role to *ordinary* if the type of *cell_i* is *Attached-Cell*. *peer_i* sets its role to *unattached* if the type of *cell_i* is *Unattached-Cell*.

- *Maintenance* **phase**: In this phase, an ever going process is executed to handle events occurs for the peer or the neighbors.

Now, we complete the description of the algorithm by describing the **(1) Local rule** and **(2) Structure updating rule** for the *CADCLA* used by activation procedure in the proposed algorithm.

1. **Local rule:** Local rule takes information of immediate neighbors of a cell *cell_i* as input and then returns reinforcement signal (β_i) and restructuring signal (ζ_i) of *cell_i*. In *cell_i*, the restructuring signal ζ_i is set to 1 if the type of that cell is equal to *Colony-Manager* and 0 otherwise. The local rule of *cell_i* returns 1 in three

cases described as follows. In the first case, the capacity of the $cell_i$ is lower than the capacity of majority of neighboring cells, and the state of $cell_i$ is equal to *Colony-Extender*. In the second case, the capacity of the $cell_i$ is higher than the capacity of majority of neighboring cells, and the state of $cell_i$ and the majority of states of immediate neighboring cells are equal to *Colony-Immobilizer*. In the third case, the unused capacity of the $cell_i$ is equal to zero, and the state of $cell_i$ is equal to *Colony-Extender*. In other cases, the reinforcement signal is equal to 0.

2. **Structure updating rule:** Structure updating rule is implemented using an operator called *Absorb* operator. Figure 4.26 shows an example of usage of *Absorb* operator. In this figure, if the restructuring signal of $cell_i$ is equal to 1, and $cell_i$ has unused capacity then the structure updating rule selects $cell_j$ using a function called *candidate-selector()* and then randomly chose some of the neighbors of $cell_j$ and uses the *Absorb* operator to transfer the chosen neighbors to $cell_i$ for filling the unused capacity of $cell_i$. Function *candidate-selector()* is described as follows. This function takes information about the neighbors of a cell and then returns one of the neighbors of that cell as output. If the state of a cell is *Colony-Immobilizer* then function *candidate-selector()* randomly selects one of neighboring cell which its state is equal to *Colony-Immobilizer*, or *Colony-Extender*, and then returns it. If the type of a cell is *Colony-Extender* then function *candidate-selector()* randomly selects one of neighboring cell which its type is equal to *Colony-Extender*, and then returns it. If function *candidate-selector()* couldn't select any cell, the neighbors of $cell_i$ remains unchanged.

Now, we give the detailed descriptions of the proposed algorithm. The pseudo code of the process executed by a $peer_i$ during joining to the network is given in Fig. 4.27. This process consists of three phases: ***initialization*** Phase, ***construction*** phase, and ***maintenance*** phase. The detailed descriptions of these phases are given in the rest of this section.

During *initialization* phase, $peer_i$ establishes its connections to other peers of the network, and initializes its corresponding cell $cell_i$. In the *initialization* phase, $peer_i$ finds some peers using Newscast protocol (Jelasity et al. 2003) to connect the network. Note that the Newscast protocol is also used in *SG-LA*, *SG-1*, and *Myconet*

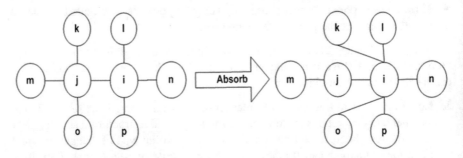

Fig. 4.26 An example for absorb operation

Algorithm peer_management()
Notation: *cell* denotes the cell corresponds to the *peer*;

01	**Begin**
02	//**initialization phase**//
03	Establish the connections of the *peer*;
04	Initialize the *cell* of the *peer*;
05	// **construction phase**//
06	Call function *Activate()* of the *cell*;
	// the pseudo code of this function is given in Figure 4-28//
07	Refine the list of the neighbors of the *peer*;
	// according to the changes made by procedure *Activate*//
08	Set the role of the *peer*;// according to the attribute
	of the *cell*
09	// **maintenance phase**//
10	Wait until some event occurs;
11	**If** (*peer* has detected that it has no neighboring peers) **Then**
12	Goto **initialization phase**;// goto line 02//
13	**EndIf**
14	**If** (*peer* has detected that *join, leave, Absorb Operation, or*
	cellular operation has been occurred in its neighborhood)
	Then
15	Perform appropriate management operation;
	// considering the operation occurred in the neighborhood //
16	Refine the list of the neighbors of the *peer*;
	// considering the effect of the management operation //
17	Goto **construction phase**;// goto line 05 //
18	**EndIf**
19	**If** (*peer* has been activated to exchange information with *peer$_j$*)
	Then
20	Exchange information with *peer$_j$*;
21	Goto **maintenance phase**;// goto line 09//
22	**EndIf**
23	**End**

Fig. 4.27 Pseudo code of the proposed algorithm

for establishing initial connections. After establishing the initial connections, $peer_i$ initializes its corresponding cell $cell_i$. During initializing $cell_i$, the capacity of $cell_i$ is computed and the type of $cell_i$ is set to *Unattached-Cell*. Then $peer_i$ goes to *construction* phase.

During *construction* phase, $peer_i$ determines its role. In the *construction* phase, $peer_i$ executes the activation procedure of its corresponding cell that is $cell_i$. Figure 4.28 shows the pseudo code of the procedure which each cell of *CLA* executes after activation. After executing the activation procedure of $cell_i$, $peer_i$ sets its role using the type of $cell_i$ and goes to *maintenance* phase. $peer_i$ sets its role to super-peer if the type of $cell_i$ is equal to *Colony-Manager*. $peer_i$ sets its role to ordinary-peer if the type of $cell_i$ is *Attached-Cell*. $peer_i$ sets its role to unattached-peer if the type of $cell_i$ is *Unattached-Cell*.

During the *maintenance* phase which is an ever going process, $peer_i$ continually waits for one of the events leaving a peer, joining a peer, request for execution of *Absorb* operation, and request for exchanging information.

If $peer_i$ has detected that it has no neighboring peers, it goes to *initialization* phase. If $peer_i$ has detected that *join, leave, Absorb operation*, or *cellular operation* has been occurred in its neighborhood, it performs appropriate management operation. The management operations are described as below:

- If $peer_i$ has received a request for joining peer from $peer_j$, $peer_i$ connects to $peer_j$.
- If $peer_i$ has detected that one of its neighboring peers has left, $peer_i$ removes information about that neighbors form the list of its neighbors.
- If $peer_i$ has been activated by a $peer_j$ to execute *Absorb* operation, then $peer_i$ executes *Absorb* operation with $peer_j$.
- If $peer_i$ has been activated by a $peer_j$ to execute a cellular operation (such as computing restructuring signal, gathering attributes and etc.) then $peer_i$ executes appropriate operation with $peer_j$.

After executing the management operation, $peer_i$ refines the list of its neighbors considering the effects of the management operations and then goes to the *construction* phase. If $peer_i$ has been activated to exchange information with its neighbors, then $peer_i$ exchanges information with its neighbors and then restarts the *maintenance phase*.

4.3.3   Experimental Results

All simulations have been implemented using *OverSim* (Baumgart et al. 2009). The performance of the proposed algorithm which we call it *X-NET* is compared with four different algorithms *Myconet* (Snyder et al. 2009), *SG-1* (Montresor 2004), *SPS* (Liu et al. 2013), and *SG-LA* (Gholami et al. 2014) among which *SG-1* is a well-known super-peer selection algorithm. The reason for selecting these

---

Algorithm Activate ()

Input: $cell_i$

Output: the list for new neighbors of the *cell*

Notations: $cell_i$ denotes the cell corresponds to the $peer_i$;

F_1 denotes the *local rule*

F_2 denotes the structure updating rule

Ψ_i denotes the attribute of $cell_i$

Φ_i denotes the state of $cell_i$

ζ_i denotes the restructuring signal of $cell_i$

N_i denotes the set of neighbors of $cell_i$

β_i denotes the reinforcement signal of the learning
automata of $cell_i$

Begin

// preparation phase//

Compute Ψ_i;// using using the statemachine given for fungal growth pattern//

Compute ζ_i^1;// using F_1//

Ask from neighboring cells of $cell_i$ to compute their *restructuring signals*;

Gather the *restructuring signals* of the Neighboring cells;

// structure updating phase//

If (the value of ζ_i is 1) **Then**

Compute N_i and Ψ_i;// using F_2 //

EndIf

//state updating phase//

Each *LA* of $cell_i$ chooses one of its actions;

Set Φ_i;// set Φ_i to be the set of actions chosen by the set of
learning automata in $cell_i$//

Compute β_i;//using F_1//

Update the action probabilities of *LAs* of $cell_i$ using β_i;

End

Fig. 4.28 Pseudo code of the procedure which each cell executes after activation

algorithms is that the concept of peer capacity used in these algorithms is similar to the concept of peer capacity used in *X-NET*.

In order to evaluate the performance of the *X-NET*, **Number of Super-Peer (NSP), Peer Transfer Overhead (PTO), Control Message Overhead (CMO),** and **Capacity Utilization (CU)** are used to compare the performance of the *X-NET* with other super-peer selection algorithms. The definitions of these metrics are given below.

- **NSP** is the sum of number of super-peers of the network [z in (4.5)]. The super-peer selection algorithms try to decrease NSP. Higher value of NSP leading to a large set of super-peers which is not appropriate. This metric is used in (Montresor 2004; Snyder et al. 2009; Liu et al. 2013; Gholami et al. 2014).
- **PTO** is the number of peers which are transferred between super-peers. This metric implicitly shows the changes which were made by the operators (such as *Absorb* operator) of the management algorithms. *PTO* can be used to study the changes occur in the configuration of the super-peer network. Higher value of *PTO* indicates higher changes in the configuration of the network which is bad. This metric is used in (Liu et al. 2013; Gholami et al. 2014).
- **CMO** is the number of extra control messages generated by the management algorithm. This metric is used in (Liu et al. 2013; Gholami et al. 2014). Higher value of *CMO* indicates higher traffic in the network.
- **CU** is the ratio of current number of attached clients to total capacity provided by super-peer as given in (4.11). In (4.11), let S denotes number of selected super-peers. This metric is used in (Snyder et al. 2009; Gholami et al. 2014). If the value of *CU* becomes one then the capacity of all super-peers is used. High value for *CU* is preferred.

$$CU = \frac{\#Attached\_OrdinaryPeer}{\sum_{i=1}^{s} c_i} \qquad (4.11)$$

Results reported are averages over 50 different runs. For both *X-NET* and *SG-LA*, each peer is equipped with a variable structure learning automaton of type L_{RP}. The reward parameter a and penalty parameter b *for* L_{RP} are set to 0.25 and 0.25, respectively. To generate the capacities of peers Pareto distribution is used. For Pareto distribution the maximum capacity is set to 100 and the parameter is set to 2.

Experiment 1 is conducted to study the performance of *X-NET* with respect to *NSP, PTO, CMO, CU, Entropy* and *potential energy*. In experiment 2 to experiment 3, *X-NET* is compared with *SG-1, SG-LA, Myconet,* and *SPS* algorithms with respect to *NSP, CMO, PTO,* and *CU*.

Experiment 1:

This experiment is conducted to study the performance of the proposed algorithm with respect to *NSP, PTO, CMO, CU, Entropy* and *potential energy*. In this experiment, the network size is 10,000 and the *power-law* distribution is used to generate the capacities of peers. The results of this experiment are given in Figs. 4.29 and 4.34. According to the results of this experiment, we may conclude the followings.

- Figure 4.29 plots the *Entropy* versus round during the execution of the algorithm. This figure shows that the value of *Entropy* is high at early rounds and gradually decreases. This means that the changes in the role taken by a peer frequently occur during the early rounds and becomes less frequent in the later rounds.
- Figure 4.30 plots the *potential energy* versus round. This figure shows that the value of *potential energy* is high during the early rounds and gradually decreases which indicates that the network approaching a fixed structure.
- Figure 4.31 plots *NSP* versus round during the execution of the algorithm. This figure shows that the value of *NSP* is high at initial rounds but gradually decreases. Lower *NSP* means smaller set of super-peers selected by the algorithm.
- Figures 4.32 and 4.33 show the value of *CMO* and *PTO* per round. At the early rounds, both *CMO* and *PTO* are high. Each *CMO* or *PTO* eventually reaches a fixed value.
- Figure 4.34 show the plot of CU versus round for proposed algorithm. This figure indicates that the value of *CU* approaches one which means that if the proposed algorithm is used as super-peers selection algorithm then every super-peer will eventually reach its full capacity.

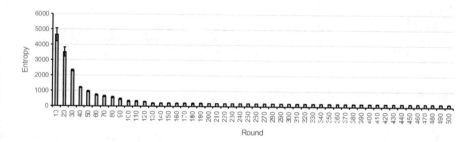

Fig. 4.29 Entropy of the proposed algorithm

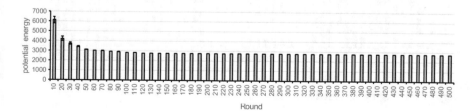

Fig. 4.30 *Potential energy* of the proposed algorithm

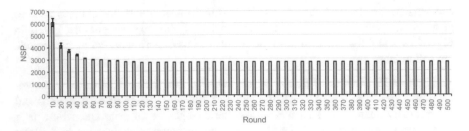

Fig. 4.31 NSP of the proposed algorithm

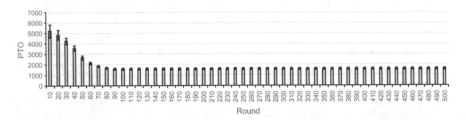

Fig. 4.32 PTO of the proposed algorithm

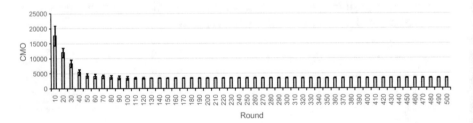

Fig. 4.33 CMO of the proposed algorithm

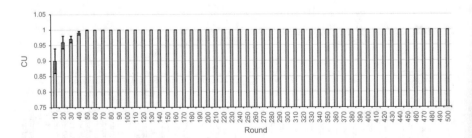

Fig. 4.34 CU of the proposed algorithm

Experiment 2:

This experiment is conducted to study the impact of catastrophic failure on the performance of the proposed algorithm. For this purpose, we removed different percentages (30 and 60) of super-peers from the network at the beginning of round 30 of the simulation. It should be noted that, the removed peers were added (at the same time of removing peers) to the network as unattached-peers. In this experiment, the network size is 10,000 and the *power-law* distribution is used to generate the capacities of peers. The results obtained are compared with the results obtained for *SG-1, SG-LA, SPS,* and *Myconet* algorithms with respect to *NSP, PTO, CMO,* and *CU*. From the result of this experiment given in Figs. 4.35, 4.36, 4.37, 4.38, 4.39, 4.40, 4.41 and 4.42, we may conclude the following:

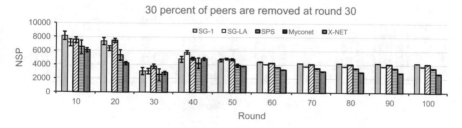

Fig. 4.35 Comparison of different algorithms with X-NET with respect to NSP when 30% of peers are removed

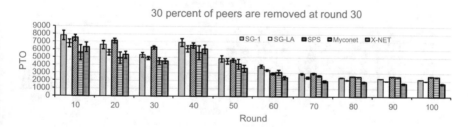

Fig. 4.36 Comparison of different algorithms with X-NET with respect to PTO when 30% of peers are removed

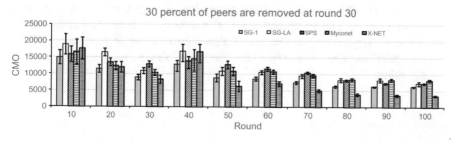

Fig. 4.37 Comparison of different algorithms with X-NET with respect to CMO when 30% of peers are removed

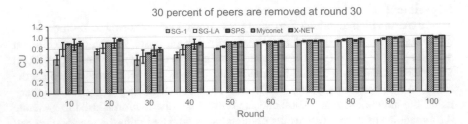

Fig. 4.38 Comparison of different algorithms with X-NET with respect to CU when 30% of peers are removed

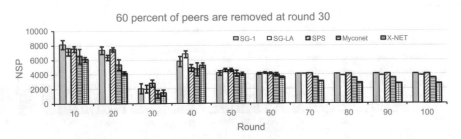

Fig. 4.39 Comparison of different algorithms with X-NET with respect to NSP when 60% of peers are removed

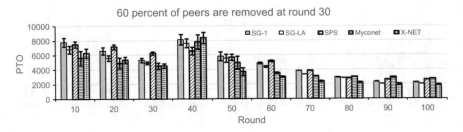

Fig. 4.40 Comparison of different algorithms with X-NET with respect to PTO when 60% of peers are removed

- In terms of *NSP* and *CU*, the proposed algorithm performs better than other algorithms under catastrophic failure. The results also have shown, the number of rounds required by the proposed algorithm in order to reach a appropriate configuration after a catastrophic failure is fewer as compared to other algorithms.

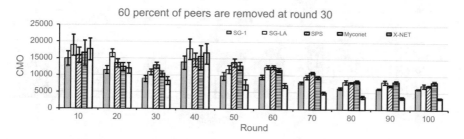

Fig. 4.41 Comparison of different algorithms with X-NET with respect to CMO when 60% of peers are removed

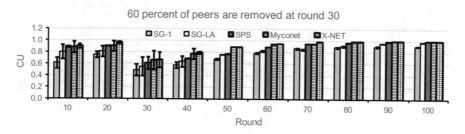

Fig. 4.42 Comparison of different algorithms with X-NET with respect to CU when 60% of peers are removed

Experiment 3:

This experiment is conducted to study the impact of different churn models on the performance of the proposed algorithm. In this experiment, the network size is 10,000 and the *power-law* distribution is used to generate the capacities of the peers. The churn models used in this experiment are described as below.

- **Random churn. 1** is designated based on Random churn model reported in (Baumgart et al. 2009). Random churn model has two parameters: joining _probability and leaving _probability. In Random churn. 1, joining _probability and leaving _probability parameters are set to 0.7 and 0.3 respectively.
- **Pareto churn. 1** is designated based on Pareto churn model reported in (Baumgart et al. 2009). Pareto churn model has two parameters: *LifetimeMean* and *DeadtimeMean*. In Pareto churn. 1, *LifetimeMean* and *DeadtimeMean* parameters are set to 50 and 20 s respectively.

The results obtained are compared with the results obtained for *SG-1, SG-LA, SPS,* and *Myconet* algorithms with respect to *NSP, PTO, CMO,* and *CU.* From the result of this experiment given in Figs. 4.43, 4.44, 4.45, 4.46, 4.47, 4.48, 4.49 and 4.50, we may conclude the following:

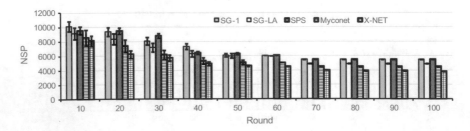

Fig. 4.43 Comparison of different algorithms with X-NET with respect to NSP when random churn.1 is used

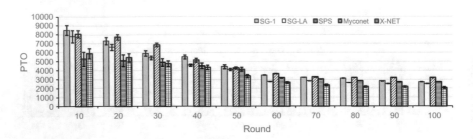

Fig. 4.44 Comparison of different algorithms with X-NET with respect to PTO when random churn. 1 is used

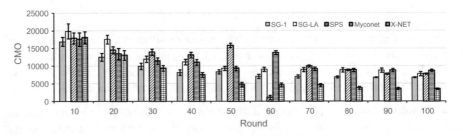

Fig. 4.45 Comparison of different algorithms with X-NET with respect to CMO when random churn. 1 is used

- In terms of *NSP* and *CU* the proposed algorithm performs better than other algorithms.
- The values *CMO* and *PTO* are high at early rounds of operation of the network during which the peers try to gather information about each other for the purpose of searching an appropriate configuration. Higher values for *CMO* and *PTO* throughout the operation of the network especially at the early rounds is the price that we need to pay if we want to find a configuration for which *NSP* attains it lowest possible value and *CU* attains its highest possible value.

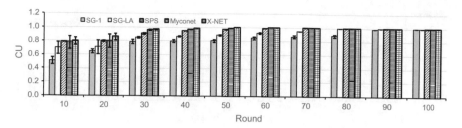

Fig. 4.46 Comparison of different algorithms with X-NET with respect to CU when random churn. 1 is used

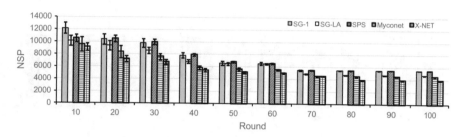

Fig. 4.47 Comparison of different algorithms with X-NET with respect to NSP when pareto churn. 1 is used

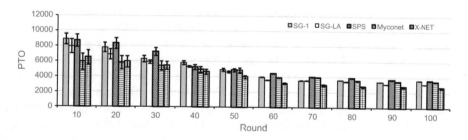

Fig. 4.48 Comparison of different algorithms with X-NET with respect to PTO when pareto churn. 1 is used

4.4   A Cognitive Engine for Solving Topology Mismatch Problem Based on Voronoi Diagrams

In this section, *a Voronoi diagram construction algorithm* is fused with *DCLA* (Saghiri and Meybodi 2017b). *Voronoi diagrams* are well-known diagrams in the field of computational geometry and have several applications in computer networks (Aurenhammer 1991). This section gives a landmark clustering algorithm for solving topology mismatch problem based on *CADCLA-VL*. To show the

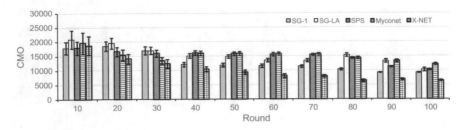

Fig. 4.49 Comparison of different algorithms with X-NET with respect to CMO when pareto churn. 1 is used

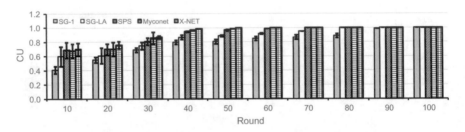

Fig. 4.50 Comparison of different algorithms with X-NET with respect to CU when pareto churn. 1 is used

superiority of the proposed landmark clustering algorithm, it is compared with *mOverlay* (Zhang et al. 2004) and *lOverlay* (Saghiri and Meybodi 2015) algorithms. In order to study the performance of *DCLA*, two metrics: entropy and *potential energy* will be used. The rest of this section is organized as follows. In Sects. 4.4.1 and 4.4.2, we briefly introduce landmark clustering algorithms, and *Voronoi* diagrams. In Sect. 4.4.2, an adaptive landmark clustering algorithm for solving topology mismatch problem for unstructured peer-to-peer networks based on *CADCLA-VL* is described. Section 4.4.3 reports the results of experiments.

4.4.1 Preliminaries

In this section, in order to provide basic information for the remainder of this chapter, we present a brief overview of learning automata, landmark clustering algorithms, and *Voronoi* diagrams.

***Voronoi* Diagrams:**   In the field of computational geometry, constructing the *Voronoi* diagram for a set of n points in the Euclidean plane is one of well-known problems. In a Euclidean plane, the *Voronoi* diagram of a set of points is a collection of cells that divide up the plane. Figure 4.51 shows an example for a

Fig. 4.51 An example for
Voronoi diagram for six
points in a plane

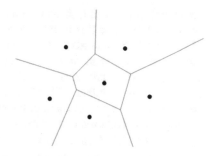

Voronoi diagram for six points in a plane. Each cell corresponds to one of the points. All of the points in one cell are closer to the corresponding cell than to any other cell. This problem has an centralized optimal algorithm with $O(nlogn)$ complexity (Aurenhammer 1991), but the optimal algorithm is not applicable to distributed systems. In a distributed system, computing the *Voronoi* diagram where the nodes are modeled as points in the plane leads to a challenging problem. This is because, in some distributed systems, a node cannot gather information about all nodes of the system and the *Voronoi* construction algorithms for building the *Voronoi* diagram among the nodes use information of all nodes. In the next paragraph, we focus on a type of *Voronoi* construction algorithms reported for distributed systems.

Several distributed algorithms for constructing *Voronoi* diagrams in distributed manner are reported in (Chen et al. 2004; Sharifzadeh and Shahabi 2004; Bash and Desnoyers 2007; Zhou et al. 2009b). In (Sharifzadeh and Shahabi 2004; Bash and Desnoyers 2007), a type of algorithms for constructing *Voronoi* diagram in distributed manner is reported for sensor networks. In this type of algorithms, each node of the network uses information about 1-hop neighbors to build initial local *Voronoi* cell. Then the node modifies it corresponding *Voronoi* cell gradually as it gathers messages containing location information from other neighboring nodes.

Landmark clustering problems: The management algorithms for solving the topology mismatch problem can be classified into three classes (Saghiri and Meybodi 2016b): the algorithms in which each peer in the network uses some services such as *GPS* to gather information about the locations of peers, Landmark clustering algorithms, and the algorithms in which each peer in the network uses local information about its neighbors. The landmark clustering algorithms are described in more details in the next paragraph.

Consider n peers which are connected to each other through an overlay network over an underlay network. The topology of the overlay network can be represented by each graph such as $G = (V, E)$ in which $V = \{peer_1, peer_2, ..., peer_n\}$ is a set of peers and $E \subseteq V \times V$ is a set of links connecting the peers in the overlay network. This graph is called as overlay graph. The topology of the underlay network can be represented by graph $G' = (V', E')$ in which $V' = \{position_1, position_2, ..., position_n\}$ is a set of positions in the underlay network and $E' \subseteq V' \times V'$ is a set of links connecting the positions in the underlay network. In peer-to-peer networks, each peer in V is mapped to a position V' according to a one-to-one function $H : V \rightarrow V'$. In landmark clustering algorithms, the overlay graph is formed by clusters in which each cluster has a landmark peer. The peers which are not selected as landmark peers are called as ordinary peers. The communication between ordinary peer should be handled by landmark peers. Let C_i (where $i \in \{1, 2, 3, 4, \ldots, p\}$) represents the set of p clusters in the network and $L_i \in V$ denotes the landmark peer of cluster C_i. The topology of the landmark peers network can be represented by a graph $G^l = \left(V^l, E^l\right)$ in which $V^l \subseteq V$ is a set of landmark peers and $E^l \subseteq V \times V$ is a set of links connecting the landmark peers in the overlay network. This graph is called landmark graph. In the peer-to-peer network which is constructed based on landmark clustering algorithms, each landmark peer in V^l is mapped to several ordinary peers in V according to an one-to-many function $H' : V^l \rightarrow V$. In other word, each landmark peer in V^l is mapped to several positions in V' using H' and H. Figure 4.52 shows an example of a peer-to-peer network which uses two clusters to manage overlay topology. In this example $peer_m$ and $peer_j$ are two landmark peers. A primary goal of landmark clustering algorithms is to improve the performance of the overlay network in term of communication delay, because topology mismatch problem causes redundant communication delay. Landmark clustering algorithms consists of two parts; landmark selection to select some peers as landmark peers and topology optimization to organize the links between peers. The problem of finding the set of peers as landmark peers which minimizes the total communication delay is similar to a version of super peer selection in super peer network which is proven to be an NP-hard problem (Wolf and Merz 2007). In (Wolf and Merz 2007), This problem was cast as a special case of the Hub Location Problem (O'kelly 1987) which is an NP-Hard problem. Landmark clustering is more complex than classic Hub Location problem, because it must respond to dynamic joins and leaves of peers. For solving landmark clustering problem the following problem should be solved.

$$\text{Min } z = \sum_{i=1}^{n} \sum_{j=1, j \neq i}^{n} \sum_{k=1}^{n} \sum_{m=1}^{n} \left(d_{ik} + d_{km} + d_{mj}\right) \times x_{ik} \times x_{jm} \qquad (4.12)$$

Fig. 4.52 An overlay topology which uses landmark clustering algorithm

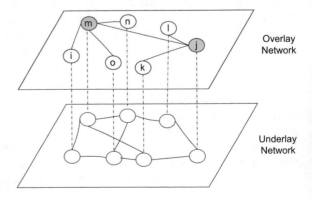

s.t,

$$x_{ij} \leq x_{jj} \quad i, j = 1, \ldots, n \tag{4.13}$$

$$\sum_{j=1}^{n} x_{ij} = 1 \quad i, j = 1, \ldots, n \tag{4.14}$$

$$\sum_{j=1}^{n} x_{jj} = \mathrm{p} \tag{4.15}$$

$$x_{ij} \in \{0, 1\} \quad i, j = 1, \ldots, n \tag{4.16}$$

In Eq. (4.12), z is the total all-pairs end-to-end communication delay of the overlay network. In (4.12), x_{ij} indicates the existence or absence of a connection between *peer$_i$* and *peer$_j$*. If $x_{ij} = 1$ then there is a connection between *peer$_i$* and *peer$_j$*. If $x_{kk} = 1$, *peer$_k$* is chosen as a landmark peer. d_{ij} is the end-to-end delay from *peer$_i$* and *peer$_j$* in the underlay network that is $d_{ij} = D\big(H(peer_i), H(peer_j)\big)$ where function $D : V' \times V' \to \mathbb{R}$ gives the end-to-end delay between pair of positions in the underlay network. Constraint (4.14) which is used in (Zhang et al. 2004) indicates that each ordinary peer connects to exactly one landmark peer. Constraint (4.15) which is used in (Ratnasamy et al. 2002) indicates that there will be exactly p cluster each with own landmark peer.

In Landmark clustering algorithms such as those reported in (Ratnasamy et al. 2002; Xu et al. 2003; Zhang et al. 2004; Tian et al. 2005; Jiang et al. 2006; Scheidegger and Braun 2007; Wolf and Merz 2007; Li and Yu 2011; Ju and Du 2012), the overlay network is organized by clusters in which each cluster has one landmark peer and several ordinary peers. In the landmark clustering algorithms, the peers organize themselves into clusters such that peers that get in within a given cluster are relatively closed to each other in terms of communication delay. These algorithms can be classified into two subclasses: algorithms that use static landmark selection methods (Ratnasamy et al. 2002; Xu et al. 2003; Wolf and Merz 2007; Ju and Du 2012) and algorithms that use dynamic landmark selection methods (Zhang et al. 2004; Scheidegger and Braun 2007; Saghiri and Meybodi 2015). The dynamic landmark selection methods can be further classified into two subclasses: algorithms that use non-adaptive landmark selection methods such *mOverlay* reported in (Zhang et al. 2004), and algorithms that use adaptive landmark selection methods such as *lOverlay* reported in (Saghiri and Meybodi 2015). A drawback of *lOverlay* is that it has many parameters that must be tuned.

4.4.2 Proposed Algorithm: xOverlay

Initially, a *CLA* isomorphic to the overlay graph is created which involves defining the initial structure, local rule, structure updating rule, automata trigger function, restructuring function, and local environments. The overlay graph is then mapped

Fig. 4.53 A snapshot of the CLA at a particular time

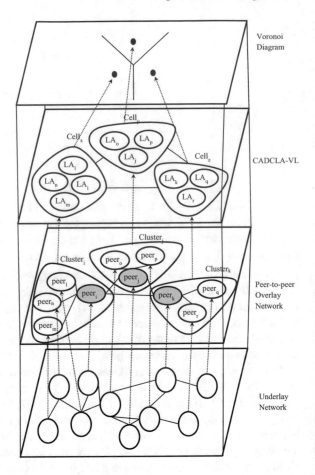

into the *CLA* in which each cell is equipped to a learning automaton. Each peer plays one of three roles: *Undecided*, *Ordinary* and *Landmark*. The initial role of each peer is set to *Undecided*. The learning automata in each cell have two actions "change the role to Landmark" and "change the role to Ordinary". We call the set of nodes in each cell a cluster of peers. Clearly, at the beginning of the algorithms each cluster contains one peer of the overlay graph and hence the number of cluster is equal to the number of cells in the *CLA*. Each cell of the *CLA* has an attribute which contains the geographical positions of the peers of the cell in the underlay graph. Figure 4.53 shows a snapshot of the *CLA* at a particular time. As shown, *cluster$_i$*, *cluster$_j$*, and *cluster$_k$* are corresponding clusters to *cell$_x$*, *cell$_y$*, and *cell$_z$* respectively. *peer$_i$*, *peer$_j$*, and *peer$_k$* are the landmark peers of *cluster$_i$*, *cluster$_j$*, and *cluster$_k$* respectively. Once the *CLA* is created, the proposed algorithm utilizes it to manage the roles of the peers. The process executed by each peer *peer$_i$* when joining to the network consists of three phases: *construction* phase, *organization* phase, and *maintenance* phase (Fig. 4.54). These phases are briefly described as below.

Algorithm *xOverlay* ()

Notation: *Cluster$_x$* denotes the cluster that the *peer* belongs to at any time.
cell denotes the cell corresponds to the *cluster Cluster$_x$*.

01	**Begin**
02	//**Construction phase**//
03	**If** (there is an appropriate cluster *cluster$_x$*) **Then**
04	Connect to the *Cluster$_x$*;
05	Set the role of the *peer* to *Ordinary*;
06	**Else**
07	Set the role of the *peer* to *Landmark*;
08	goto **Maintenance phase**; // goto line 13//
09	**EndIf**
10	//**Organization phase**
11	**Call** *Activate*(*cell*) ;
12	Determine the role of the peers according to the state of *cell*
13	//**Maintenance phase**
14	**While** (management phase is needed) **Do**
15	Broadcast management information of cluster *cluster$_x$*;
16	**If**(reorganizing in the cluster is needed) **Then**
17	Goto **Organization phase**; // goto line line 10//
18	**EndIf**
19	**If** (reconstruction in the cluster is needed) **Then**
20	Goto **Construction phase**; // goto line line 02//
21	**EndIf**
22	**EndWhile**
23	**End**

Fig. 4.54 Simplified pseudo code of the proposed algorithm

- **Construction phase**: During the *construction* phase performed by a peer, the peer tries to find an appropriate cluster and sets an initial role for itself. If the peer couldn't find any cluster, it sets its role to *Landmark*, and goes to *maintenance* phase. If the peer found a cluster, it connects to the cluster, sets its role to *Ordinary*, and goes to *organization* phase.
- **Organization phase**: During the *organization* phase performed by a peer, the peer and other peers that they have the same cluster as the cluster of that peer try to find an appropriate landmark peer for their cluster by activating their corresponding cell in the *CADCLA-VL*. In a cell, the changes were made by the activation function in the *CADCLA-VL* may change the connections of the peer and also select a new landmark peer according to feedbacks received from the network. In other word, As the Algorithm proceeds, the structure of CLA changes and as a result the number of peers in a cell may change. At the end of *organization* phase, the peer goes the *maintenance* phase.
- **Maintenance phase**: In this phase, the peer monitors (and broadcast) required information from (to) all peers of its corresponding cluster, and then execute appropriate management operation. In this phase, the peer may jump to organization or construction phases in some situations.

Now, we complete the description of the algorithm by describing the **Local Rule** and, **Structure Updating Rule** for the *CLA* used by function **Activate (cell)** called by Algorithm *xOverlay*.

- **Local Rule**: Local rule takes information of immediate neighbors of a cell $cell_i$ as input and then returns reinforcement signal (β_i) and restructuring signal (ζ_i) of $cell_i$.

  - In $cell_i$, the reinforcement signal is equal to 1, if the learning automaton of $peer_j$ (returned by the *Voronoi_center_selector* function as *central candidate* peer) has selected "set the role to landmark peer", other learning automata of $cell_i$ have selected "set the role to ordinary peer" and there is no departed peer (returned by the *Voronoi_center_selector* function as d-peer), and 0 otherwise.
  - The *restructuring signal* of $cell_i$ is set to 1 if there is a *departed* peer returned by the function *Voronoi_center_selector* for $cluster_i$, and 0 otherwise. Let $cluster_i$ be the corresponding cluster to $cell_i$ (Fig. 4.55).

Remark 4.4 in a peer-to-peer network, the information about delays among peers cannot be gathered with high accuracy. Inaccurate information results in creating inaccurate *Voronoi* cells. We solved this problem using a function called *distance_tester* which is described as follows. Function *distance_tester* takes two values (m and n), and threshold $q \in [0, 1)$ as input and then return true if $(m \geq n \text{ and } (1 - q) \times m \leq n) \text{ or } (m < n \text{ and } (1 - q) \times n \leq m)$ and false otherwise. This function determines whether two values of delays can be considered equal or not. This function is also used in landmark clustering algorithms such as those reported in (Scheidegger and Braun 2007; Saghiri and Meybodi 2015) (Fig. 4.56).

Algorithm Voronoi_center_selector()

Input:	$group_i$ // the set of nodes of $cell_i$//
	r // a threshold//
	u // a vector containing distances from all nodes of $group_i$ to the hub nodes of the groups adjacent to $group_i$//
Output:	c-node // a candidate node//
	d-nodes // set of departed nodes//
	m // mean of distances computed for node *c-node*//
	v // variance of distances computed for node *c-node*//

Begin

Find the c-node; // a c-node (candidate node) is a node that has the minimum mean and variance of distances to the hub nodes of the groups adjacent to $group_i$ and spoke nodes of $group_i$//

Find the d-nodes; // In $group_i$, a $node_d$ is a d-nodes (*departed* node) if the distances from $node_d$ to the hub nodes of the adjacent groups of $group_i$ is not equal (based on the output of function *distance_ evaluator* given in Figure 4-56) to the distances from the *c-node* to the hub nodes of the adjacent groups of $group_i$//

Return (c-node, d-node, m, v);

End

Fig. 4.55 The pseudo code for function *voronoi_center_selector*

Algorithm *distance_ evaluator*()

Input: r // a threshold//

              x // a real value//

              y // a real value//

Output: z // a value which determines whether two values x and y

                  can be considered equal or not considering parameter r//

Begin

  If $(x \geq y \ and \ (1 - r) \times x \leq y)$**or**

    $(x < y \ and \ (1 - r) \times y \leq x)$ **Then**

        z ← True;

  Else

        z ← False;

  EndIf

  Return (z);

End

Fig. 4.56 The pseudo code for function *distance_ evaluator*

Remark 4.5 In each cluster, function *Voronoi_center_selector* categories the peers of that cluster and returns some information about that cluster. In *cluster$_i$*, this function takes a threshold t, the delays among all peers of *cluster$_i$* and delays from all peers of *cluster$_i$* to the landmark peers of the clusters adjacent to *cluster$_i$* and then returns *m, v, c-peer,* and *d-peers. c-peer* is called *central candidate* peer and it has the minimum mean and variance of delays to the landmark peers of the clusters adjacent to *cluster$_i$* and ordinary peers of *cluster$_i$*. In *cluster$_i$*, the *central candidate* peer is an appropriate landmark peer. *m* and *v* denote the mean and variance of delays computed for peer *c-peer* respectively. *d-peers* is called *departed* peers set which contains *departed* peers. In *cluster$_i$*, a *peer$_d$* is called *departed* peer if the delays from *peer$_d$* to the landmark peers of adjacent clusters of the *cluster$_i$* is not equal (based on the output of function *distance_tester*) to the delays from the *c-peer* to the landmark peers of the adjacent clusters of the *cluster$_i$*. The algorithm used for finding *central candidate* peer is similar to an algorithm used for finding center point of *Voronoi* cells reported in (Sharifzadeh and Shahabi 2004; Bash and Desnoyers 2007).

- **Structure Updating Rule**: This rule is implemented using three operations called **Migrate** operation, **split** operation, and **Merge** operation which are

---

Algorithm Structure updating rule

Input: The neighbors, attributes and restructuring signals of cell$_i$ and its neighbors
Output: Immediate neighbors and attribute of cell$_i$
Notations: Let *cluster$_i$* be the set nodes of *cell$_i$*.
         Let *cluster$_j$* be the set nodes of *cell$_j$*.
         Let d-nodes be the set of departed nodes of *cell$_i$*.
         Let c-nodes be the candidate node of *cell$_i$*.

Begin

   If (the restructuring signal of cell$_i$ is equal to 1) **Then**

  Find set d-peers;// using function *Voronoi_center_selector* given in Figure 4-55.//

   If (d-peers is not empty) **Then**

      Select a peer peer$_k$ from set d-peers randomly;

      Select an appropriate cluster *cluster$_j$* for peer$_k$;

     If (*cluster$_j$* has been found) **Then**

       If (*cluster$_i$* has one peer) **Then**

       Execute **Merge** operation;

       Else

       Execute **Migrate** operation;

      EndIf

     Else

     Execute **Split** operation;

    EndIf

   EndIf

  EndIf

End

Fig. 4.57 The pseudo code of the structure updater

described below. These operations are described as bellows. This rule use the concept of *Voronoi* diagrams to appropriately divide the peers among the clusters in such a way that an appropriate clustered network with low communication delay can be obtained (Fig. 4.57).

1. *Migrate operation*: Figure 4.58 shows an example of the *Migrate* operation. In this figure, *cell$_i$* and *cell$_j$* have participated in a *Migrate* operation. Let *cluster$_i$* and *cluster$_j$* denote the clusters of peers for *cell$_i$* and *cell$_j$* respectively and *peer$_i$* and *peer$_j$* are the *landmark* peers of *cluster$_i$* and *cluster$_j$*, respectively and also *peer$_k$* is returned by function *Voronoi_center_selector* as a *departed* peer for

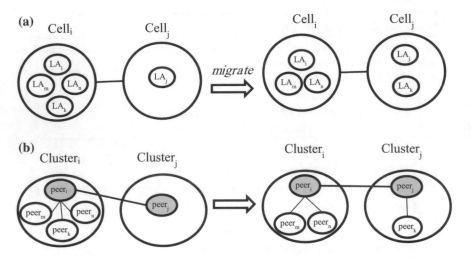

Fig. 4.58 a Example of *migrate* operation for two cells. **b** The effect of migrate operation on the structure of the peer-to-peer overlay network

cluster$_i$. In *cluster$_i$*, the *structure updating rule* of *cell$_i$* uses *Migrate* operation to assign the LA_k (the learning automaton of *peer$_k$*) of *cell$_i$* to cell *cell$_j$*. *cell$_j$* is a neighboring cell of *cell$_i$* whose landmark peer has the least delay to the *landmark* peer of *cell$_i$*. Note that, assigning the LA_k to *cell$_y$* leads to changing the connection of *peer$_k$* from *peer$_i$* to *peer$_j$*.

2. **Split operation**: Figure 4.59 shows an example of the *Split* operation. In this figure, *cell$_i$*, *cell$_j$*, *cell$_w$*, and *cell$_z$* have participated in a *Split* operation. Let *cluster$_i$* and *cluster$_j$* are the clusters of peers for *cell$_i$* and *cell$_j$* respectively, and *peer$_j$* is returned by function *Voronoi_center_selector* as a *departed* peer for *cluster$_i$*. In *cluster$_i$*, if *node$_i$* cannot find an appropriate cell for executing the *Migrate* operation with it, the *structure updating rule* of *cell$_i$* uses *Split* operation to split the *cell$_i$* into two cells; a new cell which contains the learning automaton of *peer$_j$* and a cell which contains the rest of learning automata of *cell$_i$*. Note that the neighbors of the new cell are the neighbors of *cell$_i$* (before execution of *Split* operation).

3. **Merge operation**: Figure 4.60 shows an example of the *Merge* operation. In this figure, *cell$_a$*, *cell$_b$*, *cell$_i$* and *cell$_j$* have participated in a *Merge* operation. The *Merge* operation is performed if after the execution of *Migrate* operation there is no learning automaton in *cell$_i$* (*cell$_j$*).

Detailed descriptions of the xOverlay algorithm is given in (Saghiri and Meybodi 2017b).

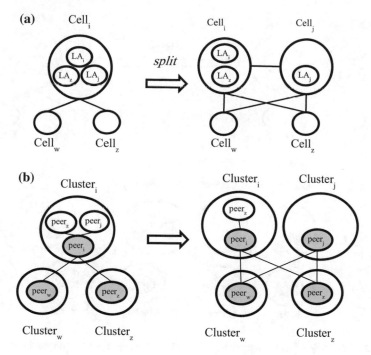

Fig. 4.59 **a** Example of *split* operation. **b** The effect of *split* operation on the structure of the peer-to-peer overlay network

4.4.3 Experimental Results

In this section, computer simulations have been conducted to evaluate the performance of the proposed algorithm (*xOverlay*) and then the results are compared with the results obtained for other existing algorithms. All simulations have been implemented using the *OverSim* which is a flexible overlay network simulation framework (Baumgart et al. 2009). Different types of underlay network models such as *Simple* and *ReaSE* are supported in *Oversim*. The *Simple model* is the most scalable model of underlay networks. We will use one underlying network topology denoted by *T.1(k)* as an example of the *Simple model* where k determines the size of the network. T.1(k): In this underlay topology, k peers are placed on a N-Dimensional Euclidean space and the Internet latencies are based on CAIDA/Skitter data (Huffaker et al. 2002; Mahadevan et al. 2005).

To evaluate the performance of the *xOverlay algorithm*, it is compared with two groups of algorithms. The first group contains the algorithms reported in the literature for solving the landmark clustering problem that is listed below.

- The *mOverlay* algorithm (Zhang et al. 2004) which is a well-known landmark clustering algorithm in unstructured peer-to-peer networks.

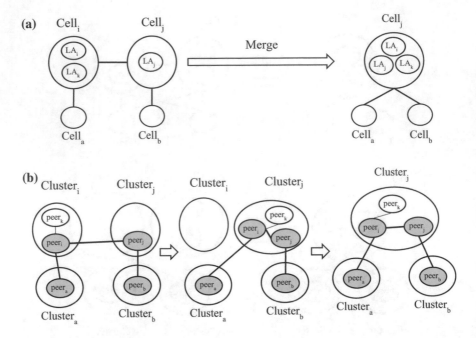

Fig. 4.60 a Example of *merge* operation. **b** The effect of *merge* operation on the structure of the peer-to-peer overlay network

- The *lOverlay* algorithm (Saghiri and Meybodi 2015) reported recently is an extension of the *mOverlay* algorithm.

The second group contains different versions of the *xOverlay algorithm*. These algorithms are used to study different aspects of the proposed learning model used in *xOverlay algorithm*. These visions of *xOverlay* are briefly described below.

- The *nOverlay* algorithm is a version of the *xOverlay algorithm* in which *CADCLA-VL* is replaced with a *pure-chance CADCLA-VL*. A pure-chance *CADCLA-VL* is a *CADCLA-VL* in which, each *learning automaton* is replaced with a *pure-chance* automaton. In this case, each peer participates in the landmark selection procedure with probability 0.5. *nOverlay* will be used to study the learning capability of the *CLA* on the performance of *xOverlay algorithm*.
- The *hOverlay* algorithm is a version of the *xOverlay algorithm* in which the *learning automata* of *CADCLA-VL* are deactivated. In this algorithm, each peer always participates in the landmark selection procedure without considering the actions selected by the *learning automata*. This means that restructuring of the network is performed based on the positions of the peers only. This is in contrast to *xOverlay* in which the restructuring of the network depends on both the positions of the peers and the actions selected by the *learning automata*. *hOverlay* will be used to study the performance of the landmark clustering algorithm when the learning automata are absent.

- *kOverlay*(p) algorithm is a version of *xOverlay* in which the information about each peer such as its position and the actions selected by the *learning automata* residing the cells containing that peer is used with probability p. *kOverlay*(p) will be used to study the effect of the mentioned information on the performance of *xOverlay* algorithm.

The parameters of the *lOverlay* algorithm are set according to Table 4.6. The *xOverlay* algorithm has two main parameters: t, and *THRSHOLD*. Parameter *THRSHOLD* must be set to a large value in order to provide enough time for the peers of the network to communicate with their neighbors. For the experiments, *THRSHOLD* is set to 5 s. For all experiments, each peer is equipped with a variable structure learning automaton of type L_{RI} with reward parameter $a = 0.1$ and the neighborhood radius of all cells of the *CADCLA-VL* is set to 2.

The results reported are averages over 60 different runs. In each run, the algorithms are executed for 100 rounds. The algorithms are compared with respect to four metrics: Total Communication Delay (*TCD*), Control Message Overhead (*CMO*), *Entropy*, and *potential energy*. The definitions of *entropy*, *Potential energy* were previously given in Sect. 2. The definitions of other metrics are given below.

- **Total Communication Delay** is the total of all-pairs end-to-end communication delay of the overlay network. This metric is measured using (4.12).
- **Control** Message **Overhead** is the number of extra control messages generated for the purpose of reconfiguration of overlay network.

The design of experiments is given as follows. Experiment 1 is conducted to study the impact of the learning capability of the learning automata of the *CADCLA-VL* on the performance of the *xOverlay* algorithm with respect to *TCD* and *CMO*. Experiment 2 is designated to study the performance of the *CADCLA-VL* of the *xOverlay* algorithm with respect to *Entropy, and Potential energy*. Experiment 3 is conducted to study the effect of network size on the *mOverlay*, the *lOverlay* and the *xOverlay* algorithms.

Experiment 1:

This experiment is conducted to study the impact of the learning capability of the learning automata of the *CADCLA-VL* on the performance of *xOverlay* algorithm with respect to *TCD* and *CMO*. For this purpose, *xOverlay* algorithm is compared with *nOverlay* and *hOverlay* algorithms. For this experiment parameter q is set to 0.3, and *T.1*(10,000) is used as underlay topology. The results are compared with respect to *TCD, and CMO*. According to the results of this experiment given in Table 4.7, we may conclude the followings.

Table 4.6 The parameters of lOverlay algorithm

Parameters	Values
A	0.1
T	0.3
MAX_ITERATION_LEARNING	1
MAX_ITERATION	1000

Table 4.7 Comparison of different versions of the proposed algorithm

	xOverlay	hOverlay	nOverlay
TCD	173980 ± 223	174300 ± 314	179980 ± 1158
CMO	170113 ± 2952	191232 ± 1521	181132 ± 2115

- In terms of *TCD*, *xOverlay* algorithm performs better than *nOverlay* because in *xOverlay* algorithm each peer decides on its role (to be an ordinary or landmark) adaptively based on the decisions made by its neighboring peers whereas in *nOverlay* at a given time the history of a peer about its role does not affect its decision regarding its role (chooses its role with probability 0.5).
- In terms of *TCD*, the *xOverlay* algorithm performs better than *hOverlay*. This is because in *hOverlay*, unlike *xOverlay* all the learning automata are deactivated and as a result each peer always participates in the landmark selection procedure. In *hOverlay* restructuring of the network is performed based on the positions of the peers only whereas in *xOverlay* restructuring of the network depends on both the positions of the peers and the actions selected by the learning automata.
- The *xOverlay* algorithm has lower *CMO* as compared to *hOverlay* algorithm. This is because in *hOverlay*, each peer always participates in the landmark clustering which results in generating higher number of control messages. *xOverlay* has lower *CMO* as compared to *nOverlay* algorithm. This is because *nOverlay* performs an exhaustive search in order to find appropriate landmarks whereas the search performed by *xOverlay* is guided by *CLA*.

Experiment 2:
This experiment is conducted to study the performance of the *CADCLA-VL* of *xOverlay* algorithm with respect to *Entropy, and potential energy*. For this purpose, *xOverlay* algorithm is compared with *nOverlay* and *kOverlay* (0.8) algorithms. For this experiment parameter q is set to 0.3, and *T.1*(10,000) is used as underlay topology. The results obtained from different rounds of the simulation are reported in Figs. 4.61 and 4.62. According to the results of this experiment, one may conclude the followings.

- In terms of *Entropy*, *xOverlay* performs better than other algorithms. Figure 4.61 indicates that the *Entropy* is high at initial rounds of the simulation and gradually decreases. This means that the changes in the states of *CLA* becomes less frequent as the set of landmark peers found by the algorithm becomes closer to the appropriate set of landmark peers.
- Figure 4.62 shows the changes in *potential energy* during the execution of *xOverlay*. This figure shows that *potential energy* is initially high and approaching zero indicating the fact that the overlay network topology is approaching to a fixed structure. Experimentations have shown that for *nOverlay* and *kOverlay*, *potential energy* shows a gradual decrease but approaching to a fixed structure is very slow.

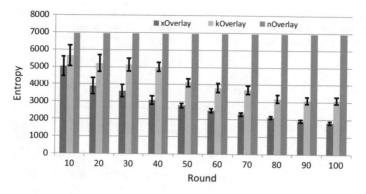

Fig. 4.61 Comparison of xOverlay, kOverlay (0.8), and nOverlay with respect to entropy

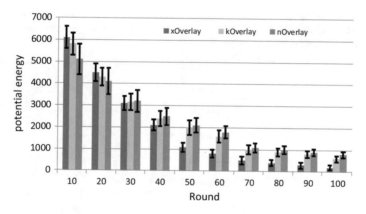

Fig. 4.62 Comparison of xOverlay, kOverlay (0.8), and nOverlay with respect to *potential energy*

Experiment 3:

This experiment is conducted to study the impact of the network size on the performance of the *xOverlay* algorithm when the underlay network topology is generated by the simple model as described before. For this purpose, we generate five topologies from $T.1(k)$ for $k = 10,000, 20,000, 30,000, 40,000$ and $50,000$. For *xOverlay* algorithm, the parameter q is set to 0.3. The results obtained are compared with the results obtained for *mOverlay* and *lOverlay* algorithms with respect to the criteria mentioned above. According to the results of this experiment that are shown in Figs. 4.63 and 4.64. one may conclude that in terms of *TCD*, *xOverlay* algorithm performs better than other algorithms because in *xOverlay* algorithm, *CADCLA-VL* gradually finds appropriate landmark peers for the clusters which results in lower *TCD*. *mOverlay* performs better than other algorithm in terms of *CMO* but worse than other algorithms in terms of *TCD*. This is because *mOverlay* unlike other algorithms is not adaptive and does not have the burden of generating messages for the sake of overlay structure adaptation.

Fig. 4.63 Comparison of
xOverlay with mOverlay and
lOverlay with respect to TCD

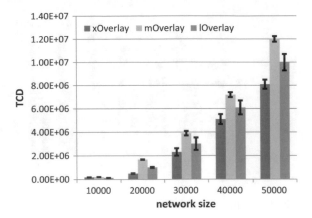

Fig. 4.64 Comparison of
xOverlay with mOverlay and
lOverlay with respect to CMO

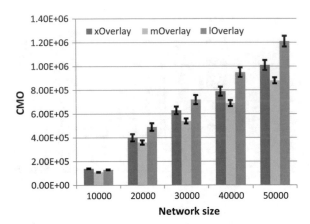

4.5   Conclusion and Future Work

In this chapter, a framework for cognitive peer-to-peer networks was introduced
and then an approach based on *DCLAs* for designing cognitive engines in the
cognitive peer-to-peer networks was proposed. The proposed approach was used to
suggest three cognitive engines for solving topology mismatch and super-peer
selection problems. In order to design the structure updating rule of the *DCLAs*, the
restructuring rules of Schelling segregation model, fungal growth model, and
Voronoi diagrams model were borrowed. Experimental results showed that the
suggested cognitive engine can compete with existing algorithms.

Chapter 5
Learning Automata for Complex Social Networks

5.1 Complex Social Networks

Many real-world complex phenomena in biological, chemical, technological, information and social systems are taken the form of networks which are represented as graphs with a set of nodes (e.g., users of social networks) and edges (e.g., a kind of relations between users of social networks). Plenty of studies have been done on the capturing the structural and dynamical characteristics of complex social networks in various applications (Zheng et al. 2015; Wang et al. 2016; Zhang 2016; Zhao et al. 2016) and lots of real networks are found to show the small-world property (Watts and Strogatz 1998), a power-law degree distribution (Watts and Strogatz 1998) or existence of community structures (Fortunato 2010) in networks. Studying and characterization of the main features of online social networks plays a significant role for analysis of online social networks. For such analysis objectives, online social networks similar to many real world networks are usually modeled and represented as deterministic graphs with a set of nodes as users of social networks and edges as kind of relations between users of social networks.

In online social networks, studying networks using understanding the central users according to social relationship between users is one of the basic analyses and can be applied for identifying influential users for maximizing the spread of influence through a social network. The central users such as influential spreaders of network may not necessarily be a user with the maximum number of friends at a particular time; rather the behavior of user activities over time is more important than the number of friends of users. For example, many users share their daily life activities such as photos, videos, news or collaborates a scientific paper with their friends or colleagues through online social network services. The shared items and activities of audiences such as their likes, comments, joining to communities and making citations to a scientific paper which may change over time and have an impact on the behavior of other users' activities to become interested in favorite items of their friends or collaborators. In online social networks, the degree of

© Springer International Publishing AG 2018
A. Rezvanian et al., *Recent Advances in Learning Automata*, Studies in
Computational Intelligence 754, https://doi.org/10.1007/978-3-319-72428-7_5

friendship among the users changes directly with the weights of their edges to one another over time, and it is clear that sometimes some edges with high weights are more important than that of some others because they are more accessible and willing to communicate or taking friendship activities. In contrast, there are sometimes the same users are in holidays and have their contributions in their social graph have a less edge weight because they are less accessible and decreased their activities in online social networks.

The existing research works for studying and analysis of social networks in the literature show that the most of graph models are usually assumed to be a deterministic graph with fixed weights for its edges or nodes, however online social networks are intrinsically non-deterministic and their structures or behaviors such as online activities of users have unpredictable, uncertain and time varying nature, thus deterministic assumptions for graph model of online social networks with fixed weights are too restrictive to solve most of the real network problems and may lose real information of network due to non-deterministic nature of real world networks. For example in online social networks, the activity behaviors of online users such as following friends, taking a comment on a wall post, liking comments on a given post, posting status posts, visiting user profiles, rating recommendations, asking questionnaires, frequency of notifying events, and frequency of sharing a news, to name a few, vary over time with unknown probabilities (Jin et al. 2013). Analyzing online social networks with deterministic graph models cannot take into consideration the continuum and variety of activities of the users occurring over time. Even modeling complex social networks with weighted graphs in which the edge or node weights are assumed to be fixed weights considers only a snapshot of a real-world network and it is not valid when the activities and behavior of users on networks vary with time.

5.1.1   Stochastic Graphs as a Graph Model for Complex Social Networks

Modeling real networks as deterministic binary networks with fixed weights simplify the metrics, models, algorithms, applications and analyses, further it results in intentionally losing much of the important information contained in the varying edge weights of network reflecting the real nature of the network. Deterministic view about the central users and their activities only focuses on the impact of activities of users at a particular time and ignore the time varying nature of such users' activities and relations. As a solution to this problem, researchers have recently suggested that since understanding the characteristics of social relationship between users is the basis for capturing the realistic analysis of online social networks, stochastic graphs in which the weights associated with the edges are random variables may be a better candidate for modeling complex social networks and stochastic network measures which are random variables provide more information for analyzing the network. Once the network model is chosen to be stochastic graphs, every aspect of the network such as path, clique, spanning tree, dominating set, network measures, sampling algorithms, to mention a few, should be treated stochastically. For example, choosing stochastic graph as the graph model of an online social network and defining community structure in terms of clique (Hao et al. 2016), and the associations among the individuals within the community as random variables, the concept of stochastic clique may be used to study community structure properties. Another example can be identifying influential nodes with respect to network centrality measures, thus, the concept of stochastic network centrality measures may be used to study the influence maximization by identifying influential nodes. Recently, several works regarding modeling social networks as stochastic graphs have been reported in the literature. In some network problems such as maximum clique, shortest paths (Mollakhalili Meybodi and Meybodi 2014) and minimum vertex covering in stochastic graphs have been discussed. In particular network measures and sampling algorithms must be redefined to deal with stochastic graphs directly in order to provide useful and real analysis for social networks.

A weighted undirected graph can be described by a triple $G = \langle V, E, W \rangle$ where $V = \{v_1, v_2, \ldots, v_n\}$ is the set of nodes, $E = \{e_{ij}\} \in V \times V$ is the edge-set in which e_{ij} indicates a connection between node v_i and node v_j, and W is a matrix in which w_{ij} is a weight associated with the edge between nodes v_i and v_j if such an edge exists. Graph G is said to be stochastic if weight w_{ij} associated with edge e_{ij} is a random variable with the probability density function (PDF) q_{ij}.

The set of nodes V can represent users in social networks, authors in citation networks, friends in friendship networks or airline terminal in transportation networks. An edge in edge-set E can represent a type of a connection, relationship, task, activity or flow which can be generated, produced, transferred or terminated through the edges of graph. The weights may be defined according to the tasks/activities on an edge between two nodes such as strength activities between two users in online social networks that it is not always deterministic, the number of collaborations (co-authoring a paper) between scientist authors that may change over time, the amount of trust between friends that varies with time or the links in a communication network that may be affected by collisions, congestions, interferences or some other factors. The weights of connections in real world networks activities for each mentioned process may vary over time. Hence, it seems better to represent these networks as stochastic graphs where the weights associated with the nodes or edges are random variables rather than deterministic graphs.

Torkestani et al. (2012a) designed a learning automata-based heuristic algorithm to solve the minimum spanning tree problem in stochastic graphs in such a way that the algorithm determines the edges that must be sampled at each stage. Furthermore, he defined a variant version of spanning tree problems as the stochastic min-degree constrained minimum spanning tree problem and then presented a decentralized learning automata-based heuristic to solve it (Torkestani 2013b). In Torkestani and Meybodi (2010b), the mobility-based multicast routing algorithm for wireless mobile Ad hoc networks are modeled with a stochastic graph and a stochastic version of the minimum Steiner connected dominating set problem in weighted networks in which the relative mobility of each host is considered as its weight is introduced. Torkestani et al. presented several learning automata-based algorithms to solve the minimum weakly connected dominating set problem in a stochastic graph (Torkestani and Meybodi 2010a). Also they introduced the stochastic version of the minimum connected dominating set problem in vertex weighted graphs, then, they presented several learning automata-based algorithms for solving the problem. They claimed that their algorithm guaranteed neither the domination of all the graph nodes nor the connection of the dominators (Torkestani and Meybodi 2012b). Ni et al. studied two typical minimum weight covering problems under stochastic environments. According to different decision criteria, they presented three stochastic programming models. They also solved these problems using a hybrid intelligent algorithm that integrates stochastic simulation with genetic algorithm (Ni 2012). In Rysz et al. (2016), A combinatorial branch-and-bound algorithm presented for the problem of detecting risk-averse low-diameter clusters in graphs under the assumption that clusters represent k-clubs and that uncertain information manifests itself in the form of stochastic vertex weights whose joint distribution is known. Moradabadi et al. modeled the time series link prediction problem in online social networks as a stochastic graph problem by considering different similarity metrics and link occurrence over time, concurrently. Their solution is established based on a team of continuous action set learning automata for solving a stochastic graph problem (Moradabadi and Meybodi 2016).

5.2   Stochastic Graph Problems

Dynamic nature of many decision-making problems (Elamvazuthi et al. 2013; Ganesan et al. 2014) in real world networks such as computer networks (Lin and Huang 2013), biochemical reaction networks (Hwang and Velázquez 2013), and social networks (Ni and Shi 2013; Liu and Zhang 2014) cause some of their structural and behavioral parameters be time varying parameters and for this reason deterministic graphs for modeling such networks may not be appropriate. In most scenarios of networking science in literatures, it is assumed that the weights associated with the edges of model graph are fixed. Such an assumption is not valid when the tasks/activities on networks vary with time. In such application stochastic version of graph problems can be used for real world application.

In this section, maximum clique and minimum vertex covering in stochastic graphs are first defined and then several learning automata-based algorithms are presented for solving these problems in stochastic graphs under unknown environment (that is, the probability distribution functions of the weights associated with the graph edges are unknown). These algorithms need to take samples from the edges of the stochastic graph in order to find the clique with maximum expected weight or vertex covering with minimum expected weight. These algorithms use a kind of guided sampling implemented with the aid of learning automata in order to reduce the number of samples needed to be taken from the edges of the stochastic graph for finding the clique with maximum expected weight or vertex covering with minimum expected weight. This is done by taking samples from the edges which are along the paths toward the edges of stochastic maximum clique or stochastic minimum vertex covering. In the LA-based algorithms, each vertex of the graph is equipped with a learning automaton whose actions correspond to choosing the edges of the corresponding vertex. The set of learning automata which forms a distributed learning automata tries to guide the sampling process in such a way that to obtain the clique with maximum expected weight or vertex covering with minimum expected weight with as fewer number of samples taken from the edges of the graph as possible. The guided sampling process implemented by distributed learning automata tends to take more samples from the edges along the paths toward the clique with maximum expected weight or the vertex covering with minimum expected weight instead of wandering around and taking unnecessary samples from non-promising edges.

The rest of this chapter is organized as follows. Section 5.2.1 will be devoted to maximum clique problem in stochastic graphs and Sect. 5.2.2 will be devoted to minimum vertex covering problem in stochastic graphs.

5.2.1 Maximum Clique Problem in Stochastic Graphs

By choosing stochastic graph as the graph model of an online social network and defining community structure in terms of clique, and the associations among the individuals within the community as random variables, the concept of stochastic clique may be used to study community structure properties (Rezvanian and Meybodi 2015b). Following are some of basic definitions for maximum clique and its variants.

Definition 5.1 (*Clique*) Let $G = \langle V, E \rangle$ be a connected and undirected graph, where V is the set of vertices, and $E \subset V \times V$ denotes the edge-set. A clique C is a subset of V such that all vertices in this subset are pair-wise adjacent in the graph.

Definition 5.2 (*Stochastic graph*) A stochastic graph G can be described by a triple $G = \langle V, E, W \rangle$, where $V = \{v_1, v_2, \ldots, v_n\}$ is the set of nodes, $E = \{e_{ij}\} \subseteq V \times V$ is the edge-set, and W is a matrix in which w_{ij} is a random variable associated with edge e_{ij} between node v_i and node v_j if such that edge exists.

Definition 5.3 (*Maximum clique*) Let in graph G, there are r distinct cliques $\Theta = \{\theta_1, \theta_2, \ldots \theta_r\}$, *clique* θ^* is maximum clique (MC) if and only if $\theta*$ has the largest cardinality among all possible cliques of G.

Definition 5.4 (*Maximum weight clique*) Let in graph G, $\Theta = \{\theta_1, \ldots, \theta_n\}$ be the set of all possible cliques of graph G and $w_{\theta_i} = \sum_{e_{jk} \in \theta_i} w_{jk}$ be the weight of clique θ_i, where w_{jk} is the edge weight of edge $e_{jk} \in E$. Therefore, clique θ^* is the *Maximum weight clique* (MWC) of graph G if and only if $w_{\theta^*} = \max_{\theta_i \in \Theta}\{w_{\theta_i}\}$.

Definition 5.5 (*Stochastic maximum clique*) Let in stochastic graph G, Θ is the set of cliques of stochastic graph, stochastic clique θ^* is called *stochastic maximum clique* if and only if $w_{\theta^*} = \max_{\theta_i \in \Theta}\{w_{\theta_i}\}$ where w_{θ_i} is the expected weight of clique θ_i $\in \Theta$. That is a *stochastic maximum clique* is a stochastic clique with the minimum expected weight.

Maximum clique (MC) problem is known to be NP-hard (Karp 1972) and for this reason several approximation algorithms have been reported in the literature for solving them. MC of users in online social networks indicates a highly connected group of users as community. Several studies are reported in the literatures for finding maximum clique in deterministic graph (Wu and Hao 2015). Gibbons et al. (Motzkin and Straus 1965) solved MC problem by minimizing a quadratic form over the standard simplex via formulating MWC problem as a standard quadratic optimization problem. A new framework for MWC problem by linear complementary problem is presented in (Massaro et al. 2002). They showed that generically, all stationary points of the MWC problem quadratic program represent strict complementary. Also branch and bound based algorithms are presented in for solving the MC problem to ' (Cheng et al. 2011). A nice review on branch and bound algorithms for the MCP is given in Carmo and Züge (2012). They implemented and compared eight different branch and bound algorithms under a unifying framework.

5.2.1.1 LA-Based Algorithm for Finding Maximum Clique in Stochastic Graphs

In this section, we introduce a distributed learning automata-based algorithm for finding the maximum weight clique (MWC) in the stochastic graph. It is assumed that the edge weight of graph is a positive random variable with unknown probability distribution function. The LA-based algorithm tries to find a clique with the maximum expected weight by sampling the edges of the graph in such a way that the number of samples taken from the edges be minimized. The sampling from the edges will be guided by a distributed learning automata isomorphic to the input graph. Let $G = \langle V, E, W \rangle$ be the input stochastic graph, where $V = \{v_1, v_2, ..., v_n\}$ is the set of vertices, E is the edge set, and $W = \{w_1, w_2, ..., w_m\}$ is a set of random variables each of which is associated to one edge of the input graph. n and m be the total number of vertices and edges in the input graph respectively. The LA-based algorithm uses a DLA which is isomorphic to the input graph. This network can be defined by 2-tuple $\langle A, \alpha \rangle$ where $A = \{A_1, A_2, ..., A_n\}$ is the set of learning automata each of which assigned to a vertex of the input graph; specifically A_i is assigned to

v_i, $\alpha = \{\alpha_1, \alpha_2, \ldots, \alpha_n\}$ is the action-set in which $\alpha_i = \{\alpha_i^1, \alpha_i^2, \ldots, \alpha_i^{r_i}\}$ is the set of actions where r_i is the number actions that can be taken by learning automata A_i for each $\alpha_i \in \alpha$. An action of an automaton corresponds to choosing an edge of the corresponding vertex. At any time each learning automaton can be in one of two modes: active and inactive. Initially, all learning automata are in inactive mode.

The LA-based algorithm consists of four phases: 1. *initialization*, 2. *finding a candidate stochastic clique*, 3. *calculating dynamic threshold*, and 4. *updating action probabilities*. After the initialization is performed the algorithm iterates phases 2, 3, and 4 until stopping conditions are met. In initialization phase learning automata $A = \{A_1, A_2, \ldots, A_n\}$ are created and assigned to the vertices. In phase two a candidate stochastic clique by performing a guided search in the graph with the help of learning automata residing in the vertices of the graph is determined. In phase 3, a threshold which is the average of the weights of all candidate stochastic cliques found in all iterations performed so far is computed and in phase 4, the action probability vectors of visited vertices during phase 1 are updated. Finally the algorithm stops if the stop conditions are met. The details of each phase are described as follows.

A. Initialization

A distributed learning automata $\langle A, \alpha \rangle$ whose topology is isomorphic to the input graph is created. $A = \{A_1, A_2, \ldots, A_n\}$ is the set of learning automata and $\alpha = \{\alpha_1, \alpha_2, \ldots, \alpha_n\}$ where α_i is the action set for automaton A_i and $\alpha_i = \{\alpha_i^1, \alpha_i^2, \ldots, \alpha_i^{r_i}\}$ is the set of actions for learning automaton A_i. Learning automaton A_i which is associated to vertex v_i has r_i actions each of which corresponds to selecting one of the outgoing edges of vertex v_i. Let $p(v_i) = \{p_i^1, p_i^2, \ldots, p_i^{r_i}\}$ be the action probability vector of learning automaton A_i. Initially $p_i^j = 1/r_i$ for all j. Each learning automaton can be in either active or inactive mode. At the beginning of the algorithm all learning automata are in inactive mode. Set θ at any time during the execution of the clique finding algorithm contains the vertices of the stochastic clique found up to that point. θ initially contains vertex v_i which is selected randomly. Weight of θ denoted by $\bar{w}(\theta)$ is defined to be the sum of the weights of the edges between every pair of vertices in θ.

B. Finding a candidate stochastic clique

In this phase, one of the vertices of the input graph is randomly selected and added to θ_k. Then learning automaton associated to this vertex is activated. Let this vertex be v_i. Learning automaton A_i chooses one of its actions according to its action probability vector. Let α_i^j be the chosen action corresponding to edge (v_i, v_j). Each inactive learning automaton connected to A_i removes action α_i^j from its action-set and then scales its action probabilities. Then automaton A_j is activated and selects an action. If adding vertex v_j to θ_k forms a clique then vertex v_j will be added to the candidate stochastic clique θ_k. The process of activating a learning automaton, choosing an action, changing the number of actions for inactive automata connected to the activated chosen automaton and adding a proper vertex to candidate clique is repeated until we cannot activate an automaton. We may not be able to activate an automaton if either the action of the activated automaton is empty or all the vertices of the graph have been visited. The reason for removing the action selected by the activated from the list of actions of automata connected to the activated automata is to avoid the formation of cycles during the traversal of the graph which may lead to a situation that the algorithm repeatedly traverses a part of the graph forever.

C. Calculating dynamic threshold

The *dynamic threshold* at iteration k, $DT(k)$, which is the average of the weights of all stochastic cliques found so far is computed according to Eq. (5.1).

$$DT(k) = \frac{1}{k}\sum_{i=1}^{k} \bar{w}(\theta_i) \tag{5.1}$$

where k is the number of iteration, $\bar{w}(\theta_i)$ is the sum of the weights of the edges between every pair of vertices in θ_i. Dynamic threshold is used to decide whether or not a clique found during an iteration can becomes a candidate for stochastic maximum clique. If the clique found during iteration has a weight greater than the dynamic threshold $DT(k)$ then that clique becomes a candidate for the stochastic maximum clique.

D. Updating action probabilities

In this phase, the probability vectors of learning automata assigned to the visited vertices of the graph are updated according to the learning algorithm on the basis of the result obtained from the previous phase. If $\bar{w}(\theta_k)$ is greater than or equal to the dynamic threshold $DT(k)$, then the actions selected by all learning automata along the traversed path are rewarded and penalized otherwise. At the end of this phase, the actions removed from the set of actions of a learning automaton will be added to the action set of that learning automaton and then its action probability is rescaled to reflect this addition. Also, all activated learning automata will become inactive.

E. Stopping conditions

The algorithm stops when the number of iterations k exceeds a predefined number K_{max} or $PC(k)$ defined by Eq. (5.2) reaches a predefined value P_{max}.

Algorithm 5-1. The LA-based algorithm for finding maximum stochastic clique

Input: Stochastic Graph $G\langle V, E, W\rangle$, Thresholds P_{max}, K_{max}
Output: The stochastic maximum clique
Assumptions:
    Assign an automaton A_i to each vertex v_i and initially set to the inactive mode;
    Let k denotes the iteration number which is initially set to 1;
    Let θ_k denotes the candidate stochastic maximum clique at iteration k;
Begin
    Let $DT(k)$ denotes the dynamic threshold at iteration k initially set to 0;
    Let $PC(k)$ is the product of the probabilities the edges of candidate stochastic maximum clique θ_k;
    Repeat
        Let $\bar{w}(\theta_k)$ is the average weight of all samples taken from θ_k;
        v_i is randomly chosen as the initial candidate stochastic clique θ_k;
        While (able to activate automaton) **Do**
            Automaton A_i is activated and then chooses an action using its action probability vector;
            Let the action chosen by A_i be (v_i, v_j);
            Visit (sample) the chosen edge (v_i, v_j);
            If adding vertex v_j to θ_k forms a clique **then**
                Add vertex v_j to the candidate stochastic clique θ_k;
                Compute the weight for the new clique $\bar{w}(\theta_k)$;
            End if
            Remove action α_j^i from action set of each inactive LA connected to A_i and scale their action probabilities;
            $v_i \leftarrow v_j$;
        End While
        Calculate dynamic threshold $DT(k)$ using equation (5-1);
        If $(\bar{w}(\theta_k) \geq DT(k))$ **Then**
            Reward the chosen actions by all the activated learning automata;
        Else
            Penalize the chosen actions by all the activated learning automata;
        End If
        Calculate $PC(k)$ using equation (5-2);
        $k \leftarrow k + 1$;
        Insert actions of all learning automata and rescale their action probabilities;
        Deactivate all learning automata;
    Until ($PC(k)$ is greater than P_{max} **or** k is greater than K_{max})
End Algorithm

Fig. 5.1 Pseudo-code for the LA-based algorithm for finding maximum stochastic clique

$$PC(k) = \prod_{v_i \in \theta_k} (max(p(v_i))) \tag{5.2}$$

$PC(k)$ is the product of maximum probabilities in probability vectors of learning automata of the vertices of candidate stochastic clique θ_k at iteration k. $p(v_i)$ is the action probability vector of learning automaton A_i residing in vertex v_i.

Figure 5.1 shows the pseudo-code for the LA-based algorithm for finding maximum stochastic clique.

The performance of any learning automata based algorithm is very sensitive to learning rate of the learning algorithm. Large value for learning rate results in increasing the speed of convergence and decreasing the accuracy of the algorithm, while small value for learning rate results in increasing the accuracy and decreasing the speed of convergence of the algorithm. In the LA-based algorithm, all learning automata use a same learning rate; it means that the algorithm considers a same

Table 5.1 Stochastic graphs
used in simulations

Graphs	Number of vertices	Number of edges
Alex1-B	8	14
Alex2-B	9	15
Alex3-B	10	21

degree of importance for the edges being traversed during the traversal of the graph. One can use different adaptive strategy for learning rate adjustment for improving results as reported in Rezvanian and Meybodi (2015b).

5.2.1.2 Experiments

In this section, performance of the LA-based algorithm for solving stochastic maximum clique problem is investigated on a number of stochastic graphs. Table 5.1 describe the stochastic graphs are used for the experimentations and their characteristics. These graphs are well-known stochastic graphs which are borrowed from Hutson and Shier (2006), with their weights of edges being random variables.

All simulations are carried out 30 times and their averages of each parameter are reported for the experiments. The stopping conditions of LA-based algorithm is either the number of iteration reached 100,000 iterations or the value of PC (k) which mentioned as Eq. (5.2) for product of maximum probabilities in probability vectors of learning automata of the vertices of candidate stochastic clique θ_k at iteration k is greater than 0.9. All experiments for the LA-based algorithm are evaluated in terms of total number of samples taken from graph. For all algorithms, the reinforcement scheme used for updating the action probability vector is L_{R-I} and the learning rate a is set to 0.1.

In the experiment, we compare the number of samples needed by the LA-based algorithm in order to achieve a given accuracy and then compare it with the number of samples required by standard sampling method (SSM) to reach the same accuracy. In order to have a fair comparison, both the LA-based algorithm and SSM must use the same confidence level $1 - \varepsilon$. The LA-based algorithm may reach the confidence level of $1 - \varepsilon$ by a proper choice of learning rate a. According to (Torkestani and Meybodi 2012b), such a learning rate can be estimated using equation $ax/(e^{ax} - 1) = max_{i \neq j}(d_j/d_i)$ where d_i is the reward probability of action α_i. Based on the SSM, to obtain a certain confidence level $1 - \varepsilon$ for a subset of edges in graph, we need to build a confidence level $1 - \varepsilon_i$ for each edge e_i such that $\sum_{i=1}^{n} \varepsilon_i = \varepsilon$. In this experiment, same confidence level $1 - \varepsilon_0$ is assumed for all edges of the stochastic graph. The minimum required number of samples for each edge of graph for SSM is calculated subject to $p[|\bar{x}_n - \mu| < \delta] \geq 1 - \varepsilon$, where $\delta = 0.001$.

For this experimentation, error rate ε varies from 0.4 to 0.1 with increment 0.1 for the LA-based algorithm. Table 5.2 shows the results of this experimentation. From the results, one may say for LA-based algorithm, the total number of samples

Table 5.2 Number of samples taken from different test graphs using the LA-based algorithms and SSM for solving stochastic maximum clique problem with different confidence levels

Confidence level	Alex1-B		Alex2-B		Alex3-B	
	LA-based algorithm	SSM	LA-based algorithm	SSM	LA-based algorithm	SSM
0.50	466	3171	687	3413	838	3802
0.60	534	3187	723	3437	982	3867
0.70	713	3118	891	3418	1141	3887
0.80	881	3152	1049	3464	1287	3953
0.90	914	3404	1131	3743	1365	4201
0.95	1079	3445	1291	3773	1447	4214

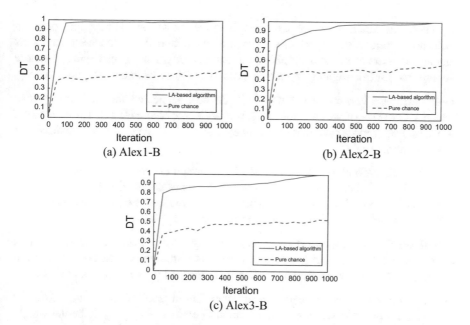

Fig. 5.2 Comparison of LA-based algorithm with LA-based algorithm in which learning automata are replaced with pure chance automata for solving stochastic maximum clique problem

taken from a graph is fewer than the number of samples taken using standard sampling method (SSM) for all cases of graph size and confidence level.

It is also conducted another experiment to compare LA-based algorithm with LA-based algorithm in which the leaning automaton residing in each vertex is replaced by a pure chance automaton. In pure chance automaton the actions are always chosen with equal probabilities (Narendra and Thathachar 1989). The comparison is made with respect to dynamic threshold (DT) scaled between (0, 1) by dividing DT by the weight of maximum clique. The plot of DT versus k given in Fig. 5.2 shows the role of learning automata in guiding the process of sampling from edges of the graph for finding the stochastic maximum clique. With the aid of learning automata the process of finding the stochastic maximum clique is done with fewer numbers of samples taken from the graph as compared to the case where learning is absent. Similar results for other algorithms can also be obtained.

Experimental results showed that the LA-based algorithm can successfully reduce the number of samples needed to be taken from the edges of the stochastic graph as compared to the number of samples needed by standard sampling method at a given confidence level.

5.2.2   Minimum Vertex Covering in Stochastic Graphs

By choosing stochastic graph as a graph model of social network and defining influential users with respect to coverage of users and the influence associations among the users as random variables, the concept of stochastic cover may be more appropriate for investigating spreading properties (Rezvanian and Meybodi 2015a).

Definition 5.6 (*Vertex cover*) Let $G = \langle V, E \rangle$ be a connected and undirected graph, where V is the set of vertices, and $E \in V \times V$ denotes the edge-set. *Vertex cover C* is a subset of vertices such that for every $e(v_i, v_j) \in E$ at least one of the vertices v_i or v_j is an element of C.

Definition 5.7 (*Stochastic graph*) A stochastic graph G can be described by a triple $G = \langle V, E, W \rangle$, where $V = \{v_1, v_2, ..., v_n\}$ is the set of nodes, $E = \{e_{ij}\} \subseteq V \times V$ is the edge-set, and W is a matrix in which w_{ij} is a random variable associated with edge e_{ij} between node v_i and node v_j if such that edge exists.

Definition 5.8 (*Minimum vertex cover*) Let in graph G, there are r distinct vertex covers $\Theta = \{\Phi_1, \Phi_2, ... \Phi_r\}$, *vertex cover Φ^** is minimum vertex cover (MVC) if and only if Φ^* has the minimum cardinality among all possible vertex covers of G.

Definition 5.9 (*Minimum weight vertex cover*) Let in graph G, $\Theta = \{\Phi_1, ..., \Phi_n\}$ be the set of all possible vertex covers of graph G and $w_{\Phi_i} = \sum_{v_j \in \Phi_i} w_j$ be the weight of vertex cover Φ_i, where w_j is the vertex weight of vertex $v_j \in V$. Therefore, vertex cover Φ^* is the *Maximum weight vertex cover* (MWVC) of graph G if and only if $w_{\Phi^*} = \min_{\Phi_i \in \Theta} \{w_{\Phi_i}\}$

Definition 5.10 (*Stochastic minimum vertex cover*) Let in stochastic graph G, Θ is the set of vertex covers of stochastic graph G, stochastic vertex cover Φ^* is called *stochastic minimum vertex cover* if and only if $w_{\Phi^*} = \min_{\Phi_i \in \Theta} \{w_{\Phi_i}\}$ where w_{Φ_i} is the expected weight of vertex cover $\Phi_i \in \Theta$. That is a *stochastic minimum vertex cover* is a vertex cover with the minimum expected weight.

Due to natural computational complexity of NP-hard problems of minimum vertex covering (MVC) (Karp 1972), several algorithms have been reported in the literature for solving it which most of them are based on heuristics for reaching near optimal solutions in a reasonable computation time. The MVC and MWVC problems are used in a wide-spread of many real-world applications such as scheduling (Kuhn and Mastrolilli 2013), facility location (Gupta et al. 2011), network flow (Bentz et al. 2012), network security (Jadliwala et al. 2013), cascading failures (Veremyev et al. 2014), belief propagation (Jin-Hua and Hai-Jun 2014), propagation of ideas through a social network (Kim et al. 2014) to name a few. MVC of users in online social networks indicates a high influential set of users. The LA-based algorithm for finding minimum vertex covering in stochastic graphs is described as following subsection.

5.2.2.1 LA-Based Algorithm for Finding Minimum Vertex Covering

Let $G\langle V, E, W \rangle$ be the input stochastic graph, where $V = \{v_1, v_2, ..., v_n\}$ is the set of vertices, E is the edge set, and $W = \{w_1, w_2, ..., w_n\}$ is a set of random variables each of which is associated to a vertex of the input graph. n and m are the total number of vertices and edges in the input graph respectively. It is supposed that the weight of each vertex is a positive random variable with unknown probability distribution function. The LA-based algorithm uses a network of learning automata isomorphic to the input graph and tries to find a vertex cover with minimum expected weight by sampling the vertices of the graph in such a way that the number of samples taken from the vertices of graph be minimized. An action of automaton A_i assigned to vertex v_i corresponds to the selection of one of the neighboring vertices of vertex v_i. The process of sampling from the vertices of graph is guided by the set of learning automata assigned to the vertices of the input graph. In the LA-based algorithm at any time each learning automaton can be in either active or inactive modes. The LA-based algorithm consists of four steps as described below.

A. Initialization

A network of learning automata defined by 2-tuple $\langle A, \alpha \rangle$ isomorphic to the input graph is created. $A = \{A_1, A_2, ..., A_n\}$ is the set of learning automata each of which assigned to a vertex of the input graph; specifically A_i is assigned to v_i, $\alpha = \{\alpha_1, \alpha_2, ..., \alpha_n\}$ is the action-set in which $\alpha_i = \{\alpha_i^1, \alpha_i^2, ..., \alpha_i^{r_i}\}$ is the set of actions where r_i is the number of actions that can be taken by learning automata A_i. Action α_i^j is the selection of vertex v_j by automaton A_i. Let $p(v_i) = \{p_i^1, p_i^2, ..., p_i^{r_i}\}$ be the action probability vector of learning automaton A_i, where p_i^j is initialized to $1/r_i$ for all j. Let Φ_k be the candidate stochastic vertex cover at iteration k and initially set empty. At first all learning automata are in inactive mode. Also a new vertex v_t is added to the input graph and then connected to all vertices of the input graph. A pure chance learning automaton with n actions each of which correspond to one of the vertices of the input graph will be assigned to this vertex. A pure chance learning automaton is a learning automaton which always selects its actions with equal probability (Narendra and Thathachar 1989).

The insertion of the new vertex v_t allows the algorithm to continue the traversal of the input graph when it encounters situations like entering a node with degree one or two which causes the current iteration of the algorithm terminates and a new iteration with a new randomly selected starting vertex begins. This causes the information gathered during that iteration to be lost and results in prolonging the process of finding the solution by the algorithm.

B. **Construction of stochastic vertex cover**

In this step, a starting vertex called v_i is randomly selected from one of the vertices of the input graph. Then v_i is inserted into candidate stochastic vertex cover Φ_k, the neighbors of v_i are inserted into N_k and learning automata A_i corresponding to v_i is activated. Learning automaton A_i chooses one of its actions α_i^j (selection of node v_j adjacent to vertex v_i) according to its action probability vector. Each inactive learning automaton connected to A_i disables action α_i^j from its action set and then scales its action probabilities and also disables those actions that cause selection of the vertices belongs to the N_k except v_t and scale its action probabilities. After that, automaton A_j corresponding to vertex v_j is activated and selects an action. Let this action be selecting vertex v_l. If selected vertex v_l is not v_t, then vertex v_l will be added to the candidate stochastic vertex cover Φ_k and their new neighboring vertices to N_k. The process of activating a learning automaton, choosing an action, changing the number of actions for inactive automata connected to the activated automaton and neighbors N_k, adding a vertex to candidate vertex cover Φ_k and its neighbors to N_k is repeated until the algorithm cannot activate another automaton. The algorithm may not be able to activate an automaton if either Φ_k is a vertex cover or action set of the activated automaton becomes empty.

C. **Computation of dynamic threshold**

Dynamic threshold is used by the algorithm to judge the quality of current solution found by the algorithm. Dynamic threshold at iteration k, $DT(k)$, which is the average of the weights of all stochastic vertex cover found so far is computed according to the following equation.

$$DT(k) = \frac{1}{k}\sum_{i=1}^{k} \bar{w}(\Phi_i) \tag{5.3}$$

where k is the number of iteration, $\bar{w}(\Phi_i)$ is the sum of the weights of the vertices in Φ_i. Dynamic threshold is used to decide whether or not a vertex cover found during an iteration of the algorithm can become a candidate for stochastic minimum vertex cover. If the vertex cover found during iteration k has a weight equal or lower than the dynamic threshold $DT(k)$ then that vertex cover becomes a candidate for the stochastic minimum vertex cover Φ^*.

D. Updating action probabilities

In this step, the probability vectors of learning automata assigned to the visited vertices of the input graph are updated according to the learning algorithm on the basis of the results obtained from the previous steps. If $\bar{w}(\Phi_k)$ is lower than or equal to dynamic threshold $DT(k)$, then the actions selected by all automata along the traversed path are rewarded and penalized otherwise. At the end of this step, the disabled actions from the action set of activated learning automata will be enabled and their action probability vectors are rescaled to reflect enabling the disabled actions. Finally, all activated learning automata will become inactive.

E. Stopping the algorithm

The algorithm stops when the number of iterations k exceeds a specified number K_{max} or $PC(k)$ defined by Eq. (5.4) reaches a given value P_{max}.

$$PC(k) = \prod_{v_i \in V, v_j \in \Phi_k} \left(max\left(p_i^j\right) \right) \tag{5.4}$$

$PC(k)$ is the product of maximum probabilities of probability vectors of learning automata of the vertices connected directly to the vertices of candidate stochastic vertex cover Φ_k at iteration k. p_i^j is the action probability vector of learning automaton A_i residing in vertex v_i which is connected to v_j.

Figure 5.3 shows the pseudo-code for the LA-based algorithm for finding stochastic minimum vertex cover.

5.2.2.2 Experiments

In this section, to evaluate the performance of the LA-based algorithm, we conducted number of experiments on synthetic stochastic graphs. The stochastic graphs used for the experimentations and their characteristics are given in Table 5.3 which are borrowed from Hutson and Shier (2006), with their weights of vertices being random variables.

All experiments are conducted 30 times and their averages of number of samples taken from graph are reported. The stopping criteria of LA-based algorithm is either the number of iteration reached 10,000 iterations or the value of $PC(k)$ as given by Eq. (5.4) is greater than 0.90. For LA-based algorithm, the reinforcement scheme used for updating the action probability vector is L_{R-I} and the learning rate a is set to 0.01.

This experiment is conducted to study the performance of the LA-based algorithm and standard sampling method (SSM) for different confidence levels in terms of total number of samples taken from graph. The results of this experiment are given in Table 5.4 for *Alex1-B*, *Alex2-B* and *Alex3-B*. From the results we may conclude that LA-based algorithm successfully finds the minimum vertex covering with fewer samples need to be taken from the graph in comparison with SSM.

Algorithm 5-2. The LA-based algorithm for finding stochastic minimum vertex cover

Input: Stochastic Graph $G\langle V, E, W\rangle$, Thresholds P_{max}, K_{max}

Output: The stochastic minimum vertex cover Φ^*

Assumptions:

   Create a network of learning automata isomorphic to graph G by assigning an automaton A_i to each vertex v_i;

   Insert a new vertex v_t to graph G and connect it to all other vertices and assign a pure chance learning automata to vertex v_t;

   Let k denotes the iteration number and initially set to 1;

   Let Φ_k denotes the candidate stochastic minimum vertex cover found at iteration k;

   Let N_k denotes the neighboring vertices of Φ_k;

   Let $\overline{w}(\Phi_k)$ denotes the average weight of all samples taken from all vertices in Φ_k;

   Let Φ^* denotes the stochastic minimum vertex cover and initially is empty;

Begin

   Let $DT(k)$ denotes the dynamic threshold at iteration k and initially set to 0;

   Let $PC(k)$ is the product of maximum probabilities of LA of the vertices connected directly to the vertices in Φ_k at iteration k;

   Inactive all learning automata;

   Repeat

      Select v_i randomly as starting vertex;

      $\Phi_k \leftarrow v_i$;

      Insert neighboring vertices of v_i into N_k

      While (a learning automaton can be activated) **Do**

         Learning automaton A_i is activated and then chooses an action according to its action probability vector;

         Let the action chosen by A_i be the action corresponding to vertex v_j;

         Take a sample from vertex v_j;

         Disable the action corresponding to selecting vertex v_i and vertex set N_k and then scale their action probabilities;

         Learning automaton A_j is activated and then selects an action using its action probability vector;

         Let the action selected by A_j be selecting vertex v_l;

         Take a sample from vertex v_l;

         If ($v_l \neq v_t$) **then**

            $\Phi_k \leftarrow \Phi_k \cup v_l$;

            Insert neighboring vertices of v_l into N_k;

         End if

         Disable the action corresponding to selecting vertex v_j and vertex set N_k and then scale their action probabilities;

         $v_j \leftarrow v_l$;

      End While

      Compute the weight of stochastic vertex cover $\overline{w}(\Phi_k)$;

      Compute dynamic threshold $DT(k)$ using equation (5-3);

      If ($\overline{w}(\Phi_k) \geq DT(k)$) **Then**

         Reward the chosen actions by all the activated learning automata;

      Else

         Penalize the actions chosen by all the activated learning automata;

      End If

      Calculate $PC(k)$ using equation (5-4);

      Enable actions of all learning automata and rescale their action probability vector according to equation (1-4);

      Deactivate all learning automata;

      If ($\overline{w}(\Phi_k) \leq \overline{w}(\Phi^*)$) **Then**

         $\Phi^* \leftarrow \Phi_k$;

      End If

      $k \leftarrow k + 1$;

   Until ($PC(k)$ is greater than P_{max} **or** k is greater than K_{max})

End Algorithm

Fig. 5.3 Pseudo-code for LA-based algorithm for finding stochastic minimum vertex cover

The number of samples required by the LA-based algorithm and the standard sampling method (SSM) in order to reach a particular accuracy is also compared. In order to have a fair comparison, both the LA-based algorithm and SSM must use the same confidence level $1 - \varepsilon$. The LA-based algorithm may reach the confidence level of $1 - \varepsilon$ by a proper choice of learning rate a. According to (Torkestani and

Table 5.3 Stochastic graphs used in simulations

Graphs	Number of vertices	Number of edges
Alex1-B	8	14
Alex2-B	9	15
Alex3-B	10	21

Table 5.4 Number of samples taken from different test graphs using the LA-based algorithm and SSM for solving minimum vertex covering with different confidence levels

Confidence level	Alex1-B		Alex2-B		Alex3-B	
	LA-based algorithm	SSM	LA-based algorithm	SSM	LA-based algorithm	SSM
0.50	863	3171	886	3413	702	3802
0.60	1257	3187	1125	3437	1089	3867
0.70	1634	3118	1842	3418	1298	3887
0.80	2246	3152	2173	3464	1679	3953
0.90	2749	3404	2459	3743	1934	4201
0.95	2961	3445	2714	3773	2387	4214

Meybodi 2012b), such a learning rate can be estimated using equation $ax/(e^{ax} - 1) = max_{i \neq j}(d_j/d_i)$ where d_i is the reward probability of action α_i. To obtain a confidence level not smaller than $1 - \varepsilon$ for the minimum vertex cover, it is sufficient to build a confidence with level $1 - \varepsilon_i$ for every vertex v_i such that $\sum_{i=1}^{n} \varepsilon_i = \varepsilon$, where n is the number of vertices of vertex cover.

Proof The required number of samples for each vertex of graph to satisfy a confidence level $1 - \varepsilon_i$ can be obtained using the vertex sampling method described below.

   Vertex sampling method Let $(x_1, x_2, \ldots, x_N)$ be N random samples of random variable X_i associated with weight of vertex v_i having unknown mean μ and variance σ^2. If $\bar{x}_N \pm \frac{\sigma}{\sqrt{N\varepsilon_i}}$, where $\bar{x}_N = \frac{1}{N}\sum_{j=1}^{N} x_j$ is a $(1 - \varepsilon_i)\%$ confidence interval for mean μ, then for each sufficiently small value of δ, there exists a positive number N_0 such that

$$\text{Prob}[|\bar{x}_N - \mu| < \delta] > 1 - \varepsilon_i \tag{5.5}$$

for all $N \geq N_0$.

   The problem is then to find a confidence region for each of the edges under which a desired confidence level $1 - \varepsilon$ is guaranteed for the minimum vertex cover. The confidence region for the minimum vertex cover is defined as the intersection $\bigcap_{i=1}^{n} C_i(\varepsilon_i)$ of the confidence regions for the edges, where $C_i(\varepsilon_i) = 1 - \varepsilon_i$ denotes the confidence region for vertex v_i and n denotes the number of vertices of the stochastic vertex cover. Using Booles-Bonferroni inequality (B Alt 1982), we have

$$\min_{1 \leq i \leq n-1} (1 - \varepsilon_i)\text{Prob}\big[\mu_i \in \cap_{i=1}^{n} C_i(\varepsilon_i)\big] \geq 1 - \sum_{i=1}^{n} \text{Prob}[\mu_i \notin C_i(\varepsilon_i)] \qquad (5.6)$$

and so

$$\min_{1 \leq i \leq n-1} (1 - \varepsilon_i)\text{Prob}\big[\mu_i \in \cap_{i=1}^{n} C_i(\varepsilon_i)\big] \geq 1 - \sum_{i=1}^{n} \varepsilon_i \qquad (5.7)$$

Hence, the confidence level of the minimum vertex cover is not smaller than $1 - \sum_{i=1}^{n} \varepsilon_i$. In this theorem, the objective is to obtain a confidence level not smaller than $1 - \varepsilon$ for the minimum vertex cover. To achieve this, according to the Bonferroni Correction (Bonferroni 1936) it is sufficient to build a confidence with level $1 - \varepsilon_i$ for each vertex v_i such that $\sum_{i=1}^{n} \varepsilon_i = \varepsilon$, and hence the proof of the theorem. $\square$

In what following, we describe estimation method of learning rate a in order to reach a given confidence level. Before describing the estimation method of learning rate, we need to prove the following theorem.

Theorem 5.1 Let $q_i(k)$ be the probability of formation the vertex cover C_i at iteration k, and $1 - \varepsilon$ is the probability with which LA-based algorithm converges to the vertex cover θ_i. if $q(k)$ is updated according to LA-based algorithm, then there exist a learning rate $a \in (\varepsilon, q)$ for every error parameter $\varepsilon \in (0, 1)$, so that

$$\frac{xa}{e^{xa} - 1} = \max_{j \neq i} \left(\frac{d_j}{d_i} \right) \qquad (5.8)$$

Where $1 - e^{-xq_i} = (1 - e^{-x})(1 - \varepsilon)$ and $q_i = [q_i(k)|k = 0]$.

Proof Let $V[u]$ is defined as

$$V[u] = \begin{cases} \frac{e^u - 1}{e^u} & u \neq 0 \\ 1 & u = 0 \end{cases} \qquad (5.9)$$

and also according to (Narendra and Thathachar 1989) we have

$$\frac{1}{V[x]} = \max_{j \neq i} \left(\frac{d_j}{d_i} \right) \qquad (5.10)$$

It has been proved in Rezvanian and Meybodi (2015a, b) that there always exists a $x > 0$ under which the Eq. (5.8) is satisfied, if $\frac{d_j}{d_i} < 1$, for all $j \neq i$. then we conclude that

$$\frac{e^{-xq_i/a} - 1}{e^{-x/a} - 1} \leq \Gamma_i(q) \leq \frac{1 - e^{-xq_i}}{1 - e^{-x}} \tag{5.11}$$

where $\Gamma_i(q)$ is the probability with which the LA-based algorithm converges to the unit vector e_i with initial vector q and defined as

$$\Gamma_i(q) = prob[q_i(\infty) = 1|q(0) = q] = prob[q^* = e_i|q(0) = q] \tag{5.12}$$

q_i is the initial probability of the stochastic minimum vertex cover C_i. It is assumed that the probability with which the LA-based algorithm with learning rate a^* converges to the stochastic minimum vertex cover θ_i is $1 - \varepsilon$, for each $0 < a < a^*$, where $a^*(\varepsilon) \in (0,1)$. So it is concluded that,

$$\frac{1 - e^{-xq_i}}{1 - e^{-x}} = 1 - \varepsilon \tag{5.13}$$

It is also assumed that for every error parameter $\varepsilon \in (0,1)$ there exists a value x under which the Eq. (5.8) is satisfied. So, we have

$$\frac{ax}{e^{ax} - 1} = \max_{j \neq i}\left(\frac{d_j}{d_i}\right) \tag{5.14}$$

It is concluded that there exists a learning rate $a \in (\varepsilon, q)$ for every error parameter $\varepsilon \in (0, 1)$ under which the probability with which the LA-based algorithm is converged to stochastic minimum vertex cover is greater than $1 - \varepsilon$ and so Theorem 5.1 is proved. □

According to the above theorem, in order to reach a confidence level $1 - \varepsilon$ by the LA-based algorithm, the learning rate a must be estimated by solving the last equation based on parameter x.

Based on the SSM, to obtain a certain confidence level $1 - \varepsilon$ for a subset of vertices in graph, we need to build a confidence level $1 - \varepsilon_i$ for each vertex v_i such that $\sum_{i=1}^{n} \varepsilon_i - \varepsilon$. In this experiment, same confidence level $1 - \varepsilon_0$ is assumed for all vertices of the stochastic graph. The minimum required number of samples for each vertex of graph for SSM is calculated subject to $p[|\bar{x}_n - \mu| < \delta] \geq 1 - \varepsilon$, where $\delta = 0.001$.

Another experiment is conducted to compare the LA-based algorithm with LA-based algorithm in which the leaning automaton residing in each vertex is replaced by a pure chance automaton for solving minimum vertex covering problem. In pure chance automaton, the actions are selected with equal probabilities (Narendra and Thathachar 1989). The comparison is made with respect to dynamic threshold (DT) scaled between (0, 1) by dividing DT by the weight of maximum vertex cover minus the weight of minimum vertex cover. The plot of DT versus iteration presented in Fig. 5.4 show the role of learning automata in guiding the process of sampling from vertices of the stochastic graph for forming the stochastic minimum vertex cover. With the aid of learning automata the process of forming the

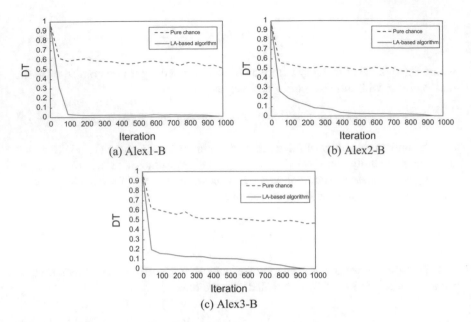

Fig. 5.4 Comparison of LA-based algorithm with LA-based algorithm in which learning automata are replaced with pure chance automata for test networks (Alex1-B, Alex2-B, Alex3-B)

stochastic minimum vertex cover is done with fewer numbers of samples taken from the graph as compared to the case where learning is absent. Similar results for other algorithms can also be obtained.

5.3 Social Network Analysis

In recent years, online social networks such as *Facebook* and *Twitter* have provided simple facilities for online users to generate and share a variety of information about users' daily life, activities, events, news and more information about their real worlds which results in the online users have become the main features of online social networks and studying how users behave and interact with their friends in online social networks play a significant role for analysis of online social networks. Online social networks similar to many real world networks are usually modeled and represented as deterministic graphs with a set of nodes as users and edges as connection between users of networks. In social networks analysis, when networks are modeled as weighted or unweight deterministic graphs, one can study, characterize and analyze the statistical characteristics and the dynamical behaviors of the network and its user behaviors by some network measures (Costa et al. 2007). Popular network measures such as degree, betweenness, closeness and clustering coefficient are originally defined for deterministic graphs. These network measures not only used in characterization of networks but also used as a part of some algorithms such as Girvan-Newman community detection algorithm using high betweenness edges (Girvan and Newman 2002), overlapping community detection using nodes' closeness (Badie et al. 2013) and finding the outstanding nodes as a set of seed nodes for maximization of the spread of influence in social network aims to introduce and promote a new products, services, innovations or technologies by ranking important humans (Li et al. 2014).

Online social networks are intrinsically non-deterministic and their structures or behaviors such as online activities of users have unpredictable, uncertain and time varying nature and for this reason modeling those using deterministic graph models with fixed weights are too restrictive to solve most of the real network problems. For example in online social networks the online activities of users such as friendship behaviors, liking a comment on a given post in *Facebook* and frequency of taking a comment on a wall post vary over time with unknown probabilities (Jin et al. 2013). Analyzing online social networks with deterministic graph models cannot take into consideration the continuum of activities of the users occurring over time. Even modeling social networks with weighted graphs in which the edge weights are assumed to be fixed weights considers only a snapshot of a real-world network and it is not valid when the activities and behavior of users on networks vary with time. Moreover, in analyzing online social networks not only understanding the structure and topology of the network is important but the degree of association among the users in network is also important for analysis of user behaviors in online social networks.

According to the aforementioned points, it seems that stochastic graphs in which weights associated with the edges are random variables is a better candidate as a graph model for real-world network applications with time varying nature. By choosing a stochastic graph as a graph model, every feature, measures and concept of the graph such as path (Beigy and Meybodi 2006), cover, clique, and spanning tree (Torkestani and Meybodi 2012a) should be treated as stochastic features. For example, choosing stochastic graph as the graph model of an online social network and defining community structure in terms of clique, and the associations among the humans within the community as random variables, the concept of stochastic clique may be used to study community structure properties.

In this section, we first present the definition of some network measures for stochastic graphs and then introduce a learning automata-based algorithm for computing them under the situation that the probability distribution functions of the weights associated with the edges are unknown. The LA-based algorithms for computing network measures by taking samples from the edges of the stochastic graph try to estimate the distribution of the network measures. The process of sampling from the edges of the graph is guided by the aid of learning automata in such a way that the number of samples needed to be taken from the edges of the stochastic graph for estimating the network measures to be reduced as much as possible. In the LA-based algorithms, the guided sampling process implemented by learning automata aims to take more samples from the promising region of the graph, the regions that reflects higher rate of changes (*i.e.*, higher rate of user activities), instead of walking around and taking unnecessary samples from non-promising region of the graph (Rezvanian and Meybodi 2016a).

5.3.1   Network Centralities and Measures for Social Network Analysis

Network measures and calculating them play a significant role in social network analysis (Borgatti 2005). Popular network measures such as degree, betweenness, closeness and clustering coefficient not only used for evaluating the node importance in actual complex network studies but also used as a part of some algorithms such as Girvan-Newman community detection algorithm using betweenness (Girvan and Newman 2002), overlapping community detection using node closeness (Badie et al. 2013). In this section some of well-known network measures for deterministic networks are introduced.

Definition 5.11 (*degree*) Degree as a basic network measure has been widely used in many studies and degree of node v_i defined in binary network (also called unweighted network) as follows

$$k_i = \sum_{j \neq i} a_{ij} \tag{5.15}$$

where j is the index of all other nodes of graph and a_{ij} is 1 if node v_i is adjacent to node v_j, and 0 otherwise. In other words, degree of node v_i is the number of nodes that directly connected to node v_i. Degree centrality is useful in the context of finding the single human which gets affected by the diffusion of any information in the network. It follows from the fact that the human with high degree centrality has the chance of getting affected from many numbers of sources (Freeman 1979).

Definition 5.12 (*strength*) Node strength of node v_i is defined as the sum of adjacent edge weights for weighted network as follows

$$s_i = \sum_{j \neq i} w_{ij} \tag{5.16}$$

where w_{ij} is greater than 0 if node v_i is adjacent to node v_j and its value indicates the weight of edge between node v_i and node v_j. Similar to degree, a human with high strength centrality is known as popular human with high strength links to other humans; however a human with high strength may not consist of necessarily the maximum number of friends. Strength centrality is useful in the context of finding a human which gets affected by the amount of spreading of any information in the network (Freeman 1979).

Definition 5.13 (*closeness*) Closeness of a node is the inverse sum of shortest paths to all other nodes from that node and defined for binary network with n nodes as follows

$$c_i = \frac{1}{\sum_{j \neq i} d_{ij}} \tag{5.17}$$

where d_{ij} is the length of shortest path between node v_i and node v_j. Closeness can be regarded as a measure of how long it will take to spread information from that node to all other nodes sequentially. Since, the spread of information can be modeled by the use of shortest paths, in applications such as spread of information, a human with high closeness centrality can be considered as the central point because that human can spread the information faster than other human with lower closeness centrality (Freeman 1979).

Definition 5.14 (*betweenness*) Betweenness of a node is the number of shortest paths from all nodes to all other nodes of graph that pass through that node. Betweenness of node v_i is defined for binary network as follows

$$b_i = \frac{g_{st}(i)}{g_{st}} \tag{5.18}$$

where g_{st} is the number of shortest paths between all pair of nodes of source node v_s and destination node v_t in a binary network and $g_{st}(i)$ is the number of those paths that pass through node v_i. Since, a node with high betweenness centrality will typically be the one which acts as a bridge between many pairs of nodes, the betweenness introduced as a measure for quantifying the control of a human on the communication between other humans in a social network. In this conception, humans that have a high probability to occur on a randomly chosen shortest path between two randomly chosen humans have a high betweenness. Betweenness can be regarded as a measure of how to control Information flows over communication links. Also, Information flows could be dominated by humans with high betweenness centrality in a network (Freeman 1979; Newman 2005).

Definition 5.15 (*Clustering coefficient*) The clustering coefficient for node v_i is defined as a proportion of the number of all edges among its neighbors over the possible number of edges between them. The local clustering coefficient for a node v_i in binary network can be defined as follows

$$CC(i) = \frac{1}{k_i(k_i - 1)} \sum_{v_j, v_h \in N_i} a_{ij} \cdot a_{ih} \cdot a_{jh} \tag{5.19}$$

where k_i is the degree of node v_i, n is the total number of nodes of graph, N_i is the set of nodes v_j adjacent to node v_i and a_{ij} is 1 if node v_i is adjacent to node v_j, and 0 otherwise. The clustering coefficient of a node is a measure of the connectivity among the neighborhood of the node (transitivity of a node). A human with high clustering coefficient indicates the high tendency of that human to form cluster with other humans. In other words, the clustering coefficient measures transitivity of a network. And also, most of the friends of a human with high clustering coefficient can collaborate with one another even if the focal human is removed from the network.

Definition 5.16 (*Stochastic path*) In a stochastic graph, a *path* π_i with weight of w_{π_i} and length of n_i from source node v_s to destination node v_d, can be defined as an ordering $\left\{ \pi_1^i, \pi_2^i, \ldots, \pi_{n_i}^i \right\} \subset V$ of nodes in such a way that $\pi_1^i = v_s$ and $\pi_{n_i}^i = v_s$ are source and destination nodes, respectively and edge $e(\pi_{j-1}^i, \pi_j^i) \in E$ for $1 \leq j \leq n_i$, where π_j^i is the jth node in path π_i.

Definition 5.17 (*Stochastic shortest path*) Let in a stochastic graph, there are r distinct paths $\Pi = \{\pi_1, \pi_2, \ldots \pi_r\}$ between source node v_s and destination node v_d, path π^* between source node v_s and destination node v_d is defined as a stochastic shortest path if and only if $w_{\pi^*} = \min_{\pi_i \in \Pi} \{w_{\pi_i}\}$ where $w_{\pi_i} = \sum_{e_{jk} \in \pi_i} w_{jk}$ is the expected weight of path $\pi_i \in \Pi$ and called *stochastic shortest path*. That is a stochastic shortest path is a stochastic path with the minimum expected weight.

Definition 5.18 (*Stochastic strength*) The strength of a node v_i in a stochastic graph is a random variable defined by following equation

$$SS(v_i) = \sum_{j \neq i} \bar{w}_{ij} \tag{5.20}$$

where $\bar{w}_{ij}$ is a random variable associated with the edge weight between node v_i and node v_j.

Definition 5.19 (*Stochastic closeness*) The closeness of node v_i in a stochastic graph is a random variable defined by Eq. (5.21)

$$SC(v_i) = \frac{1}{\sum_j ED_{ij}} \tag{5.21}$$

where ED_{ij} is a random variable associated with the weight of shortest path between node v_i and node v_j with minimum expected weight.

Definition 5.20 (*Stochastic betweenness*) The betweenness of node v_i in a stochastic graph is a random variable SB_i defined by Eq. (5.22)

$$SB(v_i) = \frac{G_{st}(v_i)}{G_{st}} \tag{5.22}$$

where $G_{st}(v_i)$ is the distribution of the sum of weight of shortest paths between node v_s and node v_t that passes through node v_i and G_{st} is the distribution of the sum of weights of all shortest paths between node v_s and node v_t.

Definition 5.21 (*Stochastic clustering coefficient*) The clustering coefficient of a node v_i in a stochastic graph is a random variable which can be defined as follows

$$SCC(v_i) = \frac{1}{(D(v_i) - 1) \cdot SS(v_i)} \sum_{j \neq i} \sum_{k \neq i,j} \frac{(w_{ij} + w_{ik})}{2} a_{ij} \cdot a_{ik} \cdot a_{jk} \qquad (5.23)$$

where $D(v_i)$ is the number of nodes adjacent to node v_i, and a_{ij} is 1 if node v_i is adjacent to node v_j, and 0 otherwise. $SS(v_i)$ is the strength of node v_i and w_{ij} is the random variabl ae associated with the weight of edge e_{ij}.

5.3.2 LA-Based Algorithm for Computing Network Centralities and Measures in Stochastic Social Networks

In the previous section, we defined some network measures for stochastic graphs. In this section, a learning automata-based algorithm is presented for computing network measures in stochastic graphs under the situation that the probability distribution functions of the weights associated with the edges of graph are unknown. The LA-based algorithm for computing network measures by taking samples from the edges of the stochastic graph tries to estimate the distribution of the network measures. The process of sampling from the edges of the graph is guided by the aid of learning automata in such a way that the number of samples needed to be taken from the edges of the stochastic graph for estimating the network measures to be reduced as much as possible. In the LA-based algorithms, the guided sampling process implemented by learning automata aims to take more samples from the promising region of the graph, the regions that reflects higher rate of activities, instead of walking around and taking unnecessary samples from non-promising region of the graph. The algorithm after initialization phase iterates the updating phase until one of the stopping conditions is met. The details of initialization phase, updating phase, and stopping conditions are given below.

A. Initialization

Let $G\langle V, E, W \rangle$ be the input stochastic graph, where $V = \{v_1, v_2, \ldots, v_n\}$ is the set of nodes, $E = \{e_{i,j}\} \subseteq V \times V$ is the edge-set, and $W = \{w_{i,j}\}$ is the set of random variables each of which is associated with an edge weight of the input stochastic graph. It is assumed that the weight of each edge is a positive random variable with unknown probability distribution function. The LA-based algorithm uses a network of learning automata isomorphic to the input graph and tries to estimate weights of edges by sampling from the edges of the graph in such a way that the number of samples taken from the edges of graph be reduced. Learning automaton A_i residing in node v_i has two actions as $\alpha_i = \{\alpha^1, \alpha^2\}$ where action α^1 is "take sample from the edges of chosen node v" and action α^2 is "do not take sample from the edges of chosen node v". Let $p(v_i) = \{p_i^1, p_i^2\}$ be the action probability vector of learning automaton A_i and initialized equally $p_i^1 = p_i^2 = 1/2$ for all $v_i \in V$. The process of taking samples from the edges on which a node is incident is governed by the learning automaton assigned to that node. At the beginning of the algorithm, the weight of each edge is initialized with some random samples in order to provide a coarse estimate of the weight of that edge.

B. Updating

In this phase, every learning automaton in the network chooses an action according to its action probability vector in parallel manner. Then a sample from every edge incident on a node whose corresponding learning automaton has chosen the action of taking sample (α^1) will be taken. Computation of the new estimates for the means and standard deviations of the unknown distributions of the edge weights are then performed. Using these new estimates, the network measure is calculated for the given network and then the difference between the computed network measure and the network measure obtained in the previous iteration is calculated using DDQC [as given in Eq. (5.24)]. Based on this difference, the probability vectors of learning automata assigned to the nodes of the input graph are updated according to the learning algorithm. That is, if this difference is lower than or equal to a given error ε_p, then action α^1 of all learning automata are rewarded proportional to the amount of improvement and penalized otherwise.

DDQC is a measure for the difference between two calculated network measures in the consecutive iterations. (Aliakbary et al. 2015). In DDQC, the given degree distribution divided into eight regions based on the following description and the sum of parameter distribution in each region is calculated as distribution percentiles. The absolute difference between these extracted values for real and estimated distributions in the original data and obtained samples data is considered as the distance between two distributions as given following equation.

$$DDQC(G, G') = \sum_{i=1}^{8} |Q_i(G) - Q_i(G')| \tag{5.24}$$

where $Q(G) = \langle IDP(I(i)) \rangle_{i=1,\dots,8}$ is the set of values for eight regions of graph G and $IDP(I)$ is the interval distribution probability and can be calculated as given below

$$IDP(I) = Pr(left(I) \leq D \leq right(I)) \tag{5.25}$$

where $Left(R)$ and $Right(R)$ is the minimum and maximum values for each interval I. The interval I can be divided in two regions as following

$$I(i) = \begin{cases} \left[lef\,t\left(R\left\lceil \frac{i}{2}\right\rceil\right), left\left(R\left\lceil \frac{i}{2}\right\rceil\right) + \frac{\left|R\left\lceil \frac{i}{2}\right\rceil\right|}{2}\right] & i\ is\ odd \\[2ex] \left[lef\,t\left(R\left\lceil \frac{i}{2}\right\rceil\right) + \frac{\left|R\left\lceil \frac{i}{2}\right\rceil\right|}{2}, right\left(R\left\lceil \frac{i}{2}\right\rceil\right)\right] & i\ is\ eve \end{cases} \tag{5.26}$$

where

$$|R(r)| = \max(right(R(r)) - left(R(r)), 0) \tag{5.27}$$

is the range length of values. The range of values is divided in four regions based on mean and standard deviation of a network measure.

$$R(r) = \begin{cases} [D_{min}, \mu - \sigma] & r = 1 \\ [\mu - \sigma, \mu] & r = 1 \\ [\mu, \mu + \sigma] & r = 1 \\ [\mu + \sigma, D_{max}] & r = 1 \end{cases} \tag{5.28}$$

where $\mu = \sum_{D_{min}}^{D_{max}} x \times Pr(x)$ and $\sigma^2 = \sum_{D_{min}}^{D_{max}} Pr(x) \times (x - \mu)^2$ are the mean and variance of the probability distribution function $Pr(x)$ respectively, which obtained from a network measure.

C. Stopping

Updating phase is repeated until one of the following stopping conditions is met.

- The number of iterations k exceeds a given number K_{max},
- The difference between the computed measure in two consecutive iterations becomes lower than a specified error E_{min}, or
- The average of entropy of probability vector of learning automata reaches a predefined value T_{min}.

In the LA-based algorithm, the entropy is used for measuring learning process. The information entropy for a learning automaton with r actions can be defined as follows (Mousavian et al. 2013):

$$H = -\sum_{i}^{r} p_i.Log(p_i) \tag{5.29}$$

where p_i is the probability of choosing ith action of a learning automaton. The entropy for a learning automaton has maximum value of one when all the actions have equal probabilities of choosing and has minimum value of zero when the action probability vector is a unit vector. The pseudo-code of the LA-based algorithm is shown in Fig. 5.5.

Algorithms 5-1. The LA-based algorithm for computing network measures in stochastic graphs

Input: Stochastic Graph $G\langle V, E, W\rangle$, Network measure M_Θ, Thresholds K_{max}, T_{min}, E_{min}, ε_p

Output: Computed network centralities/measures $\hat{\theta}_1, \hat{\theta}_2, \dots, \hat{\theta}_n$ where $\hat{\theta}_i$ is the computed network measure for node v_i

Initialization:

    Create a network of learning automata isomorphic to the graph G by assigning an automaton A_i to each node v_i;

    Let $\alpha_i = \{\alpha^1, \alpha^2\}$ be the set of actions for learning automata A_i;

    Let $P=\{p_1, p_2, \dots, p_n\}$ be the set of action probabilities where $p_i=\{p_i^1, p_i^2\}$ and initialized equally $p_i^1 = p_i^2 = 1/2$ for all $v_i \in V$;

    Let k be the iteration number and initially set to 1;

    Let $\bar{w}_{i,j}$ be the estimate for mean of edge weight of all samples taken from edge e_{ij} and initialized with some random samples;

    Let $\hat{\theta}_1^k, \hat{\theta}_2^k, \dots, \hat{\theta}_n^k$ be the computed network measure for the nodes at iteration k;

    Let G^k be the network with mean edge weight at iteration k;

Begin

    Let $H_i(k)$ be the entropy of learning automaton A_i at iteration k and computed based on equation (5-29);

    Repeat

        For all learning automata do in parallel

            Learning automaton A_i chooses an action according to its action probability vector;

            If $(\alpha_i = \alpha^1)$ **then**

                Take a sample from every edge on which node v_i is incident;

                Construct G^k by computing the new weight for every edge on which node v_i is incident;

            Else

                No action;

            End if

        End for

        Compute new estimates for the centrality/measure of the nodes $\hat{\theta}_1^k, \hat{\theta}_2^k, \dots, \hat{\theta}_n^k$

        Calculate $DDQC^k(G^k, G^{k-1})$ using equation (5-24); // Calculate DDQC between consequent estimated network measures

        If $(k > 1)$ **Then**

            Calculate $\Delta^k = DDQC^k - DDQC^{k-1}$; // Calculate the change rate between iterations k and k-1;

            If $(|\Delta^k| \geq \varepsilon_p)$ **Then**

                Reward the chosen actions of all learning automata;

            Else

                Penalize the chosen actions of all learning automata;

            End If

        End If

        $H(k) \leftarrow \frac{1}{n}\sum_{i=1}^{n} H_i(k)$;

        $k \leftarrow k + 1$;

    Until $(k < K_{max}$ **OR** $H(k) > T_{min}$ **OR** $|\Delta^k| > E_{min})$

End Algorithm

Fig. 5.5 Pseudo-code for LA-based algorithm for compute network measures in stochastic graphs

5.3.3 Experiments

In this section, to study the performance of the LA-based algorithm, several computer simulations are conducted on synthetic stochastic graphs which are generated based on well-known random graph with their weights of edges being random variables. We use well-known computer generated network model of *Barabási-Albert* model (*BA* model) as a scale-free network with heavy-tailed degree distribution (Barabási and Albert 1999) which is utilized widely in literature. We set network parameters of BA model as $N \in \{1000, 2000, 5000, 10,000\}$ and $m_0 = m = 5$. Other synthetic network as a small world network is *Watts–Strogatz* model (*WS* model) (Watts and Strogatz 1998) which reflect a common property of many real networks such as short average path length with $N \in \{1000, 2000, 5000, 10,000\}$, $k = 4$ and $p = 0.2$. Also, we use well-known random network of *Erdös–Rényi* model (*ER* model) (Erdos and Rényi 1960) which is utilized widely in literature as $N \in \{1000, 2000, 5000, 10{,}000\}$ and $p = 0.2$. The edge weights for these graphs are random variables with exponential distributions whose mean are chosen randomly from set $\mu \in \{0.8, 1, 1.2, 1.5\}$. It is noted that the chosen mean values is adopted from an empirical observations by Bild et al. (2015).

In order to evaluate the accuracy of estimations by the LA-based algorithm, it is used Kolmogorov-Smirnov distance statistic as a distance function between distribution of real parameters and estimated parameters.

Kolmogorov-Smirnov (KS) D-statistics is one of the statistical test methods commonly used for assessment the distance between two cumulative distribution functions (CDFs). *KS* measures acceptability between original distribution and estimated distribution. The result of this test is a value between 0 and 1 that as closer as it is to zero, both distributions will have a greater similarity; and as closer as it is to unit, the two distributions will show a greater discrepancy. This measure has been defined as

$$KS(P, Q) = \max_{x} |P(x) - Q(x)| \qquad (5.30)$$

where P and Q are two CDFs of original and estimated data, respectively, and x represents the range of the random variable. So it is computed as the maximum vertical distance between the two distributions.

The stopping criteria for the LA-based algorithm is either the number of iteration k reached $K_{max} = 50 \times n$ iterations where n is the number of nodes for each instance graph or the value of average information entropy $H(k)$ is lower than $T_{min} = 0.10$ or the difference between the computed measure in two consecutive iteration Δ^k becomes lower than $\varepsilon_h = 0.001$. The edge weights for synthetic stochastic graphs are random variables with exponential distributions whose mean are selected randomly from set $\mu \in \{0.8, 1, 1.2, 1.5\}$. The reinforcement scheme used for updating the action probability vector of learning automata is $L_{R\varepsilon p}$ with $a = 0.05$, $b = 0.01$. The reported results are the averages taken over 30 runs.

In this experiment, we compare the number of samples required by the LA-based algorithms in order to reach certain accuracy with the number of samples needed by the standard sampling method (SSM) to reach the same accuracy. In order to have a fair comparison, both the LA-based algorithm and SSM must use the same confidence level $1 - \varepsilon$. The LA-based algorithm may reach the confidence level of $1 - \varepsilon$ by a proper choice of learning parameter a. According to appendix C, such learning parameter can be estimated using $ax/(e^{ax} - 1) = max_{i \neq j}(d_j/d_i)$ where d_i is the reward probability of action α_i. Based on the SSM, to obtain a certain confidence level $1 - \varepsilon$ for a subset of edges in stochastic graph, we need to build a confidence level $1 - \varepsilon_i$ for each edge e_i such that $\varepsilon = \sum_{i=1}^{m} \varepsilon_i/m$. In this experiment, same confidence level $1 - \varepsilon_0$ is assumed for all edges of the stochastic graph. The minimum required number of samples for each edge of graph for SSM is calculated subject to $p[|\bar{x}_n - \mu| < \delta] \geq 1 - \varepsilon$, where $\delta = 0.01$.

For this experimentation, error rate ε varies from 0.3 to 0.05 with increment 0.05 (confidence levels 0.7 to 0.95) for the LA-based algorithms. Different synthetic stochastic graphs with size from 1000 to 10,000 are used. The results of this experimentation are averages taken over all the test networks with respect to edge sampling rate (ESR) which is defined as the average number of samples that is needed to be taken from each edge of stochastic graphs to reach a certain confidence level (CL). The average and standard deviation of ESR are given in Table 5.5 for synthetic stochastic ER, BA and WS graphs. From the results one can see that for LA-based algorithm, the average number of samples taken from each edge of graph is fewer than the number of samples taken using SSM for all synthetic stochastic graphs and different confidence levels.

According to the results of Table 5.5, one can conclude that LA-based algorithm outperforms the standard sampling method in terms of average number of samples taken from each edge.

Another experiment is conducted to study the performance of the LA-based algorithm and standard sampling method (SSM) for sampling from stochastic graphs in terms of different distance metrics. For this experimentation, different synthetic stochastic graphs (BA, WS and ER) with size from 1000 to 10,000 are used. The results of this experimentation are averages taken over all the test networks with respect to Kolmogorov-Smirnov (KS) for estimated measures including: betweenness, clustering coefficient, closeness and strength. The results of average and standard error with 95% confidence interval of mentioned metrics are given in Fig. 5.6 for synthetic stochastic graphs. From Fig. 5.6 one can observe that LA-based algorithm outperforms standard sampling method for all test graphs and standard sampling method achieves the low accuracy in terms of KS for mentioned metrics. For all test graphs, the results for betweenness, clustering coefficient and strength is more reliable than the results for closeness in terms of KS.

Following experiment is conducted to compare the LA-based algorithm with the pure chance algorithm (LA-based algorithm in which the leaning automaton residing in each node is replaced by a pure chance automaton) with respect to the number of samples taken from the edges of the graph. In pure chance automaton, the actions are chosen with equal probabilities. In order to perform this experiment, we plot average number of samples taking from each edge of graph (ESR) during the execution of the algorithms for computing network measures versus iteration number for LA-based algorithm and pure chance algorithm. The results of this experiment are taken averages for each synthetic stochastic test graphs and are given in Fig. 5.7. In Fig. 5.7, the points along the curves show the average value and the error bars represent 95% confidence interval. As it is shown, as the algorithms proceeds, the number of samples taken from each edge (ESR) abundantly increases in initial iterations for all algorithms and then reduces and gradually approaches zero for LA-based algorithm, however ESR value for pure chance algorithm remains unchanged. This implies that for the LA-based algorithm, the average number of samples taken from each edge of the graph gradually approaches zero. And also it is done with fewer numbers of samples taken from the graph as compared to the case where learning is absent that indicates the important impact of learning on the superiority of the learning automaton based algorithm for computing network measures in stochastic graphs. From the error bars shown in

Table 5.5 Average results [± standard deviation (std)] for SSM and the LA-based algorithm for synthetic stochastic graphs with different error rates in terms of ESR

Networks	Methods	Confidence level				
		0.75	0.8	0.85	0.9	0.95
ER	SSM	30.00 ± 1.08	36.98 ± 1.34	46.42 ± 1.69	60.26 ± 2.20	85.15 ± 3.13
	Algorithm 1	12.66 ± 3.62	17.46 ± 3.06	24.50 ± 5.27	31.37 ± 6.08	52.38 ± 7.49
BA	SSM	24.66 ± 3.39	30.16 ± 2.28	37.19 ± 2.82	46.65 ± 3.55	60.59 ± 4.65
	Algorithm 1	9.27 ± 1.74	12.83 ± 1.70	15.04 ± 1.78	19.19 ± 1.94	27.18 ± 2.57
WS	SSM	25.58 ± 0.80	31.29 ± 1.10	38.58 ± 1.36	48.42 ± 1.73	62.90 ± 2.24
	Algorithm 1	10.92 ± 1.42	13.43 ± 2.17	15.87 ± 2.68	18.84 ± 2.15	28.85 ± 6.78

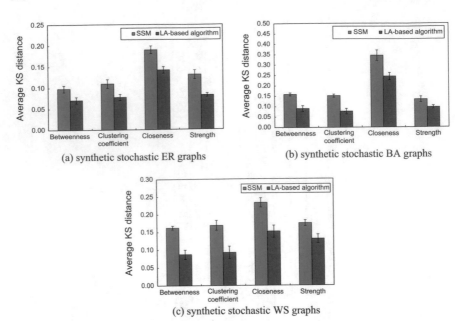

Fig. 5.6 Comparing results of average KS distances over betweenness, clustering coefficient, closeness and strength for stochastic graphs

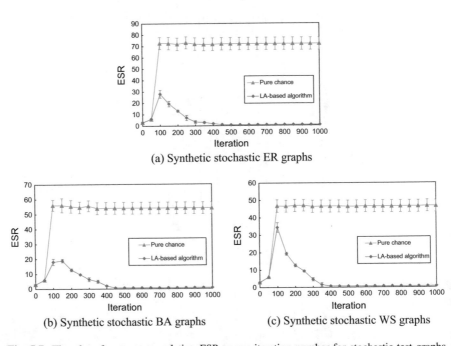

Fig. 5.7 The plot of average cumulative *ESR* versus iteration number for stochastic test graphs

Fig. 5.7, we conclude that the variances in the results decreases as the iteration number increases. According to the results for comparison among network models, we can see that for Synthetic stochastic ER graphs, the average number of samples taking from each edge of graph is much more than that of both Synthetic stochastic BA and WS graphs. Similar result can be obtained for other algorithms.

Since it is assumed that the structural and behavioral parameters of online social networks are time varying parameters and it can be observed by the network. Using the observed parameters, the network can calculate different measurements such as strength, closeness and betweenness in an ongoing fashion. The LA-based algorithm can be used by the network as a means for observing the time varying parameters of the network for the purpose of network's measurements computations. The aim of this algorithm is to collect information from the network in order to find good estimates for the network's measurements using fewer numbers of samples than that of standard sampling methods.

5.4 Network Sampling

This section consists of two set of LA-based algorithms related to network sampling for deterministic networks and stochastic networks.

5.4.1 LA-Based Algorithms for Network Sampling in Deterministic Graphs

In recent years, various researchers have focused on the analysis of dynamic behaviors of individuals in social networks. However, there are several issues such as the large scale, complexity, dynamicity and limitations on accessing these networks which make them hard or even impossible to study. Thus, network sampling algorithms have emerged to realize studying, characterizing and analysis of complex networks such as online social networks. Network sampling can provide an energy efficient way to conduct any research on the complex social networks in an offline manner in a small amount of time and lower computational cost (e.g., process, storage, bandwidth and money) (Gjoka et al. 2011; Papagelis et al. 2013; Ahmed et al. 2014). The network sampling is aimed to construct a sampled network with small scale which is preserved the properties of the original network. The properties of the networks can be studied and characterized by network sampling with some network measures and these measures of the sampled network are used to study the network instead of access to the whole network data (Murai et al. 2013; Papagelis et al. 2013).

Accuracy of most studies and analyses on real networks critically depends on the accuracy of the sampled networks obtained from the original networks. The main purpose of network sampling algorithms is to construct an appropriate sampled network that directly resembles the original network and preserves most properties of the original network. In order to reach a desirable accuracy for sampled network sometimes more processing is required for the sampling algorithm. Although recently several simple network sampling algorithms have been presented, most of these algorithms independently sample just the nodes or edges with low computational cost, which results in sampled networks that are poor in recovering the important natural characteristics of original networks. A proper sampling algorithm should be able to preserve different natural characteristics of social networks, such as the presence or absence of important, central and influential individuals.

A social network can be represented as a graph with a set of nodes such as users and a certain type of relationship between users as edges of graph. Let $G = \langle V, E \rangle$ be a given social network graph, where V is the set of nodes with size $n = |V|$ and E is the set of edges. Sampling is a function $f\colon G \to G_s$ from graph G to sampled network $G_s = \langle V_s, E_s \rangle$ as the induced subgraph of G such that $V_s \subset V, E_s \subset E$, and $|V_s| = \varphi \times n$, where $0 < \varphi < 1$ denotes the sampling rate. It is noted that the main goal of sampling algorithms is to achieve smaller subgraph similar to the original graph G. Sampling algorithms play an important role in preprocessing, characterizing, and studying real networks (Volz and Heckathorn 2008; Gjoka et al. 2011; Papagelis et al. 2013). Sampling can be used to study a small part of networks while preserving main features of the original network.

Several sampling algorithms have been introduced for sampling networks that can be categorized into three main strategies in terms of collecting samples: random sampling, crawling based sampling (also called topology based sampling, traversal based sampling or link tracing sampling), and coarse graining based sampling algorithms. One may also classify the sampling algorithms into two groups: (1) one-phase sampling algorithms which construct the sampled network by random selection of edges/nodes or using some kinds of graph traversal procedure (e.g., RDS (Gile and Handcock 2010), RWS (Yoon et al. 2007), FFS (Kurant et al. 2010), SBS (Frank 2011), to name a few); (2) two-phase sampling algorithms which construct sampled networks using a graph traversal procedure and some pre or post processing (e.g., random PageRank (RPN) (Yoon et al. 2015), DPL (Yoon et al. 2015), IFFST-PR (Tong et al. 2016), to mention a few). The first group of sampling algorithms is relatively simple and has low cost, low accuracy and fails to perform very well on all kinds of networks. The second group uses additional pre or post processing in order to get the information about the network such as classifying important nodes into groups based on Katz centrality measures (Luo et al. 2015), scoring or ordering nodes by PageRank algorithm (Salehi et al. 2012; Tong et al. 2016), extracting groups of nodes in network (Blagus et al. 2015), extracting community structures of network (Yoon et al. 2015) to achieve higher accuracy. Such a pre or post processing certainly increases the cost of the sampling algorithms, which must be paid if achieving higher accuracy is the goal. This should be also noted that the sampled network, once constructed, can be used many times for various applications and analyses.

In this section, two algorithms based on distributed learning automata and extended distributed learning automata are introduced for sampling from complex social networks. In both LA-based algorithms, a learning automaton is assigned to each node of the graph and set of distributed learning automata cooperate to each other to take promising samples from the important part of the network. In both LA-based algorithms, based on the evaluation on the sampled network during the traversal of the given network, the learning algorithm are applied on active learning automata.

5.4.1.1 Distributed Learning Automata Sampling Algorithm (DLAS)

In this algorithm (Rezvanian et al. 2014), it is assumed that each node of a social network is available by a unique ID of each user as nodes of input network and also their allowable connected users as corresponding edges between them. DLAS uses a set of learning automata in order to guide the process of visiting nodes by visiting the important parts of the input network, those parts of the network which contains important nodes. A DLA is constructed by assigning a learning automaton A_i to node v_i of the input network. Action-set of each LA corresponds to selecting the edges of the node to which the LA is assigned. Each LA in DLA initially selects its actions with equal probabilities. In an iteration of the algorithm, DLA starts from a randomly chosen node, and then follows a path of nodes according to the probability vectors of set of LA constituting DLA. If a LA selects an action which corresponds to a node which has been already visited in the previous paths (iterations) then the probability of selecting that action will be increased according to the learning algorithm. After several iterations, the visited nodes are sorted according to their probabilities of selecting. The sampled network is then constructed using a particular number (given number (sampling ratio multiply the number of nodes in the network or also a given percentage) of mostly visited nodes. Pseudo-code of DLA based sampling algorithm (DLAS) is given in Fig. 5.8.

5.4.1.2 Extended Distributed Learning Automata Sampling Algorithm (EDLAS)

In this section, a sampling algorithm for social networks which is based on extended distributed learning automata (eDLA) (Mollakhalili Meybodi and Meybodi 2014) called EDLAS is introduced. In EDLAS (Rezvanian and Meybodi 2017b), a set of learning automata which forms an eDLA isomorphic to the input network and they cooperate with each other in order to take appropriate samples from the network. The algorithm iteratively tries to identify promising nodes or edges by traversing promising parts of the network and then forms the sampled network.

In the EDLAS, it is assumed that each node of a social network is available by a unique ID of each user as nodes of input network and also their allowable

---

Algorithm 5-3. DLA based sampling algorithm (DLAS)

Input: Network $G=\langle V, E \rangle$, Maximum number of iteration K, Sampling ratio ϕ

Output: Sampled network $G_s=\langle V_s, E_s \rangle$

Assumptions:

    Construct an DLA by assigning an automaton A_i to each node v_i and initialize their action probabilities ;

    Let k denotes the iteration number of algorithm which is initially set to 1;

Begin

    Disable all learning automata;

    While ($k < K$)

        Select node v_s randomly as a starting node;

        While (number of visited nodes $<=\phi\times|V|$)

            Automaton A_s is enabled and then selects an action according to its action probability vector;

            Let the selected action by A_s be (v_s , v_i);

            If (v_i is already visited in previous paths) **then** // favorable nodes

                Reward the selected action;

            End If

            $v_s \leftarrow v_i$;

            Disable A_i;

        End While

        $k \leftarrow k + 1$;

    End While

    Sort the visited nodes according to the number of times that the nodes are visited in descending order;

    Construct sampled network G_s using $\phi\times|V|$ mostly visited nodes;

End Algorithm

Fig. 5.8 Pseudo-code of DLA based sampling algorithm (DLAS)

connected users as corresponding edges between them. The LA-based sampling algorithm uses an eDLA to guide the process of visiting nodes by visiting the important parts of the input network, those parts of the network which contains more hub and central nodes. At first, an eDLA isomorphic to the input network is constructed by assigning a learning automaton A_i to node v_i of the input network. Action-set of each LA corresponds to the selecting an edge of the node to which the LA is assigned with equal probabilities. In eDLA, at any time, a LA can be in one of four levels: *Passive, Active, Fire* and *Off*. Thus in the EDLAS, all learning automata are initially set to *Passive* level. At the first of algorithm, one of the nodes in eDLA is chosen by firing function to be the starting node and its activity level is set to *active* by governing rule. In what follows an iteration of the EDLAS is explained.

At the beginning of an iteration, one of the *active* nodes chosen randomly by firing function and then fired (its activity level changes to *fire* and the activity level of its *passive* neighboring nodes change to *active*) and chooses one of its actions which correspond to one of the edges of the *fired* node and then the level of the *fired* LA changes to *off* by governing rule. Then one of LA in eDLA with activity level *active* is chosen and fired (its activity level changes to *fire* and the activity level of its *passive* neighboring nodes change to *active*). The fired LA chooses one of its actions which correspond to an edge of the fired node and then the activity level of the fired LA changes to *off* by governing rule. The process of selecting a learning automaton from the set of learning automata with activity level *active* in eDLA by firing function, firing it, choosing an action by fired LA, changing activity levels of *passive* adjacent LA to *active* LA, and changing activity level of fired LA from *fire* to *Off* by governing rule is repeated until either the number of LA with activity level *Off* is reached a given number (sampling ratio multiply the number of nodes in the network) or the set of LA with activity level *active* is empty. In the EDLAS, the action probabilities of all fired learning automata are updated on the basis of the responses received from the environment. The probabilities are updated as follows: The probabilities of choosing actions of all fired learning automata which correspond to the edges traversed in the previous iterations are updated according to the learning algorithm. At the end of each iteration, the activity level of all learning automata in *e*DLA are set to *passive*. Thus, the EDLAS iteratively visits a set of paths and also updates its action probability vectors until it iterates the maximum number of iterations. After that, the visited nodes are sorted according to the number of times that they have been visited in descending order. The sampled network is then constructed using a given number (sampling ratio multiply the number of nodes in the network or also a particular percentage) of mostly visited nodes. Figure 5.9 is the pseudo-code of *e*DLA based sampling algorithm.

In EDLAS visiting a node causes the probability of visiting this node in later iterations to be increased. This means that mostly visited nodes have incoming edges with high probabilities. That means that the EDLAS constructs a sample of the network which includes central and hub nodes leading to a better sampling of the network.

5.4.1.3 Experiments

In this section, performance of the two LA-based sampling algorithms is investigated on several well-known real and synthetic test networks which are commonly used benchmarks for simulations. Table 5.6 describes the test networks used for the experimentations and their characteristics. We use well-known computer generated network model of *Barabási-Albert* model (BA model) for scale free networks with heavy-tailed degree distribution (Barabási and Albert 1999) which is utilized widely in literature. We set network parameters of BA model as $N = \{1000, 2000, 5000, 10,000\}$ and $m_0 = m = 5$. We also test the LA-based sampling algorithm on several real networks such as *ca-HepTh* (Collaboration network of ArXiv High Energy Physics), *ca-GrQc* (Collaboration network of ArXiv General Relativity) and *ca-CondMat* (Collaboration network of ArXiv Condense Matter Physics).

To investigate the performance of the LA-based algorithm, several simulation experiments are employed on the well-known data sets of complex networks. In the simulation experiments, the LA-based algorithms are compared with the several sampling algorithms from complex networks. The description of networks has been listed in Table 5.6. In the all experiments presented in the chapter, the learning scheme is L_{R-I} and the reward is set to 0.1. The threshold predefined iteration for stopping algorithm is set to 1000.

It is used *Kolmogorov-Smirnov* (KS) and *Relative Error* (RE) as distance measures for performance studies.

Algorithm 5-4. *e*DLA based sampling algorithm (EDLAS)

Input: Network $G=\langle V, E\rangle$ with $V=\{v_1, v_2, \dots v_n\}$, Maximum number of iteration K, Sampling ratio φ
Output: Sampled network $G_s=\langle V_s, E_s\rangle$
 Assumptions:
　Construct an *e*DLA by assigning an automaton A_i to each node v_i from input network G;
　Initialize action probabilities of automata and activity level of each LA initially set to *Passive*;
　Let k denotes the iteration number of algorithm which is initially set to 1;
　Let Pa is the set of learning automata with activity level of *Passive* and initially set $Pa \leftarrow \{v_1, v_2, \dots v_n\}$;
　Let Ac is the set of learning automata with activity level of *Active* and initially set to empty;
　Let Of is the set of learning automata with activity level of *Off* and initially set to empty;
　Let Fi is the learning automaton with activity level of *Fire* and initially set to empty;
　Let S is an array of visited nodes with size n and initially each element set to 0;
　Let $N(v_i)$ is a function that returns the all adjacent nodes of v_i in *Passive* level;
 Begin
　Select an starting node v_s at random by firing function and change its activity level to *Active* level;
　$Ac \leftarrow \{A_s\}$;
　$Pa \leftarrow Pa - \{A_s\}$;
　While $(k < K)$
　　Select one learning automata among *Active* learning automata Ac as A_s by firing function;
　　$Fi \leftarrow A_s$;
　　$Ac \leftarrow N(v_s) \setminus v_s$;
　　$Pa \leftarrow Pa \setminus Ac$;
　　Automaton A_s chooses an action using its action probability vector;
　　Let the action chosen by A_s be (v_s, v_m);
　　$S_k[v_k] \leftarrow S_k[v_k] + 1$;
　　$S_k[v_m] \leftarrow S_k[v_m] + 1$;
　　If $(|Of| >= \varphi \times |V|$ or Ac is empty) **then**
　　　If $(\Sigma(S_k)) > \Sigma S_{k-1})$ **then** // favorable traverse
　　　　Reward the actions chosen by the learning automata with activity level *Of*
　　　Else
　　　　Penalize the actions chosen by the learning automata with activity level *Of*
　　　End If
　　　$Pa \leftarrow \{v_1, v_2, \dots v_n\}$;
　　　$Fi \leftarrow \{\}$;
　　　$Ac \leftarrow \{\}$;
　　　$Of \leftarrow \{\}$;
　　　Select a starting node v_s at random by firing function and change its activity level to *Active* level;
　　　$Ac \leftarrow \{A_s\}$;
　　　$Pa \leftarrow Pa \setminus A_s$;
　　Else
　　　$Of \leftarrow Of \cup Fi$
　　　$Fi \leftarrow \{\}$
　　End If
　　$k \leftarrow k + 1$;
　End While
　Sort the visited nodes according to the number of times that the nodes are visited in descending order;
　Construct sampled network G_s using $\varphi \times |V|$ mostly visited nodes;
 End Algorithm

Fig. 5.9 Pseudo-code of the *e*DLA based sampling algorithm (EDLAS)

Table 5.6 Description of test networks

Network	Node	Edge	Description
BA-1000	1000	9952	Synthetic scale-free network
BA-2000	2000	19,952	Synthetic scale-free network
BA-5000	5000	49,944	Synthetic scale-free network
BA-10000	10,000	99,945	Synthetic scale-free network
ca-GrQc	5242	28,980	Collaboration network of ArXiv general relativity
ca-HepTh	9877	25,998	Collaboration network of ArXiv high energy physics
ca-CondMat	23,133	93,497	Collaboration network of ArXiv condense matter physics

Kolmogorov-Smirnov D-statistic (KS-D)

Kolmogorov-Smirnov D-statistic is one of the statistical test methods used for calculating the distance between two cumulative distribution functions (CDF). *KS distance* is a commonly used measure for acceptability between distribution of the original network and distribution of the sampled network. It is calculated as the maximum absolute distance between the two distributions. The result of this test is a value between 0 and 1. As closer as it is to zero, both distributions will have a greater similarity; and as closer as it is to unit, the two distributions will show a greater discrepancy (Goldstein et al. 2004). This metric has been calculated as the following equation

$$KS(P, Q) = \max_{x} |P(x) - Q(x)|$$

where P and Q are two CDFs of original and estimated data, respectively and x represents the range of the random variables. So it is computed as the maximum vertical distance between the two distributions. The result of this test can be employed for comparison of sampling algorithms. In this section, we apply KS-D on the degree distribution of the original network and the sampled network.

Relative Error (RE)

Relative Error (RE) can be applied to assess the accuracy of the results for a single parameter, which is defined by the following equation:

$$RE = \frac{|P - Q|}{P} \tag{5.32}$$

where, P and Q denote the values of a network parameter (*i.e.*, clustering coefficient) in the original network and the sampled network respectively (Papagelis et al. 2013). In this section, we calculate RE for clustering coefficient in the original network and the sampled network.

All simulations are conducted 30 times and the average results are reported. Linear learning scheme L_{R-I} with learning parameter 0.05 is used in all experiments except experiment I. For evaluation purpose, Relative Error (RE) and Kolmogorov-Smirnov (KS) statistic are used.

This experiment is performed to compare the performance of the *e*DLA based sampling algorithm (EDLAS) with Random Node Sampling (RNS) (Leskovec and Faloutsos 2006), Random Edge Sampling (RES) (Leskovec and Faloutsos 2006), Random Walk Sampling (RWS) (Yoon et al. 2007), Metropolis-Hastings Random Walk (MHRW) (Lee et al. 2012), and the only learning automata based sampling algorithm (DLAS) for sampling rates 10, 20 and 30%. The linear learning scheme L_{R-I} with learning parameter $a = 0.05$ is used for both DLAS and EDLAS. The comparison is made in terms of KS-D for degree distribution and relative error (RE) for clustering coefficient (CC), and RE for degree distribution (DD) parameter and then the results are summarized in Tables 5.7 and 5.8. Table 5.7 is for KS-D for degree distribution and Table 5.8 for RE for clustering coefficient. According to the results, we may conclude that for all the test networks, higher (lower) value for

Table 5.7 Average KS-D for degree distribution for different sampling rate

Sampling rate (%)	Sampling algorithms					
	RNS	RES	RWS	MHRW	DLAS	EDLAS
10	0.74	0.57	0.46	0.47	0.44	0.32
20	0.66	0.42	0.41	0.33	0.32	0.24
30	0.62	0.35	0.36	0.30	0.26	0.19
40	0.59	0.33	0.31	0.26	0.23	0.18
50	0.55	0.32	0.28	0.24	0.21	0.17

Table 5.8 Average RE for clustering coefficient for different sampling rate

Sampling rate (%)	Sampling algorithms					
	RNS	RES	RWS	MHRW	DLAS	EDLAS
10	0.65	0.53	0.50	0.47	0.45	0.43
20	0.58	0.46	0.43	0.43	0.40	0.36
30	0.55	0.42	0.41	0.38	0.36	0.33
40	0.51	0.4	0.38	0.35	0.33	0.29
50	0.47	0.38	0.36	0.33	0.31	0.27

sampling rate results in lower (higher) value of RE and KS distance. From the results, we also see that the EDLAS outperforms other sampling algorithms for many of the test networks and different sampling rates. The results also show that the EDLAS in comparison with the DLAS has a 26.93% improvement for the average KS value for degree distribution taken over all tested real and synthetic networks. These improvements for the average of RE values for degree distribution and clustering coefficient taken over all tested real and synthetic networks are 18.27 and 7.89% respectively. Besides, the results indicate that in terms of KS-D, the EDLAS, DLAS, MHRW, RWS, RES and RNS, has rank 1, 2, 3, 4, 5 and 6, respectively for real and synthetic networks.

Furthermore, another experiment is conducted to compare LA-based sampling algorithm (EDLAS) with EDLAS in which the leaning automaton residing in each node is replaced by a pure chance automaton. In a pure chance automaton the actions are always chosen with the same probabilities (Narendra and Thathachar 1989). These two algorithms are compared in terms of average KS-D for degree distribution and average relative error (RE) for clustering coefficient with respect to test networks. Average results of this experiment are given in Fig. 5.10. According to the results, we may conclude that LA-based algorithm perform better than LA-based algorithm which learning automata are replaced with pure chance automata. That is indication of the fact that traversal of the network with the help of learning automata leads to a better sampling of the network. Similar results from comparing DLAS and DLAS in which the leaning automaton residing in each node is replaced by a pure chance automaton are also obtained.

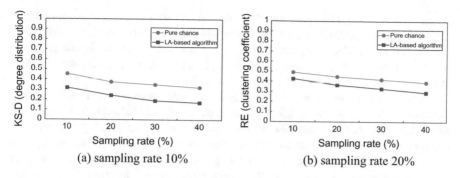

(a) sampling rate 10%   (b) sampling rate 20%

Fig. 5.10 Comparison of LA-based algorithm with LA-based algorithm which learning automata are replaced with pure chance automata for network sampling

5.4.2 LA-Based Algorithms for Network Sampling Algorithms in Stochastic Graphs

In the previous section, we introduced some LA-based algorithms for network sampling in deterministic graphs. In this section, we introduce two sampling algorithms for stochastic graphs (Rezvanian and Meybodi 2017a).

Network sampling for stochastic graphs can be redefined as follows: Let $G = \langle V, E, W \rangle$ be the input stochastic graph, $V = \{v_1, v_2, \ldots, v_n\}$ is the set of nodes, E denotes the edge-set and $W = \{w_1, w_2, \ldots, w_m\}$ is the set of edge weights of the graph in which w_{ij} associated with every edge $e_{ij} \in E$ is a random variable with unknown probability distribution function q_{ij}. Network sampling for stochastic graphs is a sampling algorithm for constructing a sampled network $G' = \langle V', E', W' \rangle$ from original network G such that $V' \subset V$, $E' \subset E$ and W' in which $\bar{w}_{ij}$ is an estimate of the edge weight associated with edge $e_{ij} \in E'$ with sampling rate $0 < \varphi < 1$, where $|V'| = \varphi \times |V|$. Hence, sampling algorithm tries to estimate the unknown distribution of the weigh associated with an edge by taking samples from that edge. The LA-based algorithms for sampling from stochastic graphs use a network of learning automata to guide the process of sampling edges of the stochastic graph in such a way that the number of samples taken from the edges to be reduced as much as possible. These algorithms are the generalization of two sampling algorithms described for binary network in the previous section.

5.4.2.1 Distributed Learning Automata Based Sampling for Stochastic Graphs (DLAS-SG)

In this section, we introduced the generalization of the distributed learning automata (DLAS) sampling algorithm to be used for sampling stochastic graphs and call it *DLAS-SG*. The DLAS-SG algorithm iteratively visits the nodes of the stochastic graph from different starting nodes several times (e.g., K_{max} starting nodes) using a distributed learning automata and then use the visited nodes for constructing a sampled graph. The DLA used for this purpose is a DLA isomorphic with the input graph. Action-set of learning automaton A_i which is assigned to node v_i is the set of outgoing edges of node v_i in the input graph. Each learning automaton in DLA initially chooses its actions with equal probabilities.

DLAS-SG consists of number of stages. In each stage a path in the graph is traversed and the average weights of the edges along this path are updated. To do this, DLA starts from a randomly chosen starting node v_i and learning automaton A_i corresponding to node v_i is activated and chooses one of its actions according to its action probability vector. Let the chosen edge be e_{ij}. The chosen edge e_{ij} is sampled (visited) and then the average weight of the sampled edge is updated. Then learning automaton A_j is activated and chooses one of its actions. The process of activating a learning automaton, choosing an action, activating another learning automaton and updating the average weight of the chosen edge is repeated until either the number of visited nodes reaches $|V| \times \varphi$ or the activated learning automaton cannot activate another automaton. Once the traversal of path π_t at stage t is terminated, the probability vectors of automata along the traversed path π_t are updated as follows. If the weight of π_t ($\bar{w}_{\pi_t}$) is equal or greater than the dynamic threshold T_t computed as $T_t = [(t-1)T_{t-1} + \bar{w}_{\pi_t}]/t$ then the probability of actions chosen by all the activated learning automata along the traversed path π_t are increased and decreased otherwise according to the learning algorithm. At the end of a stage, all learning automata in DLA are deactivated and then a new stage begins. A stage begins if the maximum number of stages (K_{max}) has not reached and the difference between two dynamic thresholds in two consecutive stages is equal or greater than a predefined threshold T_{min}.

When the execution of stages is over, the nodes visited during all stages are sorted in descending order according to the number of paths along which a node has been appeared. The sampled network is then constructed by considering an induced sub-graph of the input network whose node-set contains a given number of mostly visited nodes and its edge weights are the average weight of edges estimated during the execution of the algorithms. Pseudo-code of DLAS-SG algorithm for stochastic graphs is given in Fig. 5.11.

5.4.2.2 Extended Distributed Learning Automata Based Sampling for Stochastic Graphs (EDLAS-SG)

In this section, we introduce the generalization of the extended distributed learning automata (EDLAS) sampling algorithm for sampling stochastic graphs and call it *EDLAS-SG*. The EDLAS-SG algorithm iteratively traverses stochastic graph from different starting nodes (e.g., K_{max} starting nodes) eDLA and then use them for constructing the sampled graph. The algorithm uses an eDLA isomorphic with the input graph. Action-set of learning automaton A_i which is assigned to node v_i is the set of outgoing edges of node v_i in the input graph. Each learning automaton in

---

Algorithm 5-5. DLAS-SG(G, φ, K_{max}, T_{min})

Input: Stochastic graph $G=\langle V, E, W \rangle$, Sampling rate φ, Thresholds K_{max}, T_{min}.

Output: Sampled graph $G'=\langle V', E', W' \rangle$.

Initialization

   Construct a DLA by assigning an automaton A_i to each node v_i and initialize their action probabilities.

   t denotes the iteration number of algorithm which is initially set to 1.

   $\bar{w}_{\pi_t}$ denotes the average weight of all samples taken from edges along the path of π_t.

   Ls is an one-dimensional array with size $|V|$ which is used to store the number of times that the nodes are sampled.

Begin algorithm

   Deactivate all learning automata;

   $t \leftarrow 1;$ $T_k \leftarrow 0;$

   While $(t < K_{max}$ **OR** $|T_t - T_{t-1}| \geq T_{min})$ **Do**

   $\pi_t \leftarrow \{\};$

   Select starting node v_i randomly from non-visited nodes NV;

   $NV \leftarrow NV \setminus v_i;$

   While (number of visited nodes $\leq \phi \times |V|$ **AND** a learning automaton can be activated) **Do**

   Learning automaton A_i is activated and then chooses an action according to its action probability vector;

   Let the chosen action by A_i be edge e_{ij};

   Visit and take a sample from the chosen edge e_{ij};

   $\pi_t \leftarrow \pi_t \cup \{e_{ij}\};$

   $\bar{w}_{\pi_t} \leftarrow \bar{w}_{\pi_t} + \bar{w}_{ij};$

   $Ls[v_i] \leftarrow Ls[v_i] + 1;$

   Set A_i to A_j;

   End While

   $T_t \leftarrow [(t-1)T_{t-1} + \bar{w}_{\pi_t}]/t;$

   If $(\bar{w}_{\pi_t} \geq T_t))$ **Then**   // favorable path

   Reward the actions chosen by all the activated learning automata along path π_t;

   Else

   Penalize the actions chosen by all the activated learning automata along path π_t;

   End If

   $t \leftarrow t + 1;$

   End While

   Sort Ls in descending order;

   Construct an induced sub-graph $G'=\langle V', E', W' \rangle$ using $\phi \times |V|$ mostly visited nodes;

End Algorithm

Fig. 5.11 Pseudo-code of distributed learning automata based sampling algorithm for stochastic graphs

$eDLA$ initially chooses its actions with equal probabilities. In $eDLA$, at any time, a LA can be in one of four levels: *Passive, Active, Fire* and *Off*. All learning automata are initially set to *Passive* level.

The EDLAS-SG consists of number of stages. In each stage a sub-graph in the graph is traversed and the average weights of the edges along this sub-graph are updated. To do this, one of the nodes in $eDLA$ is chosen by firing function as randomly to be the starting node and its activity level is set to *active* by governing rule. Then learning automaton A_i is fired (its activity level changes to *fire* and the activity level of its *passive* neighboring nodes change to *active*) and chooses one of its actions which correspond to one of the edges of the *fired* node and at the same time the level of the *fired* LA A_i changes to *off* by governing rule. Let the chosen action (edge) be e_{ij}. Edge e_{ij} is sampled (visited) and the average weight of the sampled edge is recomputed. Then one of LA in $eDLA$ with activity level *active* is chosen and fired (its activity level changes to *fire* and the activity level of its *passive* neighboring nodes change to *active*). The fired LA chooses one of its actions which correspond to an edge of the fired node and then the activity level of the fired LA changes to *off* by governing rule. The process of selecting a learning automaton from the set of learning automata with activity level *active* in *e*DLA by firing function, firing it, choosing an action by fired LA, changing activity levels of *passive* neighboring LA to *active* LA, computing the average weight of the chosen edge, and changing activity level of fired LA from *fire* to *Off* by governing rule is repeated until either the number of LA with activity level *Off* is reaches $|V| \times \varphi$ or the set of LA with activity level *active* is empty.

Once the traversal of sub-graph τ_t at stage t is terminated the probability vectors of fired learning automata along the traversed path τ_t are updated as follows. If the weight of τ_t ($\bar{w}_{\tau_t}$) is equal or greater than the dynamic threshold T_t computed as $T_t = [(t-1)T_{t-1} + \bar{w}_{\pi_t}]/t$, then the probability of actions chosen by all the fired learning automata along the traversed path τ_t are increased and decreased otherwise according to the learning algorithm. Before a new stage begins, activity levels of all learning automata in *e*DLA are changed and the activity level of all learning automata in *e*DLA are set to *passive*. A stage begins if the maximum number of stages (K_{max}) has not reached and the difference between two dynamic thresholds in two consecutive stages is equal or greater than a predefined threshold T_{min}.

When the execution of stages is over, the nodes visited during all the stages are sorted in descending order according to the number of sub-graph along which a node has been appeared. The sampled network is then constructed by considering an induced sub-graph of the input network whose node-set contains a given number of mostly visited nodes and its edge weights are the average weight of edges estimated during the execution of the algorithms. Pseudo-code of ESLAS-SG algorithm for stochastic graphs is given in Fig. 5.12.

5.4.2.3 Experiments

In this section, the performance of the LA-based sampling algorithms for stochastic graphs is studied on several well-known real and synthetic stochastic networks. Table 5.9 describes the characteristics of test networks used for the experimentations. These networks are *HT09* (Isella et al. 2011), *Facebook-like-OPSAHL-UCSOCIAL* (Opsahl and Panzarasa 2009), *Cit-HepTh* (Leskovec et al. 2007), *Facebook-wall* (Viswanath et al. 2009), and *LKML-Reply* (KONECT 2016) as real networks and also synthetic networks are: *ER-SG* is generated based on *Erdős–Rényi* model (Erdos and Rényi 1960) which is utilized widely in literature as random networks, *WS-SG* is made based on *Watts–Strogatz* model (Watts and Strogatz 1998) which is a synthetic small world network and *BA-SG* is created based on *Barabási–Albert* model (Barabási and Albert 1999) which is a synthetic scale-free network. The nodes for all synthetic networks is $N = 10,000$ and $p = 0.15$ for ER-SG, $p = 0.2$ for WS-SG, and $m_0 = m = 5$ for BA-SG. The edge weights of these real networks are random variables which are the number of activities among individuals during a specified timestamp for each network.

Algorithm 5-6. EDLAS-SG(G, φ, K_{max}, T_{min})

Input: Stochastic graph $G=\langle V, E, W\rangle$, Sampling rate φ, Thresholds K_{max}, T_{min}.

Output: Sampled graph $G'=\langle V', E', W'\rangle$.

Initialization

   Construct an eDLA by assigning an automaton A_i to each node v_i.

   Initialize action probabilities of automata and set the activity level of each LA to *Passive*.

   t is the iteration number of algorithm which is initially set to 1.

   Pa is the set of learning automata with activity level of *Passive* which is initially set to $\{v_1, v_2, ..., v_n\}$.

   Ac is the set of learning automata with activity level of *Active* which is initially set to empty.

   Of is the set of learning automata with activity level of *Off* which is initially set to empty.

   Fi is the learning automaton with activity level of *Fire* which is initially set to empty.

   $\bar{w}_{\tau_t}$ is the average weight of all samples taken from edges along subgraph of τ_t.

   Ls is a one dimensional array with size $|V|$ which is used to store the number of times that the nodes are sampled.

   $N(v_i)$ is a function that returns the all adjacent nodes of v_i with *Passive* level.

Begin algorithm

   $t \leftarrow 1$; $T_t \leftarrow 0$;

   While ($t \leq K_{max}$ **OR** $|T_t - T_{t\text{-}1}| \geq T_{min}$) **Do**

     $Pa \leftarrow \{v_1, v_2, ...v_n\}$; $Fi \leftarrow \{\}$; $Ac \leftarrow \{\}$; $Of \leftarrow \{\}$;

     Select a starting node v_i randomly by firing function and change its activity level of A_i to *Active* level;

     $Ac \leftarrow Ac \cup \{A_i\}$; $Pa \leftarrow Pa \setminus A_i$;

     $\tau_t \leftarrow \tau_t \cup \{v_i\}$;

     $Ls[v_i] \leftarrow Ls[v_i] + 1$;

     While ($|Of| \leq \phi{\times}|V|$ **AND** $|Ac| \geq 0$) **Do**

       Select one *Active* LA A_i by firing function and then chooses an action according to its action probability vector;

       Let the action chosen by A_i be edge e_{ij};

       Take a sample from the chosen edge e_{ij};

       $\tau_t \leftarrow \tau_t \cup \{e_{ij}\}$;

       $\bar{w}_{\tau_t} \leftarrow \bar{w}_{\tau_t} + \bar{w}_{ij}$;

       $Ls[v_j] \leftarrow Ls[v_j] + 1$;

       $Fi \leftarrow A_i$; $Ac \leftarrow N(v_i) \setminus v_i$; $Pa \leftarrow Pa \setminus Ac$; $Of \leftarrow Of \cup \{Fi\}$; $Fi \leftarrow \{\}$;

     End While

     $T_t \leftarrow [(t\text{-}1)T_{t\text{-}1} + \bar{w}_{\tau_t}]/t$;

     If ($\bar{w}_{\tau_t} \geq T_t$)) **Then**

       Reward the actions chosen by all the fired learning automata with activity level of *Off* in subgraph τ_t;

     Else

       Penalize the actions chosen by all the fired learning automata with activity level *Off* in subgraph τ_t;

     End If

     $t \leftarrow t + 1$;

   End While

   Sort list of visited node Ls in descending order;

   Construct an induced sub-graph $G'=\langle V', E', W'\rangle$ using $\phi{\times}|V|$ mostly visited nodes;

End Algorithm

Fig. 5.12 Pseudo-code of extended distributed learning automata based sampling algorithm for stochastic graphs

The edge weights of synthetic networks are random variables with *Weibull* distribution whose parameters are $a = 0.32$, $b = 0.17$. The chosen parameters is adopted from an empirical observation from *Twitter* for the distribution of lifetime tweets (Bild et al. 2015).

Distance between a property of original networks and that of sampled network is often calculated for evaluating the quality of sampled network. In this section, we use the Kolmogorov-Smirnov (KS) D-statistic and relative error (RE) as distance measures to compare different sampling algorithms in this section for stochastic graphs. These evaluation measures as evaluating criteria are described below.

Table 5.9 Description of the test networks for the experimentation

Network	Node	Edge	Type	Directed	Description
HT09 (Isella et al. 2011)	113	2196	Face-to-face conversation	N	ACM Hypertext 2009 dynamic contact network, Torin, IT, during 3 days
Facebook-like-OPSAHL-UCSOCIAL (Opsahl and Panzarasa 2009)	1899	59,835	User-user messaging	Y	Network of sent messages between the users of an online community of students from the University of California, Irvine
Cit-HepTh (Leskovec et al. 2007)	27,770	352,807	Author–author collaborations	Y	The collaboration graph of authors of scientific papers from the arXiv's high energy physics—theory (Hep-Th) section
Facebook-wall (Viswanath et al. 2009)	63,731	1,545,684	User-user wall post	Y	Network of the wall posts from the Facebook new orleans networks
LKML-Reply (KONECT 2016)	63,399	1,096,440	User-user reply	Y	Networks of the communication network of the linux kernel mailing list
ER-SG	10,000	8,060,206	Synthetic ER stochastic graph	N	Synthetic random graph based on *Erdős–Rényi* model, the edge weights are *Weibull* random variables with $a = 0.32$, $b = 0.17$
WS-SG	10,000	10,001,633	Synthetic WS stochastic graph	N	Synthetic small world random graph generated using watts–strogatz model, the edge weights are *Weibull* random variables with $a = 0.32$, $b = 0.17$
BA-SG	10,000	58,749	Synthetic BA stochastic graph	N	Synthetic scale free random graph based on *Barabási–Albert* model graph, the edge weights are *Weibull* random variables with $a = 0.32$, $b = 0.17$

Kolmogorov-Smirnov D-statistic (KS-D) is one of the statistical test methods commonly used for assessment the distance between two cumulative distribution functions (CDFs). *KS*-D computes the maximum vertical distance between the cumulative distribution function of the original distribution from original graph and that of estimated distribution from sampled graph. This measure is defined as

$$D(P, Q) = \max_{x}\{|P(x) - Q(x)|\} \qquad (5.33)$$

where P and Q are two CDFs of original and estimated data, respectively, and x represents the range of the random variable. KS-D is sensitive to both locations and shapes of distributions, and it is an appropriate measure for the similarity of the distribution. If the value of KS-D between original network and the sampled network as closer as it is to zero, it means the both networks have a greater similarity and as closer as it is to unit, the two networks have a greater difference.

Relative error (RE) can be applied to assess accuracy of the results for a single parameter, which is defined by the following equation

$$RE = \frac{|P - Q|}{P} \qquad (5.34)$$

where, P and Q denote the values of real and sampled parameters (*i.e.*, real clustering coefficient and estimated clustering coefficient) in the original data and obtained samples data respectively (Papagelis et al. 2013).

For LA-based algorithms, the maximum number of iteration K_{max} is set $n \times \varphi$ where n is the number of nodes for each instance graph. The reinforcement scheme used for updating the action probability vector of learning automata is L_{R-I} with a = 0.05 and threshold T_{min} is set to 0.05. The reported results are the averages taken over 30 runs.

The experiment is carried out to study the performance of the LA-based sampling algorithms for stochastic graphs (DLAS-SG and EDLAS-SG) with respect to stochastic strength and stochastic clustering coefficient and the results are compared with pure chance automata. In this experiment, the sampling rate is varied from 10% to 30% with increment 5% and the average results of all algorithms over all test networks are given in terms of average KS-D for strength distribution and average RE for average clustering coefficient in Fig. 5.13. According to the results, we may conclude that for all the test networks, the performance of the sampling algorithms in terms of mentioned measures increases as the sampling rate increases. From the results shown in Fig. 5.13a, it is clear that in terms of KS-D for strength distribution, EDLAS-SG outperforms DLAS-SG algorithm. Also from the results shown in Fig. 5.13b, in terms of RE for clustering coefficient, EDLAS-SG outperforms DLAS-SG algorithm. For all test networks, LA-based algorithms are significantly better than pure chance automata. The results also demonstrate the role of learning automata in guiding the process of taking samples from nodes of the stochastic graph for estimation of edge weights. With the aid of learning automata the process of sampling the stochastic

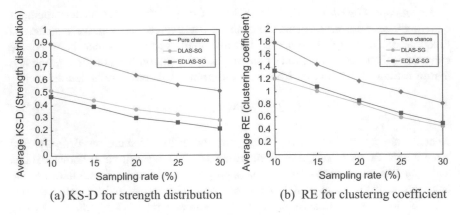

(a) KS-D for strength distribution   (b) RE for clustering coefficient

Fig. 5.13 Comparing LA-based sampling algorithms for different sampling rate in terms of average KS-D for strength distribution and average RE for clustering coefficient

graph is done with fewer numbers of samples taken from the graph as compared to the case where learning is absent and also LA-based algorithms produce better sampled network.

5.5 Community Detection

It is shown that in variants of real world networks, there are common universal characteristics such as small world phenomena (Watts and Strogatz 1998), small shortest path lengths in average manner (Kleinberg 2000), power low degree distribution (Clauset et al. 2009) and specially existence of community structures in networks (Fortunato 2010). Community structure refers to a set of vertices whose edges inside are more densely than edges outside (Rabbany et al. 2013). Finding community in a network known as a community detection problem (also called community structure identification or cluster finding). Community detection plays a significant role for studying and understanding the structure and function of real world networks including several various domains such as clustering web stations having similar functionality and are geographically near to each other may improve the performance of services provided on the World Wide Web (Krishnamurthy and Wang 2000). Online social networks often consist of a set of communities based on common properties of users such as locations, hobbies, interests, activities and carriers (Ranjbar and Maheswaran 2014). Citation networks form communities on the basis of similar research interest topics between authors (Girvan and Newman 2002). Communities (modules) in protein–protein interaction (PPI) networks may correspond to known or even unknown functional modules/protein complexes with significant functionality in cellular biological systems (Gao et al. 2009). Therefore, the detection of the community structure in a network has important practical applications and can help researchers to understand the organization and function of complex network systems.

Detecting the communities in complex networks due to the wide spread of applications have been received a great attentions in literature by scholars. A good review for community detection consists of techniques and applications is the one presented by Fortunato (2010). He classified community detection methods into five categories including traditional algorithms, hierarchical algorithms, modularity-based methods, spectral algorithms and dynamic algorithms. Among the all types of community detection approaches, hierarchical clustering techniques are widely used techniques which put similar vertices into larger communities. Hierarchical clustering algorithms including two categories of divisive and agglomerative form the communities gradually in a hierarchical manner. Several scholars improved the hierarchical algorithms using some metrics to select a suitable partition or a proper set of partitions that satisfies particular metrics such as the number of desired communities, the maximum (or minimum) number of vertices in each community and optimize an objective function (Fortunato 2010). Divisive techniques try to find the edges that connect vertices of different communities and iteratively eliminate them, so that the communities separated from each other. *Girvan* and *Newman* have introduced a famous divisive method (Girvan and Newman 2002) which includes the removal of the edges based on their values of edge betweenness. An agglomerative method, however, tries to form communities in a bottom up manner. In general, the common notion of agglomerative methods is to partition vertices into communities iteratively starting from a partition in which communities are composed of a single vertex. The process of partitioning vertices continues until a single community consists of all vertices of input network is achieved (Newman 2006). Most of agglomerative algorithms select the best partition that maximizes a typical quality objective function. One of the best known quality function and the most widely used quality function is a *modularity* metric which is proposed by Newman et al. (2006). Using modularity, *Clauset,* et al. proposed a fast greedy modularity optimization method (Clauset et al. 2004) which starting from a set of isolated vertices and then a pair of vertices iteratively connected to each other such that it achieves the maximum possible value of modularity at each step. Although using this measure by modularity optimization achieves many promising results for community detection, it is shown that it has some limitations such as resolution limit. For example in the extreme case, the modularity optimization algorithms are failed for a network with several cliques connected by a single edge (Kumpula et al. 2007).

5.5.1 LA-Based Algorithm for Community Detection

In this section, we describe the LA-based algorithm using distributed learning automaton for finding communities in complex social networks (Khomami et al. 2016b). It is assumed that the input network $G = \langle V, E \rangle$ is an undirected and un-weighted network where $V = \{v_1, v_2, \ldots, v_n\}$ is the set of vertices and $E \subseteq V \times V$ is the set of edges in the given network. The distributed learning automaton algorithm including four steps tries to find iteratively a set of communities that are more densely connected internally to each other than to the rest of the network. After the initialization step is performed, by assigning a learning automaton to the vertices of input network, the LA-based algorithm repeats community finding by doing a guided traversal in the network with the help of distributed learning automata, evaluates the set of found communities and updates their action probability vectors iteratively until stopping criteria are satisfied. We describe four steps of the LA-based algorithm in the following subsections in details.

5.5.1.1 Initialization

In the first step, a distributed learning automata $\langle A, \alpha \rangle$ which is isomorphic to the input network is constructed. The resulting network can be defined by 2-tuple $\langle A, \alpha \rangle$ where $A = \{A_1, A_2, \ldots, A_n\}$ is the set of learning automata corresponding to the set of vertices, $\alpha = \{\alpha_1, \alpha_2, \ldots, \alpha_n\}$ denotes the set of actions in which $\alpha_i = \{\alpha_i^1, \alpha_i^2, \ldots, \alpha_i^{r_i}\}$ are the set of actions that can be taken by learning automaton A_i and r_i is the number of actions that can be taken by learning automaton A_i. An action of a learning automaton A_i corresponds to choosing an adjacent vertex of the corresponding vertex v_i. Let $p(v_i) = \{p_i^1, p_i^2, \ldots, p_i^{r_i}\}$ be the action probability vector of learning automaton A_i and $p_i^j = 1/r_i$ equally initialized for all j. At all iterations each learning automaton can be in either active or inactive mode. At the beginning of the LA-based algorithm all learning automata initially are set in inactive mode. Let C_k be the set of vertices in the kth community and initially is set to be empty, G' represent the set of unvisited vertices in the execution of algorithm and is initially equal to G and also π^t is the path of visited vertices at the iteration t.

5.5.1.2 Finding Communities

At the tth iteration of this step, the algorithm finds k communities in such a way that the LA-based algorithm starts with randomly selecting vertex v_i among unvisited vertex set G' and the selected vertex v_i is inserted in the set of current community C_k and current path π^t. Then, learning automaton A_i corresponds to the starting vertex v_i is activated and chooses one of adjacent vertex v_i according to its action probability vector. Let the chosen action by learning automaton A_i be vertex v_j. If the number of internal connections for union of selected vertex v_j and current community C_k is greater than the number of internal connections for current community C_k then v_j is inserted to the set of current community C_k, C_k is removed from set G' and also visited vertex v_j is updated in path π^t. The process of activating an automaton, choosing an action, checking the condition of inserting chosen vertex v_j

in the current community C_k, inserting new vertex v_j to C_k, updating visited vertex v_j in path π' and removing C_k from set G' is repeated until total number of edges inside the current community C_k is more than the total number of edges outside the current community C_k or active learning automaton could not select any action. The process of finding new communities according to the above description and updating path of visited vertices at the current iteration π' is continued when the union of all vertex-set of found communities is equal to the input network G.

5.5.1.3 Computing Objective Function

Let $C' = \{C_1, C_2, \ldots, C_k\}$ be the set of k communities found at the iteration t. The quality of the set of communities found at the iteration t is evaluated via normalized cut as objective function (Dhillon et al. 2004) by following equation

$$NC(C') = \frac{1}{k}\sum_{i=1}^{k}\frac{cut(C_i, \bar{C}_i)}{vol(C_i)} \tag{5.35}$$

where $cut(C_i, \bar{C}_i)$ denotes the number of edges between communities C_i and $\bar{C}_i = G\backslash C_i$, $vol(C_i)$ is the total degree of vertices that are the members of community C_i and also k is the number of communities. Since mentioned in Shi and Malik (2000), normalized cut with low complexity considers extracting the global impression of the network, instead of local features and measures both the total dissimilarity between the different communities as well as the total similarity within the communities. So, using the LA-based algorithm can gradually decreased normalized cut which means that the algorithm gradually converges to the minimum normalized cut and approach to the proper set of communities.

5.5.1.4 Updating Action Probability Vectors

In this step, the set of k communities found at the iteration t is evaluated via corresponding normalized cut and if the value of the normalize cut at current iteration $NC(C')$ is less than or equal to the value of normalized cut at previous iteration $NC(C^{t-1})$, then the chosen action along the path π' by all the activated learning automata are rewarded according to the learning algorithm described in Sect. 5.2 and penalized otherwise.

5.5.1.5 Stopping Criteria

The LA-based algorithm iterates steps 2, 3 and 4 until the number of iterations exceeds a given threshold T or $P_t = \prod_{v_i \in C^t} \max_{v_j \in N(v_i)} \left(p_i^j\right)$ at iteration t becomes greater than a particular threshold τ where p_i^j is the probability of choosing neighboring vertex v_j by learning automaton A_i residing in vertex v_i and $N(v_i)$ is the set of neighboring vertex v_i.

Figure 5.14 shows the pseudo-code for the LA-based community detection algorithm for complex networks.

5.5.2 Experiments

In order to study the performance of the LA-based algorithm for community detection (DLACD), we conducted a number of experiments on the well-known real and synthetic modular networks. Table 5.10 describes the set of real test networks that are used for experiments including the popular real networks: *Karate* (Zachary 1977), *Dolphins* (Lusseau et al. 2003), *Books* (Newman 2015), *Football* (Girvan and Newman 2002) *Reactome*(Joshi-Tope et al. 2005) and *Les-miserables* (Newman 2015) and also *LFR1* and *LFR2* as LFR benchmark networks (Lancichinetti et al. 2008) for synthetic modular networks.

To investigate the performance of the LA-based algorithm, we employed several simulations on the well-known real and synthetic networks. In all experiments presented in this paper, the learning scheme is L_{R-I} and the learning rate is set to be 0.02. The maximum threshold τ is set to 0.9 and maximum iteration T is set to $n \times 1000$ where n is the number of vertices of graph.

We provide an experiment on the known synthetic modular networks based on LFR benchmark. In this experiment, the LFR benchmark network parameters are set as $N = 5000$, $K = 15$, $max_k = 50$, $\mu = \{0.1, 0.5\}$, $min_c = 10$, $max_c = 50$ and the mixing parameter is changed from 0.05 to 0.5 with 0.05 increment. Several community detection algorithms including walk trap (WT) (Pons and Latapy 2005), label propagation (LP) (Raghavan et al. 2007), DANON greedy optimization (DGO) (Arenas et al. 2010), Fuzzy community detection (FCD) (Reichardt and Bornholdt 2004) and Multi-resolution community detection (MRCD) (Ronhovde and Nussinov 2009) are conducted on mentioned synthetic modular networks and compared with respect to NMI results. NMI is defined as follows

Normalized Mutual Information (NMI) (Danon et al. 2005) measures the similarity between known set of communities and set of communities found by the algorithm, where A and B are two partitions of the input network, and NMI is a value between $[0, 1]$, the higher value indicates the partitions A and B are totally independent.

$$NMI(A, B) = \frac{-2 \sum_{a \in A} \sum_{b \in B} |a \cap b| log\left(\frac{|a \cap b|n}{|a||b|}\right)}{\sum_{a \in A} |a| log\left(\frac{|a|}{n}\right) + \sum_{b \in B} |b| log\left(\frac{|b|}{n}\right)}. \qquad (5.36)$$

This measure is useful for artificial networks with a prior knowledge about built-in communities.

Algorithm 1. LA-based algorithm for community detection in complex social networks

Input: A network $G=(V, E)$, Thresholds τ, T // *τ: stopping threshold for product of probabilities, T:maximum iteration number*

Output: Set of found communities C^*

Assumptions

Assign an automaton A_i to each vertex v_i;

Let k is the number of communities;

Let C_k is the set of kth community and initially set to empty;

Let t is the iteration number of algorithm and initially set to 0;

Let $NC(C^t)$ is the normalized cut value for set of communities found at iteration t and initially set to 0;

Let P_t is the product of maximum probabilities in probability vector of LA the vertices of a set of communities at iteration t and initially set to 0;

Let π^t is the path of visited vertices by the algorithm at iteration t and initially set to empty;

Begin

$G' \leftarrow G$; // set of unvisited vertices

All learning automata initially are set inactive mode;

While ($t<T$ **or** $P_t<\tau$)

  Repeat

    $t \leftarrow t+1$;

    $k \leftarrow 1$;

    Vertex v_i is selected randomly from G';

    $C_k \leftarrow C_k \cup v_i$;

    $\pi^t \leftarrow \pi^t \cup v_i$;

    While ($d_{in}(C_k) < d_{out}(C_k)$ **AND** $|\alpha_i|\neq0$) **Do** // *Finding k^{th} community*

      Automaton A_i is activated and then choose an action using its action probability vector;

      Let the chosen action by A_i be v_j;

      If ($d_{in}(C_k \cup v_j) > d_{in}(C_k)$ **AND** $d_{out}(C_k \cup v_j) < d_{out}(C_k)$) **then** //*checking internal and external connections*

        $C_k \leftarrow C_k \cup v_j$;

        $\pi^t \leftarrow \pi^t \cup v_j$;

        $v_i \leftarrow v_j$;

      End If

    End while

    $G' \leftarrow G' \setminus C_k$;

    $k \leftarrow k + 1$;

  Until ($|G'|\neq0$)

  Compute $NC(C^t)$ according to equation (5.35);

  If ($NC(C^t) < NC(C^{t-1})$)

    Reward the actions chosen along the path π_t by all the activated learning automata;

  Else

    Penalize the actions chosen along the path π_t by all the activated learning automata;

  End If

  Compute $P_t = \prod_{v_i \in C^t} \max_{v_j \in N(v_i)}(p_i^j)$;

  Set all learning automata in inactive mode;

End while

$C^* \leftarrow C^t$;

End Algorithm

Fig. 5.14 Pseudo-code of the LA-based algorithm for community detection in complex social network

Table 5.10 Description of the test networks used for the experiments

Networks	Vertex	Edge	Description
Karate	34	78	Zachary karate club network
Dolphins	62	159	The network of dolphins
Les-miserables	77	254	Co-appearance network of characters in the novel Les Miserables
Books	105	441	The network of American politics books
Football	115	615	Network of American college football teams
Reactome	6327	147,547	Network of protein–protein interactions in humans
LFR1	5000	38,160	Synthetic modular benchmark
LFR2	5000	250,000	Synthetic modular benchmark

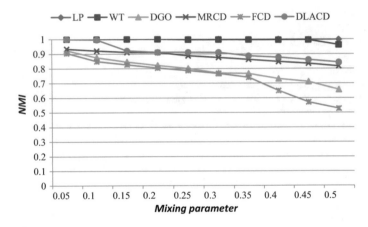

Fig. 5.15 The comparison results on synthetic LFR benchmark with different mixing parameter μ

As shown in Fig. 5.15, for the small size mixing parameter, all algorithms obtained almost same and high NMI values. As the complexity of the network is increased by the value of mixing parameter, the quality of each algorithm with respect to NMI is decreased. This figure also shows that the LA-based community detection algorithm is better than MRCD, DGO and FCD in term of NMI.

Another experiment is done to study the performance of the LA-based community detection algorithm in comparison with other community detection algorithms including walk trap (WT) (Pons and Latapy 2005), label propagation (LP) (Raghavan et al. 2007), DANON greedy optimization (DGO) (Arenas et al. 2010), Fuzzy community detection (FCD) (Reichardt and Bornholdt 2004) and Multi-resolution community detection (MRCD) (Ronhovde and Nussinov 2009) in term of modularity, Min-max-cut, Coverage, Performance value for real world networks. The results of this experiment are given in Table 5.11 for modularity (Newman 2006) as defined follows.

Table 5.11 Comparison of the community detection algorithms in term of modularity

Methods	Networks					
	Karate	Dolphins	Books	Les-miserable	Football	Reactome
MRCD	0.267	0.376	0.337	–	0.504	0.621
DGO	0.383	**0.522**	0.523	–	0.566	0.652
FCD	0.360	0.389	0.442	–	0.476	0.524
WT	**0.407**	0.519	0.52	0.521	**0.602**	0.733
LP	0.317	0.514	0.497	0.353	0.583	**0.741**
DLACD	0.384	0.349	**0.526**	**0.524**	0.584	0.651

Note The best results are highlighted in boldface

$$Q = \frac{1}{2m} \sum_{C \in P} \sum_{v_i, v_j \in C} \left[A_{i,j} - \frac{k_i k_j}{2m} \right] \qquad (5.37)$$

where A is the adjacency matrix that $A_{i,j}$ is equal to 1 if there is an edge between vertex v_i and vertex v_j and zero otherwise. $k_i = \sum_j A_{ij}$ is the degree of vertex v_i and m is the total number of edges in the network. The summation is over all pairs of vertices that member of the same community C of partitioning P.

According to the results of Table 5.11, for Books and Les-Miserables, the LA-based community detection algorithm outperforms the other algorithms in terms of modularity. Also, for other networks, the modularity of LA-based algorithm is approximately in the same range as of the other algorithms. It is noted that the DANON is designed for only modularity as objective function and thus it is biased toward high modularity's; therefore, it is not surprising that DANON reaches higher modularity in some networks, and this means that some more criteria must be considered to analyze the quality of algorithms accurately. The modularity of DLACD is low in some of the datasets, but it is still much higher than those of many metrics that shown in the following. The obtained results are negligible for other network datasets and for most of networks are in the same range. Note that the LA-based community detection algorithm tries to find optimal normalized cut value in the network and it is independent of some restrictions such as resolution limits.

5.6 Conclusion

The focus of this chapter was on the introducing learning automata based algorithms for solving stochastic graph problems, social network analysis, network sampling algorithms for deterministic and stochastic graphs and community detection. The learning automata based algorithms for stochastic graphs on network of learning automata tried to estimate the unknown distribution of the weigh associated with an edge by taking samples from the edges of stochastic graphs. For the sake of social network analysis when stochastic graph are used as graph model of network, the learning automata can be used by the network as a means for observing the time varying parameters of the network for the purpose of analyzing network. The aim of the LA-based algorithms are to collect information from the social network in order to find good estimates for the network parameters and measurements using fewer numbers of samples than that of standard sampling methods. The performances of the LA-based algorithms were tested through experimentations on a number of real and synthetic stochastic networks. The experimental results showed the simplicity, superiority and efficiency of the LA-based algorithms for social network analysis.

Chapter 6
Adaptive Petri Net Based on Learning Automata

6.1 Introduction

Petri Nets (PNs) are graphical tools, useful for formal description of systems, whose dynamics are characterized by concurrency, synchronization, mutual exclusion, and conflict, which are typical features of distributed environments. PNs incorporate a notion of (distributed) state and a rule for state change that allow them to capture both the static and the dynamic characteristics of a real system (Marsan et al. 1994).

A PN comprises of three structural components; places, transitions, and arcs. Places are used to describe possible local system states, which can be either conditions or situations. Transitions are used to describe events that may modify the system state. Arcs specify the relation between local states and events; they indicate the local state in which an event can occur, and local state transformations induced by that event.

The evolution of a Petri net can be described in terms of its initial state and a number of firing rules (Murata 1989). A state, or marking, of a PN is specified by tokens which are indistinguishable markers that reside in places. If a place describes a condition, that condition is true if a token is present in the place and is false otherwise. If a place defines a situation, the number of tokens contained in the place is used to specify the situation. Firing rules define the marking modification induced by transition firings. A transition can fire only if all of its input places contain at least one token. In this case the transition is said to be enabled. The firing of an enabled transition removes one token from each of its input places, and generates one token in each of its output places. More tokens are required/generated in an input/output place if the weight of its arc is greater than one. The firing of a transition is an atomic operation. Tokens are removed from input places, and deposited into output places with one indivisible action (Marsan et al. 1994).

In the evolution of a PN, two or more enabled transitions may be in conflict; this is occurred when firing of one transition results in disabling the other(s). Thus, in a PN, a controlling mechanism is needed which specifies the order for firing of

© Springer International Publishing AG 2018
A. Rezvanian et al., *Recent Advances in Learning Automata*, Studies in
Computational Intelligence 754, https://doi.org/10.1007/978-3-319-72428-7_6

transitions so that probable conflicts among transitions will be resolved. Different controlling mechanisms have been proposed in literature so far such as random selection (Peterson and Net 1981), queue regimes (Burkhard 1981), priority (Bause 1996, 1997), and external controller (Holloway and Krogh 1994) to mention a few. Among these mechanisms, random selection is the mostly used mechanism.

Despite its simplicity and effectiveness in many situations, random selection is not an appropriate mechanism for resolving conflicts. Sets of conflicting transitions may change when markings change and hence the random selection for selecting the transition to be fired is inappropriate when PN is used to model the dynamics of real world problems. This is mainly due to the fact that random selection does not consider the dynamics of the environment under which it operates and hence, it cannot adapt itself appropriately. To tackle this problem, one approach is to add a mechanism to PN to adaptively resolve conflicts among the enabled transitions.

Petri nets can also be used to represent algorithms of different classes (Reisig 2013a). One class of algorithms is cellular algorithm (Terán-Villanueva et al. 2013) which commonly follows the so-called distributed computing model (Linial 1992). In this model, each cell executes the same algorithm, referred to as the local algorithm, and the whole problem is solved by a mean of cooperation between neighboring cells. Petri nets can be used to represent such cellular algorithms as well. For this to be made possible, two steps must be taken: (1) a Petri net must be designed for representing the local algorithm to be executed in each cell. This Petri net is then replicated into all cells; and (2) a mechanism must be devised to handle the required cooperation between any two neighboring cells. A common mechanism for cooperating between any two neighboring cells is representing the cooperation within the structure of the two Petri nets, designed for the cells; that is by connecting the two Petri nets using ordinary elements of Petri nets such as common places and inhibitor arcs (Reisig 2013b). This results in a large Petri net for the whole algorithm, which is extremely hard to analyze and simulate (Chiola and Ferscha 1993; Fujimoto 2001). To tackle this problem, one approach is to handle the cooperation, between the two neighboring cells, within the controlling mechanisms of their Petri nets, instead of their structures.

The aim of this chapter is to fuse Petri nets with learning automata in order to make several hybrid machines, suitable for representing adaptive algorithms. A learning automaton (LA) is an adaptive decision-making unit that improves its performance by learning how to choose the optimal action from a finite set of allowed actions through repeated interactions with a random environment (Thathachar and Sastry 2004; Narendra and Thathachar 1989). An attractive feature of learning automata approach is that it could be regarded as a simple unit from which complex systems such as games, cellular and distributed could be constructed to handle the complicated learning problems. In proposed hybrid machines, learning automata are used as adaptive controlling mechanisms which resolve conflicts among transitions in PNs.

Adaptive Petri net based on Learning Automata (APN-LA) is introduced by the fusion of learning automata and ordinary Petri nets (Vahidipour et al. 2015). A generalization of this machine, referred to as ASPN-LA, is represented

(Vahidipour et al. 2015) and resulted from the fusion of learning automata with Stochastic Petri nets (SPN). To be able to represent cellular algorithms, adaptive Petri net based on Irregular Cellular Learning Automata (APN-ICLA) (Vahidipour et al. 2017a) and Cellular Adaptive Petri Nets based on Learning Automata (CAPN-LA) (Vahidipour et al. 2017c) are introduced. The APN-ICLA is a Petri net in which a cellular structure of learning automata, i.e., ICLA (Esnaashari and Meybodi 2015), is utilized as the adaptive controlling mechanism. On the other hand, in the CAPN-LA, a number of APN-LAs are organized into a cellular structure. This cellular structure represents a neighborhood relationship between the APN-LAs. Each APN-LA represents a local algorithm to be executed within a cell and the cooperation between neighboring cells is handled by the cooperation between the LAs reside in the Petri nets of the cells.

Applications of the proposed hybrid machines to the following three problems are also presented: (1) Priority Assignment in Queuing systems with unknown characteristics (Vahidipour et al. 2015; Vahidipour and Esnaashari 2017), (2) finding the shortest path in stochastic graphs (Vahidipour et al. 2017b), and (3) the vertex-coloring problem in graphs (Vahidipour et al. 2017a, c). Since the notations used in this chapter may be different from those used in other chapters, the notations and their descriptions used in this chapter are listed in Table 6.1.

This chapter is organized as follows: In Sect. 6.2 a brief introduction to Petri nets and adaptive Petri nets will be presented. Four adaptive Petri nets, designed atop of the fusion of learning automata and Petri nets will be introduced in Sect. 6.3 and their evolution will be studied. In Sect. 6.4, the ASPN-LA will be applied to solve the problem of priority assignment in a queuing system with unknown characteristics. The queuing system consists of a server and m queues. Jobs are processed from the queues in such a way that the average waiting time of the system will be minimized. Extensive computer simulations, reported in this section, justify the effectiveness of the proposed ASPN-LA. Section 6.5 will review two applications of the ASPN-LA and CAPN-LA in graphs. In the first application, the ASPN-LA is used to analyze an algorithm which is proposed to solve the shortest path problem in stochastic graphs. In the second application, the CAPN-LA is used to represent a cellular algorithm for solving the vertex coloring problem. For each of the applications, computer simulations are conducted to show the effectiveness of the proposed adaptive Petri nets. Section 6.6 concludes this chapter and presents suggestions for further studies.

6.2   Petri Nets

Petri nets (PNs) are graphical and mathematical modeling tools, which have been applied to many different systems. They are used to describe and study information processing systems with concurrent, asynchronous, distributed, parallel, non-deterministic and/or stochastic characteristics (Reisig 2013b). In such

Table 6.1 List of symbols used in this chapter

Symbols	Description
LA_i	ith learning automaton
m_i	The number of actions in LA_i
β_i	The reinforcement signal for LA_i
q_i	The action probability vector of LA_i
$\hat{q}_i$	Scaling action probability vector of LA_i (an LA with changing number of actions)
P	The set of places in the PN
T	The set of transitions in the PN
ω_t	The weight assigned to the immediate transition t in the SPN
M_0	The initial marking of a PN system
$\bullet x$	The pre-set of x
$\bullet x$	The post-set of x
$M[t >$	Transition t is enabled in marking M
$M[t > M'$	Firing the transition t in marking M yields a new marking M'
$\mathcal{N}$	An adaptive Petri net
$\hat{P}$	The set of places in the APN
$\hat{T}$	The set of transitions in the APN
T^u	The set of updating transitions in the APN
t_i^u	ith updating transition in the APN
S	The set of clusters in the APN
s_0	Cluster of concurrent transitions in the APN
ω_t^0	The weight assigned to the immediate transition t in the cluster s_0 in the ASPN-LA
$s_i, i > 0$	A cluster of conflicting transitions in the APN
s_{-1}	Cluster of timed transitions in the ASPN-LA
A_i	ith APN-LA in the CAPN-LA
$\mu_{i,j}$	jth marking in A_i
$s_{i,0}$	Cluster of concurrent transitions in A_i
$s_{i,1}$	Cluster of conflicting transitions in A_i
$\hat{F}$	The set of local rules
F_i	The local rule of ith cell in APN-ICLA and CAPN-LA
$\underline{q}(k)$	The conflicting transitions configuration in CAPN-LA
$\underline{q}_i$	The selection probability of transitions in the cluster $s_{i,1}$
N_i	The neighboring set of any cluster $s_{i,1}$
N_i^j	jth neighboring cluster of the cluster $s_{i,1}$
Ft_i	The last fired transition of the cluster $s_{i,1}$
$\pi_\mu(k)$	The probability of the APN-LA A_i being in the marking μ_i at time instant k
t_i^j	jth transition in the cluster $s_{i,1}$
$q_i^j(k)$	The probability of choosing the transition t_i^j for firing at time instant k
$\hat{q}_i^j(k,\mu)$	The scaled probability of choosing the transition t_i^j for firing in marking μ

(continued)

Table 6.1 (continued)

Symbols	Description
$d_i^j\left(\underline{q}(k)\right)$	The average reward of firing the transition t_i^j at time instant k
$D_i\left(\underline{q}(k)\right)$	The average reward for the APN-LA A_l at time instant k
$D\left(\underline{q}\right)$	The total average reward for the CAPN-LA at configuration $\underline{q}$
M_E	Measure of expediency

information processing systems, controlling mechanisms may be needed with respect to synchronization and conflict resolution. For PNs, control can be thought as the mechanism which specifies the order, by which the transitions are fired, so that probable conflicts among transitions will be resolved. Different controlling mechanisms have been proposed in literature so far for resolving conflicts among a set of enabled transitions in a PN. A controlling mechanism is an adaptive one, which adapts itself to the changes in markings caused by changes of conflicts among the transitions. Such an adaptation is needed, when Petri nets are used to model dynamics of real world problems.

As a generalization of state machines, Petri nets have been around for almost fifty years [developed by Carl Adam Petri in his doctoral thesis (Petri)]. They can be used graphically and mathematically to communicate between technical and non-technical audiences and construct behavioral models of process-oriented systems. In (van der Aalst 1998), Van der Aalst suggests the following as main reasons for using Petri nets as a process modelling technique:

- Their graphical nature has underlying formal semantics,
- They are state-based and event-based,
- A range of analysis techniques is available to study properties of the system represented by a Petri net.

6.2.1 Definitions and Notations

A Petri net (PN) is graphically represented by a directed bipartite graph, in which the two types of nodes (places and transitions) are drawn as circles and either bars or boxes, respectively. Arcs of the graph are classified (with respect to transitions) as input arcs (arrow-headed arcs from places to transitions) and output arcs (arrow-headed arcs from transitions to places). A Petri net can be formally defined in the following definition.

Definition 6.1 A Petri net (PN) is a 4-tuples $\{P, T, F, W\}$ (Popova-Zeugmann 2013), where

- P and T are finite, non-empty, and disjoint sets. P is the set of places and T is the set of transitions with $P \cup T \neq \varnothing$ and $P \cap T = \varnothing$,
- $F \subseteq (P \times T) \cup (T \times P)$ is called a flow relation of the net, represented by arcs with arrows from places to transitions or from transitions to places, and
- $W : ((P \times T) \cup (T \times P)) \rightarrow \mathbb{N}$ is a mapping that assigns a weight to an arc: $W(x, y) > 0$ *iff* $(x, y) \in F$ and $W(x, y) = 0$ otherwise, where $x, y \in P \cup T$. $\square$

Places can contain tokens, which are drawn as black dots within places. The state of a PN is called marking, and is defined by the number of tokens in each place. As in classical automata theory, there is a notion of initial state (initial marking) in the PN. The number of tokens, initially found in the PN places, can be conveniently represented by symbols that are parameters of the system. A marking of PN can be defined as follows.

Definition 6.2 A marking M of a Petri net N is a mapping from P to $\mathbb{N}$. $M(p)$ denotes the number of tokens in place p. A place p is marked by a marking M *iff* $M(p) > 0$ (Li and Zhou 2009). $\square$

The initial marking is denoted by M_0. A Petri net along with an initial marking M_0 creates a PN system:(N, M_0). The initial marking of a PN system can generally change into a successor marking according to certain rules and this can itself transform in turn into successor markings. The rules describing the possible changes from one marking to the next one are called firing rules, the occurring change itself is called a firing. Throughout such firings, the distribution of tokens over the places of a Petri net can change and thereby the whole view of the net changes. In other words, the Petri net also has a dynamic aspect, or evolution, which is defined by the firing rules.

Before we define the firing rules for PNs, some basic notions need to be explained. For each element x, either a place or a transition, its pre-set is defined as $\bullet x = \{y \in P \cup T | (y, x) \in F\}$ and its post-set is defined as $x \bullet = \{y \in P \cup T | W(x, y) > 0\}$. For example, for every place p, the pre-set (or pre-transitions) is $\bullet p = \{t | t \in T \wedge W(t, p) > 0\}$ and post-set (or post-transitions) is $p \bullet = \{t | t \in T \wedge W(p, t) > 0\}$ (Popova-Zeugmann 2013).

The dynamic aspect, or evolution, of a Petri net is defined by the firing rules. The firing rules reflect causal relations within a permanently changing system: the events of the real system are modeled by transitions of the Petri net. The causes or preconditions of an event are represented by the pre-places of the transition modeling the event. The places in the post-set of the transition describe the post-conditions of the event, which of course in turn can be preconditions of other events. Whenever a place in the pre-set is marked, the respective condition is considered to be fulfilled. In the real system an event can take place, when all preconditions of the event are fulfilled. In the Petri net the occurrence of the event is represented by firing of the corresponding transition. After an event has taken place, its preconditions (in general) are not fulfilled any more. Therefore, the corresponding places in the pre-set are no longer marked. Instead, the post-conditions of the event are fulfilled and places in the post-set of the transition are marked. This atomic process in the real system provides the basic idea for the firing rule in the Petri net. In classic Petri nets, this basic idea is carried out with the multiplicity $W \equiv 1$ (these are the so-called ordinary Petri nets) and each place holds at most one token, i.e., for each reachable marking M it holds that $M(p) \in \{0, 1\}$ (these are the so called 1-safe Petri nets). Ordinary, 1-safe Petri nets are also called condition/event nets (Popova-Zeugmann 2013).

If not all multiplicities of arcs in a Petri net are 1, the firing rule is extended consistently. The preconditions of an event are fulfilled if, for each place, the number of tokens it holds is no smaller than the multiplicity of the arc from this place to the corresponding transition. After an event has taken place, every place in the post-set of the transition obtains the number of tokens equivalent to the multiplicity of the arc from the transition to the corresponding place. A transition t can therefore fire if the Petri net is in a marking M, which assigns at least as many tokens as t needs in each place in the pre-set of t. All preconditions can be regarded as fulfilled. We can now formally introduce the notions enabled and firing.

Definition 6.3 A transition $t \in T$ is *enabled* at a marking M *iff* $\forall p \in \bullet t, W(p, t) \leq M(p)$. This fact is denoted by $M[t >$. Firing t yields a new marking M' such that $\forall p \in P, M'(p) = M(p) - W(p, t) + W(t, p)$, as denoted by $M[t > M'$. Marking M' is called a directly reachable marking from M. Marking M'' is said to be reachable from M if there exists a sequence of transitions $\sigma = t_0 t_1 \cdots t_n$ and markings $M_1, M_2, \cdots$, and M_n such that $M[t_0 > M_1[t_1 > M_2 \cdots M_n[t_n > M''$ holds (Li and Zhou 2009).   □

This firing rule defines the firing of a single transition, i.e., it is a mono-firing rule. It is also possible to define a rule firing a set (step) of transitions. This firing rule, also called a step firing rule, is usually used in time-dependent Petri nets (Popova-Zeugmann 2013). In this chapter, we will always use the mono-firing rule.

A special kind of arc is inhibitor arc. An inhibitor arc is a directed arc which leaves a place P_i to reach a transition t_j. Its end is marked by a small circle as shown in Fig. 6.1. The inhibitor arc between P_2 and t_1 means that transition t_1 is only enabled if place P_2 does not contain any tokens. In this figure, transition t_1 cannot fire, because the token in P_2 inhibits it. PNs with inhibitor arcs have the computational power of Turing machines and in the general case they cannot be transformed into ordinary PNs (Reisig 2013b).

6.2.1.1 Reachability Set and Reachability Graph

The firing rule defines the evolution of the PN systems. Starting from the initial marking, it is possible to compute the set of all markings reachable from it (*state space* of the PN system) and all the paths that the system may follow to move from state to state.

Fig. 6.1 Transition t_1 cannot
fire, because the token in P_2
inhibits it

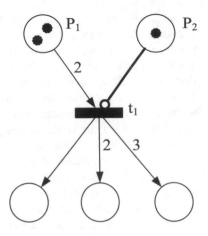

Definition 6.4 The Reachability Set of a PN system with initial marking M_0 is denoted $RS(M_0)$, and it is defined as the smallest set of markings such that

- $M_0 \in RS(M_0)$,
- $M_1 \in RS(M_0) \,\&\, \exists t \in T : M_1[t > M2 \Rightarrow M_2 \in RS(M0)$ (Reisig 2013b). $\square$

When there is no possibility of confusion, we indicate with RS the set $RS(M_0)$. We also indicate with $RS(M)$ the set of markings reachable from a generic marking M (Reisig 2013b).

The RS contains no information about the transition sequences fired to reach each marking. This information is contained in the reachability graph, where each node represents a reachable state, and there is an arc from M_1 to M_2 if the marking M_2 is directly reachable from M_1. If $M_1[t > M_2$, the arc is labelled with t. Note that more than one arc can connect two nodes (it is indeed possible for two transitions to be enabled in the same marking and to produce the same state change), so that the reachability graph is actually a multi-graph.

Definition 6.5 Consider a PN system and its reachability set RS. Reachability Graph, referred to as $RG(M_0)$, is the labelled directed multi-graph, in which M_0 is taken as the initial node, whose set of nodes is RS and whose set of arcs A is defined as follows:

- $A \subseteq RS \times RS \times T$,
- $\langle M_i, M_j, t \rangle \in A \Leftrightarrow M_i[t > M_j$. $\square$

We use the notation $\langle M_i, M_j, t \rangle$ to indicate that a directed arc connects the node corresponding to marking M_i to the node corresponding to marking M_j, and a label t is associated with the arc. When the RS is not finite, the RG is not finite as well. This happens when some place in the PN system can contain an unlimited number of tokens.

6.2.1.2 Examples

In this section, we describe two examples of Petri net systems, borrowed from (David and Alla 1994). The first example is referred to as *billiard balls* and is illustrated in Fig. 6.2. Two billiard balls, A and B, move on the same line parallel to one of the bands. In Fig. 6.2a, A moves to the right while B moves to the left (We assume they have the same speed). The conditions for the event hitting the balls to occur are satisfied. When this event occurs, the balls set off again in the opposite direction at the same speed (Fig. 6.2c). A ball which strikes a band sets off in the other direction at the same speed (Fig. 6.2e).

This example is modeled by the Petri net in Fig. 6.2b, d, f. The event hitting is associated with transition T_1. When places P_1 and P_2 are marked, the conditions for

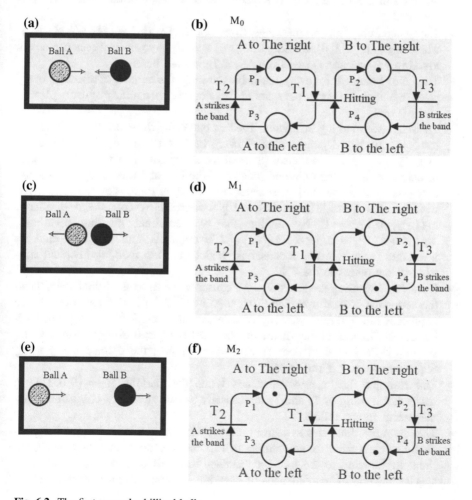

Fig. 6.2 The first example: billiard balls

this event to occur are fulfilled. This appears in Fig. 6.2b because transition T_1 is enabled. We do not know when this event will occur, but we know that it will occur (because T_1 is the only enabled transition) for the current marking M_0. After hitting, i.e., firing of transition T_1, the marking is $M_1 = (0, 0, 1, 1)$, which is illustrated in Fig. 6.2d. Then, there are two enabled transitions, namely T_2 and T_3. Transition T_2 corresponds to the event A strikes the left band, and transition T_3 corresponds to the event B strikes the right band. We do not know which of these two events will occur first. If T_2 is fired before T_3, the marking becomes $M_2 = (1, 0, 0, 1)$ as shown in Fig. 6.2f. In Fig. 6.2f, only transition T_3 is enabled, then the next event will be B strikes the right band; after firing of T_3 the initial marking M_0 is obtained again.

From Fig. 6.2, one may conclude the following concepts (David and Alla 1994):

- This is an autonomous Petri net; neither time nor external synchronization is involved in this model. This is a qualitative description of the system which is observed.
- The reachability set of markings from M_0 is $RS(M_0) = \{M_0, M_1, M_2, M_3\}$, where $M_3 = (0, 1, 1, 0)$. Markings M_0, M_1, and M_2 have already been presented. Marking M_3 can be reached from M_1, if T_3 is fired before T_2.
- The PN is bounded; for every reachable marking, the number of tokens in every place is bounded. Furthermore, the PN is safe, i.e., the marking of every place is either 0 or 1.
- There are two marking invariants. The first one is $M_i(P_1) + M_i(P_3) = 1$; for any reachable marking M, the number of tokens in the set of places $\{P_1, P_3\}$ is equal to 1. This invariant has a clear physical meaning: ball A may have two states, namely moving to the right and moving to the left, and it is always in one and only one state. Place P_1, is associated with the first state, and P_3 is associated with the second one. The set $\{P_1, P_3\}$ is a conservative component. Similarly, $M_i(P_2) + M_i(P_4) = 1$. By adding the two minimal marking invariants, $M_i(P_1) + M_i(P_3) + M_i(P_2) + M_i(P_4) = 2$ is obtained. This is a new marking invariant. The whole PN is conservative since the last marking invariant contains all the places in the PN.
- From M_0, the firing sequence $T_1 T_2 T_3$ causes a return to the initial state. Then this sequence is repetitive. The firing sequence $T_1 T_3 T_2$ is also repetitive. These sequences are different, but they contain the same number of firings for each transition. There is a firing invariant whose meaning is: if every transition in the set $\{T_1, T_2, T_3\}$ is fired once from a state M, then the corresponding firing sequence causes a return to M_0.
- This example illustrates concurrency. When the marking $M_2 = (0, 0, 1, 1)$ is reached, the firing of T_2 and T_3 are causally independent (i.e., concurrent, they may occur in any order).
- This example illustrates synchronization. Although the balls behave independently of each other for some time, ball A cannot change from 'moving to the right' to 'moving to the left' independently of ball B (and vice versa). This synchronization of both changes of direction is illustrated by transition T_1

   The second example is referred to as *common memory* and is illustrated in
Fig. 6.3. Computer CP_1 has three possible states: either it requests the memory
(place P_1), or it uses it (place P_2), or it does not need it (place P_3). Similarly,
computer CP_2 has three possible states.

   When the memory is free (place P_7 is marked) and CP_1 and CP_2 request it,
transitions T_1 and T_4 are enabled (Fig. 6.3a). If transition T_1 is fired, then CP_1 uses the
memory (Fig. 6.3b). When CP_1 has finished, transition T_2 is fired, then the marking in
Fig. 6.3c is reached (the memory is released, and may be re-used either by CP_1 or
CP_2). From, one may conclude the following concepts (David and Alla 1994):

- There is a conflict: the place P_7 is an input of both transitions T_1 and T_4. This is
  a structural conflict (the structural conflict will be defined in the Sect. 6.2.2).
  Now, when there is one token in every place P_1, P_4 and P_7 (Fig. 6.3a), there is
  an effective conflict between transition T_1 and T_4 (the effective conflict is also

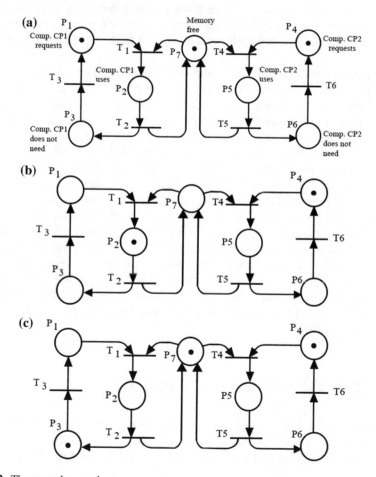

Fig. 6.3 The second example: two computers use a common memory

defined in Sect. 6.2.2). As a matter of fact both transitions are enabled, but only one can be fired. If T_1 is fired, then T_4 is no longer enabled, and vice versa. We will define conflict later in Sect. 6.2.2.

• Resource sharing: the memory may be used by two computers, but not at the same time (this implies a conflict). One can observe that there is a marking invariant $M_i(P_2) + M_i(P_5) + M_i(P_7) = 1$. This means that if there is a token in P_2, there is no token in P_5, and vice versa. This property expresses that the memory cannot be used by both computers at the same time.

6.2.1.3 Abbreviations and Extensions of Petri Nets

Petri nets cab be divided into three main classes (David and Alla 1994): ordinary Petri nets, abbreviations and extensions. In the ordinary Petri net, all the arcs have the same weight which is 1, there is only one kind of token, the place capacities are infinite (i.e., the number of tokens is not limited by place capacities), the firing of a transition can occur if each its pre-place contains at least one token, and no time is involved.

The abbreviations correspond to simplified representations, useful in order to lighten the graphical representation, to which an ordinary Petri net can always be made to correspond. Generalized PN (Peterson and Net 1981), finite capacity PN (David and Alla 1994), and colored PN (Jensen 1981) are abbreviations. Then they have the same power of description as the ordinary Petri nets.

The extensions correspond to PNs to which functioning rules have been added, in order to enrich the initial model, enabling a greater number of applications to be treated. Three main subclasses may be considered (David and Alla 1994). The first subclass corresponds to PNs which have the description power of Turing machines: inhibitor arc PN (Peterson and Net 1981) and priority PN (Bause 1996, 1997). The second subclass corresponds to extensions allowing modeling of continuous and hybrid systems: continuous PN (David and Alla 1994) and hybrid PN (Bail et al. 1991). The third subclass corresponds to non-autonomous Petri nets, which describe the functioning of systems whose evolution is considered by external events and/or time: synchronized PN (di Mascolo et al. 1991), timed PN (Ramchandani 1974), interpreted PN (David and Alla 1994) and stochastic PN (Marsan et al. 1994).

6.2.2 Controlling Mechanisms in Petri Nets

The evolution of a Petri net system can be described in terms of its initial marking and a number of firing rules. At any given time during the evolution of a Petri net system, current marking of the system evolves to a new marking by applying the firing rules. Firing rules of an ordinary Petri net system, for any marking M, consist of three steps (Murata 1989):

1. Determining the set of enabled transitions from available transitions,
2. Selecting a transition from the set of enabled transitions for firing, and
3. Generating a new marking by firing the selected transition.

Mechanisms which determine the set of enabled transitions from the available transitions in PNs try to reduce the number of enabled transitions by introducing concepts such as inhibitor arc (Peterson and Net 1981), priority (Bause 1996, 1997), time (Marsan et al. 1994), and color (Jensen 1981).

In a Petri net, a decision point is raised when two or more transitions are simultaneously enabled in a given marking. The decision to be made at such a point is the selection of an enabled transition for firing. Decision making in Petri nets is accomplished by so called *controlling mechanisms*. Different controlling mechanisms have been proposed in literature so far such as random mechanism (Peterson and Net 1981), queue regimes (Burkhard 1981), priority (Bause 1996, 1997), and external controller (Holloway and Krogh 1994). Before giving a more details about these different control mechanisms, a number of required definitions are given.

Definition 6.6 Let $N = (P, T, F, W, M_0)$ be a Petri net system (Li and Zhou 2009).

- Two transitions t_1 and t_2 from T are in a static (structural) conflict, if they have at least one common place in their pre-sets, i.e., $\bullet t_1 \cap \bullet t_2 \neq \varnothing$.[1]
- Two transitions t_1 and t_2 from T are in a dynamic (effective) conflict in the marking M if they are in a static conflict and by firing one of the transitions in marking M the other one may become disabled, i.e., $\bullet t_1 \cap \bullet t_2 \neq \varnothing$ and $W(p, t_1) \leq M(p)$ and $W(p, t_2) \leq M(p)$ but $W(p, t_1) + W(p, t_2) \nleq M(p)$.
- Two transitions t_1 and t_2 from T are said to be concurrent in marking M if they are both enabled in M, and they are not in effective conflict, i.e., $\bullet t_1 \cap \bullet t_2 = \varnothing$ and $W(p, t_1) \leq M(p)$ and $W(p, t_2) \leq M(p)$ and $W(p, t_1) + W(p, t_2) \leq M(p)$. □

Control for simultaneous processes may be needed with respect to synchronization and conflict resolution. For Petri nets, controlling mechanism can be considered as a mechanism that specifies the order, by which transitions are fired so that probable conflicts among transitions can be resolved. The most used controlling mechanism in the literature is the random selection which is inappropriate in conflicting situations (Peterson and Net 1981). The queue regime (Burkhard 1981) is another set of mechanisms used to control the order of firing in the PN. In this mechanism, enabled transitions are put in a queue. To select a transition from the queue of transitions, different approaches have been addressed. None of these approaches is appropriate to conflicting situations when the environment is dynamic.

---

[1] It is noted that, in this chapter, we will always use conflict instead of the static conflict unless stated otherwise. In addition, term "a set of conflicting transitions" indicates a set of transitions, each transition of which is in a static conflict with at least another transition from this set.

In priority net, a priority level is assigned to each transition. In any marking, an enabled transition with the highest priority will be fired (Bause 1996, 1997). The priority net can be either static (Bause 1996) or dynamic (Bause 1997; Kounty 1992). In the static priority net, the priority level of a transition does not change, whereas in the dynamic priority net, the priority level of a transition, even though different for different markings, for a specific marking is fixed and does not change as priority net evolve. For this reason, neither static priority net nor dynamic priority net is a good candidate to control firing of transitions, when the environment is dynamic.

A Petri net is introduced in (Burkhard 1985), in which the firing of transitions is controlled by a set of finite automata. In this PN, transitions are partitioned into a number of disjoint sets. Each automaton is responsible for controlling transition firing in the corresponding set of transitions. This mechanism is able to handle conflicts among transitions, but due to the deterministic nature of finite automaton, it cannot adapt itself to the environmental changes.

Controlled Petri nets (CtlPNs) are a class of PNs with a new kind of places called control places (Holloway and Krogh 1994). A control place allows an external controller to influence the evolution of the CtlPN. In most cases, the external controller must control the evolution of CtlPN so that the system always remains in a specified set of allowed states, or equivalently, a set of forbidden markings are never reached by the CtlPN. Such a control policy can be designed by general approaches such as linear integer programming (Li and Wonham 1994) and path-based approaches (Krogh et al. 1991). Controlled Petri nets, within the framework of supervisory control (Murata 1989) is used as a discrete event model (Chen et al. 2013; Holloway et al. 1997). A drawback of these mechanisms is the lack of learning capability, which prevents them from being adapted to the changes in the environment.

A controlling mechanism must be able to select an enabled transition for firing in any of the following two situations: (1) All enabled transitions are concurrent and they are causally independent, which means one transition may fire before, after or in parallel with the others (Murata 1989), and (2) some of the enabled transitions are in conflict, which means if any of them is fired, none of the others can be fired (Murata 1989). Considering these two situations, a controlling mechanism in Petri net can be divided into two sub-mechanisms: a mechanism for selecting an enabled transition among a set of concurrent transitions and the other mechanisms for selecting an enabled transitions among a set of conflict transitions. In this chapter, these two sub-mechanisms are referred to as *concurrent selector* and *conflict resolver*, respectively.

6.2.3   Time in Petri Nets

The (ordinary) Petri net systems include no notion of time. The concept of time was intentionally avoided in the original work by Petri (1962), because of the effect that timing may have on the behavior of PNs. In fact, the association of timing constraints with the activities represented in PN systems may prevent certain transitions from firing, thus destroying the important assumption that all possible behaviors of a real system are represented by the structure of the PN.

Very soon PNs were recognized as possible models of real simultaneous systems, capable of coping with all aspects of parallelism and conflict in asynchronous activities with multiple actors. In this case, timing is not important, when considering only the logical relationships among the entities that are part of the real system. The concept of time becomes instead of paramount importance, when the interest is driven by real applications whose efficiency is always a relevant design problem. Indeed, in areas like hardware and computer architecture design, communication protocols, and software system analysis, timing is crucial even to define the logical aspects of the dynamic operations.

The pioneering works in the area of timed PNs were performed by Merlin and Farber (1976), and Noe and Nutt (1973). In the former work, the authors applied PNs augmented with timing to communication protocol modelling; in the latter case, a new type of timed nets was used for modelling a computer architecture. In both cases, PNs were not viewed as a formalism to statically model the logical relationships among the various entities that form a real system, but as a tool for the description of the global behavior of complex structures. PNs were used to narrate all the possible stories that the system can experience, and the temporal specifications were an essential part of the picture.

Different ways of incorporating timing information into PN systems were proposed by many researchers during the last two decades (Ramchandani 1974; Merlin and Farber 1976; Zuberek 1980; Holliday and Vernon 1987; Ramamoorthy and Ho 1980; Sifakis 1978; Wong et al. 1985); the different proposals are strongly influenced by the specific application fields. In this chapter, we do not discuss and compare the different approaches, but only the temporal semantics used in Stochastic Petri nets (SPN).

6.2.3.1 Stochastic Petri Nets

Before giving the definition of stochastic Petri net (SPN), the two types of transitions, i.e., timed and immediate transitions, which both of them are used in the SPN, are introduced.

- **Timed Transitions**: the firing of a transition in a PN system corresponds to the event that changes the state of the real system. This change of state can be due to one of two reasons: it may either result from the verification of some logical condition in the system, or be induced by the completion of some activity. Considering the second case, we note that transitions can be used to model activities, so that transition enabling periods correspond to activity executions and transition firings correspond to activity completions. Hence, time can be naturally associated with transitions.

A timed transition t can be associated with a local clock or timer. When this transition is enabled, the associated timer is set to an initial value. The timer is then decremented at constant speed and the transition fires when the timer reaches the value zero. The timer associated with the transition can thus be used to model the duration of an activity whose completion induces the state change that is represented by the change of marking produced by the firing of the timed transition t.

The type of activity associated with the transition, whose duration is measured by the timer, depends on the modelled real system: it may correspond to the execution of a task by a processor, or to the transmission of a message in a communication network, or to the work performed on a part by a machine tool in a manufacturing system. It is important to note that the activity is assumed to be in progress while the transition is enabled. This means that in the evolution of more complex nets, an interruption of the activity may take place if the transition loses its enabling condition before it can actually fire. The activity may be resumed later on, during the evolution of the system in the case of a new enabling of the associated transition. This may happen several times until the timer goes down to zero and the transition finally fires.

It is possible to define a timed transition sequence or timed execution of a timed PN system as a transition sequence (as defined in Definition 6.3) augmented with a set of non-decreasing real values describing the epochs of firing of each transition. Such a timed transition sequence is denoted as follows: $[(\tau_{(1)}, t_{(1)}), \ldots, (\tau_{(i)}, t_{(i)}), \ldots]$. The time intervals $[\tau_{(i)}, \tau_{(i+1)}]$ between consecutive epochs represent the periods during which the timed PN system sojourns in marking $M_{(i)}$. This sojourn time corresponds to a period, in which the execution of one or more activities is in progress and the state of the system does not change.

- **Immediate Transitions**: it is noted that not all the events that occur in a real system model correspond to the end of time-consuming activities (or to activities that are considered time-consuming at the level of detail at which the model is developed). For instance, a model of a multiprocessor system described at a high level of abstraction often neglects the durations of task switching, since these operations require a very small amount of time, compared with the durations of task executions. The same can be true for bus arbitration compared with bus use. In other cases, the state change induced by a specific event may be quite complex, and thus difficult to obtain with the firing of a single transition. Moreover, the state change can depend on the present state in a complex manner. As a result, the correct evolution of the timed PN system can often be conveniently described with subnets of transitions that consume no time and describe the logics or the algorithm of state evolution induced by the complex event. To cope with both these situations in timed PN systems, it is convenient to introduce a second type of transition called immediate. Immediate transitions fire as soon as they become enabled (with a null delay), thus acquiring a sort of precedence over timed transitions, thus leading to the choice of giving priority to immediate transitions in the definition of stochastic Petri nets. In this chapter, immediate transitions are depicted as thin bars, whereas timed transitions are depicted as boxes or thick bars.

Several reasons suggest the introduction of the possibility of using immediate transitions into PN systems together with timed transitions. The firing of a transition may describe either the completion of a time-consuming activity, or the verification of a logical condition. It is thus natural to use timed transitions in the former case, and immediate transitions in the latter. Moreover, by allowing the use of immediate transitions, some important benefits can be obtained in the model solution. We only mention here the fact that the use of immediate transitions may significantly reduce the cardinality of the reachability set, and may eliminate the problems due to the presence in the model of timed transitions with rates that differ by orders of magnitude. On the other hand, the introduction of immediate transitions in an SPN does not raise any significant complexity in the analysis.

In the SPN, markings in the reachability set can be classified as tangible or vanishing. A tangible marking is a marking in which (only) timed transitions are enabled. A vanishing marking is a marking in which (only) immediate transitions are enabled (the "only" is in parentheses since the different priority level makes it impossible for timed and immediate transitions to be enabled in the same marking, according to the definition of the enabling condition for PN systems with priority (Marsan et al. 1994). A tangible marking is a marking, in which no transition is enabled. The time spent in any vanishing marking is deterministically equal to zero, while the time spent in tangible markings is positive with probability one.

To describe the SPN dynamics, we separately observe the timed and the immediate behavior, hence referring to tangible and vanishing markings, respectively. Let us start with the timed dynamics (hence with tangible markings). We can assume that each timed transition possesses a timer. The timer is set to a value that is sampled from the negative exponential pdf associated with the transition, when the transition becomes enabled for the first time after firing. During all time intervals in which the transition is enabled, the timer is decremented. Transitions fire when their timer reading goes down to zero. With this interpretation, each timed transition can be used to model the execution of some activity in a distributed environment; all enabled activities execute in parallel (unless otherwise specified by the PN structure) until they complete. At completion, activities induce a change of the system state, only as regards their local environment. No special mechanism is necessary for the resolution of timed conflicts: the temporal information provides a metric that allows the conflict resolution.

In the case of vanishing markings, the SPN dynamics consumes no time: everything takes place instantaneously. This means that if only one immediate transition is enabled, it fires, and the following marking is produced. If several immediate transitions are enabled, a metric is necessary to identify which transition will produce the marking modification. Actually, the selection of the transition to be fired is relevant only in those cases in which a conflict must be resolved: if the enabled transitions are concurrent, they can be fired in any order. For this reason, SPNs associate weights with immediate transitions belonging to the same conflict set.

We can thus observe a difference between the specification of the temporal information for timed transitions and the specification of weights for immediate transitions. The temporal information associated with a timed transition depends only on the characteristics of the activity modelled by the transition. Thus, the temporal specification of a timed transition requires no information on the other (possibly conflicting) timed transitions, or on their temporal characteristics. On the contrary, for immediate transitions, the specification of weights must be performed considering at one time all transitions belonging to the same conflict set. Indeed, weights are normalized to produce probabilities by considering all enabled transitions within a conflict set, so that the specification of a weight, independent of those of the other transitions in the same conflict set, is not possible. We can now formally define the SPN. It should be noted that the above definition of SPN coincides with the definition of Generalized SPN (GSPN) given in Marsan et al. (1994).

Definition 6.7 A Stochastic Petri net (SPN) system is a triple $\{N, R, \omega\}$, where

- $N = (P, T, F, W, M_0)$ is a Petri net system,
- $\forall t \epsilon T, R_t \in \mathbb{R}^+ \cup \{\infty\}$ is the rate of exponential distribution for the firing time of transition t. If $R_t = \infty$, the firing time of t is zero; t is an immediate transition. On the other hand, a transition t with $R_t < \infty$ is a timed transition, and
- $\forall t \epsilon T, \omega_t \in \mathbb{R}^+$ is the weight assigned to the firing of the enabled transition t, whenever its rate R_t is equal to ∞. The firing probability of transition t enabled in a vanishing marking M is computed as $\dfrac{\omega_t}{\sum_{M|t' > } \omega_{t'}}$. $\qquad\square$

6.2.4 Adaptive Petri Nets

Several approaches are used in the literature to make PNs adaptive by fusing intelligent techniques such as neural networks and fuzzy logic systems with PNs (ul Asar et al. 2005; Rutkowski and Cpalka 2003; Khosla and Dillon 1997; Jain and Martin 1999). But none of these mechanisms is adaptive and cannot be used to resolve conflicts among transitions, when the environment is dynamic. In the following paragraphs some of these adaptive PNs are briefly described.

In Kadjinicolaou et al. (1990), a feed-forward Neural Petri Net (NPN) obtained by fusing PNs with neural networks is introduced. The NPN is constructed using the basic concepts of PNs (place, transition and arc). Places in the NPN are decomposed into three layers: input, hidden and output. In order to construct connections between places in hidden and output layers, thresh-holding transitions and arcs with trainable weights are introduced. A thresh-holding transition is enabled, when the summation of its weighted inputs is equal to or greater than its predefined threshold. Using a set of input-output training examples, the weights of arcs are trained according to the back propagation rules. Another example of fusion between PNs and neural networks has been recently reported (Ding et al. 2014).

In this PN, a special transition, called Adaption transition (A-transition), is introduced which is associated with a multilayer neural network. The inputs of the neural network are the markings of its input places whereas the outputs are the markings of its output places. The inputs of all A-transitions create an environment for the system; thus, when the environmental changes happen, A-transitions can adapt themselves to the changes. Based on the markings of the output places of a fired A-transition, a sub-graph of the reachability graph will be traversed. Other types of fusion between PNs and neural networks are also reported in Chen et al. (1990) and Zha and Fok (1998).

An adaptive PN obtained from the fusion of PNs and Fuzzy logic systems is reported in Gao and Wu (2003), where a PN is used to represent a fuzzy rule-based system. In this PN, places represent propositions and tokens represent states of those propositions. A value between 0 and 1 is assigned to each token representing the truth degree of a proposition state. Firing a transition expresses a rule reasoning process (Looney 1988; Cardoso and Dubois 1999). A combination of the LA with Color Fuzzy Petri Net (CFPN) has been recently reported (Barzegar et al. 2011). The LA is used to adjust adaptively the membership functions defined on the input parameters.

6.2.5 Applications of Petri Nets

To name just a few, Petri nets have a wide variety of applications in modeling, verification, manufacturing systems, scheduling systems, sequence controllers, communication networks, and software design. In what follows, we provide a very brief review for cellular applications of Petri nets, which are useful to follow this chapter.

Petri nets are suitable for modeling cellular applications (Lin and Chan 2009). In Teng and Black (1990), PNs are used to model and represent the dynamic behavior of a cell's operations in a Cellular Manufacturing System (CMS). In each cell of CMS, a cell control system is required to determine the cell's operations, as well as handling machine breakdowns in a real time manner. Such a control system can be reliably modeled by designing a Petri net for the CMS. Other applications of Petri nets for CMS are reported in literature (Bruno and Biglia 1985; Merabet 1986; Ravichandran and Chakravarty 1986).

PNs can be used to model different strategies in cellular networks such as channel assignment. For instance, in James-Romero et al. (1997), PNs are used to model a centralized Dynamic Channel-Assignment (DCA), as well as a distributed DCA (DDCA) strategy. A kind of handshaking mechanism is embedded in the PN so that no two cells within the channel reuse distance can use the same channel. PNs are also used in resource management of cellular networks (Ma et al. 2002), bandwidth allocation in IEEE 802.16 networks (Geetha and Jayaparvathy 2010), handoff process in cellular mobile telephone systems (Dharmaraja et al. 2003), and fairness analysis in cellular networks (Schoenen et al. 1983).

Petri nets are widely used in other fields of computer networks. In Haines et al. (2007), an extended PN is introduced as a formal verification technique of IEEE 802.11 MAC protocol. This extended PN is applied as a case study to the centralized control mechanism of IEEE 802.11, which is used to support delay sensitive streams and fading data traffics. In Salah and Mustafa (2004), a colored PN is used to model and analyze the station to station (STS) protocol in computer networks. In Pengand et al. (2007), a PN is used to model the Session Initiation Protocol (SIP) and to accomplish its verification. Another interesting work is presented in Liu et al. (2008) where the 802.11i 4-way handshaking mechanism is analyzed utilizing high-level PNs. The results of the analysis confirmed that this mechanism is vulnerable to the Denial-of-Service attack.

6.3   Hybrid Machines Based on LAs and PNs

This Section contributes toward proposing adaptive Petri nets, which evolve under the control of a number of learning agents. To this end, we utilize learning automata as learning agents and introduce a number of hybrid machines, which are constructed by the fusion of Petri nets and learning automata.

The first and basic hybrid machine, which is introduced in this Section, is referred to as APN-LA (Vahidipour et al. 2015). In the APN-LA, learning automata are used to provide an adaptive controlling mechanism for the Petri net. The second machine, constructed atop of the APN-LA, is ASPN-LA (Vahidipour et al. 2015). The ASPN-LA is an APN-LA where the underlying Petri net is a Stochastic Petri net (SPN). Another hybrid machine, introduced in this Section, is APN-ICLA (Vahidipour et al. 2017a). The APN-ICLA can be regarded as a Petri net, in which the controlling mechanism is cooperative. The last introduced hybrid machine is CAPN-LA (Vahidipour et al. 2017c). The CAPN-LA consists of a number of identical APN-LAs organized into a cellular structure.

6.3.1   APN-LA: Adaptive PN Based on LA

In this section, an adaptive Petri net based on learning automata (APN-LA) will be introduced. The APN-LA uses learning automata to resolve conflicts among the transitions of the PN. In this model, at first, transitions are partitioned into several clusters of conflicting transitions. Then, each cluster is equipped with an LA, which is responsible for controlling the conflicts among transitions in the corresponding transition cluster. Each LA has a number of actions, each of which corresponds to the selection of one of the enabled transitions. During the evolution of the APN-LA, unlike standard PN, a cluster, instead of a transition, is selected for firing at each marking. A cluster can be selected for firing only if it has an enabled transition. Then, the LA associated with the cluster is activated to select one of the enabled transitions within that cluster. The selected transition will be fired and a new marking will be generated. Upon the next activation of this LA, a reinforcement signal will be generated using the sequence of markings generated between the two activations of the LA. Using the reinforcement signal produced by the environment, the internal structure of the LA will be updated to reflect the markings' changes.

In the rest of this section, first the APN-LA will be formally defined and second, a procedure used by the designer to construct an APN-LA will be given. Finally, to describe the evolution of the APN-LA, its firing rules will be presented. It should be noted that in this chapter the notation 'q' is used as the action probability vector of learning automata and the notation 'p' is used as the place in Petri nets.

6.3.1.1 Formal Definition

Before we give the formal definition of the APN-LA, three required definitions are given.

Definition 6.8 Potential Conflict is a set of transitions $\{t_1, t_2, \ldots, t_n\}$ *iff* every $t_k \in \{t_1, t_2, \ldots, t_n\}$ shares at least one input place with one $t_j \in \{t_1, t_2, \ldots, t_n\} \setminus \{t_k\}$. The potential conflict only depends on the structure of the PN graph. □

Definition 6.9 Maximal Potential Conflict is a set of transitions $\rho = \{t_1, t_2, \ldots, t_n\}$ *iff* the following conditions are held:

- $\{t_1, t_2, \ldots, t_n\}$ are in potential conflicts,
- For every $t \in T \setminus \rho$, $\{t\} \cup \rho$ are not in potential conflict. □

We refer to a maximal potential conflict set as a *cluster* hereafter. Informally, for each transition t_k in the cluster ρ, there exists at least one transition $t_j \neq t_k$ in the cluster ρ, such that t_k and t_j have at least one input place in common and also two transitions in two different clusters do not have any input place in common.

Definition 6.10 An updating transition t^u, is an immediate transition, except when it fires, the action probability vector of the LA that fused with the PN is updated. □

An APN-LA can be formally defined as follows:

Definition 6.11 An Adaptive Petri Net based on Learning Automata (APN-LA) is a quintuple $(\hat{P}, \hat{T}, \hat{W}, S, L)$, where

- $\hat{P}$ is a finite set of places,
- $\hat{T} = T \cup T^U$ is a finite set of ordinary transitions and updating transitions $T^U = \{t_1^u, \ldots t_n^u\}$,
- $\hat{W} : ((\hat{P} \times \hat{T}) \cup (\hat{T} \times \hat{P})) \to \mathbb{N}$ defines the interconnection of $\hat{P}$ and $\hat{T}$,
- $L = \{LA_1, \ldots, LA_n\}$ is a set of learning automata with varying number of actions, and
- $S = \{s_0, s_1, \ldots, s_n\}$ denotes a set of clusters, each consists of a set of transitions:

- $s_i, i = 1, \ldots, n$ are sets of clusters, each is a maximal potential conflict set. Each cluster s_i is equipped with a learning automaton, denoted by LA_i. Number of actions of LA_i is equal to the number of transitions in s_i; each action corresponds to a transition,
- s_0 is the set of remaining transitions in $\hat{T}$.   □

Definition 6.12 An *APN-LA system* is a triple $(\hat{N}, M_0, \hat{F})$, where $\hat{N}$ is an APN-LA, M_0 is the initial marking, $\hat{F} = \{f_1, \ldots, f_n\}$ is the set of reinforcement signal generator functions. $f_i : M \rightarrow \beta_i$ is the reinforcement signal generator function related to LA_i. Sequence of markings in the APN-LA is the input of f_i and the output of f_i is reinforcement signal β_i. Upon the generation of β_i, LA_i updates its action probability vector using the learning algorithm.   □

Figure 6.4 shows the operation of the learning automaton LA_i in APN-LA. LA_i selects an enabled transitions in the cluster s_i by choosing the action α_i. When updating transition t_i^u in the APN-LA is fired, the signal generator function f_i is executed and the reinforcement signal β_i is generated for LA_i.

6.3.1.2 Construction Procedure

An APN-LA is constructed by fusing a Petri net with learning automata according to the following steps (Fig. 6.5):

1. Determine all sets $s_i, i = 1, \ldots, n$ of maximal potential conflicts, which call them clusters in the PN. The remaining transitions in this PN, which are not in any of the maximal potential conflicts sets $s_i, i = 1, \ldots, n$, form a cluster called s_0.
2. For any cluster $s_i, i = 1, \ldots, n$ do the following steps:

   (a) Assign LA_i with a variable number of actions to s_i. Each action of LA_i corresponds to a transition in s_i.
   (b) Insert a newly updating transition t_i^u with one input and one output place.
   (c) Connect the newly inserted output place to t_i^u with an inhibitor arc.
   (d) Add the newly inserted output place to the preset of all transitions in the cluster s_i.
   (e) Find all shared pre-places of transitions in cluster s_i. Let the union of the presets of these places be the preset of the newly inserted input place.

Fig. 6.4 The operation of LA in the APN-LA system

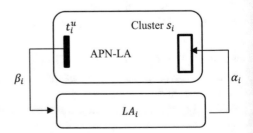

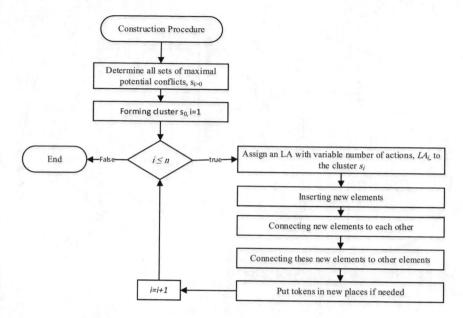

Fig. 6.5 Construction procedure

(f) Put a token in newly inserted output place if at least one token exists in one of the shared input places of transitions in cluster s_i.

To illustrate the construction of APN-LA according to the above steps, we will use these steps to form an APN-LA from a Petri net given in Fig. 6.6. The resulting APN-LA is given in Fig. 6.7.

1. The maximal potential conflict is $\{t_1, t_2\}/\{t_4, t_5\}/\{t_6, t_7\}$. Following step 1 of the construction procedure, three clusters s_1, s_2, and s_3 are formed. We illustrate these clusters with dotted-rectangles in Fig. 6.6. Remaining transitions t_3, t_8, and t_9 form the cluster s_0.
2. For cluster $s_1/s_2/s_3$ we have:

   (a) In the resulted APN-LA, three LAs with two actions, i.e., $LA_1/LA_2/LA_3$ are responsible for controlling the conflicts among the transitions in $s_1/s_2/s_3$. One action of $LA_1/LA_2/LA_3$ corresponds to the selection of $t_1/t_4/t_6$ and the other action corresponds to the selection of $t_2/t_5/t_7$.
   (b) The updating transition $t_1^u/t_2^u/t_3^u$, input place $p_9/p_{11}/p_{13}$ and output place $p_{10}/p_{12}/p_{14}$ are inserted into PN (Fig. 6.7).
   (c) The newly output place $p_{10}/p_{12}/p_{14}$ is connected to updating transition $t_1^u/t_2^u/t_3^u$ with an inhibitor arc.
   (d) $p_{10}/p_{12}/p_{14}$ is added to the preset of $t_1/t_4/t_6$ and $t_2/t_5/t_7$.

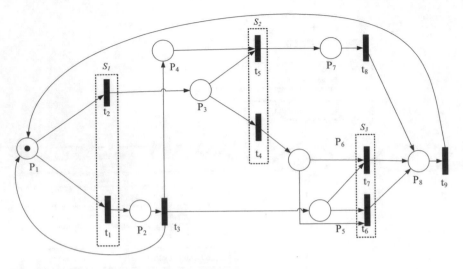

Fig. 6.6 A PN system of a concurrent model

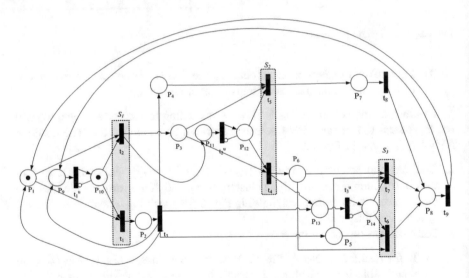

Fig. 6.7 The yielded PN-layer from a PN shown in Fig. 6.6

(e) The place $p_1/p_3/p_6$ is the shared input place of transitions $t_1/t_4/t_6$ and $t_2/t_5/t_7$. Therefore, we add preset of this place, i.e., t_3 and $t_9/t_2/t_4$, to preset of $p_9/p_{11}/p_{13}$.

(f) One/No/No token is appeared in $p_{10}/p_{12}/p_{14}$ because there exists one/no/no token in $p_1/p_3/p_6$.

6.3.1.3 Evolution of an APN-LA System

The evolution of a Petri net system can be described in terms of its initial marking and a number of firing rules (Murata 1989). At any given time during the evolution of a Petri net system, the current marking of the system evolves to a new marking by applying the firing rules. Firing rules of an ordinary Petri net system, for any marking M, are defined as follows (Murata 1989):

- A transition $t \in T$ is said to be enabled if each input place p of t is marked with at least $W(p, t)$ tokens.
- An enabled transition may or may not fire (only one enabled transition can fire in each marking).
- A firing of an enabled transition t removes $W(p, t)$ tokens from each input place p of t, and adds $W(t, p)$ tokens to each output place p of t.

Firing rules of an APN-LA system differ from those of an ordinary PN due to the fusion with the LA. In the remaining of this section, we will introduce the firing rules for the APN-LA. We first give a number of definitions, which will be used later for defining the firing rules of the APN-LA.

Definition 6.13 Cluster s_i is *enabled* in marking M when at least one transition in s_i is enabled. □

Definition 6.14 *Fired cluster* is an enabled cluster which is selected as the fired cluster, from which a transition will be fired in marking M. □

Definition 6.15 LA_i is *activated* in marking M *iff* the cluster s_i is selected as the fired cluster in marking M. □

Definition 6.16 A subset E_M^s of a cluster s is said to be in *effective conflict* in marking M *iff* these conditions are held: $\forall t \in E_M^s$ is enabled in M and $\{\forall t' \in s | t' \notin E_M^s, \sim M[t']\}$. □

Considering the above definitions, the firing rules of an APN-LA in any marking M can be described as follows (Fig. 6.8):

1. A transition $t \in \hat{T} - T \cup T^U$ is said to be enabled if each input place p of t is marked with at least $\hat{W}(p, t)$ tokens.
2. A cluster s_i is said to be enabled if at least one transition in s_i is enabled.
3. If s_0 is enabled, then s_0 is selected as fired with probability $\frac{|t' \in s_0, M[t']|}{|t \in \hat{T}, M[t]|}$, where $\| \|$ stands for norm operator; otherwise, an enabled cluster $s_i, i = 1, \ldots, n$ is selected as fired with probability $\frac{|E_M^{s_i}|}{|t \in \hat{T}, M[t]|}$ (only one enabled cluster can be selected as the fired cluster in each marking).
4. Only one enabled transition $t \in \hat{T}$ from the fired cluster can fire in each marking M.
5. If s_0 is the fired cluster, then an enabled transition t is randomly selected from s_0 for firing; otherwise, an LA associated with the fired cluster is activated. The available action set of the activated LA consists of the actions corresponding to

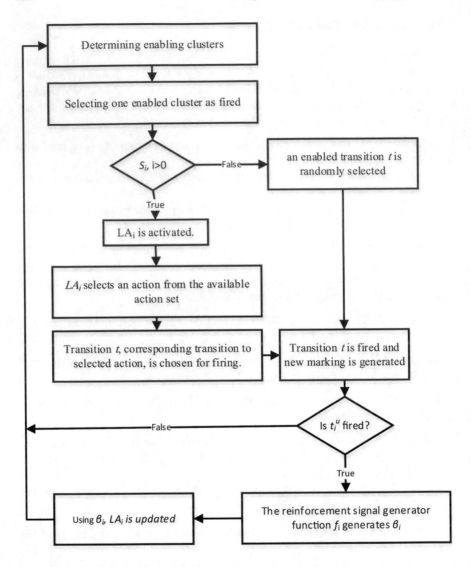

Fig. 6.8 Firing rules flowchart

the enabled transitions $(E_M^{s_i})$ in the fired cluster. The activated LA selects an action from its available action set according to its action probability vector and then the corresponding transition is selected for firing.

6. Firing of a transition $t \in \hat{T}$ removes $\hat{W}(p, t)$ tokens from each input place p of t, and adds $\hat{W}(t, p)$ tokens to each output place p of t.

7. The reinforcement signal generator function $f_i \in \hat{F}$ is executed when the updating transition $t_i^u \in T^u$ fires.

Fig. 6.9 Two transitions t_2 and t_3 can be enabled at the same time, without structural conflict

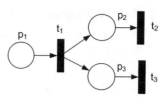

8. Using the generated reinforcement signal, the internal structure of the LA will be updated to reflect the markings' changes.

It is worth mentioning that two or more transitions may be enabled at the same time without sharing any input places. For example, two transitions t_2 and t_3 shown in Fig. 6.9, are enabled at the same time without sharing any input places. Although these transitions are in actual conflict in marking $\{p_2 + p_3\}$, the potential conflict condition in Definition 6.8 does not hold for these transitions; our proposed APN-LA does not resolves this actual conflict. As a consequence, the proposed APN-LA as defined above, is only applicable for conflict resolution among the effective conflicts.

6.3.2   ASPN-LA: Generalization of APN-LA

In this section, we generalize the proposed APN-LA and obtain ASPN-LA, which is a fusion between the LA and the Stochastic Petri net (SPN). The most important difference between PN and SPN is the addition of timed transitions. A timed transition has a countdown timer, which is started whenever the transition becomes enabled. This timer is set to a value that is sampled from the negative exponential probability density function associated with the transition. An enabled transition can be fired whenever its timer reaches zero. Thus, no special mechanism is necessary for the resolution of timed conflicts: the temporal information provides a metric that used in the conflict resolution mechanism (Marsan et al. 1994). As such, to generalize the APN-LA to the ASPN-LA, it is sufficient to put all timed transitions of the SPN into a single cluster, denoted by s_{-1}, in which the temporal information is used to select an enabled timed transition for firing.

6.3.2.1   Formal Definition

Formally, an ASPN-LA can be defined as follows:

Definition 6.17 An Adaptive Stochastic Petri Net based on Learning Automata (ASPN-LA) is a 6.tuples $\left(\hat{P}, \hat{T}, \hat{W}, \hat{S}, L, R, \omega^0\right)$, where

- $\hat{P}, \hat{T}, \hat{W}, L$ are defined as in Definition 6.11,
- $\hat{S} = \{s_{-1}, s_0, s_1, \ldots, s_n\}$ denotes a set of clusters, each contains of a set of transitions:

  - s_{-1} contains all timed transitions in the ASPN-LA,
  - $s_i, i = 0, \ldots, n$ are defined as in Definition 6.11.

- $\forall t \epsilon T, R_t \in \mathbb{R}^+ \cup \{\infty\}$ is the rate of exponential distribution for the firing time of transition i. If $R_t = \infty$, the firing time of t is zero; such a transition is called immediate. On the other hand, a transition t with $R_t < \infty$ is called timed transition. A marking M is said to be vanishing if there is an enabled immediate transition in this marking; otherwise, M is said to be tangible (Marsan et al. 1994), and
- $\forall t \epsilon s_0, \omega_t^0 \in \mathbb{R}^+$ is the weight assigned to an enabled transition t in the cluster s_0. $\square$

Note that the definition of ω_t in the SPN (Definition 6.7) is changed into ω_t^0 in the ASPN-LA. In other words, in the ASPN-LA, weights are only assigned to the transitions in cluster s_0. In the ASPN-LA, conflicts among immediate transitions in $s_i, i = 1, \ldots, n$ are resolved by LAs; hence, no weigh is required for these sets of transitions.

6.3.2.2 Evolution of an ASPN-LA System

Firing rules of an ASPN-LA system differs from those of an APN-LA because of the addition of the cluster s_{-1}. Therefore, the following rules must be added to the firing rules of the APN-LA (described in Sect. 6.3.1.3):

1. Cluster s_{-1} is said to be enabled if at least one timed transition in s_{-1} is enabled and none of other clusters $s_i, i \geq 0$ is enabled.
2. Cluster s_{-1} is selected as fired cluster if it is enabled.
3. If s_{-1} is the fired cluster, the temporal information associated with the enabled transitions within s_{-1} is used to select a transition for firing.

In addition if s_0 is selected as the fired cluster, then instead of selecting a transition at random for firing, an enabled transition t is selected for firing with probability $\frac{\omega_t^0}{\sum_{M[t']} \omega_{t'}^0}$.

It is worth mentioning that in an SPN, if several immediate transitions are enabled, a metric will be necessary to identify which transition produces the marking's change. However, this is required only in those cases, in which a conflict must be resolved; if the enabled transitions are concurrent, they can be fired in any order (Marsan et al. 1994). In conflicting situations, weights assigned to immediate transitions are used to determine which immediate transition will actually fire. A practical approach for assigning weights to conflicting transitions is given in Marsan et al. (1994). The main difference between the ASPN-LA and the GSPN (Marsan et al. 1994) is that, the conflict resolution mechanism in the ASPN-LA is adaptive to the environmental changes, whereas weights and priorities, assigned to immediate transitions in the GSPN, are constant.

6.3.3 APN-ICLA: Adaptive PN Based on Irregular Cellular LA

In the APN-LA, the controllers that control different sets of conflicting transitions in the PN act independently from each other. But there could be situations, where resolving the conflict within a set conflicting transitions affects another set of conflicting transitions by possibly enabling or disabling some of the transitions within that set. In such situations, it seems it is more appropriate to let the controllers within the PN cooperate with each other, instead of operating independently. In other words, if resolving conflicts among two different sets of transitions affects each other, then their controllers must be aware of the decisions made by each other to make better decisions.

Irregular cellular learning automata (ICLA), which is a network of learning automata (Esnaashari and Meybodi 2015), is utilized to construct a new adaptive Petri net, called APN-ICLA. The APN-ICLA consists of two layers: PN-layer and ICLA-layer, both of which are constructed according to the problem to be solved. The PN-layer is a Petri net, in which conflicting transitions are partitioned into several clusters. A cluster in the PN-layer is mapped into a cell of ICLA-layer and hence, the LA residing in that cell acting as the controller of that cluster. The connections among the cells in the ICLA-layer are determined by the locality defined in the application, for which the APN-ICLA is designed. Two clusters in the PN-layer are said to be neighbors if their corresponding cells in the ICLA-layer are neighbors. We will also define the firing rules for the proposed adaptive Petri net and hence, the definition of the APN-ICLA will be completed.

6.3.3.1 Formal Definition

An APN-ICLA consists of two layers: PN-layer and an ICLA-layer (Fig. 6.10). To construct the PN-layer, a Petri net is needed. This PN is designed according to the problem to be solved. Actually, this Petri net is used to design an algorithm for solving that problem. To create this PN-layer, we first determine all sets $s_i, i = 1, \ldots, n$ of maximal potential conflicts which call them clusters in the PN. Then, the remaining transitions in this PN, which are not in any of the maximal potential conflicts sets $s_i, i = 1, \ldots, n$, form the cluster s_0. The construction procedure of a PN-layer from a Petri net is similar to the construction procedure of an APN-LA, which is described in Sect. 6.3.1.2.

Every cluster $s_i, i = 1, \ldots, n$ in the PN-layer is mapped into a cell in the ICLA-layer. The connections among the cells in the ICLA-layer are determined by

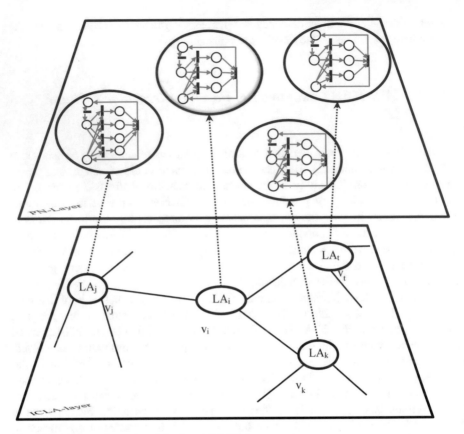

Fig. 6.10 The two-layered model of an APN-ICLA

the locality defined in the application, for which the APN-ICLA is designed. Two clusters in the PN-layer are said to be neighbors if their corresponding cells in the ICLA-layer are neighbors. The LA resides in a particular cell in the ICLA-layer acts as the controller of the corresponding cluster in the PN-layer.

Like ICLA, there is a local rule that APN-ICLA operates under. The local rule of the APN-ICLA and the actions selected by the neighboring LAs of any particular LA determine the reinforcement signal to that LA. The neighboring LAs of any particular LA constitute the local environment of that LA. The local environment of an LA in the APN-ICLA is non-stationary because the action probability vectors of the neighboring LAs vary during the evolution of APN-ICLA. An APN-ICLA can be formally defined as follows (Vahidipour et al. 2017a).

Definition 6.18 An Adaptive Petri Net based on Irregular Cellular Learning Automata (APN-ICLA) with n cells is a tuple $\mathcal{N} = (PN - layer = (\hat{P}, \hat{T}, \hat{W}, S), ICLA - layer = (G\langle E, V\rangle, \Phi, A, F))$, where

1. PN-Layer

   - $\hat{P}, \hat{T}, \hat{W}, S$ are defined as in Definition 6.11.

2. ICLA-layer

   - Φ is a finite set of states. The state of the cluster $s_i, i = 1, \ldots, n$ is denoted by φ_i,
   - $A = \{LA_1, \ldots, LA_n\}$ is a set of learning automata with varying number of actions,
   - G is an undirected graph, with V as the set of vertices (clusters) and E as the set of edges (adjacency relations), and
   - $\hat{F} = \{F_1, \ldots, F_n\}$ is the set of local rules. $F_i : \underline{\Phi}_i \rightarrow \beta_i$ is the local rule of ICLA in the cluster $s_i, i = 1, \ldots, n$, where $\underline{\Phi}_i = \{\varphi_j | (i,j) \in E\} \cup \{\varphi_i\}$ is the set of states of all neighbors of s_i and β_i is the reinforcement signal. Upon the generation of β_i, LA_i updates its action probability vector using the learning algorithm.

3. Mapping between layers

   - Each cluster $s_i, i = 1, \ldots, n$ in the PN-layer is mapped into a cell c_i in the ICLA-layer,
   - Two clusters s_i and s_j in the PN-layer are neighbors if the cells c_i and c_j in the ICLA-layer are neighbors, and
   - LA_i, resides in the cell c_i of the ICLA-layer, becomes the controller of the cluster s_i of the PN-layer. Number of actions of LA_i is equal to the number of transitions in the cluster s_i.   □

Definition 6.19 An APN-ICLA system is $(\mathcal{N}, M_0)$, where $\mathcal{N}$ is an APN-ICLA and M_0 is the initial marking.   □

6.3.3.2 Evolution of an APN-ICLA

The evolution of an APN-ICLA can be described in terms of its initial marking and a number of firing rules. At any given time during the evolution of an APN-ICLA, the current marking of the APN-ICLA evolves to a new marking by applying the firing rules. The firing rules of an APN-ICLA in any marking M can be described as follows:

1. A cluster s_i of PN-layer is said to be enabled if at least one transition in s_i is enabled.

2. If s_0 is enabled, then s_0 is selected as fired with probability $\frac{|t' \in s_0, M[t')|}{|t \in \hat{T}, M[t)|}$, where $||$ stands for norm operator; otherwise, an enabled cluster $s_i, i = 1, \ldots, n$ is selected

as fired with probability $\frac{|E_M^{s_i}|}{|t \in \hat{T}, M(t)|}$ (only one enabled cluster can be selected as the fired cluster in each marking).

3. Only one enabled transition $t \in \hat{T}$ from the fired cluster can fire in each marking M.
4. If s_0 is the fired cluster, then an enabled transition t is randomly selected from s_0 for firing; otherwise, the LA in corresponding cell with the fired cluster is activated. The available action set of the activated LA consists of the actions corresponding to the enabled transitions in the fired cluster. The activated LA selects an action from its available action set according to its action probability vector and then the corresponding transition is selected for firing.
5. Firing of a transition $t \in \hat{T}$ removes $\hat{W}(p, t)$ tokens from each input place p of t, and adds $\hat{W}(t, p)$ tokens to each output place p of t.
6. The local rule $F_i \in \hat{F}$ is used to generate the reinforcement signal β_i, when the updating transition $t_i^u \in T^u$ fires.
7. Using β_i, the action probability vector of LA_i is updated by the learning algorithm.

6.3.4 CAPN-LA: Cellular Adaptive PN Based on LA

Petri nets can also be used to represent algorithms of different classes (Reisig 2013a). One class of algorithms is cellular algorithms (Terán-Villanueva et al. 2013), which commonly follows the so-called distributed computing model (Linial 1992). In this model, each cell executes the same algorithm, referred to as the local algorithm, and the whole problem is solved by a mean of cooperation between neighboring cells. Petri nets can be used to represent such cellular algorithms as well. For this to be made possible, two steps must be taken:

1. A Petri net must be designed for representing the local algorithm to be executed in each cell. This Petri net is then replicated into all cells,
2. A mechanism must be devised to handle the required cooperation between any two neighboring cells.

In the first step, when a Petri nets is designed for the local algorithm, a local solution is formed. This solution entails a number of decision points, at which there needs to be mechanisms for making decisions. Each decision point in a Petri net is aroused when two or more transitions are enabled in a given marking. At such point, the decision to be made is the selection of an enabled transition for firing among the set of enabled transitions. In Petri net literature, such decision making mechanisms are referred to as controlling mechanisms. Different controlling mechanisms have been proposed in literature so far such as random mechanism (Peterson and Net 1981), queue regimes (Burkhard 1981), priority (Bause 1996, 1997), and external controller (Holloway and Krogh 1994). Having a local solution, any of these controlling mechanisms may result in a specific local algorithm.

For the cooperation mechanism between any two neighboring cells, which is required in the second step above, a common way, which usually is adopted in the Petri net literature, is representing the cooperation within the structure of the two Petri nets, designed for the cells; that is by connecting the two Petri nets using ordinary elements of Petri nets such as common places and inhibitor arcs (Reisig 2013b). This results in a large Petri net for the whole algorithm, which is extremely hard to analyze and simulate (Chiola and Ferscha 1993; Fujimoto 2001).

To overcome the issue of large Petri nets, we suggest that the cooperation between the two neighboring cells to be handled within the controlling mechanisms of their Petri nets, instead of their structures. This can be possible only if the controlling mechanisms are capable of cooperating with each other. Among the controlling mechanisms proposed for Petri nets so far, the APN-LA mechanism is the only one which includes such property. This is due to the fact that in this mechanism, controlling is handled using a Learning Automaton (LA), and LAs have been shown to be able to cooperate with each other (Narendra and Thathachar 1989). Therefore, a cellular adaptive Petri net, called CAPN-LA, in which a number of APN-LAs are organized into a cellular structure, is proposed. This cellular structure represents a neighborhood relationship between the APN-LAs. Each APN-LA represents a local algorithm to be executed within a cell and the cooperation between neighboring cells is handled by the cooperation between the LAs reside in the Petri nets of the cells.

For the proposed CAPN-LA, the concept of expediency can be introduced. Informally, a CAPN-LA is considered to be expedient if, in the long run, each of its APN-LAs performs better than its equivalent APN-Pure-Chance-Automaton (APN-PCA). The equivalent APN-PCA of an APN-LA can be formed by replacing the LA in that APN-LA with a pure-chance automaton; that is, in APN-PCA, the controlling mechanism randomly selects an enabled transition for firing among the set of enabled transitions. Expediency is a notion of learning. The steady-state behavior of CAPN-LA is studied and then conditions, under which a CAPN-LA becomes expedient, are presented.

6.3.4.1 Formal Definition

A Cellular Adaptive Petri net based on Learning Automata, called CAPN-LA and shown in Fig. 6.11, consisting of a number of identical APN-LAs organized into a cellular structure. This cellular structure consists of a number of cells, in each of which an APN-LA is assigned. The neighborhood relationship between cells in the cellular structure represents the neighborhood relationship between APN-LAs. Each APN-LA represents a local algorithm to be executed within a cell of the cellular structure and the cooperation among these local algorithms specifies the behavior of the CAPN-LA.

The APN-LA assigned to c_i is denoted by A_i. Like other APN-LAs, each A_i has a controlling mechanism with two parts: (1) a concurrent selector and (2) a conflict resolver. The concurrent selector in the APN-LA is a random mechanism and its

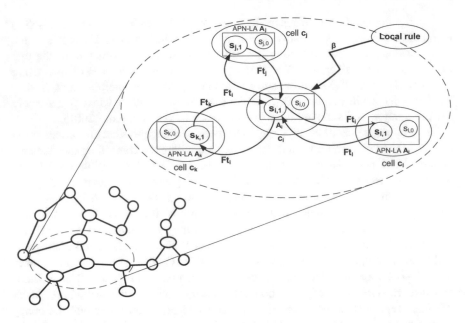

Fig. 6.11 The cellular adaptive Petri net based on learning automata (CAPN-LA) (Vahidipour et al. 2017c)

conflict resolver is an LA with varying number of actions. This LA (LA_i) is assigned to a cluster of transitions ($s_{i,1}$), which some of these transitions may be in conflict with each other in some markings. For such a marking, LA_i selects an enabled transition, denoted by Ft_i, among the enabled transitions in $s_{i,1}$ for firing.

Two clusters $s_{i,1}$ and $s_{j,1}$ in the CAPN-LA are neighbors if their corresponding APN-LAs, A_i and A_j, are neighbors. The neighboring clusters of any particular cluster, indicated by $s_{i,1}$, constitute the local environment of the LA in that cluster (LA_i), which provides the reinforcement signal to LA_i. In the CAPN-LA, a local rule takes the last firing transitions of cluster $s_{i,1}$ and its neighboring clusters as inputs and then, returns the reinforcement signal as output. The local rule provides a mean of cooperation between neighboring clusters of conflicting transitions, which in turn, results in cooperation between neighboring APN-LAs. This indeed provides a mean of cooperation between local algorithms represented by neighboring APN-LAs.

Formally, a CAPN-LA can be defined according to the following definition.

Definition 6.20 A cellular Adaptive Petri net based on Learning Automata (CAPN-LA) is a structure $\mathcal{N} = \left(G\langle E, V\rangle, \{A_1, \ldots, A_n\}, \hat{F}\right)$, where

- G is an undirected graph, with V as the set of vertices (cells) and E as the set of edges (adjacency relations),

- $\{A_1, \ldots, A_n\}$ is a finite set of APN-LAs, where $n = |V|$. Each A_i is assigned to cell c_i and is defined by a quintuple[2] $A_i = \left(\widehat{P}_i, \widehat{T}_i, \widehat{W}_i, S_i, LA_i \right)$, in which

  - $\widehat{P}_i$ is the set of places,
  - $\widehat{T}_i = T \cup \left\{ t_{i,1}^u \right\}$ is a finite set of ordinary transitions and an updating transition $t_{i,1}^u$,
  - $\widehat{W}_i : \left(\left(\widehat{P}_i \times \widehat{T}_i \right) \cup \left(\widehat{T}_i \times \widehat{P}_i \right) \right) \rightarrow \mathbb{N}$ defines the interconnection of $\widehat{P}_i$ and $\widehat{T}_i$,
  - LA_i is a learning automaton with varying number of actions, and
  - $S_i = \left\{ s_{i,0}, s_{i,1} \right\}$ is a set of transitions where

- $s_{i,0}$ contains the concurrent transitions from $\widehat{T}_i$. No pair of transitions in this cluster has any input places in common, i.e., $\{ \forall t_k \in s_{i,0}, \forall t_j \in s_{i,0} | \bullet t_k \cap \bullet t_j = \varnothing \}$. A random mechanism is used as the concurrent selector in this cluster.
- $s_{i,1}$ contains the remaining transitions in $\widehat{T}_i$ which are not in $s_{i,0}$. The set of transitions in the cluster $s_{i,1}$ may conflict with each other in some markings, and thus a conflict resolver is required to select an enabled transition for firing in this cluster.
- $\widehat{F} = \{F_1, \ldots, F_n\}$ is a set of local rules which implies the cooperation among clusters of conflicting transitions. $F_i : \underline{s}_{i,1} \rightarrow \beta_i$ is the local rule of the cell c_i, where $\underline{s}_{i,1} = \{ s_{j,1} | (i,j) \in E \} \cup \{ s_{i,1} \}$ is the set of conflicting transitions of all neighbors of A_i, and β_i is the reinforcement signal related to LA_i. Upon the generation of β_i, A_i updates its action probability vector using its learning algorithm. $\quad\square$

Definition 6.21 A CAPN-LA system is a triple $(\mathcal{N}, M_0)$, where

- $\mathcal{N}$ is a CAPN-LA structure,
- $M_0 = \left[\mu_{1,0}, \ldots, \mu_{n,0} \right]$ is the initial marking vector of $\mathcal{N}$, such that $\mu_{i,0}$ is the initial marking of A_i. M is used to represent the set of markings of $\mathcal{N}$ and μ_i to represent the set of markings of A_i; $\mu_{i,j}$ represents the jth marking in A_i. To represent an arbitrary marking in the APN-LA, the notation μ will be used. $\quad\square$

---

[2]A_i, assigned to cell c_i, is an APN-LA, in which there exists only one updating transition.

6.3.4.2 Evolution of a CAPN-LA

A CAPN-LA system evolves from the current marking to a new marking according to the firing rules of it constituting APN-LAs. Each APN-LA evolves independently from the other APN-LAs in CAPN-LA. The firing rules of A_i, the APN-LA assigned to the cell c_i, in any marking M can be described as follows:

1. A transition $t \in \widehat{T_i}$ is said to be enabled if each input place p of t is marked with at least $\widehat{W_i}(p,t)$ tokens.

2. Two probability values $P_{i,0}$ and $P_{i,1}$ are calculated: $P_{i,0} = \dfrac{\left|t' \in s_{i,0}, M[t'\rangle\right|}{\left|t \in \widehat{T_i}, M[t\rangle\right|}$ and $P_{i,1} = \dfrac{\left|t' \in s_{i,1}, M[t'\rangle\right|}{\left|t \in \widehat{T_i}, M[t\rangle\right|}$ where $\|$ stands for norm operator and $P_{i,0} + P_{i,1} = 1$.

3. One of the two clusters $s_{i,0}$ and $s_{i,1}$ is selected as the fired cluster according to the calculated probabilities $P_{i,0}$ and $P_{i,1}$. $P_{i,0}$ (or $P_{i,1}$) is the probability of selecting the cluster $s_{i,0}$ (or $s_{i,1}$) as the fired cluster.

4. Only one enabled transition $t \in \widehat{T_i}$ from the fired cluster can fire in each marking M.

5. If $s_{i,0}$ is the fired cluster, then an enabled transition t is randomly selected from $s_{i,0}$ for firing, using concurrent selector mechanism; Otherwise, LA_i is activated. The available action set of the activated LA_i consists of the actions corresponding to the enabled transitions. The activated LA selects an action from its available action set according to its action probability vector and then the corresponding transition is selected for firing.

6. Firing of a transition $t \in \widehat{T_i}$ removes $\widehat{W_i}(p,t)$ tokens from each input place p of t, and adds $\widehat{W_i}(t,p)$ tokens to each output place p of t.

7. The local rule $F_i \in \hat{F}$ is used to generate the reinforcement signal β_i for LA_i, when the updating transition $t_{i,1}^u$ fires. For generating the reinforcement signal, the last selected actions of LA_i and its neighboring LAs (or the last fired transitions of $s_{i,1}$ and its neighboring APN-LAs) are used.

8. Using β_i, the learning algorithm of LA_i updates the action probability vector of LA_i.

6.3.4.3 Behavior of CAPN-LA

In this section, the steady-state behavior of the proposed CAPN-LA system will be analyzed. Before that, some definitions will be provided, which will be used later for the analysis of the CAPN-LA.

Definition 6.22 At time instant k, in the CAPN-LA, the conflicting transitions configuration is denoted by $\underline{q}(k) = \left(\underline{q}_1, \underline{q}_2, \ldots, \underline{q}_n\right)^T$, where $\underline{q}_i$ is the selection probability of transitions in the cluster $s_{i,1}$ and T denotes the transpose operator. $\square$

Definition 6.23 In the CAPN-LA, the set of probabilistic configurations $\mathcal{K}$ is defined by the Eq. (6.1), as given below.

$$\mathcal{K} = \left\{ \underline{q} \big| \underline{q} = \left(\underline{q}_1, \ldots, \underline{q}_n \right)^T, \underline{q}_i = \left(q_i^1, \ldots, q_i^{m_i} \right)^T, \forall y, \forall i, 0 \leq q_i^y \leq 1, \sum_y q_i^y = 1 \right\}$$

(6.1)

Definition 6.24 Evolution of the conflicting transitions is defined by a sequence of configurations $\left\{ \underline{q}(k) \right\}_{k \geq 0}$, with initial configuration $\underline{q}(0) \in \mathcal{K}$. In this sequence, $\underline{q}(k+1) = \mathcal{G}(\underline{q}(k))$, where $\mathcal{G}$ is a mapping $\mathcal{G} : \mathcal{K} \to \mathcal{K}$. Whenever an updating transition is fired, this mapping will be performed. □

Definition 6.25 A Pure-Chance Automaton (PCA) is an automaton that chooses each of its actions with equal probability. Therefore, for an pure-chance automaton with m_i-actions $q_i^j, j = 1, \ldots, m_i$ is equal to $\frac{1}{m_i}$. □

Definition 6.26 An Adaptive Petri Net based on Pure-Chance Automaton (APN-PCA) is an APN-LA, in which, instead of an LA, the conflicts among transitions are resolved by a PCA. □

Definition 6.27 The neighboring set of any cluster $s_{i,1}$, denoted by N_i, is defined as the set of all clusters with conflicting transitions in the neighborhood cells of the cell c_i, that is,

$$N_i = \{ s_{j,1} | (i,j) \in E \}$$

(6.2)

Let $\mathcal{N}_i$ be the cardinality of N_i. In addition, assume the following notations:

- $N_i^j, (1 \leq j \leq \mathcal{N}_i)$ is the jth neighboring cluster of the cluster $s_{i,1}$,
- Ft_i denotes the last fired transition of the cluster $s_{i,1}$,
- $ED(\mu, s_{i,1})$ indicates the number of enabled transitions within $s_{i,1}$ in marking μ,
- $\pi_\mu(k)$ represents the probability of the APN-LA A_i being in the marking μ_i at time instant k,
- t_i^j indicates the jth transition in the cluster $s_{i,1}$,
- $E(t, \mu_i)$ is a Boolean function defined in Eq. (6.3). This function indicates whether or not the transition t is enabled in marking μ_i.

$$E(t, \mu_i) = \begin{cases} 1, & if \ \mu_i[t) \\ 0, & otherwise \end{cases}.$$

(6.3)

- $q_i^j(k)$ is the probability of choosing the transition t_i^j from cluster $s_{i,1}$ for firing at time instant k, and
- $\hat{q}_i^j(k, \mu)$ is the scaled probability of choosing the enabled transition t_i^j for firing in marking μ. $\hat{q}_i^j(k, \mu)$ is calculated, when transition t_i^j is enabled in marking μ, according to the Eq. (6.4), as given below. $\hat{q}_i^j(k, \mu) = 0$, when transition t_i^j is not enabled in marking μ. □

$$\hat{q}_i^j(k, \mu) = \frac{q_i^j(k)}{\sum_{t_i^l | En(t_i^l, \mu)} q_i^l(k)} \tag{6.4}$$

Definition 6.28 The average reward of firing the transition t_i^j at time instant k is defined in Eq. (6.5) and the average reward for the APN-LA A_i at time instant k is defined in (6.6). □

$$d_i^j\left(\underline{q}(k)\right) = \sum_{t_1 \in N_i^1, \ldots, t_{N_i} \in N_i^{N_i}} F_i\left(t_1, \ldots, t_{N_i}; t_i^j\right) \prod_{t_l \in t_1, \ldots, t_{N_i}, t_l \in s_{L,1}} q_L^{t_l} \tag{6.5}$$

$$D_i\left(\underline{q}(k)\right) = \sum_{\mu \in \mu_i} \left(\sum_{t_i^j} d_i^j(\hat{q}(k)) \times \hat{q}_i^j(k, \mu) \right) \times \pi_\mu(k) \tag{6.6}$$

In Eq. (6.5), $F_i\left(t_1, \ldots, t_{N_i}; t_i^j\right)$, which is the local rule of the cell c_i, generates the reinforcement signal $\beta_i(k)$ for a set of transitions $\{t_1, \ldots, t_{N_i}; t_i^j\}$; this set consists of transitions $t_l, 1 \leq l \leq N_i, l \neq i$ from the neighboring clusters of $s_{i,1}$ as well as transition t_i^j from the cluster $s_{i,1}$. The average reward of transition t_i^j, calculated by the Eq. (6.5), is a weighted sum of F_i, in which the weights are the products of the probabilities of firing transitions $\{t_1, \ldots, t_{N_i}\}$. The average reward for all transitions in the cluster $s_{i,1}$ is then calculated for a marking set μ_i according to the Eq. (6.6). Note that by definition, $\hat{q}_i^j(k, \mu) \neq 0$ only when the transition t_i^j is enabled. In other words, enabled transitions are only considered for calculating the average reward in the cluster $s_{i,1}$. It has been assumed that $d_{i,1}^j\left(\underline{q}\right) \neq 0$ for all i, j, and $\underline{q}$. In other words, in any configuration, any transition has a non-zero chance of receiving reward.

$$D\left(\underline{q}\right) = \sum_i D_i\left(\underline{q}\right) \tag{6.7}$$

Definition 6.29 The total average reward for the CAPN-LA at configuration $\underline{q} \in \mathcal{K}$ is the sum of the average rewards for all APN-Las, as shown in Eq. (6.7). Atop of this definition, notation $D\left(\underline{q}^{PC}\right)$ is referred to as the total average reward for a CAPN-PCA. □

Expediency is a notion of learning. Any model that is said to learn must then do at least better than its equivalent pure-chance model. In the rest of this section, we define the expediency concept for the CAPN-LA.

Definition 6.30 A CAPN-LA is assumed to be expedient with respect to the cluster $s_{i,1}$ if $\lim_{k\to\infty} \underline{q}(k) = \underline{q}^*$ exists and the following inequality holds:

$$\lim_{k\to\infty} E\left[D_i\left(\underline{q}(k)\right)\right] > \sum_{\mu \subset \mu_i} \left[\frac{1}{ED(\mu, s_{i,1})} \times \sum_{t_i^j} d_i^j\left(\underline{q}^*\right)\right] \times \pi_\mu^* \qquad (6.8)$$

where π_μ^* is the steady-state probability of marking μ. □

In other words, a CAPN-LA is expedient with respect to the APN-LA A_i if, in the long run, A_i performs better (receives more reward) than an APN-PCA.

Definition 6.31 A CAPN-LA is assumed to be expedient if it is expedient with respect to any of its constituting APN-LAs. □

Theorem 6.1 *A CAPN-LA, in which the learning automata with the SLRP learning algorithm are the conflict resolvers, regardless of the local rule being used, is expedient.* □

Proof It has to be shown that CAPN-LA with *SLRP* learning automata is expedient with respect to all of its APN-LAs, that is, $\lim_{k\to\infty} \underline{q}(k) = \underline{q}^*$ exists and the following inequality holds for every i:

$$\lim_{k\to\infty} E\left[D_i\left(\underline{q}(k)\right)\right] > \sum_{\mu \in \mu_i} \frac{1}{ED(\mu, s_{i,1})} \times \left[\sum_{t_i^j} d_i^j\left(\underline{q}^*\right)\right] \times \pi_\mu^*, \quad \textit{for every } i \quad (6.9)$$

Lemma 6.1 *Consider a probabilistic configuration $q \in \mathcal{K}$, which evolves with SLRP learning algorithm. Regardless of the initial configuration $\underline{q}(0)$, $\lim_{k\to\infty} \underline{q}(k) = \underline{q}^*$ and $\lim_{k\to\infty} E\left[\underline{q}(k)\right] = \underline{q}^*$ exist.* □

Proof They both have been proven in (Esnaashari and Meybodi 2015), in which q represents a probabilistic matrix constructed by the action probability vector of a team of LAs. Theses LAs form an irregular cellular learning automata and all of them evolve with *SLRP* learning algorithm. □

Lemma 6.2 *Consider a probabilistic configuration $q \in \mathcal{K}$, which evolves with SLRP learning algorithm. Let q_i^{j*} be the selection probability of the transition $t_{i,1}^j$ in the cluster $s_{i,1}$ in the long run and $d_{i,1}^j\left(\underline{q}^*\right)$ be the generated reward value by*

Eq. (6.5). *Regardless of the local rule being used, the value of* q_i^{j*} *is proportional to the value of* $d_{i,1}^j\left(\underline{q}^*\right)$.   □

Proof It has been shown in Esnaashari and Meybodi (2015) that the value of q_i^{j*} is inversely proportional to the received penalty ($rp_{i,1}^j\left(\underline{q}^*\right)$) from the environment, that is $q_i^{j*} \propto \frac{1}{rp_{i,1}^j\left(\underline{q}^*\right)}$. Since the reward signal, $d_{i,1}^j\left(\underline{q}^*\right)$ is equal to $1 - rp_{i,1}^j\left(\underline{q}^*\right)$, we get

$$q_i^{j*} \propto \frac{1}{1 - d_{i,1}^j\left(\underline{q}^*\right)} \tag{6.10}$$

In Eq. (6.10), if the value of $d_{i,1}^j\left(\underline{q}^*\right)$ is increased, then the denominator decreases and consequently, the q_i^{j*} value increases, and vice versa. This is an indication of the fact that the q_i^{j*} value is proportional to the $d_{i,1}^j\left(\underline{q}^*\right)$ value and hence the lemma.   □

By taking the Lemma 6.1 into consideration, it is only required to be shown that the inequality (6.9) holds. By using the average reward definition in Eq. (6.6), the left hand side of this inequity can be rewritten as follows:

$$\lim_{k\to\infty} E\left[D_i\left(\underline{q}(k)\right)\right] = \sum_{\mu\in\mu_i} \lim_{k\to\infty} E\left[\left(\sum_{t_i^j}\left[d_i^j\left(\underline{q}(k)\right) \times \hat{q}_i^j(k,\mu)\right]\right) \times \pi_\mu(k)\right] \tag{6.11}$$

Since $d_{i,1}^j\left(\underline{q}(k)\right)$, $q_i^j(k)$, and $\pi_{\mu_i}(k)$ are independent, the above equation can be simplified to

$$\lim_{k\to\infty} E\left[D_i\left(\underline{q}(k)\right)\right] = \sum_{\mu\in\mu_i}\sum_{t_i^j}\left(\lim_{k\to\infty} E\left[d_i^j\left(\underline{q}(k)\right)\right] \times \lim_{k\to\infty} E\left[\hat{q}_i^j(k,\mu)\right]\right)$$
$$\times \lim_{k\to\infty} E\left[\pi_\mu(k)\right] \tag{6.12}$$

Considering Lemma 6.1, $\lim_{k\to\infty} E\left[\underline{q}(k)\right] = \underline{q}^*$, and hence $\lim_{k\to\infty} E\left[q_i^j(k)\right] = q_i^{j*}$ for all i and j. In addition, $\lim_{k\to\infty} E\left[\pi_\mu(k)\right] = \pi_\mu^*$. Using Eq. (6.5), $\lim_{k\to\infty} E\left[d_i^j\left(\underline{q}(k)\right)\right]$ can be computed as follows:

$$\lim_{k \to \infty} E\left[d_i^j\left(\underline{q}(k)\right)\right] = \lim_{k \to \infty} E\left[\sum_{t_1 \in N_i(1),\dots,t_{N_i} \in N_i(\mathcal{N}_i)} F_i(t_1,\dots,t_{N_i}; t_i^j) \prod_{t_l \in t_1,\dots,t_{N_i}, t_l \in S_{L,1}} q_L^{t_l}\right]$$

$$= \sum_{t_1 \in N_i(1),\dots,t_{N_i} \in N_i(\mathcal{N}_i)} F_i(t_1,\dots,t_{N_i}; t_i^j) \prod_{t_l \in t_1,\dots,t_{N_i}, t_l \in S_{L,1}} \lim_{k \to \infty} E[q_L^{t_l}]$$

$$= d_i^j\left(\underline{q}^*\right)$$

$$(6.13)$$

Thus,

$$\lim_{k \to \infty} E\left[D_i\left(\underline{q}(k)\right)\right] = \sum_{\mu \in \mu_i} \left(\sum_{t_i^j} d_i^j\left(\underline{q}^*\right) \times \hat{q}_i^{j*}\right) \times \pi_\mu^* \qquad (6.14)$$

Now, for every i, it needs to be shown that:

$$\sum_{\mu \in \mu_i} \left(\sum_{t_i^j} d_i^j\left(\underline{q}^*\right) \times \hat{q}_i^{j*}\right) \times \pi_\mu^* > \sum_{\mu \in \mu_i} \left(\frac{1}{ED\left(\mu, s_{i,1}\right)}\right) \times \sum_{t_i^j} d_i^j\left(\underline{q}^*\right) \times \pi_\mu^* \quad (6.15)$$

For every $\{\mu_i | ED(\mu_i, s_{i,1})\rangle 0\}$, the above equation can be simplified to

$$\sum_{t_i^j} d_i^j\left(\underline{q}^*\right) \times \hat{q}_i^{j*} > \frac{1}{ED\left(\mu, s_{i,1}\right)} \times \sum_{t_i^j} d_i^j\left(\underline{q}^*\right) \qquad (6.16)$$

In the above inequity, each side is a convex combination of $d_i^j\left(\underline{q}^*\right)$. In the right hand side convex combination of (6.16), all of the $d_i^j\left(\underline{q}^*\right)$ have the same weights, which equals to $\frac{1}{ED(\mu, S_{i,1})}$, whereas in the left hand side convex combination, the weight of each $d_i^j\left(\underline{q}^*\right)$ equals to $\hat{q}_i^{j*}$. It was shown that in Lemma 6.2, the q_i^{j*} value is proportional to the $d_i^j\left(\underline{q}^*\right)$ value. This indicates that the greater the value of $d_i^j\left(\underline{q}^*\right)$, the greater its weight will be. As a result, the right hand side convex combination of the inequity (6.16) is smaller than the one on the left hand side, and thus the theorem is proved. $\qquad \square$

6.3.4.4 Norms of Behavior

To judge the behavior of the proposed CAPN-LA, it is necessary to set up quantitative norms of behavior. Here, we define a new metric, namely measure of expediency, in terms of the firing probabilities of the transitions in the CAPN-LA.

Definition 6.32 Measure of expediency is defined according to Eq. (6.17). $M_E < 0$ indicates that the CAPN-LA performs worse than its equivalent CAPN-PCA, $M_E = 0$ indicates that the CAPN-LA is the pure-chance, and $M_E > 0$ indicates that the CAPN-LA performs better than its equivalent CAPN-PCA. A higher value for M_E means that the CAPN-LA is more expedient.

$$M_E = \left(\frac{\lim_{k \to \infty} E\left[D\left(\underline{q}(k)\right)\right]}{D\left(\underline{q}^{PC}\right)} \right) - 1 \tag{6.17}$$

6.3.5 Concluding Remarks

One may argue why PNs are utilized in the proposed hybrid machines, while methods for representing a system are not limited to PNs. To elaborate on this matter, we can mention the following advantages of PNs (Agerwala 1979) (Aalst and C. Stahl 2011):

1. The graphical representation facilitates an easy understanding of the system,
2. Petri nets theory can analyze system behavior,
3. It is possible to systematically design systems from a Petri net model,
4. There are many software tools that support modeling and analysis of PNs,
5. PNs represent the state space of the system in a more implicit and compact way.

Similar argument may arise regarding the usage of PNs for proposing cellular hybrid machines, like APN-ICLA and CAPN-LA. There are other methods, such as finite-state automata, Markov chain, and queuing-type models, which can be used in this context. In order to clarify on our intuition in using PNs we have to note that the state space of a cellular system is a combination of the state spaces of its constituting cells, which could be too large and complex to analyze. As an example, assuming that X_i represents the state space of a cell i, then the state space of a cellular network consisting of N cells can possibly be as large as $X_1 \times X_2 \times \cdots \times X_N$ if modeled by finite-state automata. PNs possess an ability to decompose or modularize a complex interacting system, such that the interactions between neighboring cells can be resolved by adding a few places and transitions. Then, from a PN, one can conveniently see the dynamics of individual cells, discern the level of their interaction, and ultimately decompose a cellular network into logical distinct cell modules. Thus, PN is an appropriate choice for modeling a cellular system (Lin and Chan 2009).

Yet another question is that what the impacts of LA on the properties of ASPN-LA are. To answer this question, we have to note that the fusion of LA with SPN does not change the reachability graph of the underlying SPN. To see this, recall that in the construction procedure of ASPN-LA two places and one immediate transition are inserted to the underlying SPN per cluster of conflicting transitions. Since all of the inserted transitions are immediate, all of the new states in the state space of the ASPN-LA are vanishing. This means that these new inserted places and transitions do not have any impact on the reachability graph of the SPN. Therefore, any method which can be used for analyzing the underlying SPN is also applicable for analyzing the fused version of that SPN with LA. This is why we have been able to analyze ASPN-LA using its underlying CTMC, which is used for analyzing SPNs.

6.4 Priority Assignment Application

The Priority Assignment (PA) problem in the queuing system with unknown parameters, consisting of a server and m queues of jobs, is defined in (Meybodi 1983); how to select jobs from the queues so that the total waiting time of the system is minimized. The shortest total waiting time of the queuing system is achieved, when the server assigns the highest selection priority to the jobs from the queue with the highest service rate (Cobham 1954; Meybodi 1983). However, if the service rates of the queues are unknown, the total waiting time of the system can be shortened by assigning the highest selection probability, rather than priority, to the jobs from the queue with the highest service rate (Jiang et al. 2002).

In this Section, a model of adaptive stochastic Petri net (ASPN-LA) will be used to solve PA problem in the queuing systems with unknown parameters. The ASPN-LA provides a controlling mechanism to select jobs from different queues with the aim of minimization of total waiting time. This is obtained by updating the probabilities of selections of the queues in the system. In this Section, using the ASPN-LA, the problem of priority assignment in a queuing system with multiple queues and a server, when the parameters of the system are unknown, will be solved.

In this Section, we will establish a necessary condition for a PA mechanism. A PA mechanism assigns the highest priority to the queue with the lowest average service time when this necessary condition is satisfied. In addition, we will show that ASPN-LA can satisfy this necessary condition. Drawing upon the results of the conducted theoretical study, we will propose an enhance priority assignment mechanism by reducing its required learning time. Finally, we will conduct an extensive set of computer simulations to study the ASPN-LA mechanism, which support the theoretical results.

6.4.1 Priority Mechanisms

An important subject in queuing theory is that of priorities (Cobham 1954). In a priority queuing system, there are m $(m > 1)$ different classes of jobs and an arriving job belongs to one of these m different classes. These classes may be distinguished according to some measures of importance, and to indicate the relative measure of importance, a priority index i is associated to each class. It is conventional that the larger the value of the index associated with a class, the lower is the priority associated with that class. The mechanism according to which the server selects the next jobs from these classes, is called a priority mechanism (Jiang et al. 2002).

Priority mechanism is specified by two rules (Meybodi 1983). The first rule indicates the manner, in which a job is selected to give a service by the server. This rule may depend only on the knowledge of the priority class, to which a job belongs, or it may depend solely or partially on other considerations relating to the existing state of the system, which is the type of the last served job or the waiting time of present jobs. The former mechanisms are called exogenous priority mechanisms, while the latter are called endogenous priority mechanisms (Jaiswal 1968). In exogenous priority mechanisms, the decision of selecting the next job to give a service, depends only on the priority class: a job of the ith class, if present, is always selected for giving a service prior to the job of the jth class $(i < j)$ (Meybodi 1983).

The second rule specifies the manner, in which the unit is served after the service is started. In a preemptive priority mechanism, a job of higher priority takes, on arrival, immediate precedence over units of lower priority (Jaiswal 1968; Kittsteiner and Moldovanu 2005). The job whose service is interrupted returns to the service point, only when there are no higher priority jobs remaining in the system. Under non-preemptive priority mechanisms, a job, once at the service, remains there until its service is complete, then the next job to perform a service is the one with the highest priority in the classes (Kilhwan and Chae 2010).

There are some other types of priority mechanisms, which have been reported in the literature, such as alternating priority mechanism (Jurczyk 2000), priority with balking and reneging (Jaiswal 1968), round-robin mechanism (Epema 1991; Rasch 1970), dynamic priority mechanism (Chen and Lin 1990), probabilistic priority mechanism (Jiang et al. 2002), and adaptive priority mechanism.

The evaluation of various characteristics of a priority system under a given priority mechanism is an important objective in any priority system; the queue-length distribution may be important from the design point of view, the waiting time from the job point of view, and the busy period from the server point of view. It is shown that as long as the queuing mechanism selects the jobs in a way that is independent of their service time or any measure of their service time, then the distribution of the number of jobs in the system and the average waiting time will be invariant to the order of service (Bolch et al. 2006; Kleinrock 1975).

A priority system with m independent Poisson streams and arbitrary service time distributions for each of the classes is described as follows: assume a waiting cost c_i for class i, the mean service requirements μ_i^{-1} for class i, and a non-preemptive service mechanism. It is shown that a policy that ranks the classes according to the "μc" rule, indicated in the following inequality, will minimize the average waiting cost (Bolch et al. 2006).

$$\mu_1 c_1 \geq \mu_2 c_2 \geq \cdots \geq \mu_i c_i \geq \cdots \geq \mu_m c_m$$

The priority assignment is the optimal, when priorities are in the decreasing order of the $\mu_i c_i$ corresponding to the decreasing cost of system; it means that the highest priority is assigned to the class with the highest value of $\mu_i c_i$. This assignment will minimize the average cost of the system, in which service time distributions are exponentials (Jaiswal 1968).

To develop a meaningful priority schema, the probabilistic characteristics of the jobs in various classes, such as the arrival and service time distributions, are required to be known. In many practical situations, however, this information may be unknown in advance (Meybodi 1983). An interesting question rising here is that if the probabilistic (arrival and service time) characteristics of jobs in various classes are unknown in advance, how to go about developing a meaning priority assignment among the various classes. In other words, how to design a priority queuing system, in which the priorities are worked out directly without a prior knowledge of the input and service characteristics of the system.

This problem is first addressed in Varshavskii et al. (1968) and a solution is provided using fixed structure learning automata, in which one automaton is used for each class. This fixed structure is developed in the context of modeling of collective behavior (Tsetlin 1973). The states of transitions in the automaton are indirectly controlled by the service requirements of jobs in each class. At any given instance, the class corresponding to the automaton with highest state number is chosen to perform a service. This approach leads to a Markov chain with a very complex transition structure, which its analysis is impossible with the formal methods. Due to arise of this difficulty, the simulation to evaluate efficiency of this approach is resorted in Viswanathan and Narendra. The simulation shows that if the initial number of states of each automaton is chosen to be large enough, the introduced algorithm in Viswanathan and Narendra will asymptotically approach the priority assignment, in which the probabilistic characteristics of the system are known.

A simple fixed-structure finite memory learning scheme is considered to effectively learn the optimal priority assignment at a single-server queue (Viswanathan and Narendra; Kumar 1986). The algorithm parameters are well chosen so that the optimal policy (corresponding to known service time distributions) is achieved by the algorithm, asymptotically, with a probability as close to one as desired. Compared to the methods of Meybodi (1983), in which the variable structure learning schemes are used, the fixed-structure scheme exhibits a better convergence rate. In addition, the proof of convergence to the optimal priority assignment is much simpler in the fix-structure scheme (Viswanathan and Narendra; Kumar 1986).

6.4.2 ASPN-LA and Priority Assignment

In a queuing system with a server and m queues (Fig. 6.12), there are m classes of jobs. Jobs from class i arrive into queue Q_i at constant Poisson rate $\lambda_i, i = 1, \ldots, m$. The service times of these classes are independent and identically distributed with exponential distribution with the mean $\mu_i^{-1}, i = 1, \ldots, m$.

When the server becomes idle, it has to select a new job from a queue. In other words, the server must prioritize one class of jobs to other classes when they have jobs, waiting for service. This prioritization affects the total waiting time of the system (Meybodi and Lashmivarahan 1983) and is denoted by Priority Assignment (PA) problem. Thus, the PA problem in a queuing system with unknown parameters is the problem of how to select jobs from the queues so that the total waiting time of the system is minimized.

In a queuing system with m queues $Q_i, (i = 1, \ldots, m)$, if $\mu_j > \mu_k, 1 \leq j < k \leq m$, then it means that the average service time of the jobs arriving into Q_j is shorter than that of arriving into Q_k. In this queuing system, a lower total waiting time of the system is obtained when the server selects the jobs from Q_j with a higher priority than that of Q_k, for all $1 \leq j < k \leq m$ (Cobham 1954; Kleinrock 1975). If $\mu_i, i = 1, \ldots, m$ are known to the server, then the PA problem can be solved by a simple mechanism such as strict priority (Kleinrock 1975); the server always has to select the jobs from Q_j with a higher priority than that the jobs from of Q_k, for all $1 \leq j < k \leq m$. However, if the service rates of queues are unknown, the total waiting time of the system can be shortened by assigning the highest selection probability, rather than priority, to the jobs from the queue with the highest service rate. Using selection probability instead of selection priority is called Probabilistic Priority (PP) mechanism (Jiang et al. 2002).

The simple PP mechanism to solve the PA problem can be described using the stochastic Petri nets (SPN), denoted by SPN-[m]PP (Fig. 6.13), where server selects jobs from either of the queues with equal probability of $\frac{1}{m}$. In this SPN-[m]PP, place p_1 represents the server and $p_{2i}, i = 1, \ldots, m$ represents Q_i. If the server is idle $(M(p_1) = 1)$ and more than two queues have jobs $(\sum_i M(p_{2i}) > 1)$, then one of the jobs is selected randomly for execution. In other words, in this SPN-[m]PP, no priority assignment scheme is used in order to reduce the total waiting time of the

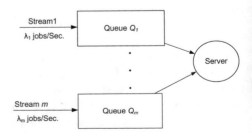

Fig. 6.12 Queuing system with one server and m queues

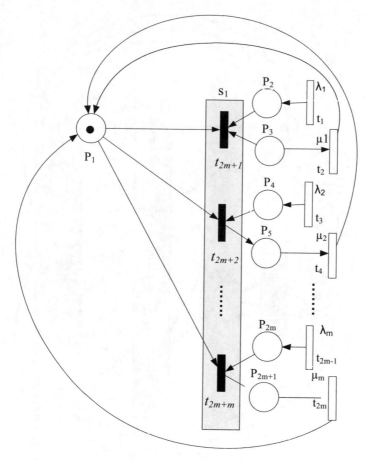

Fig. 6.13 A model of PP mechanism to solve the PA problem using an SPN

system. In the following section, an adaptive PP mechanism based on the ASPN-LA will be described.

6.4.2.1   Modeling a PA Mechanism with an ASPN-LA

An ASPN-LA is constructed based on the SPN, illustrated in Fig. 6.13, according to the construction procedure of ASPN-LA described in Sect. 6.3.1.2. Note that the set $\{t_{2m+k}, k = 1, \ldots, m\}$ is a maximal potential conflict and hence, forms the cluster s_1. The learning automaton LA_1 with m actions is assigned to the cluster s_1 to control the conflict of its transitions; each action corresponds to select one transition for firing. The updating transition t_1^u, input place p_{2m+2} and output place p_{2m+3} are inserted into the constructed ASPN-LA an then output place p_{2m+3} is connected to updating transition t_1^u with an inhibitor arc. The place p_{2m+3} is added to the preset

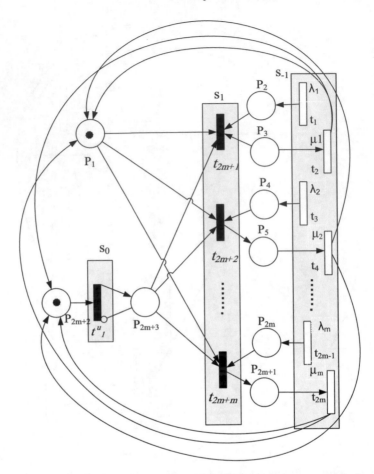

Fig. 6.14 ASPN-LA-[m]PP represents a PA mechanism (a system with m queues and one server)

of transitions $t_{2m+1}, \ldots, t_{2m+m}$. Since the place p_1 is the shared input place of transitions $t_{2m+1}, \ldots, t_{2m+m}$, we add preset of this place to preset of p_{2m+2}. Since a token exists in place p_1 (shared input place of the transitions $t_{2m+1}, \ldots, t_{2m+m}$) a token will be placed in p_{2m+2}. Transitions $t_1, \ldots, t_{2m}$ create the cluster s_{-1} and t_1^u creates cluster s_0. The ASPN-LA, in which the above modifications are made, is called ASPN-LA-[m]PP shown in Fig. 6.14.

To achieve the ASPN-LA-[m]PP system, the reinforcement signal generator function set $\hat{F} = \{f_1\}$ is needed. Upon firing of t_1^u, function f_1 is executed, which generates the reinforcement signal β_1 for LA_1. To specify how f_1 generates β_1, we first note that when t_1^u fires for the nth time, the following parameters are known to the system:

- The queue from which the last job (job n) was selected for execution,
- The execution time of job n, referred to as $\delta(n)$,

- Number of jobs from Q_j given service so far, referred to as k_j ($\sum_j k_j = n$),
- Total service time for the jobs in Q_j given service so far, referred to as $\Delta_j(n)$, and
- Average service time for jobs in Q_j given service so far, which can be calculated as $\Gamma_j(n) = \frac{1}{k_j} \times \Delta_j(n)$.

Considering above parameters f_1 can be described using Eq. (6.18), in which $\beta_1 = 1$ is considered as a penalty signal and $\beta_1 = 0$ is considered as a reward signal. Using β_1, generated according to Eq. (6.18), LA_1 updates its action probability vector, $q(n)$, according to the L_{R-I} learning algorithm.

$$\begin{cases} \beta_1 = 1; \delta(n) \geq \frac{1}{m} \times \sum_i \Gamma_i(n) \\ \beta_1 = 0; \delta(n) < \frac{1}{m} \times \sum_i \Gamma_i(n) \end{cases} \tag{6.18}$$

6.4.2.2   The Analysis of ASPN-LA-[M]PP System

In this section, we will show that LA_1, associated with the cluster s_1, gradually learns to assign a higher probabilistic probability to the class j of jobs rather than the class k, if $\mu_j > \mu_k$. To better clarify the idea, we first analyze a simple case of an ASPN-LA-[m]PP with three number of queues ($m = 3$) denoted by ASPN-LA-3PP (Fig. 6.15). We will then generalize our analysis of the ASPN-LA-3PP to the general form of ASPN-LA-[m]PP with m queues. Finally, we will show that LA solves the PA problem in the ASPN-LA-[m]PP system, if L_{R-I} is used as its learning algorithm.

To analyze the ASPN-LA-3PP, we use its underlying Continuous-Time Markov Chain (CTMC) (Bolch et al. 2006). Having derived this CTMC, we will first partition its states into four groups: S_1) the set of states in which a job is selected from Q_1 while other jobs are waiting in other queues; S_2) the set of states in which a job is selected from Q_2 while other jobs are waiting in other queues; S_3) the set of states in which a job is selected from Q_3 while other jobs are waiting in other queues; and S_4) remaining states. Assuming $\mu_1 > \mu_2 > \mu_3$, we will reach a number of conditions if hold, in the steady-state, the accumulated sojourn time of S_1 will be longer than that of S_2, and for S_2 it will be longer than that of S_3.

Deriving CTMC: an ASPN-LA-3PP is an SPN and contains immediate and timed transitions and hence, an extended reachability graph (ERG) can be generated from it (Bolch et al. 2006). Each node in this ERG is a reachable marking of ASPN-LA-3PP, and each directed arc from M_i to M_j represents the probability or rate of reaching to M_j from M_i. The ERG can be transformed into a reduced reachability graph (RG) by eliminating the vanishing marking and the corresponding transitions (Bolch et al. 2006). A CTMC will be developed from an RG (Marsan et al. 1994).

Let L_i denote the number of tokens in the place p_i and $M_j = (L_1, \ldots, L_9)$ denote the jth marking of the ASPN-LA-3PP. The ERG, achieved from the initial marking

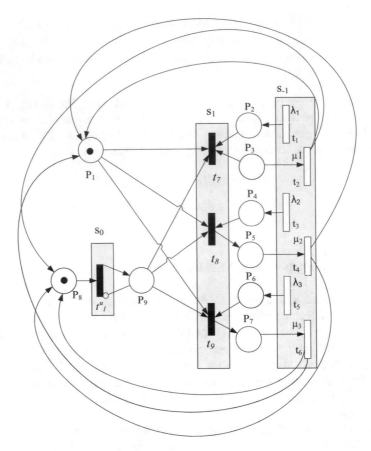

Fig. 6.15 ASPN-LA-3PP system

$M_0 = (110101010)$, is shown in Fig. 6.16. To reduce the number of markings in this ERG, we assume that places p_2, p_4 and p_6 are one-bounded (Peterson and Net 1981). This ERG contains fifteen vanishing markings {0, 1, 8, 9, 14, 15, 20, 21, 26, 27, 30, 31, 34, 35, 38} represented by rectangles and 25 tangible markings {2–7, 10–13, 16–19, 22–25, 28, 29, 32, 33, 36, 37, 39} represented by ovals. To draw reachability graph with enough precision, we use the repeated markings in figures with dash-bounded line.

The probability $q_j(n), j = 1, 2, 3$ is used to select queue j for giving service at time instant n, when there are jobs in all three queues. In other words, the action j is selected by LA_1 at time instant n with the probability $q_j(n)$, when the available action set of LA_1 has three actions. When this set consists of only two actions, the scaled action probabilities $\hat{q}(n)$ are used (defined in the LA with Changing Number of Actions). There are three possible cases: (1) no job is available at queue Q_1; (2) no job is available at queue Q_2; and (3) no job is available at queue Q_3. The set of scaled probabilities in these three cases are given in Table 6.2.

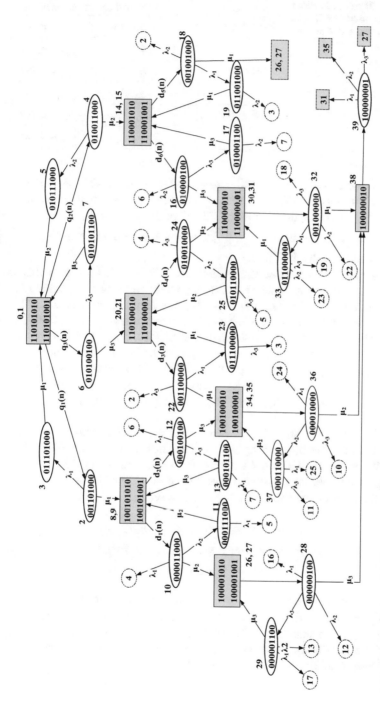

Fig. 6.16 The extended reachability graph of ASPN-LA-3PP shown in Fig. 6.15

Table 6.2 The scaled actions probabilities used to select an action from the set of actions $\mathcal{A}(n)$

Set of actions	Scaled actions probabilities
$\mathcal{A}(n) = \{\alpha_2, \alpha_3\}$	$\hat{q}_1(n) = \frac{q_2(n)}{q_2(n)+q_3(n)} = d_1(n)$
	$\hat{q}_2(n) = \frac{q_3(n)}{q_2(n)+q_3(n)} = d_2(n)$
$\mathcal{A}(n) = \{\alpha_1, \alpha_2\}$	$\hat{q}_1(n) = \frac{q_1(n)}{q_1(n)+q_2(n)} = d_3(n)$
	$\hat{q}_2(n) = \frac{q_2(n)}{q_1(n)+q_2(n)} = d_4(n)$
$\mathcal{A}(n) = \{\alpha_1, \alpha_3\}$	$\hat{q}_1(n) = \frac{q_1(n)}{q_1(n)+q_3(n)} = d_5(n)$
	$\hat{q}_2(n) = \frac{q_3(n)}{q_1(n)+q_3(n)} = d_6(n)$

The finite and irreducible CTMC, corresponding to the ASPN-LA-3PP, is shown in Fig. 6.17. In this CTMC, each state j corresponds to the tangible marking M_j of the ERG (Fig. 6.16). In drawing of the CTMC of the ASPN-LA-3PP, we have used ⬤ to denote the set of states in S_1, ◉ to denote the set of states in S_2, and ⬚ to denote the set of states in S_3.

Steady-State Analysis of the ASPN-LA-3PP: Fig. 6.16 shows that in six tangible markings, a job is selected from Q_1 while other jobs are waiting in other queues, i.e., markings $M_2, M_3, M_{18}, M_{19}, M_{22}$, and M_{23}, shown in Fig. 6.17 by **bold circle with sold fill pattern**. In other words, in these markings, the ASPN-LA-3PP system assigns the highest PP to the first class of jobs. The set of these markings is defined by $S_1 = \{s_2, s_3, s_{18}, s_{19}, s_{22}, s_{23}\}$. In the set of markings $S_2 = \{M_4, M_5, M_{10}, M_{11}, M_{24}, M_{25}\}$, the ASPN-LA-3PP system gives the highest PP to the second class and in the set of markings $S_3 = \{M_6, M_7, M_{12}, M_{13}, M_{16}, M_{17}\}$, this system gives the highest PP to the third class of jobs.

With the CTMC in Fig. 6.17, the steady-state probability vector π can be derived using equation $\pi Q = 0$, where Q is the infinitesimal generator matrix (Bolch et al. 2006). We define P_1, P_2, and P_3 according to equations defined in Eq. (6.19), in which π_k denotes the steady-state probability of the CTMC being in state k and P_i denotes the steady-state probability of marking set $S_i, i = 1, 2, 3$.

$$P_1 = \pi_2 + \pi_3 + \pi_{18} + \pi_{19} + \pi_{22} + \pi_{23},$$
$$P_2 = \pi_4 + \pi_5 + \pi_{10} + \pi_{11} + \pi_{24} + \pi_{25}, \tag{6.19}$$
$$P_3 = \pi_6 + \pi_7 + \pi_{12} + \pi_{13} + \pi_{16} + \pi_{17}.$$

Under the assumption $\mu_1 > \mu_2 > \mu_3$, we analyze our proposed ASPN-LA-3PP system by comparing the values of $P_i, i = 1, 2, 3$, when the system reaches to the steady state; if inequalities (6.20) hold, then PP mechanism assigns the highest probabilistic priority to first class of jobs and a higher probabilistic priority to the second class of jobs rather than the third class.

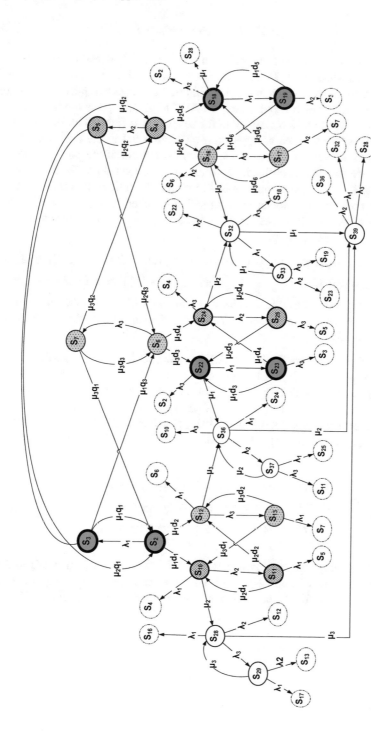

Fig. 6.17 CTMC for ASPN-LA-3PP shown in Fig. 6.15

$$\mathcal{P}_1 > \mathcal{P}_2, \mathcal{P}_1 > \mathcal{P}_3 \text{ and } \mathcal{P}_2 > \mathcal{P}_3 \qquad (6.20)$$

We propose three theorems to prove that the ASPN-LA-[m]PP reaches to the steady states, in which $\mathcal{P}_j > \mathcal{P}_k$ for $j < k; j = 1, \ldots, m - 1; k = 1, \ldots, m$ are held. In Theorem 6.2, using the steady-state probabilities, we derive two general conditions on $q_i(n), i = 1, 2, 3$ which if hold, then the ASPN-LA-3PP gives the highest PP to the first class of jobs and a higher PP to the second class rather than the third class of jobs. In Theorem 6.3, we will generalize Theorem 6.2 to the ASPN-LA-[m]PP system with m queues. In Theorem 6.4, we will show that if the L_{R-I} algorithm is used to update the action probability vector of LA_1, then $\mathcal{P}_j > \mathcal{P}_k$ for $j < k; j = 1, \ldots, m - 1; k = 1, \ldots, m$ when n goes to infinity. To follow the CTMC analysis by ordinary methods, the stochastic information attached to the arcs should be fix values. This is why we have assumed fixed values $q_1^*, q_2^*, \ldots, q_m^*$ rather than $q_1(n)$, $q_2(n), \ldots, q_m(n)$, respectively.

Theorem 6.2 Let $q_1^*, q_2^*,$ and q_3^* be the selection probabilities of $Q_1, Q_2,$ and Q_3 in the ASPN-LA-3PP. If $q_1^* > \frac{\mu_1}{2} \left(\frac{q_2^*}{\mu_2} + \frac{q_3^*}{\mu_3} \right)$ and $q_2^* > \mu_2 \left(\frac{q_3^*}{\mu_3} \right)$, then the ASPN-LA-3PP assigns the highest probabilistic priority to the first class of jobs and a higher probabilistic priority to the second class rather than the third class of jobs, i.e., inequalities (6.20).

Proof The finite and irreducible CTMC corresponding to the ASPN-LA-3PP (Fig. 6.15) is shown in Fig. 6.17, where each state j corresponds to the tangible marking M_j of the ERG (Fig. 6.16). Considering this CTMC, the steady-state probability vector π can be derived using equation $\pi \mathbb{Q} = 0$, where $\mathbb{Q}$ is the infinitesimal generator matrix (Cobham 1954; Bolch et al. 2006). Let π_k denote the steady-state probability of the CTMC being in state k and $\mathcal{P}_i, i = 1, 2, 3$ denote as the following.

$$
\begin{aligned}
\mathcal{P}_1 &= \pi_2 + \pi_3 + \pi_{18} + \pi_{19} + \pi_{22} + \pi_{23}, \\
\mathcal{P}_2 &= \pi_4 + \pi_5 + \pi_{10} + \pi_{11} + \pi_{24} + \pi_{25}, \qquad (6.21) \\
\mathcal{P}_3 &= \pi_6 + \pi_7 + \pi_{12} + \pi_{13} + \pi_{16} + \pi_{17}.
\end{aligned}
$$

To follow steady-state analysis to reach at $\mathcal{P}_i$, we drive equation $\pi \mathbb{Q} = 0$ for state 2–7, 10–13, 16–19, and 22–25 and achieve equations defined in Eq. (6.22). By simplification of Eq. (6.22), new equations are formed, defined in Eq. (6.23).

$$\pi_2(\lambda_1 + \mu_1) = q_1^*(\mu_1\pi_3 + \mu_2\pi_5 + \mu_3\pi_7) + \lambda_2\pi_{18} + \lambda_3\pi_{22}$$

$$\pi_3\mu_1 = \lambda_1\pi_2 + \lambda_2\pi_{19} + \lambda_3\pi_{23}$$

$$\pi_4(\lambda_2 + \mu_2) = q_2^*(\mu_1\pi_3 + \mu_2\pi_5 + \mu_3\pi_7) + \lambda_1\pi_{10} + \lambda_3\pi_{24}$$

$$\pi_5\mu_2 = \lambda_1\pi_{11} + \lambda_2\pi_4 + \lambda_3\pi_{25}$$

$$\pi_6(\lambda_3 + \mu_3) = q_3^*(\mu_1\pi_3 + \mu_2\pi_5 + \mu_3\pi_7) + \lambda_1\pi_{12} + \lambda_2\pi_{16}$$

$$\pi_7\mu_3 = \lambda_1\pi_{13} + \lambda_2\pi_{17} + \lambda_3\pi_6$$

$$\pi_{10}(\lambda_1 + \lambda_2 + \mu_2) = \frac{q_2^*}{q_2^* + q_3^*}(\mu_1\pi_2 + \mu_2\pi_{11} + \mu_3\pi_{13}) + \lambda_3\pi_{36}$$

$$\pi_{11}(\lambda_1 + \mu_2) = \lambda_2\pi_{10} + \lambda_3\pi_{37}$$

$$\pi_{12}(\lambda_1 + \lambda_3 + \mu_3) = \frac{q_3^*}{q_2^* + q_3^*}(\mu_1\pi_2 + \mu_2\pi_{11} + \mu_3\pi_{13}) + \lambda_2\pi_{28}$$

$$\pi_{13}(\lambda_1 + \mu_3) = \lambda_2\pi_{29} + \lambda_3\pi_{29}$$

$$\pi_{16}(\lambda_2 + \lambda_3 + \mu_3) = \frac{q_3^*}{q_1^* + q_3^*}(\mu_1\pi_{19} + \mu_2\pi_4 + \mu_3\pi_{17}) + \lambda_1\pi_{28}$$

$$\pi_{17}(\lambda_2 + \mu_3) = \lambda_1\pi_{29} + \lambda_3\pi_{16}$$

$$\pi_{18}(\lambda_2 + \lambda_1 + \mu_1) = \frac{q_1^*}{q_1^* + q_3^*}(\mu_1\pi_{19} + \mu_2\pi_4 + \mu_3\pi_{17}) + \lambda_3\pi_{32}$$

$$\pi_{19}(\lambda_2 + \mu_1) = \lambda_1\pi_{18} + \lambda_3\pi_{33}$$

$$\pi_{22}(\lambda_1 + \lambda_3 + \mu_1) = \frac{q_1^*}{q_1^* + q_2^*}(\mu_1\pi_{23} + \mu_2\pi_{25} + \mu_3\pi_6) + \lambda_2\pi_{32}$$

$$\pi_{23}(\lambda_3 + \mu_1) = \lambda_1\pi_{22} + \lambda_2\pi_{33}$$

$$\pi_{24}(\lambda_2 + \lambda_3 + \mu_1) = \frac{q_2^*}{q_1^* + q_2^*}(\mu_1\pi_{23} + \mu_2\pi_{25} + \mu_3\pi_6) + \lambda_1\pi_{36}$$

$$\pi_{25}(\lambda_3 + \mu_2) = \lambda_1\pi_{37} + \lambda_2\pi_{24}$$

$$(6.22)$$

$$\mu_1(\pi_2 + \pi_3) = q_1^*(\mu_1\pi_3 + \mu_2\pi_5 + \mu_3\pi_7) + \lambda_2(\pi_{18} + \pi_{19}) + \lambda_3(\pi_{22} + \pi_{23})$$

$$\mu_2(\pi_4 + \pi_5) = q_2^*(\mu_1\pi_3 + \mu_2\pi_5 + \mu_3\pi_7) + \lambda_1(\pi_{10} + \pi_{11}) + \lambda_3(\pi_{24} + \pi_{25})$$

$$\mu_3(\pi_6 + \pi_7) = q_3^*(\mu_1\pi_3 + \mu_2\pi_5 + \mu_3\pi_7) + \lambda_1(\pi_{12} + \pi_{13}) + \lambda_2(\pi_{16} + \pi_{17})$$

$$(\lambda_1 + \mu_2)(\pi_{10} + \pi_{11}) = \frac{q_2^*}{q_2^* + q_3^*}(\mu_1\pi_2 + \mu_2\pi_{11} + \mu_3\pi_{13}) + \lambda_3(\pi_{36} + \pi_{37})$$

$$(\lambda_1 + \mu_3)(\pi_{12} + \pi_{13}) = \frac{q_3^*}{q_2^* + q_3^*}(\mu_1\pi_2 + \mu_2\pi_{11} + \mu_3\pi_{13}) + \lambda_2(\pi_{28} + \pi_{29})$$

$$(\lambda_2 + \mu_3)(\pi_{16} + \pi_{17}) = \frac{q_3^*}{q_1^* + q_3^*}(\mu_1\pi_{19} + \mu_2\pi_4 + \mu_3\pi_{17}) + \lambda_1(\pi_{28} + \pi_{29}) \qquad (6.23)$$

$$(\lambda_2 + \mu_1)(\pi_{18} + \pi_{19}) = \frac{q_1^*}{q_1^* + q_3^*}(\mu_1\pi_{19} + \mu_2\pi_4 + \mu_3\pi_{17}) + \lambda_3(\pi_{32} + \pi_{33})$$

$$(\lambda_3 + \mu_1)(\pi_{22} + \pi_{23}) = \frac{q_1^*}{q_1^* + q_2^*}(\mu_1\pi_{23} + \mu_2\pi_{25} + \mu_3\pi_6) + \lambda_2(\pi_{32} + \pi_{33})$$

$$(\lambda_3 + \mu_2)(\pi_{24} + \pi_{25}) = \frac{q_2^*}{q_1^* + q_2^*}(\mu_1\pi_{23} + \mu_2\pi_{25} + \mu_3\pi_6) + \lambda_1(\pi_{36} + \pi_{37})$$

Therefore,

$$\mathcal{P}^1 = \frac{1}{\mu_1}\left[q_1^*A + \frac{q_1^*}{q_1^* + q_2^*}B + \frac{q_1^*}{q_1^* + q_3^*}C + (\lambda_2 + \lambda_3)(\pi_{32} + \pi_{33})\right]$$

$$\mathcal{P}^2 = \frac{1}{\mu_2}\left[q_2^*A + \frac{q_2^*}{q_1^* + q_2^*}B + \frac{q_2^*}{q_2^* + q_3^*}D + (\lambda_1 + \lambda_3)(\pi_{36} + \pi_{37})\right] \qquad (6.24)$$

$$\mathcal{P}^3 = \frac{1}{\mu_3}\left[q_3^*A + \frac{q_3^*}{q_1^* + q_3^*}C + \frac{q_3^*}{q_2^* + q_3^*}D + (\lambda_1 + \lambda_2)(\pi_{28} + \pi_{29})\right]$$

where

$$\begin{aligned}
A &= (\mu_1\pi_3 + \mu_2\pi_5 + \mu_3\pi_7) \\
B &= (\mu_1\pi_{23} + \mu_2\pi_{25} + \mu_3\pi_6) \\
C &= (\mu_1\pi_{19} + \mu_2\pi_4 + \mu_3\pi_{17}) \\
D &= (\mu_1\pi_2 + \mu_2\pi_{11} + \mu_3\pi_{13})
\end{aligned} \qquad (6.25)$$

We know that if $\mathcal{P}_1 > \mathcal{P}_2$ and $\mathcal{P}_1 > \mathcal{P}_3$, then the PP mechanism assigns the highest probabilistic probability to the first class of jobs and if $\mathcal{P}_2 > \mathcal{P}_3$ then the PP mechanism assigns a higher probabilistic probability to the second class rather than the third class of jobs. The necessary conditions for these inequalities are illustrated in Eq. (6.26).

$$q_1^* > \frac{\mu_1}{2}\left(\frac{q_2^*}{\mu_2} + \frac{q_3^*}{\mu_3}\right) \qquad \text{a}$$

$$q_2^* > \mu_2\left(\frac{q_3^*}{\mu_3}\right) \qquad \text{b}$$

$$B > C \rightarrow \mu_1\pi_{23} + \mu_2\pi_{25} + \mu_3\pi_6 > \mu_1\pi_{19} + \mu_2\pi_4 + \mu_3\pi_{17} \qquad \text{c}$$

$$C > D \rightarrow \mu_1\pi_{19} + \mu_2\pi_4 + \mu_3\pi_{17} > \mu_1\pi_2 + \mu_2\pi_{11} + \mu_3\pi_{13} \qquad \text{d} \qquad (6.26)$$

$$(\lambda_2 + \lambda_3)(\pi_{32} + \pi_{33}) > (\lambda_1 + \lambda_3)(\pi_{36} + \pi_{37}) \qquad \text{e}$$

$$(\lambda_2 + \lambda_3)(\pi_{32} + \pi_{33}) > (\lambda_1 + \lambda_2)(\pi_{28} + \pi_{29}) \qquad \text{f}$$

$$(\lambda_1 + \lambda_3)(\pi_{36} + \pi_{37}) > (\lambda_1 + \lambda_2)(\pi_{28} + \pi_{29}) \qquad \text{g}$$

where

$$(\lambda_2 + \lambda_3)(\pi_{32} + \pi_{33}) = \mu_1\pi_{33} + \mu_2\pi_{24} + \mu_3\pi_{16} + \lambda_1\pi_{39} - \mu_1(\pi_{32} + \pi_{33})$$

$$(\lambda_1 + \lambda_3)(\pi_{36} + \pi_{37}) = \mu_1\pi_{22} + \mu_2\pi_{37} + \mu_3\pi_{12} + \lambda_2\pi_{39} - \mu_2(\pi_{36} + \pi_{37}) \qquad (6.27)$$

$$(\lambda_1 + \lambda_2)(\pi_{28} + \pi_{29}) = \mu_1\pi_{18} + \mu_2\pi_{10} + \mu_3\pi_{29} + \lambda_3\pi_{39} - \mu_3(\pi_{28} + \pi_{29})$$

The inequalities (6.26)-a and (6.26)-b present our proposed necessary conditions but it is necessary to show that the inequalities form (6.26)-c to (6.26)-g are held as well. To this end, we add these inequalities and replace equations defined in (6.27) and then arrive in Eq. (6.28).

$$\mu_1\pi_{23} + \mu_2(\pi_{24} + \pi_{25}) + \mu_3(\pi_6 + \pi_{16} + \pi_{28}) + \lambda_1\pi_{29}$$
$$> \mu_1(\pi_2 + \pi_{18} + \pi_{32}) + \mu_2(\pi_{10} + \pi_{11}) + \mu_3\pi_{13} + \lambda_3\pi_{29} \qquad (6.28)$$

We can assume that $\lambda_j = \lambda, j = 1, 2, 3$. Therefore, the terms $\lambda_1\pi_{29}$ and $\lambda_3\pi_{29}$ can be removed from Eq. (6.28). We rewrite the Eq. (6.28) by three independent inequalities in Eq. (6.29).

$$\pi_{23} > \pi_2 + \pi_{18} + \pi_{32}$$
$$\pi_{24} + \pi_{25} > \pi_{10} + \pi_{11} \qquad (6.29)$$
$$\pi_6 + \pi_{16} + \pi_{28} > \pi_{13}$$

By considering Fig. 6.16, we can mix markings of the states in inequalities (6.29) and rewrite these inequalities in based on the markings of the ASPN-LA-3PP where $\times$ indicates 'do not care' value. If $M(0) = 0$, i.e., (3.27) the first entry of the ASPN-LA-3PP marking is zero, then the server is busy in the marking (state) M. In all states shown in inequalities (3.27), the server is busy. Elements $M(2i), M(2i+1), i = 1, 2, 3$ are related to ith class of jobs. If a job is waiting in Q_i for service, $M(2i)$ becomes one[3] and if a job from Q_i is getting service, $M(2i+1)$ becomes one. Below we describe the inequalities (3.27) in order to find conditions if hold, inequalities (3.27) hold too.

---

[3]Since we suppose that queues in the ASPN-LA-3PP are one bounded, the value of $M(2i)$ an be set as 0 or 1.

$$\pi_{(0,1,1,1,0,0,0,0,0)} > \pi_{(0,0,1,\times,0,\times,0,0,0)} \qquad \text{A}$$

$$\pi_{(0,1,0,\times,1,0,0,0,0)} > \pi_{(0,0,0,\times,1,1,0,0,0)} \qquad \text{B} \qquad\qquad (6.30)$$

$$\pi_{(0,\times,0,\times,0,0,1,0,0)} > \pi_{(0,0,0,1,0,1,1,0,0)} \qquad \text{C}$$

- Consider $M(2)$ and $M(3)$ in inequalities (3.27). The inequalities (3.27) hold when the steady-state probability of markings with waiting jobs for service is higher than that of markings with no waiting jobs. This situation is occurred when the proper rate λ is chosen by the ASPN-LA-3PP system. We know that $Q_i, i = 1, 2, 3$ is a stable queue if the relation $\frac{\lambda}{\mu_i}$ holds. In the other hand, there must be enough jobs to learn the LA_1 assigned to the ASPN-LA-3PP. Considering these two points, λ must be chosen.
- The former point rises again when $M(3), M(4)$ or $M(5), M(6)$ are considered.

It is concluded that the inequalities form (6.26)-c to (6.26)-g hold when the proper value of λ is set. Therefore, the Theorem 1 is proved.   □

Corollary 6.1 Assuming $\mu_1 > \mu_2 > \mu_3$, if selection probability of the first queue passes the lower bound $\frac{\mu_1}{2}\left(\frac{q_2^*}{\mu_2} + \frac{q_3^*}{\mu_3}\right)$ and the selection probability of the second queue passes the lower bound $\mu_2\left(\frac{q_3^*}{\mu_3}\right)$, then the shortest total waiting time in the ASPN-LA-3PP is reached. That is, when the action probability $q_1(n)$ is higher than $\frac{\mu_1}{2}\left(\frac{q_2(n)}{\mu_2} + \frac{q_3(n)}{\mu_3}\right)$ and the action probability $q_2(n)$ is higher than $\mu_2\left(\frac{q_3(n)}{\mu_3}\right)$, the shortest total waiting time in the ASPN-LA-3PP is reached.

Proof The proof is immediately followed by the Theorem 6.2.   □

Theorem 6.3 Let $q_j^*, j = 1, \ldots, m$ be the selection probabilities of $Q_j, j = 1, \ldots, m$ in an ASPN-LA-[m]PP. If $\frac{q_j^*}{\mu_j} > \frac{q_k^*}{\mu_k}, 1 \le j, k \le m, j \ne k$, then the ASPN-LA-[m]PP assigns a higher probabilistic priority to the jth class rather than the kth class of jobs, i.e., $\mathcal{P}_j > \mathcal{P}_k$.   □

Proof We prove this property by contradiction. Suppose that $\frac{q_j^*}{\mu_j} \le \frac{q_k^*}{\mu_k}$. Since $\frac{\mu_k}{\mu_j} < 1$ and $q_j^* \times \frac{\mu_k}{\mu_j} < q_k^*$, the relation $q_j^* < q_k^*$ must hold true, which is a contradiction. Due to selection of the class j of jobs with a higher probabilistic priority than the class k, q_j^* must be absolutely greater than q_k^*. Therefore, the Theorem 6.3 is proved.   □

Corollary 6.2 Suppose that the inequality (6.31) holds for $j = 1, \ldots, m - 1$. Then, the ASPN-LA-[m]PP assigns a higher probabilistic priority to the jth class rather than the kth class of jobs for $1 \le j, k \le m, j < k$.

$$q_j^* > \frac{\mu_j}{(m-j)} \left(\sum_{\ell=j+1}^{m} \frac{q_\ell^*}{\mu_\ell} \right) \tag{6.31}$$

Proof By Theorem 6.2, if there are $m-j$ inequalities for queue j of the form $\frac{q_j^*}{\mu_j} > \frac{q_k^*}{\mu_k}, k = j+1, \ldots, m$, then the ASPN-LA-[$m$]PP assigns a higher probabilistic priority to the jth class rather than the kth class of jobs for $1 \le j, k \le m, j < k$. Summing up the two sides of these inequalities together, we get $q_j^* > \frac{\mu_j}{(m-j)} \left(\sum_{\ell=j+1}^{m} \frac{q_\ell^*}{\mu_\ell} \right)$. Therefore, the Corollary 6.2 is proved. $\square$

Corollary 6.3 Assuming $\mu_1 > \mu_2 > \cdots > \mu_m$, when all inequalities $q_j^* > \frac{\mu_j}{(m-j)} \left(\sum_{\ell=j+1}^{m} \frac{q_\ell^*}{\mu_\ell} \right)$ for $j = 1, \ldots, m-1$ hold, the shortest total waiting time in the ASPN-LA-[m]PP is reached. That is, when inequalities (6.32) are held, the shortest total waiting time in the ASPN-LA-[m]PP is reached.

$$q_j(n) > \frac{\mu_j}{(m-j)} \left(\sum_{\ell=j+1}^{m} \frac{q_\ell(n)}{\mu_\ell} \right), \quad j = 1, \ldots, m-1 \tag{6.32}$$

Proof The proof is immediately followed by the Theorem 6.3. $\square$

Theorem 6.4 Let ASPN-LA-[m]PP be a queuing system with m queues such that $\mu_1 > \mu_2 > \cdots > \mu_m$. If the learning algorithm L_{R-I} is used to update the action probability vector $q(n)$ of LA_1, then the inequality (6.32) holds for $j = 1$, when n goes to infinity.

Proof The shortest total waiting time is achieved, when the highest probabilistic priority is assigned to class of jobs with the shortest average service time (Cobham 1954) (Meybodi 1983). Based on learning procedure L_{R-I}, if action 1 is attempted at instant n, the probability $q_1(n)$ is increased at instant $n+1$ by an amount proportional to $1 - q_1(n)$ for a favorable response and fixed for an unfavorable response. By this, it follows that $\{q(n)\}_{n>0}$ can be described by a Markov-process whose state space is the unit interval $[0, 1]$, when automaton operates in an environment with penalty probabilities $\{c_1, c_2, \ldots, c_m\}$. The schema L_{R-I} consists of m absorbing states: $\{e_i(j) = 0 \text{ and } e_i(i) = 1\}, i, j = 1, \ldots, m$. Since the probability $q_i(n)$ can be decreased only when α_i is chosen and results in a favorable response, the probability $q(k) = e_i$ holds, if $q(n) = e_i, i = 1, \ldots, m$ for all $k \ge n$. Thus, $V \triangleq \{e_1, e_2, \ldots, e_m\}$ represents the set of all absorbing states and the Markov process $\{q(n)\}_{n>0}$ generated by the schema L_{R-I} converges to the set V with probability one.

To study the asymptotic behavior of the process $\{q(n)\}_{n>0}$, a common method is to compute the conditional expectation of $q_1(n+1)$ given $q_1(n)$. For the schema L_{R-I}, this computation shows that the expected value of $q_i(n)$ increases or decreases monotonically with n depending on the values of $c_i, i = 1, \ldots, m$. Study on the asymptotic behavior shows that $q_i(n)$ converges to 0 with a higher probability, when the value of c_i is the largest value among $c_j, j = 1, \ldots, m, j \neq i$ and to 1 with a higher probability, when the value of c_i is the lowest value among $c_j, j = 1, \ldots, m, j \neq i$, if the initial probability is $q_i(0) = \frac{1}{m}$ (Narendra and Thathachar 1989).

To prove Theorem 6.4, it is enough to show that penalty signal generated by ASPN-LA-[m]PP system for the first class of jobs, is the lowest. In our queuing system, each class of job has a service time distribution with the exponential density function as Eq. (6.33).

$$f_i(t) = \mu_i e^{-\mu_i t} \tag{6.33}$$

The reinforcement signal generator function produces the penalty signal $\beta_i(n) = 1$ for class i in time instant n by probability value $c_i(n), i = 1, \ldots, m$. Let $\Gamma(n) = \frac{1}{m} \times \sum_{i=1}^{m} \Gamma_i(n)$ is the average waiting time of the system up to time instant n. Therefore, $c_i(n)$ is defined as the probability of the service time of nth job will exceed $\Gamma(n)$ and is defined by Eq. (6.34) (Meybodi 1983).

$$c_i(n) = prob[\delta_i(n) > \Gamma(n)] = e^{-\mu_i \Gamma(n)}, i = 1, \ldots, m \tag{6.34}$$

In Eq. (6.34), $\delta_i(n)$ is the execution time of job n, which is selected from ith class for service. Based on our assumption $\mu_1 > \mu_2 > \cdots > \mu_m$, it is clear that $c_1(n) < c_2(n) < \cdots < c_m(n)$ with probability one for all n (Meybodi 1983). Therefore, in ASPN-LA-[m]PP system, $q_1(n)$ converges to 1 with probability one when n goes to infinity. Therefore, the Theorem 6.4 is proved.   □

Corollary 6.4 Let ASPN-LA-[m]PP be a queuing system with m queues such that $\mu_1 > \mu_2 > \cdots > \mu_m$, in which the learning algorithm L_{R-I} is used. The ASPN-LA-[m]PP assigns the highest probabilistic priority to the first class of jobs.

Proof The proof is immediately followed by the Theorem 6.4 and Corollary 6.2. □

6.4.3   Simulation Results

In this section, we conduct a set of computer simulations to study the ASPN-LA-[m]PP mechanism compared to a SPN-based mechanism in terms of two criteria of the average waiting time and the steady-state behavior of the queuing system. In the SPN-based mechanism, SPN-[m]PP model, the server selects jobs from any of m queues with equal probability of $\frac{1}{m}$. To this end, we consider 64 different queuing systems with three different queues and one server, shown in Table 6.3. In these

Table 6.3 Set of parameters for a number of queuing systems with three queues and one server

Set of parameters

ID	$\lambda_i, \mu_i, i = 1,2,3$			ρ_i	ID	$\lambda_i, \mu_i, i = 1,2,3$			ρ_i
1	.8, 1	.6, .8	.2, .6	.8, .75, .33	33	.8, 1	.4, .6	.2, .4	.80, .67, .5
2	.8, 1	.6, .8	.2, .6	.8, .75, .33	34	.8, 1	.2, .6	.2, .4	.80, .33, .5
3	.8, 1	.4, .8	.4, .6	.8, .50, .67	35	.6, 1	.4, .6	.2, .4	.60, .67, .5
4	.8, 1	.4, .8	.2, .6	.8, .50, .33	36	.6, 1	.2, .6	.2, .4	.60, .33, .5
5	.8, 1	.2, .8	.2, .6	.8, .25, .33	37	.4, 1	.4, .6	.2, .4	.40, .67, .5
6	.6, 1	.6, .8	.4, .6	.6, .75, .67	38	.4, 1	.2, .6	.2, .4	.40, .33, .5
7	.6, 1	.6, .8	.2, .6	.6, .75, .33	39	.2, 1	.4, .6	.2, .4	.20, .67, .5
8	.6, 1	.4, .8	.4, .6	.6, .50, .67	40	.2, 1	.2, .6	.2, .4	.20, .33, .5
9	.6, 1	.4, .8	.2, .6	.6, .50, .33	41	.6, .8	.4, .6	.2, .4	.75, .67, .5
10	.6, 1	.2, .8	.2, .6	.6, .25, .33	42	.6, .8	.2, .6	.2, .4	.75, .33, .5
11	.4, 1	.6, .8	.4, .6	.4, .75, .67	43	.4, .8	.4, .6	.2, .4	.50, .67, .5
12	.4, 1	.6, .8	.2, .6	.4, .75, .33	44	.4, .8	.2, .6	.2, .4	.50, .33, .5
13	.4, 1	.4, .8	.4, .6	.4, .50, .67	45	.2, .8	.4, .6	.2, .4	.25, .67, .5
14	.4, 1	.4, .8	.2, .6	.4, .50, .33	46	.2, .8	.2, .6	.2, .4	.25, .33, .5
15	.4, 1	.2, .8	.2, .6	.4, .25, .33	47	.1, 1	.08, .8	.06, .6	.10, .10, .10
16	.2, 1	.6, .8	.4, .6	.2, .75, .67	48	.2, 1	.16, .8	.12, .6	.20, .20, .20
17	.2, 1	.6, .8	.2, .6	.2, .75, .33	49	.3, 1	.24, .8	.18, .6	.30, .30, .30
18	.2, 1	.4, .8	.4, .6	.2, .50, .67	50	.4, 1	.32, .8	.24, .6	.40, .40, .40
19	.2, 1	.4, .8	.2, .6	.2, .50, .33	51	.5, 1	.4, .8	.30, .6	.50, .50, .50
20	.2, 1	.2, .8	.2, .6	.2, .25, .33	52	.6, 1	.48, .8	.36, .6	.60, .60, .60
21	.8, 1	.6, .8	.2, .4	.8, .75, .50	53	.7, 1	.56, .8	.42, .6	.70, .70, .70
22	.8, 1	.4, .8	.2, .4	.8, .50, .50	54	.8, 1	.64, .8	048, .6	.80, .80, .80
23	.8, 1	.2, .8	.2, .4	.8, .25, .50	55	.9, 1	.72, .8	054, .6	.90, .90, .90
24	.6, 1	.6, .8	.2, .4	.6, .75, .50	56	.08, .8	.06, .6	.04, .4	.10, .10, .10
25	.6, 1	.4, .8	.2, .4	.6, .50, .50	57	.16, .8	.12, .6	.08, .4	.20, .20, .20
26	.6, 1	.2, .8	.2, .4	.6, .25, .50	58	.24, .8	.18, .6	.12, .4	.30, .30, .30
27	.4, 1	.6, .8	.2, .4	.4, .75, .50	59	.32, .8	.24, .6	.16, .4	.40, .40, .40
28	.4, 1	.4, .8	.2, .4	.4, .50, .50	60	.40, .8	.30, .6	.20, .4	.50, .50, .50
29	.4, 1	.2, .8	.2, .4	.4, .25, .50	61	.48, .8	.36, .6	.24, .4	.60, .60, .60
30	.2, 1	.6, .8	.2, .4	.2, .75, .50	62	.56, .8	.42, .6	.28, .4	.70, .70, .70
31	.2, 1	.4, .8	.2, .4	.2, .50, .50	63	.64, .8	.48, .6	.32, .4	.80, .80, .80
32	.2, 1	.2, .8	.2, .4	.2, .25, .50	64	.72, .8	.54, .6	.36, .4	.90, .90, .90

three queues, without loss of generality, we assume $\mu_1 > \mu_2 > \mu_3$. To have stable queues, in all simulations, we consider $\rho_i = \frac{\lambda_i}{\mu_i}, i = 1, \ldots, 3$ being less than 1. Let λ_i and μ_i take one of the following values: .2, .4, .6, .8, or 1. In the last 18 rows of Table 6.3, we consider the queuing systems with equal $\rho_i = \frac{\lambda_i}{\mu_i}, i = 1, 2, 3$. Thus, in these systems, λ_i will get different values. All the reported results are averaged over 1500 independent runs. Each run of simulation consists of servicing 10,000 jobs.

6.4.3.1 Experiment 1

In this experiment, considering Table 6.3, we study the average waiting times of different queuing systems resulted from the ASPN-LA-3PP and the SPN-3PP. The results of this experiment are reported in Fig. 6.18. Regarding this figure, when the LA involves in the priority assignment, that is the ASPN-LA-3PP, the average waiting time of the system is shorter than in the SPN-3PP, where no learner participates in the priority assignment process.

6.4.3.2 Experiment 2

This experiment is conducted to study the steady-state probabilities of marking sets. Let $\mathcal{P}_i, i = 1, 2, 3$ denotes the steady-state probability of marking set $\mathcal{S}_i, i = 1, 2, 3$ respectively, defined in Sect. 6.4.2.2. The steady-state probabilities $\mathcal{P}_i, i = 1, 2, 3$ are defined by the Eq. (6.35), in which π_k denotes the steady-state probability of the CTMC being in state k.

$$\begin{aligned}
\mathcal{P}_1 &= \pi_2 + \pi_3 + \pi_{18} + \pi_{19} + \pi_{22} + \pi_{23}, \\
\mathcal{P}_2 &= \pi_4 + \pi_5 + \pi_{10} + \pi_{11} + \pi_{24} + \pi_{25}, \\
\mathcal{P}_3 &= \pi_6 + \pi_7 + \pi_{12} + \pi_{13} + \pi_{16} + \pi_{17},
\end{aligned} \tag{6.35}$$

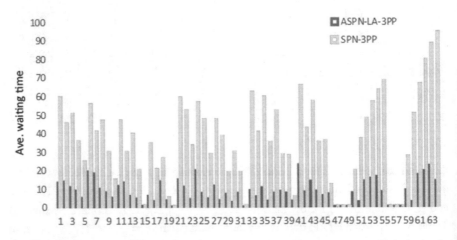

Fig. 6.18 The average waiting times of different queuing systems resulted from the ASPN-LA-3PP and the SPN-3PP

A marking M belongs to $\mathcal{S}_i$ if a job is selected from Q_i, while other jobs are waiting in other queues. In this experiment, the values of $\mathcal{P}_i$'s calculated in iteration 10,000 are used to calculate a measure called S, denoted in Eq. (6.36). Increasing the value of the measure S leads to improve the performance of algorithm in terms of the steady-state probability summation.

$$S = (\mathcal{P}_1 - \mathcal{P}_2) + (\mathcal{P}_1 - \mathcal{P}_3) + (\mathcal{P}_2 - \mathcal{P}_3) \tag{6.36}$$

The values of S related to the ASPN-LA-3PP and the SPN-3PP mechanisms are illustrated in Fig. 6.19. Regarding this figure, when the LA involves in the priority assignment, that is the ASPN-LA-3PP, the value of the measure S is larger than in the SPN-3PP, where no learner participates in the priority assignment process.

6.4.3.3 Experiment 3

This experiment is conducted to study the priority assignment behavior of the ASPN-LA-3PP. To this end, we use our proposed necessary conditions as some lower bounds defined in inequalities (6.37).

$$q_j(n) > LB_j(n), j = 1, 2 \tag{6.37}$$

$$LB_j(n) = \frac{\Gamma_j(n)}{(3-j)} \left(\sum_{\ell=j+1}^{m=3} \frac{q_\ell(n)}{\Gamma_\ell(n)} \right) \tag{6.38}$$

We define n^+ the number of jobs, after which the inequalities (6.37) hold: $q_j(n) > LB_j(n)$ for $n > n^+$ and $j = 1, 2$. We also define n^* the number of jobs, after

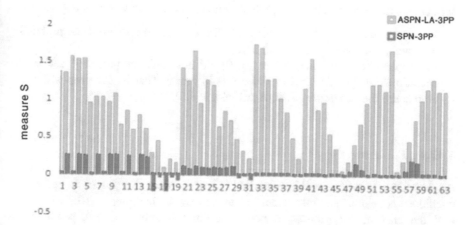

Fig. 6.19 The values of the measure S related to the ASPN-LA-3PP and the SPN-3PP mechanisms

Table 6.4 The values of n^* and n^+ for different Queuing Systems

ID	n^*	n^+	ID	n^*	n^+	ID	n^*	n^+
1	1	62	23	1	196	45	–	–
2	1	87	24	–	–	46	–	–
3	335	2357	25	1	62	47	858	3144
4	1	1	26	1	39	48	199	150
5	353	1155	27	83	1282	49	1	522
6	1	6322	28	1428	7800	50	1	43
7	142	6528	29	1	67	51	110	590
8	119	2158	30	–	–	52	1	3398
9	1	1	31	–	–	53	1	4331
10	1	172	32	–	–	54	1	233
11	–	–	33	1	58	55	1	2632
12	–	–	34	1	435	56	425	1357
13	1	645	35	1	38	57	180	297
14	–	–	36	124	139	58	1	108
15	387	229	37	1	2875	59	1	51
16	–	–	38	1	78	60	1	63
17	–	–	39	–	–	61	1	96
18	–	–	40	–	–	62	1	53
19	–	–	41	132	149	63	128	3100
20	–	–	42	1	112	64	1	74
21	1	29	43	597	6415			
22	1	48	44	1	164			

which the inequalities (6.20) hold: $P_1(n) > P_2(n) > P_3(n), n > n^*$. We use character '-' to denote that inequalities (6.37) does not hold. The following points are concluded from the Table 6.4, which illustrates the results of this experiment:

- In samples with $n^* > 1$, the ASPN-LA-3PP changes the steady-state probabilities such that inequalities (6.37) hold,
- In samples, in which inequalities (6.20) did not hold, the proposed necessary conditions, inequalities (6.37), did not hold either. That is, the proposed conditions are the necessary to learn ASPN-LA-3PP.

6.4.3.4 Experiment 4

An important question in this experiment is that when the proposed necessary conditions is held at time instance n^+, can we stop the learning in ASPN-LA? To find the answer, we propose a new algorithm to solve the PA problem in ASPN-LA-3PP system. In this new mechanism, called ASPN-LAwithNC-3PP, upon holding the proposed necessary conditions time instance n^+, updating of the action probability vector of LA is stopped, and then ASPN-LAwithNC-3PP uses the fixed action probability vector. To verify statistical differences of the ASPN-LA-3PP and the ASPN-LAwithNC-3PP, a statistical test is also run over the mechanisms. The statistical results of comparing algorithms by two-tailed t-Test with 1498 degrees of freedom at a 0.05 level of significance are given in Table 6.5 and Table 6.6. In Table 6.5, the performance of ASPN-LAwithNC-3PP vis-à-vis ASPN-LA-3PP is denoted as " + ", "-" and " ~ " when ASPN-LAwithNC-3PP is significantly better, worse than or similar to ASPN-LA-3PP, respectively. These results make the t-Test of the average waiting time of the system. Similarly, results coming from the t-Test of the measure S, defined in Eq. (6.36), are also shown in Table 6.6. Based on the test results one may conclude that

- The ASPN-LAwithNC-3PP mechanism is significantly better than or equal to the ASPN-LA-3PP, over the average time of the system, in 86% of cases.
- The ASPN-LAwithNC-3PP mechanism is significantly better than or equal to the ASPN-LA-3PP, the measure S in Eq. (6.36), in 89% of cases.
- In general, the ASPN-LAwithNC-3PP is very similar (at least 86%) to the ASPN-LA-3PP mechanism; however, in the first mechanism calling the learning algorithm of LA is lower than that in the second one.

6.4.3.5 Experiment 5

In this Section, we conduct a set of computer simulations to study the behavior of ASPN-LA-$[m]$PP ($[m] = 2$ and $[m] = 4$) in comparison to a simple mechanism, called SPN-$[m]$PP, in terms of the average waiting time of the queuing system. To have stable queues, in all simulations, we consider $\rho_i = \frac{\lambda_i}{\mu_i}, i = 1, \ldots, m$ being less than 1. Without loss of generality, we assume $\mu_j > \mu_k, 1 \leq j < k \leq m$. Let $\mu_i, \lambda_i, i = 1, \ldots, m$ take one of the following values: .2, .4, .6, .8, or 1. All the reported results are averaged over 100 independent runs of simulation. Each run of simulation consists of servicing 10,000 jobs.

The results of this experiment are reported in Table 6.7. Each row in this table represents a queuing system with specific values of λ_i and μ_i. The average waiting times of each queuing system resulted from SPN-$[m]$PP and ASPN-LA-$[m]$PP are given in the last two columns of the corresponding row of that system. As it can be seen from Table 6.7, when learning automata takes part in priority assignment, which is in ASPN-LA-$[m]$PP, the average waiting time of the system is lower than in SPN-$[m]$PP, where no learner presents in priority assignment process.

Table 6.5 Comparing mechanisms in terms of waiting time for 10,000 jobs (A1: ASPN-LA-3PP mechanism, A2: ASPN-LAwithNC-3PP mechanism, P–V: P–Value, Per: Performance)

ID	A1	A2	P–V	Per	ID	A1	A2	P–V	Per
1	11.798 ± .158	11.416 ± .140	1.34E−05	+	33	8.018 ± .053	8.040 ± .056	0.671415	
2	11.032 ± .288	10.790 ± .286	0.046681	+	34	5.358 ± .005	5.407 ± .007	0.004721	−
3	9.859 ± .039	9.659 ± .046	2.04E−05	+	35	10.454 ± .218	10.688 ± .234	0.03089	−
4	9.114 ± .131	8.810 ± .127	0.000198	+	36	3.110 ± .008	3.092 ± .008	0.388666	
5	4.050 ± .022	4.034 ± .022	0.633506		37	6.260 ± .035	6.413 ± .038	0.000465	−
6	16.203 ± .248	15.442 ± .162	1.82E−13	+	38	8.065 ± .008	8.054 ± .008	0.59442	
7	15.516 ± .670	15.047 ± .596	0.009585	+	39	7.535 ± .009	7.526 ± .009	0.674003	
8	8.257 ± .055	8.145 ± .056	0.035739	+	40	3.074 ± .022	3.098 ± .020	0.473737	
9	6.911 ± .037	6.848 ± .038	0.145667		41	14.431 ± .0257	14.450 ± .292	0.874632	
10	4.836 ± .009	4.844 ± .009	0.705582		42	7.134 ± .033	7.289 ± .046	0.000673	−
11	11.177 ± .161	11.141 ± .170	0.701404		43	13.413 ± .444	13.063 ± .461	0.022002	+
12	11.584 ± .069	11.614 ± .069	0.605686		44	6.951 ± .137	6.897 ± .140	0.527001	
13	5.709 ± .011	5.698 ± .012	0.65981		45	5.804 ± .015	5.779 ± .014	0.375289	
14	4.646 ± .006	4.674 ± .006	0.112893		46	6.989 ± .031	6.983 ± .029	0.890109	
15	0.542 ± .001	0.552 ± .001	0.077868		47	0.030 ± .000	0.030 ± .000	0.185847	
16	4.340 ± .117	4.106 ± .072	0.000817	+	48	0.050 ± .000	0.050 ± .000	0.768089	
17	3.285 ± .010	3.319 ± .012	0.152417		49	0.166 ± .000	0.166 ± .000	0.76798	
18	13.330 ± .011	13.329 ± .012	0.965074		50	7.009 ± .006	7.007 ± .006	0.892433	
19	2.507 ± .014	2.50314	0.894799		51	1.845 ± .005	1.843 ± .005	0.920045	
20	0.091 ± .000	0.091 ± .000	0.025049	+	52	11.967 ± .053	11.686 ± .060	1.99E−07	+
21	13.530 ± .521	12.645 ± .457	2.74E−08	+	53	14.707 ± .190	14.292 ± .176	2.03E−05	+
22	10.668 ± .197	9.645 ± .131	0	+	54	12.018 ± .120	11.636 ± .106	9.43E−07	+
23	4.009 ± .000	3.923 ± .022	0.008935	+	55	7.181 ± .066	6.709 ± .044	0	+

(continued)

Table 6.5 (continued)

ID	A1	A2	P–V	Per	ID	A1	A2	P–V	Per
24	17.274 ± .448	16.392 ± .411	3.48E−09	+	56	0.040 ± .000	0.040 ± .000	0.809788	
25	7.282 ± .033	7.069 ± .035	4.66E−07	+	57	0.067 ± .000	0.067 ± .000	0.435645	
26	4.669 ± .003	4.691 ± .003	0.093866		58	0.213 ± .000	0.212 ± .000	0.609269	
27	11.740 ± .048	11.750 ± .050	0.835168		59	8.021 ± .007	8.019 ± .007	0.931253	
28	3.785 ± .004	3.778 ± .004	0.602971		60	2.145 ± .006	2.153 ± .007	0.659965	
29	6.257 ± .007	6.220 ± .007	0.052742		61	14.340 ± .090	14.116 ± .106	0.001738	+
30	2.450 ± .005	2.463 ± .005	0.421046		62	17.202 ± .234	16.661 ± .243	1.11E−06	+
31	7.452 ± .018	7.424 ± .019	0.36493		63	13.808 ± .125	13.331 ± .139	8.4E−09	+
32	0.338 ± .000	0.341 ± .000	0.192869		64	8.221 ± .071	7.702 ± .061	0	+

Table 6.6 Comparing mechanisms in terms of the measure S for 10,000 jobs (A1: ASPN-LA-3PP mechanism, A2: ASPN-LAwithNC-3PP mechanism, P–V: P–Value, Per: performance)

ID	A1	A2	P–V	Per	ID	A2	A2	P–V	Per
1	1.388 ± .003	1.370 ± .004	0.194588		33	1.627 ± .000	1.627 ± .000	1.47E-05	?
2	1.427 ± .003	1.427 ± .003	0.987621		34	1.610 ± .000	1.610 ± .000	1.76E-05	?
3	1.457 ± .001	1.457 ± .001	0.194751		35	1.274 ± .001	1.274 ± .001	0.405297	
4	1.500 ± .000	1.500 ± .000	0.919298		36	1.280 ± .000	1.280 ± .000	0.557477	
5	1.513 ± .000	1.513 ± .000	0.009331	+	37	0.967 ± .000	0.967 ± .000	0.000315	?
6	0.931 ± .003	0.939 ± .003	0.533096		38	0.736 ± .000	0.736 ± .000	0.415129	
7	1.009 ± .002	1.013 ± .002	0.669407		39	0.433 ± .000	0.433 ± .000	0.053348	
8	1.159 ± .000	1.159 ± .000	0.011723		40	0.224 ± .000	0.224 ± .000	0.860133	
9	1.125 ± .001	1.125 ± .001	0.084544		41	1.414 ± .002	1.418 ± .001	0.584129	
10	0.994 ± .000	1.001 ± .000	0.147202		42	1.476 ± .000	1.454 ± .000	6E-15	?
11	0.598 ± .002	0.615 ± .002	0.11808		43	0.934 ± .002	0.934 ± .002	0.787566	
12	0.702 ± .001	0.728 ± .001	0.002526	+	44	0.866 ± .001	0.866 ± .001	0.481592	
13	0.684 ± .001	0.664 ± .001	0.013058	−	45	0.501 ± .000	0.501 ± .000	0.002839	?
14	0.758 ± .000	0.758 ± .000	0.460357		46	0.285 ± .000	0.286 ± .000	0.606219	
15	0.548 ± .000	0.548 ± .000	0.741347		47	0.016 ± .000	0.016 ± .000	0.020942	?
16	0.325 ± .001	0.340 ± .000	0.006157	+	48	0.139 ± .000	0.140 ± .000	0.162963	
17	0.385 ± .000	0.385 ± .000	0.095185		49	0.352 ± .000	0.353 ± .000	0.446912	
18	0.035 ± .003	0.035 ± .003	0.440591		50	0.676 ± .000	0.676 ± .000	0.89351	
19	0.190 ± .000	0.190 ± .000	0.112239		51	0.900 ± .002	0.900 ± .002	0.53942	
20	0.090 ± .000	0.090 ± .000	0.014795	?	52	1.027 ± .002	1.027 ± .003	0.103806	
21	1.431 ± .002	1.450 ± .002	0.056615		53	1.148 ± .004	1.152 ± .004	0.765787	
22	1.520 ± .001	1.530 ± .001	0.078277		54	1.362 ± .004	1.362 ± .004	0.965692	
23	1.568 ± .000	1.568 ± .000	7.07E-06	?	55	1.571 ± .004	1.571 ± .004	0.428867	

(continued)

Table 6.6 (continued)

ID	A1	A2	P-V	Per	ID	A2	A2	P-V	Per
24	1.041 ± .001	1.067 ± .001	0.000665	+	56	0.019 ± .000	0.019 ± .000	0.265347	
25	1.183 ± .000	1.183 ± .001	0.047744	~	57	0.158 ± .000	0.158 ± .000	0.869523	
26	1.151 ± .000	1.151 ± .000	0.695096		58	0.401 ± .000	0.402 ± .000	0.277907	
27	0.724 ± .001	0.756 ± .001	1.84E−07	+	59	0.727 ± .000	0.727 ± .000	0.800393	
28	0.805 ± .000	0.805 ± .000	0.923654		60	1.004 ± .001	1.004 ± .001	0.470342	
29	0.696 ± .000	0.696 ± .000	0.556815		61	1.136 ± .001	1.136 ± .001	0.55996	
30	0.417 ± .000	0.417 ± .000	1.96E−08	~	62	1.291 ± .002	1.292 ± .002	0.941835	
31	0.257 ± .000	0.257 ± .001	0.213921		63	1.495 ± .002	1.495 ± .001	0.57879	
32	0.173 ± .000	0.173 ± .000	0.725214		64	1.657 ± .002	1.645 ± .002	0.278113	

Table 6.7 Set of parameters for a number of queuing systems with two queues and one server

$\lambda_i,\mu_i, i=1,2$		SPN-2PP	ASPN-2PP	$\lambda_i,\mu_i, i=1,2,3,4$				SPN-4PP	ASPN-4PP
.8, 1	.6, .8	24.59	10.19	.8, 1	.6, .8	4, .6	.2, .4	198.8	30.57
.8, 1	.4, .8	14.64	7.53	.8, 1	.6, .8	.2, .6	.2, .4	162.29	28.91
.8, 1	.2, .8	3.56	2.74	.8, 1	.4, .8	.4, .6	.2, .4	179.16	24.56
.6, 1	.6, .8	18.6	11.53	.8, 1	.4, .8	.2, .6	.2, .4	153.27	29.15
.6, 1	.4, .8	6.04	5.11	.8, 1	.2, .8	.4, .6	.2, .4	154.39	14.37
.6, 1	.2, .8	0.12	0.11	.8, 1	.2, .8	.2, .6	.2, .4	118.70	16.10
.4, 1	.6, .8	8.3	8.28	.6, 1	.6, .8	.4, .6	.2, .4	195.24	40.46
.4, 1	.4, .8	0.19	0.18	.6, 1	.6, .8	.2, .6	.2, .4	166.43	43.84
.4, 1	.2, .8	0.05	0.04	.6, 1	.4, .8	.4, .6	.2, .4	181.51	26.87
.2, 1	.6, .8	0.065	0.064	.6, 1	.4, .8	.2, .6	.2, .4	145.51	22.71
.2, 1	.4, .8	0.034	0.033	.6, 1	.2, .8	.4, .6	.2, .4	152.25	19.07
.2, 1	.2, .8	26.05	11.3	.6, 1	.2, .8	.2, .6	.2, .4	182.31	38.02
.6, .8	.4, .6	6.72	4.68	.4, 1	.6, .8	.4, .6	.2, .4	147.75	40.59
.6, .8	.2, .6	12.8	10.51	.4, 1	.6, .8	.2, .6	.2, .4	171.59	18.12
.4, .8	.4, .6	0.15	0.13	.4, 1	.4, .8	.4, .6	.2, .4	136.85	9.85
.4, .8	.2, .6	0.25	0.24	.4, 1	.4, .8	.2, .6	.2, .4	140.31	30.67
.2, .8	.4, .6	19.64	9.63	.4, 1	.2, .8	.4, .6	.2, .4	103.95	4.32
.4, .6	.2, .4	25.38	7	.4, 1	.2, .8	.2, .6	.2, .4	152.68	11.53
.8, 1	.4, .6	9.04	3.93	.2, 1	.6, .8	.4, .6	.2, .4	120.27	10.19
.8, 1	.2, .6	25.03	4.8	.2, 1	.6, .8	.2, .6	.2, .4	137.14	37.27
.8, 1	.2, .4	18.7	8.68	.2, 1	.4, .8	.4, .6	.2, .4	182.31	38.02
.6, 1	.4, .6	0.3	0.26	.2, 1	.4, .8	.2, .6	.2, .4	147.75	40.59
.6, 1	.2, .6	10.45	3.68	.2, 1	.2, .8	.4, .6	.2, .4	171.59	18.12
.6, 1	.2, .4	5.3	4.16	.2, 1	.2, .8	.2, .6	.2, .4	136.85	9.85
.4, 1	.4, .6	0.08	0.07						

6.5   Graph Applications

In this section, we review two applications of the ASPN-LA and CAPN-LA in graphs. In the first application, the ASPN-LA is used to analyze an algorithm, which is proposed to solve the shortest path problem in stochastic graphs. This problem has been recently studied in the literature and a number of algorithms has been provided to find it using varieties of learning automata models. However, all these algorithms suffer from two common drawbacks: low speed and lack of a clear termination condition. In this Section, we propose a novel learning automata-based algorithm for this problem, which can speed up the process of finding the shortest path using parallelism. For this parallelism, several traverses are initiated, in parallel, from the source node towards the destination node in the graph. During each traverse, required times for traversing from the source node up to any visited node are estimated. The time estimation at each visited node is then given to the learning automaton residing in that node. Using different time estimations provided by different traverses, this learning automaton gradually learns which neighbor of the node is on the shortest path. To set a condition for the termination of the proposed algorithm, we analyze the algorithm using an ASPN-LA. The results of this analysis enable us to establish a necessary condition for the termination of the algorithm. To evaluate the performance of the proposed algorithm in comparison to the existing algorithms, we apply it to find the shortest path in six different stochastic graphs. The results of this evaluation indicate that the time required for the proposed algorithm to find the shortest path in all graphs is substantially shorter than that required by similar existing algorithms.

In the second application, the CAPN-LA is used to represent several algorithms for solving the vertex coloring (VC-) problem. This problem is a well-known coloring problem, in which a color is assigned to each vertex of the graph. A legal vertex coloring of graph is to assign distinct colors to each vertex of the graph in such a way that no two endpoints of any edge are given the same color. Using the APN-LA, one may be able to design adaptive algorithms to solve this problem. There are issues, for which designing a single APN-LA is a tedious work and results in a large and complex model. To avoid having large and complex APN-LAs for the VC problem, in this Section, a CAPN-LA, referred to as CAPN-LA-VC, is utilized to solve this problem. The CAPN-LA-VC, like other CAPN-LAs, consists of a cellular structure and a number of identical APN-LAs. The cellular structure of the CAPN-LA-VC is a graph, which is to be colored. Each APN-LA in the CAPN-LA-VC represents the algorithm, which must be executed in each vertex and the required cooperation between the neighboring vertices will be handled by means of cooperation between the APN-LAs in those vertex. The CAPN-LA-VC is then used to design different algorithms for the classic problem of vertex coloring. The measure of expediency is calculated for these algorithms and results of using them for coloring vertices of different graphs are also included.

6.5.1 The Shortest Path Problem in Stochastic Graphs: Application of ASPN-LA

The deterministic shortest path problem in graphs has been studied extensively and many algorithms have been reported in the literature to solve it (Beigy and Meybodi 2006). In this problem, one looks for a path joining source and destination nodes while minimizing the summation of the traversed edges' lengths. However, there are many applications where the underlying graph is a stochastic graph and hence, the lengths of edges are random variables; a graph with stochastic edge lengths. If the probability distribution functions of these random variables are unknown, then finding the shortest path cannot be possible using the algorithms introduced for the deterministic shortest path problem. Recently, some algorithms have been proposed to find the shortest path in stochastic graphs with unknown characteristics using Distributed Learning Automata (DLA) or extended DLA (eDLA) (Mollakhalili Meybodi and Meybodi 2014).

All of the DLA- or the eDLA-based algorithms introduced to solve the shortest path problem in stochastic graphs with unknown characteristics suffer from two common drawbacks. The first drawback is the low speed of these algorithms since these algorithms try to find the shortest path by sequentially traversing different paths along the graph from the source node towards the destination node, sampling the lengths of the traversed paths, and finally, identify the shortest path by comparing these sampled values. In this Section, we argue that this sequential process can be performed in a parallel manner, thus increasing the speed.

The second drawback is that in these algorithms, no clear condition is given for terminating the sequential traversing and sampling process. They use a simple condition of maximum number of traversing, which indeed is not a suitable condition; the specified maximum number could be too low, resulting in inadequate number of samples, or could be too high, resulting in the wasting of time, collecting non-necessary samples. In this Section, we propose a necessary condition which if it does not hold, the number of samples collected so far will still not be enough. Thus, using AND operator between this condition and the maximum number of traversing condition can at least prevent the algorithm from being stopped before inadequate numbers of samples are collected.

In the proposed algorithm, each node i, except for the source node, is equipped with a learning automaton (LA). The learning automaton residing in the node i is responsible to find the neighbor j of that node, which is on the shortest path from the source node to the node i. To this end, we let that several tokens, in parallel, start to traverse different paths between the source and the destination nodes, hop by hop, in such a way that each token estimates the required time for its traverse. Using the times provided by different tokens passed by, LA in each node gradually learns which neighbor is on the shortest path.

To establish a necessary condition for the termination of the proposed algorithm, it is required to first analyze the steady-state behavior of the algorithm. A suitable way of analyzing the steady-state behavior of an algorithm is to model it using a Petri net and then analyze the resulted Petri net. Therefore, in this Section, we first represent the proposed algorithm by an ASPN-LA, and then analyze the steady-state behavior of the yielded ASPN-LA. The reason as to why ASPN-LA, among different kinds of Petri nets, is selected to represent the proposed algorithm is that in this model, like in the algorithm, there exists several decision points at each of which an LA is used for making decisions. Therefore, representing the algorithm by the ASPN-LA can be simply accomplished by mapping the operation of LAs in the algorithm into the operation of LAs in the decision points of ASPN-LA.

6.5.1.1 Problem Statement

We first define a stochastic graph and then, explain the shortest path problem in stochastic graphs.

Definition 6.33 A *stochastic graph* is defined by a triple $G = (V, E, \mathcal{F})$, where $V = \{v^1, v^2, \ldots, v^n\}$ is set of nodes, $E \subset V \times V$ specifies set of edges, and $n \times n$ matrix $\mathcal{F}$ is the probability distribution describing the statistics of edge lengths, where n is the number of nodes. ☐

Here, we consider the length of an edge from node v^i to node v^j (d_{ij}) to be the time required for traversing from v^i to v^j. Since the graph is stochastic, d_{ij} is a positive random variable with f_{ij} as its probability density function; each f_{ij} is assumed to be an exponential distribution with an unknown rate parameter.

In a stochastic graph G, a path τ_i, consists of κ_i nodes and with expected length of $\bar{L}_{\tau_i}$, from a source node to a destination node, is defined as an ordering $\{v_{\tau_i,1}, v_{\tau_i,2}, \ldots, v_{\tau_i,\kappa_i}\} \subset V$ of nodes in such a way that $v_{\tau_i,1}$ and v_{τ_i,κ_i} are source and destination nodes, respectively, and $(v_{\tau_i,j}, v_{\tau_i,j+1}) \in E$ for $1 \leq j < k_i$, where $v_{\tau_i,j}$ is the jth node in path τ_i.

Let $v^s, v^d \in V$ be the source and destination nodes, respectively. In the stochastic graph, there are $\mathcal{M}$ different paths $\Omega = \{\tau_1, \tau_2, \ldots, \tau_M\}$ between v^s and v^d. Let $\Omega^l \subseteq \Omega$ be the set of loop-free paths between v^s and v^d. The shortest path between v^s and v^d denoted by $\tau^* \in \Omega^l$, is defined as a path with minimum expected length, that is $\bar{L}_{\tau^*} = \min_{\tau \in \Omega^l}\{\bar{L}_\tau\}$. The shortest path problem is thus to find such a path assuming no knowledge about the rate parameters of probability density functions $\mathcal{F}$.

6.5.1.2 LA-Based Algorithms for Solving the Shortest Path Problem

One available approach to solve the shortest path problem in stochastic graphs is to construct a DLA from the given stochastic graph (Beigy and Meybodi 2006) (Mollakhalili Meybodi and Meybodi 2014). The root LA is assigned to v^s and starts an operation of this DLA as follows:

- One of the outgoing edges of the root (one action of the root LA) is chosen using the corresponding action probability vector.
- The selected edge activates the LA at its other end. This LA also selects an action that results the activation of another LA.
- This process is repeated until the destination node v^d is reached.
- The time elapsed for this traverse from v^s to v^d is a sample of the time required for traversing from v^s to v^d.
- The sample time is compared with a quantity called 'dynamic threshold', which is an estimate of the required time for traversing from v^s to v^d. If the sample time is shorter than or equal to dynamic threshold, then all activated LAs get reward signal and if the sample's time is longer than the dynamic threshold or the

destination node is not reached, then all activated LAs get penalty signal. Upon the generation of reward or penalty signals, the learning algorithm L updates the action probability vectors of activated LAs.

- The value of the dynamic threshold is updated using the new sample time.

Beigy and Meybodi in (2006) solved the shortest path problem in a stochastic graph given a particular pair of source-destination by introducing different variations of the DLA. For example, in one variation, they assumed that the reward parameter of learning algorithm is different in LAs assigned to different nodes of the DLA; the closer the node to the destination, the larger the value of the reward parameter will be.

In (Beigy and Meybodi 2002), a version of the DLA has been defined by introducing a new definition for the dynamic threshold. The dynamic threshold is generally calculated by averaging the lengths of the paths traversed so far (Meybodi and Beigy 2001). But in Beigy and Meybodi (2002), the authors assumed the dynamic threshold as the minimum of the sequences of the averages, each computed at the end of one full traverse from the source to the destination node. It is argued that this new definition requires fewer number of samples to be taken from the edges of the graph to decide which path from the source to the destination node is the shortest (Meybodi and Beigy 2003).

Mollakhalili and Meybodi in (2014) solved the shortest path problem in stochastic graphs using extended DLA (eDLA). To traverse a path, the activation levels of all LAs, except for the LA in v^s, are initially set to passive. The activation level of the LA in v^s is set to active. This only active LA upgrades to fire level and selects an action. The selected action corresponds to one of the edges of v^s. This edge is added to the list of edges for the current traverse. The process of adding edges to the current traverse continues until all LAs downgrades to off. At this point, the traverse is completed, that is, a path is formed from v^s to v^d. Rest of the algorithm is the same as Beigy and Meybodi (2006).

Misra and Oommen in Misra and Oommen (2004a), introduced two different versions of an algorithm based on LA to solve the shortest path problem in stochastic graphs. The first version, called LASPA-RR, uses Ramalingam and Reps' scheme (Ramalingam and Reps 1996) in its iterations. The second, called LASPA-FMN, is an algorithm which uses the scheme introduced by Frigioni et al. (2000). Generally, an LASPA algorithm consists of two steps: initialization and iteration. To begin the initialization step, the algorithm obtains a snapshot of the directed graph with each edge having a random weight. Next, Dijkstra (1959) algorithm is run once to determine the shortest path edges on the graph snapshot. The algorithm maintains an action probability vector for each node of the graph. This vector contains the probability values to choose different actions; each possible outgoing edge corresponds to a probable action that can be selected for calculating the shortest path tree. Based on the shortest path computed using Dijkstra's algorithm, the action probability vector of each node is updated in that the outgoing edge from a node taken as belonging to the shortest path edge, has an increased probability than before the update. In an iteration of LASPA, first, a node is randomly chosen from the current graph and an action of the associated LA is selected. Second, a new sample of weight from the edge related to selected action is determined. Based on this new value, the new shortest path's tree is recalculated using either RR or FMN algorithm. Finally, the action probability vector for the node whose edge was just selected, is updated in that the edge now potentially belonging to the shortest path's tree has more likelihood of being selected than it before the update. The iteration step of LASPA is repeated for many times until the algorithm converges. Another variation of LASPA has been introduced in (Misra and Oommen 2004b) where the principles of the Generalized Pursuit (GP) method (Narendra and Thathachar 1989) is used to learn LAs. In (Misra and Oommen 2006), a similar algorithm is presented capable of computing shortest paths for all possible pairs of source-destination in the graph. In this algorithm, the Floyd Warshall's all-pairs static algorithm (Floyd 1962) has been used rather than Dijkstra's algorithm to find shortest paths in a snapshot of the stochastic graph and the Demetrescu and Italiano's algorithm (Demetrescu and Italiano 2004), rather than RR and FMN, has been used to recalculate shortest paths.

LA-based algorithms to solve the shortest problem have been successfully used in many real world applications such as finding maximum clique in stochastic graphs (Rezvanian and Meybodi 2015b), grid resource discovery (Hasanzadeh and Meybodi 2014), link prediction in adaptive web sites (Mollakhalili Meybodi and Meybodi 2008), and dynamic channel assignment (Beigy and Meybodi 2009).

6.5.1.3 Proposed Algorithm

In the proposed algorithm, we keep traversing all available paths from v^s to v^d within G repeatedly so as to find the shortest path τ^*. To traverse different paths, we use the concept of *token*. Tokens traverse the paths between v^s and v^d, hop by hop, and estimate the time required for traversing different paths. A node v^i within the graph G, receives tokens from all of its neighbors. Assigned to a token received from neighbor v^j is a time which is an estimate of the time required for traversing from v^s to v^i through v^j ($\Gamma_j^{s \sim i}$). Two neighbors v^j and v^k of v^i can be ranked among their estimated times, $\Gamma_j^{s \sim i}$ and $\Gamma_k^{s \sim i}$, that is, if $\Gamma_j^{s \sim i} < \Gamma_k^{s \sim i}$, it means that the path from v^s to v^i which goes through v^j is shorter than the path going through v^k. It is notable since traversing from v^s to v^d is performed repeatedly, the estimates $\Gamma_j^{s \sim i}$ can be changed or even improved over time. The estimates will become stable to some degree when the shortest path between v^s and v^i is the path going through neighbor v^l of v^i, for which $\Gamma_l^{s \sim i}$ is the minimum. The question here is that when it is possible to stop traversing different paths and select the shortest path using Γ estimates. The proposed algorithm uses learning automata and its theory to provide a necessary condition, which if not met, we still need more traversing.

6.5.1.3.1   Detailed Description of the Algorithm

The proposed algorithm, referred to as VDLA, seeks the shortest path τ^* between v^s and v^d within a stochastic graph. Each node v^i, except for the source node, is equipped with a learning automaton LA^i. The number of actions of LA^i is equal to the number of neighbors of v^i; each action is assigned to one neighbor. The probability of selecting each action is initially set to $\frac{1}{m^i}$, where m^i is the number of neighbors of v^i. In addition, each v^i maintains a list $\Gamma^{s \sim i}$, in which each entry $\Gamma_j^{s \sim i}$ is the estimation of the time required for traversing from v^s to v^i through neighbor v^j. For simplicity in notation, we omit the index s and use Γ^i rather than $\Gamma^{s \sim i}$ hereafter. Γ_j^i is initially set to zero for all is and js.

VDLA algorithm consists of two phases: learning phase and selection phase. In what follows, we will first describe the learning phase, and then give the detailed description of the selection phase.

- **Learning Phase**

Source node v^s initiates the learning phase by starting to repeatedly send out tokens to all of its neighbors according to an exponential distribution with rate λ. These tokens, *Learning-Tokens (LT)*, are used to sample lengths of different paths from v^s towards v^d within the graph. Assigned to each LT are an ID number and a timer. When v^s sends out a new learning-token, it creates the token with a new ID number and sets its timer to zero. This newly created LT is then sent out to neighbors of v^s.

Upon the arrival of an LT, sent out by a node v^j, at any neighbor node v^i of v^j, except for v^s, the following steps will be taken:

- Number of LTs arrived at v^i denoted by n^i is increased by one,
- Number of LTs arrived at v^i from v^j denoted by k_j^i is increased by one. Note that
  $\sum_{j=1}^{m^i} k_j^i = n^i$,
- LA^i is activated and selects the action corresponding to the neighbor v^j,
- Current estimation of the time required for traversing from v^s to v^i through neighbor v^j, Γ_j^i, is updated according to the equation
  $\Gamma_j^i = \frac{1}{k_j^i} \times \left(\left(k_j^i - 1 \right) * \Gamma_j^i + LT.timer \right)$,
- The reinforcement signal for LA^i, i.e., β^i, is generated using Eq. (6.39),

$$\begin{cases} \beta^i = 1; LT.timer \geq \frac{1}{m^i} \times \sum_k \Gamma_k^i \\ \beta^i = 0; LT.timer < \frac{1}{m^i} \times \sum_k \Gamma_k^i \end{cases} \qquad (6.39)$$

- In other words, if the time required for traversing from v^s to v^i through neighbor v^j is shorter than the average time required for traversing from v^s to v^i through

any of the neighbors of v^i, then the selected action of LA^i is rewarded (i.e., $\beta^i = 0$). Otherwise, it is penalized (i.e., $\beta^i = 1$),
- Upon the generation of β^i, LA^i updates q^i according to Eq. (6.40) where the learning algorithm L_{R-I} is used, and

$$
\begin{cases}
q_k^i(n^i) = \begin{cases} q_k^i(n^i - 1) + a[1 - q_k^i(n^i - 1)] & , k = j \\ (1-a)q_k^i(n^i - 1) & , \forall k \neq j \end{cases} & ; \beta^i = 0 \\
q^i(n^i) = q^i(n^i - 1); & \beta^i = 1
\end{cases}
\tag{6.40}
$$

- If this is the first time that an LT-with this ID number is seen in v^i, then v^i will send out the LT to all its neighbors, except for v^j. Repetitive LT tokens (tokens with repetitive ID numbers) will be discarded to avoid loops.

- **Selection phase**

In VDLA, v^d is responsible to start the selection phase by sending a token, *Selection-Token* (ST), towards v^s. This ST token is used to find the shortest path τ^*. The ST traverses a path τ_k within the graph towards v^s. When it arrives at v^s, v^s stops sending out LT tokens. At this time, τ^* is the reverse of the path τ_k. Upon the arrival of the ST at any node v^i, except for v^s, the following steps will be taken by v^i :

- It sets the neighbor node v^j, from which it receives ST, as its destination node in the shortest path τ^*.
- It sends out ST to the neighbor corresponding to the action with the maximum value of q^i.

The important question here is that when v^d can stop the learning phase and start selection phase. One common answer to this question is to set a maximum number of iterations *maxIter* for the learning phase. But using such a condition to terminate the learning phase does not take the learning process into consideration; It is possible that (1) learning does not occur at all at the terminating iteration; (2) learning occurs, but is not matured enough; or (3) learning is completed far sooner than the *maxIter*. In VDLA, we propose a necessary condition on the values of q^i, which if it does not hold, the learning has not occurred yet. Thus, using AND operator between this condition and the *maxIter* condition can at least prevent the learning phase to be stopped before any learning occur. This condition can be stated using inequality (6.41), where $LB_\varepsilon^i(n)$ is defined in Eq. (6.42). In inequality (6.41), $\varepsilon = \text{argmin}_j \Gamma_j^i(n^i)$. The above condition is achieved, when VDLA is modeled by the ASPN-LA to analyze its learning ability (see Sect. 6.5.1.4).

$$
q_\varepsilon^d(n^d) > LB_\varepsilon^d(n)
\tag{6.41}
$$

$$LB_{\varepsilon}^i(n) = \frac{\sum_{k=2}^{m^i}\left[q_k^i(n^i)\Gamma_k^i(n^i)\right]}{\Gamma_{\varepsilon}^i(n^i)(m^i - 1)} \tag{6.42}$$

Upon receiving an *LT* token, in addition to the steps explained above, v^d takes the following steps:

- If $n^d > maxIter$ Then

  - If condition specified by inequality (6.41) holds then

    Create an *ST* token and sends it out to the neighbor corresponding to $\text{argmax}_j q_j^d(n^d)$.

6.5.1.4 The Analysis of the Proposed Algorithm

For analysis of learning phase of the proposed algorithm, each node in the graph is modeled by an ASPN-LA. For this modeling, each node v^i, with $m^i > 0$ edges, can be considered as a queuing system with $m = m^i$ different classes of jobs and one server. A learning-token arriving at a node v^i from a neighboring node v^j is considered as a job from class C_j which arrives into queue Q_j. The service time of a job (learning-token) is assumed to be the required time for that token to traverse from the source node v^s to the node v^i. Two jobs arrive into a queue Q_j have different service times due to the following two reasons: (1) their traversing path from v^s to v^i may differ; and (2) the underlying graph of the problem is stochastic. As a result, the queuing system has unknown parameters, i.e., service times. Since all probability distribution functions describing the statistics of edge lengths of the graph are assumed to be exponential with unknown rate parameters, service time in a queue Q_j is also exponentially distributed with an unknown rate μ_j (Freiheit and Billington 2003). The input rates of jobs into queue Q_j is also exponentially distributed with rate $\lambda_j = \lambda, j = 1, .., m$. In other words, the input rates of all queues are equal to λ, which is the rate by which the source node v^s sends out learning-tokens into the graph. By this modeling, finding the shortest path in a stochastic graph is mapped into finding the shortest total waiting time in a queuing system with unknown parameters.

The problem of finding the shortest total waiting time in queuing systems with unknown parameters has been solved in Sect. 6.4 using ASPN-LA. Analyzing this ASPN-LA, a necessary condition has been established which can be used to terminate the algorithm. Using the mapping between the shortest path problem in stochastic graphs and the priority assignment problem in a queuing system with unknown parameters, we are now able to use the results of Sect. 6.4 to set a necessary condition for termination of the proposed VDLA algorithm. The current estimation of the time required for traversing from v^s to v^i through neighbor v^j, i.e.,

Γ_j^i, is an estimation of μ_j^{-1}. Without loss of generality, in node v^i, in time instant n, we can suppose that $\frac{1}{\Gamma_1^i(n)} > \frac{1}{\Gamma_2^i(n)} > \ldots > \frac{1}{\Gamma_m^i(n)}$. By Corollary 6.2 if

$$q_j^i(n) > LB_j^i(n), j = 1, \ldots, m - 1 \tag{6.43}$$

$$LB_j^{(i)}(n) = \frac{\sum_{\ell=j+1}^{m} \left[q_\ell^i(n^i) \Gamma_k^i(n^i) \right]}{\Gamma_j^{(i)}(n^i)(m^i - j)}, j = 1, \ldots, m - 1 \tag{6.44}$$

then node v^i significantly prefers the path from v^s to v^i, which passes through neighbor v^j to the paths passing through other neighbors. That is, a necessary condition for termination of VDLA algorithm is that in the destination node v^d, the inequalities (6.43) hold for $i = d$ whereas the selection phase, described in Section *Selection Phase*, considers inequalities (6.43) only for $j = \varepsilon$.

6.5.1.5 Simulation Results

In this section, we conduct a set of computer simulations to study the VDLA performance compared to a DLA-based algorithm introduced in (Beigy and Meybodi 2006) in terms of the two criteria of simulation time and number of samplings. To this end, we consider six different graphs constructed form two directed stochastic graphs, Graph A and Graph B, taken from Beigy and Meybodi (2002, 2006), shown in Figs. 6.20 and 6.21, respectively. Graph A is a directed stochastic graph with 10 nodes, 23 arcs, $v^s = 1$, $v^d = 10$, and $\tau^* = [1, 4, 9, 10]$. Edge cost distribution of Graph A is given in Table 6.8. Graph B is a stochastic graph with 15 nodes, 42 arcs, $v^s = 1$, $v^d = 15$, and $\tau^* = [1, 2, 5, 15]$. Edge cost distribution of Graph B is given in Table 6.9. To evaluate our proposed necessary conditions, we generate three different stochastic graphs based on Graph A, denoted by A(1), A(2), and A(3). These three graphs only differ in the edge cost distributions of the edges involved the shortest path of the Graph A (Table 6.10). In other words, the probabilistic lengths of the shortest path in these three graphs differ from each other. Similarly, three different stochastic graphs B(1), B(2), and B(3), with different lengths of the shortest path, are generated based on Graph B (Table 6.11). In the simulations, LA utilize the L_{R-I} learning algorithm, where α, the reward rate, is set to 0.01. The value of *maxIter* is set to 4000 for graphs A(1), A(2), and A(3) and to 7000 for graphs B(1), B(2), and B(3).

We consider three different versions of the proposed VDLA in the simulations. The first version, denoted by VDLA(1), is similar one described in Sect. 6.5.1.3.1. In the second version, denoted by VDLA(2), we consider all conditions given by inequalities (6.43) instead of using condition given by inequality (6.41) as the necessary condition for the termination of the algorithm. Therefore, in the learning phase of VDLA(2), v^d takes the following steps:

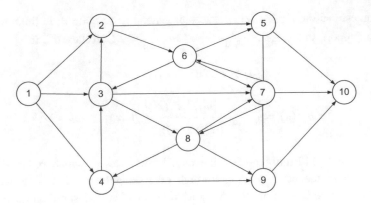

Fig. 6.20 Graph A

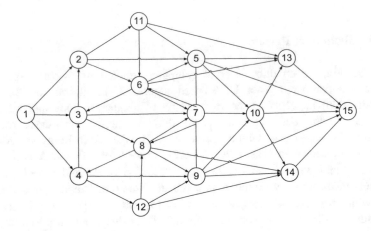

Fig. 6.21 Graph B

- If $n^d > maxIter$, then

  – Define a temporary vector $\bar{\Gamma} = sort\left(\Gamma^d\left(n^d\right)\right)$ and construct another tempo-
    rary vector $\bar{q}$ such that the action probability $\bar{q}_j$ is the probability of the action
    corresponding to $\bar{\Gamma}_j$.
  – If inequality (6.45) holds for all $j = 1, \ldots, (m^d - 1)$ then

    Create an *ST* token and sends it out to the neighbor corresponding to the
    action with the maximum value of q^d.

$$\bar{q}_j > \overline{LB}_j^d\left(n^i\right) \tag{6.45}$$

Table 6.8 Weight distribution of graph A (Fig. 6.20)

Edges	Lengths				Probabilities				Edges	Lengths				Probabilities			
(1, 2)	3.0	5.3	7.4	9.4	0.2	0.2	0.3	0.2	(6, 3)	6.8	7.7	8.5	9.6	0.4	0.1	0.1	0.4
(1, 3)	3.5	6.2	7.9	8.5	0.3	0.3	0.2	0.2	(6, 5)	0.6	1.5	3.9	5.8	0.2	0.2	0.3	0.3
(1, 4)	4.2	6.1	6.9	8.9	0.2	0.3	0.2	0.3	(6, 7)	2.1	4.8	6.6	7.5	0.2	0.4	0.2	0.2
(2, 5)	2.6	4.1	5.5	9.0	0.2	0.2	0.4	0.2	(7, 6)	4.1	6.3	8.5	9.7	0.2	0.3	0.4	0.1
(2, 6)	5.8	7.0	8.5	9.6	0.3	0.3	0.2	0.2	(7, 8)	1.6	2.8	5.2	6.0	0.2	0.3	0.3	0.2
(3, 2)	1.5	2.3	3.6	4.5	0.2	0.2	0.3	0.3	(7, 9)	3.5	4.0	5.0	7.7	0.1	0.2	0.4	0.3
(3, 7)	6.5	7.2	8.3	9.4	0.5	0.2	0.2	0.1	(7, 10)	1.6	3.4	8.2	9.3	0.2	0.3	0.3	0.2
(3, 8)	5.9	7.8	8.6	9.9	0.4	0.3	0.1	0.2	(8, 4)	7.0	8.0	8.8	9.4	0.2	0.2	0.2	0.4
(4, 3)	2.1	3.2	4.5	6.8	0.2	0.2	0.3	0.3	(8, 7)	2.1	4.6	8.5	9.6	0.4	0.2	0.2	0.2
(4, 9)	1.1	2.2	3.5	4.3	0.2	0.3	0.4	0.1	(8, 9)	1.7	4.9	6.5	7.8	0.2	0.2	0.2	0.4
(5, 7)	3.2	4.8	6.7	8.2	0.2	0.2	0.3	0.3	(9, 10)	4.6	6.4	7.6	8.9	0.4	0.1	0.2	0.3
(5, 10)	6.3	7.8	8.4	9.1	0.2	0.2	0.4	0.2									

$$\overline{LB}_j^d(n) = \frac{\sum_{\ell=j+1}^{m^d}[\bar{q}_k \bar{\Gamma}_k]}{\bar{\Gamma}_j(m^d - j)}, \quad j = 1, \ldots, (m^d - 1) \tag{6.46}$$

In the last version, denoted by VDLA(3), not only the destination node v^d, but also any node receiving an ST token, except for v^s, checks the condition given by the inequality (6.41) and sends the ST towards v^s, only after this condition holds. In VDLA(3), upon the arrival of the ST at any node v^i, except for v^s, the following two steps will be taken by v^i:

1. It sets the neighbor node v^j, from which it receives ST, as its destination node in the shortest path τ^*.
2. It sends out the ST to the neighbor corresponding to the action with the maximum value of q^i if Eq. (6.47) hold where $LB_\varepsilon^i(n^i)$ is defined in Eq. (6.40) for $= \arg\min_j \Gamma_j^i(n^i)$.

$$q_\varepsilon^i(n^i) > LB_\varepsilon^i(n^i) \tag{6.47}$$

6.5.1.5.1 Experiment One

This experiment is conducted to study the ability of the proposed algorithm in finding the shortest path in different graphs. We define n^* as the iteration number, at which the proposed necessary conditions hold for the first time. We also define $\mathbb{N}$ as the number of iterations, after n^*, at which the proposed algorithm cannot find the shortest path τ^*. An iteration here is defined as the number of unique LTs arrived at the destination node v^d.

Table 6.9 Weight distribution of Graph B (Fig. 6.21)

Edges	Lengths				Probabilities				Edges	Lengths				Probabilities			
(1, 2)	16	25	36		0.6	0.3	0.1		(7, 8)	12	15	22	24	0.3	0.3	0.3	0.1
(1, 3)	21	24	25	39	0.5	0.2	0.2	0.1	(7, 10)	19	23	37		0.6	0.2	0.2	
(1, 4)	11	13	26		0.4	0.4	0.2		(8, 4)	13	23	34		0.4	0.3	0.3	
(2, 5)	11	30			0.7	0.3			(8, 7)	14	34	39		0.6	0.2	0.2	
(2, 6)	13	37	39		0.6	0.2	0.2		(8, 9)	13	31	32		0.8	0.1	0.1	
(2, 11)	24	28	31		0.5	0.3	0.2		(8, 14)	14	15	27	32	0.3	0.3	0.2	0.2
(3, 2)	11	20	24		0.6	0.3	0.1		(9, 7)	10	17	20		0.6	0.3	0.1	
(3, 7)	23	30	34		0.4	0.3	0.3		(9, 10)	16	18	36	39	0.3	0.3	0.2	0.2
(3, 8)	14	23	34		0.5	0.4	0.1		(9, 14)	19	24	29		0.4	0.3	0.3	
(4, 3)	22	30			0.7	0.3			(9, 15)	12	23	25	32	0.4	0.3	0.2	0.1
(4, 9)	35	40			0.6	0.4			(10, 13)	14	20	25	32	0.3	0.3	0.2	0.2
(4, 12)	16	25	37		0.5	0.4	0.1		(10, 14)	23	34			0.9	0.1		
(5, 7)	15	17	19	26	0.3	0.3	0.3	0.1	(10, 15)	15	19	25		0.4	0.3	0.3	
(5, 10)	27	33	40		0.4	0.3	0.3		(11, 5)	18	19	20	23	0.3	0.3	0.3	0.1
(5, 13)	28	35	37	40	0.4	0.3	0.2	0.1	(11, 6)	10	19	39		0.5	0.4	0.1	
(5, 15)	25	32			0.7	0.3			(11, 13)	13	31	25		0.6	0.3	0.1	
(6, 3)	18	24			0.7				(12, 8)	15	36	39		0.5	0.3	0.2	
(6, 5)	18	25	29		0.5	0.3	0.2		(12, 9)	16	22			0.7	0.3		
(6, 7)	11	31	37		0.5	0.4	0.1		(12, 14)	10	13	18	34	0.3	0.3	0.3	0.1
(6, 13)	21	23			0.5	0.5			(13, 15)	12	31			0.9	0.1		
(7, 6)	12	23	31		0.5	0.3	0.2		(14, 15)	14	19	32		0.5	0.3	0.2	

All the reported results are averaged over 50 independent runs of simulation. In each run, node v^s repeatedly sends out learning-tokens to all its neighbors according to an exponential distribution with rate $\lambda = 0.1$. The average results for Experiment One are reported in Table 6.12. From this table, one may conclude the following remarks:

- The shorter the length of the shortest path, the higher the values of $LB^d(n^*)$ and n^* will be. To account for this phenomenon, recall that Γ_ε^d is an estimate of the length of the shortest path in the graph in Eq. (6.42). Considering the Eq. (6.42), a decline in the value of Γ_ε^d results in a rise in the value of LB_ε^d. As a result, when the length of the shortest path in the graph decreases, Γ_ε^d decreases and subsequently LB_ε^d increases. When LB_ε^d increases, the number of tokens required by LA^d residing in v^d to pass that LB_ε^d increases, and hence n^* will increase as well.
- The value of n^* in VDLA(3) is higher than that of VDLA(1) and VDLA(2). This is due to the fact that in VDLA(3), the algorithm terminates condition given by Eq. (6.47) must hold in all nodes v^i along the shortest path.

Table 6.10 The cost distributions of the edges along τ^* and the probabilistic length of τ^* for A(1), A(2), and A(3)

Edges of τ^*	A(1) Probabilities				A(2) Probabilities				A(3) Probabilities			
(1, 4)	0.2	0.3	0.2	0.3	0.5	0.3	0.1	0.1	0.7	0.1	0.1	0.1
(4, 9)	0.2	0.3	0.4	0.1	0.5	0.3	0.1	0.1	0.7	0.1	0.1	0.1
(9, 10)	0.4	0.1	0.2	0.3	0.5	0.3	0.1	0.1	0.7	0.1	0.1	0.1
Prob. length of τ^*	16.1				13.37				12.41			

The lengths of edges are shown in Table 6.8

Table 6.11 The cost distributions of the edges along τ^* and the probabilistic length of τ^* for B(1), B(2), and B(3)

Edges of τ^*	B(1) Probabilities			B(2) Probabilities			B(3) Probabilities		
(1, 2)	0.6	0.3	0.1	0.8	0.1	0.1	0.9	0.1	0.0
(2, 5)	0.7	0.3		0.8	0.2		0.9	0.1	
(5, 15)	0.7	0.3		0.8	0.2		0.9	0.1	
Prob. length of τ^*	64.5			60.1			55.5		

The lengths of edges are shown in Table 6.9

Table 6.12 Results of simulations on graphs

Alg.	G	n^*	$\mathbb{N}$	$LB_\varepsilon^d(n^*)$, Eq. (6.42)
VDLA(1)	A(1)	571	0	0.3500
	A(2)	748	0	0.3821
	A(3)	808	0	0.3960
	B(1)	488	0	0.2051
	B(2)	494	0	0.2089
	B(3)	499	0	0.2076

Alg.	Graph	n^*	$\mathbb{N}$	$\overline{LB}_j^d(n^*), j = 1, \ldots, m^d$, Eq. (6.46)
VDLA(2)	A(1)	624	0	[0.3499, 0.2894]
	A(2)	739	0	[0.3819, 0.2554]
	A(3)	820	0	[0.3961, 0.2487]
	B(1)	489	0	[0.2049, 0.1764, 0.1459, 0.0800]
	B(2)	491	0	[0.2088, 0.1709, 0.1414, 0.0763]
	B(3)	497	0	[0.2075, 0.1580, 0.1218, 0.0792]

Alg.	Graph	n^*	$\mathbb{N}$	$LB_\varepsilon^i(n^*)$, v^i along τ^*, Eq. (6.42)
VDLA(3)	A(1)	2319	0	0.9558, 0.4837, 0.3499
	A(2)	2789	0	0.9941, 0.5882, 0.3820
	A(3)	2954	0	0.9952, 0.6060, 0.3961
	B(1)	4782	0	0.6399, 0.4542, 0.2050
	B(2)	4983	0	0.6858, 0.4838, 0.2089
	B(3)	4983	0	0.6972, 0.4997, 0.2074

- In all simulations, for time instant $n > n^*$, the path constructed by the VDLA is equal to the shortest path τ^*. In other words, although the proposed termination condition is proved to be a necessary condition, our experiments suggest that it may be possible to consider it as a sufficient condition for the termination of the algorithm.

6.5.1.5.2 Experiment Two

We conduct a set of simulations to compare VDLA(1) and a DLA-based algorithm, introduced in (Beigy and Meybodi 2006) to solve the shortest path problem. To compare the results, three performance measures are calculated: (1) the average number of sampling taken from the edges of graph, denoted by AS, (2) the average required time for all traversing into graph, denoted by AT, and (3) the average required time for taking a sample from the edges of graph, which can be calculated by the division of AT to AS denoted by ATS (ATS $= \frac{AT}{AS}$). All reported results are averaged over 50 independent runs of simulations. In each run, we calculate the measures until the number of updates in LA^d reaches a specific number. In Table 6.13, this number is reported in a column with "Number of Updating". For example, in A(1), to update LA^d for 1000 times, the DLA algorithm takes 6351 samples from the graph in 37,602 ms whereas VDLA(1) takes 23024 samples in 9638 ms.

From Table 6.13, following points can be concluded:

- The average number of samplings for DLA algorithm is significantly lower than that of VDLA(1).
- The average required time for all traversing into the graph for VDLA(1) is substantially shorter than that of DLA algorithm. In other words, VDLA(1) takes more samples from the graph in lower time than DLA does.
- The average required time for taking a sample from edges in the graph for VDLA(1) is significantly shorter than that of DLA algorithm. In other words, the speed of taking a sample from the graph in VDLA(1) is higher than that of DLA.

Since the VDLA takes samples from all edges of the graph, we can extend VDLA algorithm to solve the single-source shortest path problem. We propose an extension of VDLA to solve this problem in stochastic graphs below.

Table 6.13 Results of simulations on graphs A(1), A(2), and A(3)

G	Number of updating	DLA			VDLA(1)		
		AS	AT(ms)	ATS (ms)	AS	AT (ms)	ATS (ms)
A(1)	1000	6351	37,602	5.92	23,024	9638	0.42
	2000	12,764	75,439	5.91	46,029	19,927	0.43
	3000	18,786	111,035	5.91	69,023	30,347	0.44
A(2)	1000	6549	36,972	5.65	23,024	10,083	0.44
	2000	12,858	72,808	5.66	46,043	19,949	0.43
	3000	18,752	104,465	5.57	69,016	29,782	0.43
A(3)	1000	6291	34,803	5.53	23,003	10,188	0.44
	2000	12,559	69,543	5.54	46,003	18,854	0.41
	3000	18,646	100,647	5.40	69,025	29,948	0.43
B(1)	1000	7840	164,870	21.03	23,519	5518	0.23
	2000	15,586	328,710	21.10	47,105	11,320	0.24
	3000	23,261	489,229	21.03	70,408	16,831	0.24
B(2)	1000	7900	163,788	20.73	23,551	5451	0.23
	2000	15,309	315,755	20.63	47,013	11,352	0.24
	3000	23,512	485,602	20.65	70,474	17,615	0.25
B(3)	1000	7628	155,027	20.32	23,565	5621	0.24
	2000	15,392	312,617	20.31	47,010	11,352	0.24
	3000	23,136	473,529	20.47	70,369	16,284	0.23

6.5.1.6 Algorithm Extension: The Single-Source Shortest Path

Here, we extend our proposed algorithm to find the shortest paths from the source node v^s to all other nodes known as the single-source shortest path problem (Cherkassky et al. 1996). In the extended algorithm, denoted by eVDLA, each node v^i, except for v^s, is responsible for starting the selection phase by sending a selection-token, denoted by ST^i, to v^s. ST^i is used to find the shortest path from the v^s to v^i denoted by τ^{*i}. ST^i traverses this path within the graph towards v^s. When v^s gets all ST^i, v^s stops sending out LT tokens.

Upon the arrival of ST^j at any node v^i, except for v^s, the following steps will be taken by v^i:

- It sets the neighbor node v^k, from which it receives ST^j, as its next node in the shortest path τ^{*j}.
- It sends out ST^j to the neighbor corresponding to the action with the maximum value of q^i.

A set of computer simulations is conducted to study the proposed necessary condition over all nodes of graphs. To do this, we consider the inequality (6.47) as the necessary condition in eVDLA and report the average value of n^* for graphs A

Table 6.14 Results of different simulation by eVDLA on graphs A(1) and B (1)

Destination	Graph	
	A(1)	B(1)
	n^*	n^*
v^2	1417	4782
v^3	2493	6417
v^4	2319	5982
v^5	1163	1872
v^6	889	2124
v^7	1172	1723
v^8	769	2086
v^9	1185	1850
v^{10}	571	936
v^{11}	–	3982
v^{12}	–	3879
v^{13}	–	1018
v^{14}	–	975
v^{15}	–	488

(1) and B(1) in Table 6.14, which summarizes the results for 50 different simulations.

In all simulations for time instant $n > n^*$, the path constructed by the VDLA for v^i is equal to the shortest path τ^{*i}. In other words, although the proposed termination condition is proved to be a necessary condition, our experiments suggest that it may be possible to consider it as a sufficient condition for the termination of the algorithm.

6.5.1.7 Algorithm Improvement: Using Discrete Learning Automata

There is always a need to improve the speed of the operation of a learning automaton (Thathachar and Sastry 2004). One of the way, in which such an improvement can be fulfilled, is discretizing the action probability space of the LA (Thathachar and Oommen 1979). In discretized automata models we restrict the action probabilities to a finite number of values in the interval [0, 1]. The number of such values denotes the level of discretization and is a design parameter. The values are generally spaced equally in [0, 1]. Every linear learning algorithm considered by a variable structure LA, can be discretized in this manner (Thathachar and Sastry 2004).

In this experiment, we use a discretized L_{R-I} algorithm to update the action probability vectors of LAs in VDLA. In the improved version of VDLA, denoted by dVDLA, the following learning algorithm is used instead of the ordinary L_{R-I} algorithm. Let $\mathcal{N}$ be the resolution parameter indicating the level of discretization. The smallest change in any action probability is then chosen as $\Delta = \frac{1}{r\mathcal{N}}$, where r is

Table 6.15   Results of simulations on graphs

G	n^*					
	VDLA(1)	dVDLA(1)	VDLA(2)	dVDLA(2)	VDLA(3)	dVDLA(3)
A(1)	571	377	624	424	2319	1531
A(2)	748	524	739	488	2789	1952
A(3)	808	566	820	549	2954	2038
B(1)	488	317	489	333	4782	3491
B(2)	494	341	491	354	4983	3438
B(3)	499	344	497	353	4983	3239

the number of actions. Let $\alpha(n) = \alpha_i$ be the action chosen by the learning automaton at instant n

$$q_j(n+1) = \begin{cases} \max\{q_j(n) - \Delta, 0\}, \forall j \neq i \\ 1 - \sum_{j \neq i} q_j(n), j = i \end{cases} \tag{6.48}$$

when the taken action is rewarded by the environment (i.e., $\beta(n) = 0$) and

$$q_j(n+1) = q_j(n), \forall j \tag{6.49}$$

when the taken action is penalized by the environment (i.e., $\beta(n) = 1$).

Utilizing discretized LAs, instead of ordinary LAs, the newly versions of VDLA (1), VDLA(2), and VDLA(3) are denoted by dVDLA(1), dVDLA(2), and dVDLA (3) respectively. Simulation settings of this experiment are completely identical to that used in Experiment one. The resolution parameter $\mathcal{N}$ is set to 100. The results for this experiment are reported in Table 6.15. As it was anticipated, using discretized learning automata increases the speed of the algorithm by a factor of 31% on average.

6.5.2   Vertex Coloring Problem: Application of CAPN-LA

In the rest of this section, a CAPN-LA will be utilized to present different algorithms to solve vertex coloring (VC) problem (Jensen and Toft 1995). To this end, first, an APN-LA will be designed for representing a local algorithm to be executed in any vertex of the graph. This APN-LA, called APN-LA-VC, is then replicated into all vertices of the graph which is to be colored; the yielded structure is a CAPN-LA and is denoted by CAPN-LA-VC. In CAPN-LA-VC, the controlling mechanism handles the required cooperation between APN-LA-VCs. The controlling mechanism of each APN-LA-VC has a number of controlling parameters; thus, using different sets of parameters in APN-LA-VCs of the CAPN-LA-VC results in different algorithms for solving VC problem. In this section, we use 6 different sets of controlling parameters for APN-LAs to form 6 different algorithms for solving VC problem. To be able to compare the results of these algorithms to each other, an evaluative measure by analyzing the APN-LA-VC in each vertex will be proposed. A number of simulation studies have been conducted, in which the results of the presented algorithms are compared to each other in terms of the proposed evaluative measure.

6.5.2.1   Problem Statement

Vertex coloring problem is a well-known coloring problem (Jensen and Toft 1995; Chen et al. 2008), in which a color is assigned to each vertex of the graph. A legal vertex coloring of graph $\mathbb{G}\langle E, V\rangle$, where $V(\mathbb{G})$ is the set of $|V| = n$ vertices and $E(\mathbb{G})$ is the edge set, is to assign distinct colors to each vertex of the graph in such a way that no two endpoints of any edge are given the same color. VC problem can be modeled by a quadruple $(V, E, \mathbb{C}, H)$, where V denotes the vertex-set of graph $\mathbb{G}$, E denotes the edge-set of graph $\mathbb{G}$, $\mathbb{C} = \mathbb{C}_1, \ldots, \mathbb{C}_m$ denotes the set of colors assigned to the vertices, and $H : V \to \mathbb{C}$ is the coloring function that assigns a color to each vertex such that $F(u) \neq F(v)$ for every $(u, v) \in E$. VC problem can be considered as a decision making problem that aims at deciding for a given graph whether or not the graph is P-colorable, and is called P-coloring problem. Graph $\mathbb{G}\langle E, V\rangle$ is P-colorable, if it can be legally colored with at most P different colors. The chromatic number $\mathcal{X}(G)$ is the minimum number of colors required for coloring the graph, and a graph $\mathbb{G}$ is said to be P-chromatic, if $\mathcal{X}(\mathbb{G}) = P$. It is shown that an arbitrary graph can be colored with at most $\Delta + 1$ colors, where Δ denotes the maximum degree of the vertices in the graph. i.e., $\mathcal{X}(\mathbb{G}) \leq \Delta + 1$.

6.5.2.2   LA-Based Algorithms for Solving the Vertex Coloring Problem

The problem of Vertex Coloring (VC) is a variation of the Graph Coloring (GC) problem (Torkestani 2013a). The problem of Graph Coloring is a classical problem of combinatorial optimization in the graph theory. The number of Graph Coloring forms is incredible. For instance, alpha coloring, lambda coloring, T-coloring, set coloring, list coloring, bandwidth coloring, multi-coloring, vertex coloring, and so on. the graph coloring is broadly used in real life applications such as computer air traffic flow management (Barnier and Brisset 2002), timetabling (Carter et al. 1996; Lewis and Paechter 2007), and light wave lengths assignment in optical networks (Zymolka et al. 2003).

In Vertex Coloring problem, each vertex of the graph is assigned a color in a way that there are no two adjacent vertices with the same color (Torkestani 2013a). Since the vertex coloring problem is NP-hard for the general graphs (Karp 1972), the exact algorithms can merely be used for the small graphs, while very large graphs usually come up in a wide range of applications (Torkestani 2013a). On the other hand, in the real life applications, a near optimal coloring of the graph will usually suffice. Therefore, a host of algorithms with polynomial time approximation has been introduced to find the near optimal solution for the coloring problem (Jensen and Toft 1995). The approximation methods, which have been addressed in the literature can be categorized as genetic algorithms (Mabrouk et al. 2009), fuzzy-based optimizations (Asmuni et al. 2009), evolutionary algorithms (Malaguti and Toth 2008), neural network approaches (Talavan and Yanez 2008), and so on.

A number of learning automata-based algorithms are addressed in the literature for the coloring of the benchmarks which are hard to color. In a major number of these algorithms, as the process of learning reaches to the end, each vertex learns how to select a color in order to color the graph with the smallest number of colors, with no neighbors with the same color (Torkestani 2013a). For example, an approximation algorithm has been addressed in (Torkestani 2010), in which each vertex is assigned by an LA. In the LA, the number of assigned auctioned to a vertex, equals the number of non-neighbor vertices of the same vertex. Any iteration of the algorithm is divided into a number of stages. In each stage, the learning automata select a subset of the non-neighbor graph vertices and assign the same color to them. In the next stages, the rest of the uncolored vertices will be selected and colored. An iteration is considered as a completed one, if the entire vertices are colored. In here, if the size of the color-set (the number of the selected colors) is smaller than the number of the selected colors, which were chosen earlier, then the coloring is rewarded; otherwise it is penalized. When the algorithm is reaching to the end, each LA learns the way of selecting one of the non-neighbor vertices, so that the number of selected vertices in a particular stage is maximized. The other LA applications in the graph coloring are addressed in the literature (Beigy and Meybodi 2002; Torkestani and Meybodi 2009).

There are a number of algorithms for the problem of vertex coloring, in which complex models are used based on LA. For example, in order to solve the problem of vertex coloring, an irregular cellular learning automata ICLA-base algorithm has been addressed in (Torkestani and Meybodi 2011). In this algorithm, the graphs are modeled by an ICLA to be colored. Each vertex in the graph corresponds to an ICLA cell. Then each cell will be equipped with an LA. The activation of the LA is automatic. When activated, a color will be selected for the assigned cell. For a proper selection of colors by an LA, the LAs can cooperate with the neighboring LAs, which are located in the neighboring cells. If the number of the colors in a coloring is larger than the number of the best coloring found so far, or if the coloring is not allowed, the selected coloring will be penalized. If none of these two conditions occur, the selected coloring will be rewarded. When the algorithm is reaching to its end, the LAs learn how to choose the colors in a way that the graph is entirely colored in a legal way with the smallest number of colors. The other ICLA applications in the problem of Graph Coloring are addressed in the literature (Eraghi et al. 2009) (Enayatzare and Meybodi 2009).

6.5.2.3   Solving the VC Problem with a CAPN-LA

To solve the VC problem with the CAPN-LA, first, an APN-LA will be constructed, referred to as APN-LA-VC, to represent a local algorithm to VC problem for each cell (or vertex) of graph G, which is to be colored. Finally, a CAPN-LA will be constructed to represent the whole algorithm to solve VC problem by replicating the APN-LA-VC into all cells (or vertices) of graph G. This yielded CAPN-LA is referred to as CAPN-LA-VC hereafter. Note that, since the cell c_i is the vertex v_i, hereafter (in some cases), vertex v_i may be referred to as cell c_i, and vice versa.

6.5.2.3.1   The APN-LA-VC

Each vertex v_i of the graph G is modeled by the simple PN depicted in Fig. 6.22. In this PN, $P_{i,0}$ is the place of making decision for the color which must be assigned to v_i and each place $P_{i,j} \in \{P_{i,1}, \ldots, P_{i,m}\}$ represents a possible color for this vertex. Using the construction procedure of the APN-LA (refer to Sect. 6.3.1.2), this PN is converted into an APN-LA, called APN-LA-VC, given in Fig. 6.23. Note that the set $\{t_{i,1}, \ldots, t_{i,m}\}$ is a maximal potential conflict and forms the cluster $s_{i,1}$. Transitions in this cluster are in effective conflict in the marking $M = P_{i,0} + P_{i,m+2}$. In this marking, the conflict resolver, i.e., LA_i, is responsible for resolving effective conflicts, and as a consequence, is responsible for determining the color of the vertex v_i. This is done as follows: the set of actions of LA_i referred to as $\underline{\alpha}_i$, contains m actions. When LA_i is activated, it selects one of its actions, say $\alpha_{i,r}$, according to its action probability vector. As a result, the transition $t_{i,r}$, corresponding to the selected action, is fired and a token is appeared in the place $P_{i,r}$. This way, the APN-LA-VC assigns a color to the vertex.

6.5.2.3.2   The CAPN-LA-VC

The CAPN-LA-VC is a graph of a number of APN-LA-VC such that the APN-LA-VC is assigned in each vertex of the graph. The LCAPN-LA-VC system is defined as below:

- $G = \langle V, E \rangle$ is an undirected graph which is to be colored.
- $\{A_1, \ldots, A_n\}$ is a finite set of APN-LA-VCs as shown in Fig. 6.23.
- $M_0 = [\mu_{1,0}, \ldots, \mu_{n,0}]$ is the initial marking vector of $\mathcal{N}$, where $\mu_{i,0} = p_{i,0} + p_{i,m+2}$ is the initial marking of A_i.
- $\hat{F} = \{F_1, \cdots, F_n\}$ is the set of local rules. F_i, the local rule related to LA_i, is executed upon the firing of $t_{i,1}^u$ in the A_i and generates the reinforcement signal β_i for LA_i using the set of last firing transitions of all neighboring clusters of $s_{i,1}$. A simple local rule can be stated as follows:

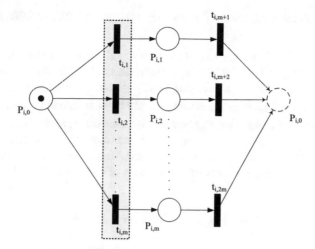

Fig. 6.22 A PN represents c_i; place $P_{i,0}$ has been duplicated (indicated by dash-line)

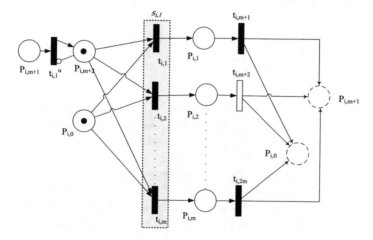

Fig. 6.23 An APN-LA represents c_i; places $P_{i,0}$ and $P_{i,m+1}$ have been duplicated (indicated by dash-line)

- If the selected transition of $s_{i,1}$ is different from the last selected transitions of all of its neighbors, then it is rewarded;
- Otherwise, it is penalized.

It should be noted that the local rule, stated above, is just a sample. In real scenarios, any other local rules can be used instead.

Based on the proposed CAPN-LA-VC, different algorithms can be designed by:

- Substituting different rules in the definition of local rules, or
- Defining different action sets for LAs, or

- Using different mechanisms for selecting action from the available action set of LAs.

A solution of VC problem is obtained from the configuration of the conflict resolvers in the CAPN-LA-VC (refer to **Definition** 6.22). To this end, *max* operator has been utilized over the action probability vector $\underline{q}_i$ to determine the color of vertex v_i; the color related to the action with the highest value in $\underline{q}_i$ is assigned to v_i.

To be able to compare such algorithms with each other, an evaluative measure is required. For this purpose, we utilize the measure of expediency, defined in Definition 6.32. In the next subsection, this measure will be calculated for CAPN-LA-VC using the results of the analysis of the proposed APN-LA-VC.

6.5.2.4 Measure of Expediency for the CAPN-LA-VC

To calculate the measure of expediency for the CAPN-LA-VC, we must determine the average reward for the vertex v_i at time instant k according to the following equation, which is a repetition of the Eq. (6.6).

$$D_i\left(\underline{q}(k)\right) = \sum_{\mu \in \mu_i}\left(\sum_{t_i^j|En\left(t_i^j,\mu\right)} d_i^j\left(\underline{q}(k)\right) \times \hat{q}_i^j(k,\mu)\right) \times \pi_\mu(k) \qquad (6.50)$$

Note that according to Eq. (6.5), $d_i^j\left(\underline{q}(k)\right)$ depends on the local rule of the CAPN-LA-VC. In other words, different algorithms result in different average rewards in the vertex v_i (D_i) due to the differences in their local rules.

To achieve $\pi_\mu(k)$ in Eq. (6.50), reachability analysis method for APN-LA-VC can be used. To this end, first, the reachability graph of APN-LA-VC related to the vertex v_i is obtained and then, its equivalent DTMC, Discrete Time Markov Chain, will be achieved. Finally, this DTMC will be used to achieve $\pi_\mu(k)$. Let first define a number of notations which is utilized in the rest of this section:

- L_k denotes the number of tokens in the place $P_{i,K}$,
- Marking $\mu_{i,j} = (L_0, L_1 \cdots L_{m_i}, L_{m_i+1}, L_{m_i+2})$ denotes the jth marking of Petri net shown in Fig. 6.22. This Petri net is related to APN-LA-VC A_i which is assigned to vertex v_i, and
- The initial marking is $\mu_{i,0} = (1,0,\ldots,0,0,1)$ in which two values L_0 and L_{m_i+2} are equal to 1 and other values are equal to 0.

Figure 6.24 shows the reachability graph of A_i (Fig. 6.22), obtained from the initial marking $\mu_{i,0}$. Figure 6.25 also shows the DTMC related to this reachability graph. The marking $\mu_{i,j}$ in the reachability graph is represented by state $\sigma_{i,j}$ in the DTMC. State $\sigma_{i,j}, (j = 1, \ldots, m)$ states that the color j is assigned to vertex v_i.

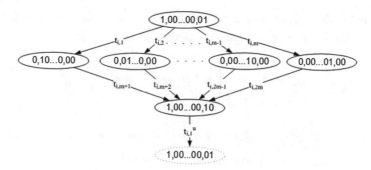

Fig. 6.24 The reachability graph of A_i; $\mu_{i,0}$ has been duplicated (indicated by dash-line)

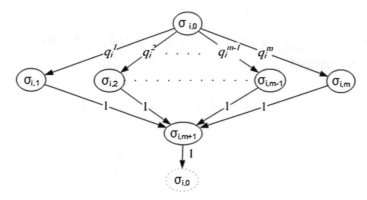

Fig. 6.25 The DTMC of A_i; $\sigma_{i,0}$ has been duplicated (indicated by dash-line)

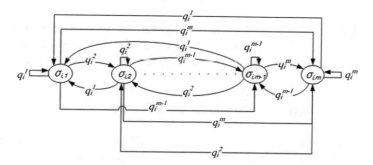

Fig. 6.26 Reduced DTMC for A_i

To reduce the number of states in the DTMC, useless states $\sigma_{i,0}$ and $\sigma_{i,m+1}$ are eliminated from the DTMC (Fig. 6.26). From DTMC shown in this figure, $\pi_{\sigma_{i,l}}(k) = q_i^l(k)$. Therefore, the Eq. (6.50) is rewritten as Eq. (6.51) given below, in which $q_i^j(k)$ is used instead of $\hat{q}_i^j(k)$ because all transitions in the cluster $s_{i,1}$ are enabled in any state σ and therefore $q_i^j(k) = \hat{q}_i^j(k, \sigma)$.

$$D_i\left(\underline{q}(k)\right) = \sum_{\sigma \in \{\sigma_{i,1}, \dots, \sigma_{i,m}\}} \left(\sum_{t_i^j} d_i^j\left(\underline{q}(k)\right) \times q_i^j(k) \right) \times q_i^l(k) \qquad (6.51)$$

Since $\sum_{t_i^j} d_i^j\left(\underline{q}(k)\right) \times q_i^j(k)$ is independent of the set of states $\sigma \in \{\sigma_{i,1}, \dots, \sigma_{i,m}\}$, we can rewrite Eq. (6.51) as below

$$D_i\left(\underline{q}(k)\right) = \left(\sum_{t_i^j} d_i^j\left(\underline{q}(k)\right) \times q_i^j(k) \right) \sum_{\sigma \in \{\sigma_{i,1}, \dots, \sigma_{i,m}\}} q_i^l(k) \qquad (6.52)$$

Since $\sum_{\sigma \in \{\sigma_{i,1}, \dots, \sigma_{i,m}\}} q_i^l(k) = 1$, the average reward for the vertex v_i at time instant k is simplified to

$$D_i\left(\underline{q}(k)\right) = \left(\sum_{t_i^j} d_i^j\left(\underline{q}(k)\right) \times q_i^j(k) \right) \qquad (6.53)$$

According to the above equation, the measure of expediency for the CAPN-LA-VC at time instant k is defined as Eq. (6.54) given below

$$M_E(k) = \left(\frac{E\left[\sum_i D_i\left(\underline{q}(k)\right) \right]}{\sum_i \frac{1}{m_i} \sum_{t_i^j} d_i^j\left(\underline{q}(k)\right)} \right) - 1 \qquad (6.54)$$

The value of $M_E(k)$ is used for comparing different VC-algorithms represented by the CAPN-LA-VC, such that a VC-algorithm with higher value of $M_E(k)$ is more expedient than a VC-algorithm with higher value of $M_E(k)$ at time instant k.

6.5.2.5   A New Algorithm for Solving the VC Problem

As mentioned before, different algorithms in CAPN-LA-VC can be designed by substituting different rules in the definition of local rules. In this section, we propose a new local rule for CAPN-LA-VC. To achieve that, we first analyze the DTMC of CAPN-LA-VC using the reduction of its states and then, using the analysis's results, we propose a new local rule. The efficiency of this new local rule will be shown in simulation results, described in the next section.

The states of DTMC shown in Fig. 6.26 can be divided into two disjoint sets:
(1) the set of allowable-coloring states ($Y_{i,1}$) and (2) the set of unallowable-coloring
states ($Y_{i,2}$). $Y_{i,1}$ is the set of states in which the color of the vertex v_i differs from the
colors of all neighboring vertices, whereas $Y_{i,2}$ is the set of states in which the color
of the vertex v_i is the same as the color of at least one of the neighboring vertices.
As a result, the graph in Fig. 6.26 is reduced to a new DTMC with two states which
is shown in Fig. 6.27. In the reduced DTMC, the transition-probability matrix $\mathbb{P}_i$ is
as follows:

$$\mathbb{P}_i = \begin{bmatrix} Z_{i,2} & Z_{i,1} \\ Z_{i,2} & Z_{i,1} \end{bmatrix}$$

$$Z_{i,1} = \sum_{S_{i,j} \in Y_{i,2}} q_i^j, Z_{i,2} = \sum_{S_{i,j} \in Y_{i,1}} q_i^j. \tag{6.55}$$

The probability of being in the state $Y_{i,1}$ at time instant k, i.e., $\pi_{Y_{i,1}}(k)$, is equal to
the probability of selecting an allowable color in the vertex v_i at k. This probability
can be stated using the following equation:

$$\pi_{Y_{i,1}}(k) = \sum_{l=1}^{m} \left[q_i^l(k) \cdot \prod_{j \in \underline{N_i}} \left(1 - \Gamma_{j,l}(k) \right) \right] \tag{6.56}$$

where $\underline{N_i}$ is the set of neighbors of the vertex v_i and $\Gamma_{i,l}(k)$ is defined according to
Eq. (6.57). Informally, q_i^l, the probability of assigning color l to vertex v_i, takes part
in $\pi_{Y_{i,1}}(k)$ if no neighbor of the vertex v_i is colored with color l.

$$\Gamma_{i,l}(k) = \begin{cases} 1, \alpha_i(k) = \alpha_{i,l} \\ 0, \alpha_i(k) \neq \alpha_{i,l} \end{cases} \tag{6.57}$$

Now, we define a measure according to the following equations in which
$|Y_{i,1}(k)|$ indicates the number of allowable colors in v_i at time instant k:

$$\omega_i(k) = \frac{\pi_{Y_{i,1}}(k)}{|Y_{i,1}(k)|} = \frac{\sum_{l=1}^{m} \left[q_i^l(k) \cdot \prod_{j \in \underline{N_i}} \left(1 - \Gamma_{j,l}(k) \right) \right]}{\sum_{l=1}^{m} \prod_{j \in \underline{N_i}} \left(1 - \Gamma_{j,l}(k) \right)} \tag{6.58}$$

Using the proposed measure $\omega_i(k)$, the proposed new local rule is defined as
follows:

Fig. 6.27 Reduced DTMC
for A_i

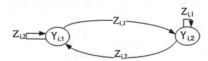

- The selected action of LA_i is rewarded if it is different from the selected actions of all of its neighbors and $\omega_i(k) \geq \hat{\omega}_i(k)$, where $\hat{\omega}_i(k)$ is defined as $\hat{\omega}_i(k) = \frac{1}{k}\sum_{j=1}^{k} \omega_i(j)$.
- Otherwise, it is penalized.

In other words, the selected action of LA_i is rewarded if (1) it is different from the selected actions of its neighbors and (2) this selection results in better ω_i than the average of ω_i resulted from previous selections of this LA.

6.5.2.6 Simulation Results

In this section, it will be shown that the CAPN-LA-VC can represent different algorithms for solving the VC problem. Theses algorithms are compared with each other in terms of the following criteria:

- CN: Number of colors required for coloring the graph,
- $\mathcal{U}$: Total number of updates on the learning automata, up to the time instant , where   is defined to be the smallest time instant at which the colors of all vertices in the graph are fixed,
- M_E: The value of the measure of expediency defined according to Eq. (6.54).

To study the efficiency of algorithms, a number of simulation studies have been conducted on a subset of hard-to-color benchmarks reported in DIMACS.[4] The rest of this section is divided into two sections:

- First, in the experiment One, the APN-ICLA-VC is used to implement and compare five different VC-algorithms. The first four algorithms are the state of the art algorithms introduced in (Torkestani 2013a); in each of which an ICLA is used to solve the VC problem. The fifth algorithm is proposed in the following section by introducing a novel local rule.
- Then, the experiment Two compares two scenarios of cooperation between APN-LAs to solve the VC problem.

6.5.2.6.1 Experiment One: Simulation on the CAPN-LA-VC

In this section, four different algorithms introduced in (Torkestani 2013a), namely APN1 to APN4, and a new algorithm proposed in this section, namely APN5, are described.

APN1: In the first algorithm, which we call APN1, it is assumed that the number of available colors for coloring each vertex v_i or equivalently the number of actions in the set of actions of LA_i is equal to the number of vertices in the graph $\mathbb{G}$; that is

---

[4]ftp://dimacs.rutgers.edu/pub/challenge/graph/.

$m = |\alpha_i| = |V|$. In the APN1, when LA_i is activated, an action from its set of actions is selected. All learning automata in the CAPN-LA-VC update their action probability vectors using an L_{R-I} learning algorithm. The local rules of the CAPN-LA-VC in APN1 algorithm are actually just simple local rules, defined as follows:

- If the selected action of LA_i is different from the selected actions of all of its neighbors, then it is rewarded;
- Otherwise, it is penalized.

APN2: In the first algorithm, the number of available actions for each LA is high. This significantly prolongs the time required by each LA to find a suitable action which in turn, prolongs the total running time of the algorithm. On the other hand, as mentioned before, it is shown that an arbitrary graph can be colored with at most $\Delta + 1$ colors, where Δ denotes the maximum degree of the vertices in the graph. To obtain a faster algorithm, in APN2, the number of available actions for each LA is equal to $\Delta + 1$. Due to the reduction in number of actions of each LA, it is expected that the APN2 colors the graph in less time and with a smaller number of colors compared to the APN1.

APN3: In APN2, all LAs have the same number of actions. But it is obvious that a vertex v_i and its neighbors can be colored using at most $\Delta_i + 1$ different colors, where Δ_i is the degree of vertex v_i. Therefore, the number of actions of each vertex v_i can be reduced to $\Delta_i + 1$. This algorithm is referred to as APN3.

APN4: Since a suitable mechanism for solving VC problem needs to decrease the number of required colors for coloring the graph, an enhanced algorithm reduces the number of available actions in each LA, where possible. In this enhanced algorithm, called APN4, learning automata with varying number of available action set are used, which is introduced in (Zhang and Zhou 2014); if the action probability of an action $\alpha_{i,l}$ falls below a certain threshold ε, then $\alpha_{i,l}$ is removed from the available set of actions of the LA_i.

APN5: All above algorithms, i.e., APN11 to APN4, utilize the simple local rule defined for APN1. However, we propose a new local rule in Sect. 6.5.2.5. Here, the APN4 algorithm with this proposed local rule is referred to as APN5.

The termination condition of all aforementioned algorithms in each vertex v_i is that the action probability of one of the actions of the learning automaton LA_i, say action $\alpha_{i,l}$, reaches a predetermined threshold τ. Then, from this time on, the color of the vertex v_i will be fixed to the corresponding color $\mathbb{C}_l$.

One problem with algorithms APN1 to APN4 is that they cannot guarantee to find a legal coloring of the graph. To eliminate the problem of improper coloring, the following modification is applied to all of the mentioned algorithms: All LAs are considered to be learning automata with varying number of available action set. The available actions for an LA at any time instant k are those which are not selected by any neighboring LAs at that time instant. This property allows each LA to pick only legal colors.

In this simulation study, the time instant k is increased whenever the LA corresponding to the vertex with the maximum degree is updated. Simulation parameters are set as follows: $a = b = .1$, $\varepsilon = .009$, and $\tau = .95$. The results of this simulation study, which are reported in Table 6.16, indicate that:

- By moving from APN1 to APN4, CN and $\mathcal{U}$ criteria decrease and $M_E(\mathbb{K})$ criterion increases. In other words, APN4 is the best and APN1 is the worst algorithm for coloring graphs. The reason behind this is that by decreasing the number of available actions for each learning automaton, its convergence rate as well as its accuracy increase.
- APN 5 is always better or at least equal to APN 4 in in terms of all the mentioned criteria. This efficiency is the results of considering the proposed measure ω_i in the local rule of the CAPN-LA-VC system.

6.5.2.6.2   Experiment Two: Cooperation Scenarios

This simulation study is conducted to compare the following two scenarios: (1) Handling the required cooperation between the neighboring cells within the controlling mechanisms, (2) Handling the required cooperation between the neighboring cells within the structure of Petri net. To this end, a Petri net consisting of multiple APN-LAs will be introduced, in which the required cooperation between any two neighboring vertices is handled within the structure of their APN-LAs. To handle the cooperation within the structure, a common way is to use inhibitor arcs. An inhibitor arc disables using of a color for a vertex when one of its neighbors uses this color for coloring. Therefore, to solve the VC problem, the APN-LA of the neighboring vertices are connected together by inhibitor arcs in order to disable using the illegal colors in the coloring of vertices. Figure 6.28 illustrates an APN-LA with its inhibitor arcs corresponding to the vertex v_i. The presented APN-LA in this figure is replicated in all vertices of the graph $\mathbb{G}$ and this way, a large Petri net will be achieved. This Petri net is referred to as APN-LA-IA, hereafter.

The number of required inhibitor arcs in the APN-LA-IA for coloring the graph $\mathbb{G}$ with n vertices is equal to $\sum_{i=1}^{n} m_i \times \Delta_i$ where m_i is the number of available colors for coloring in vertex v_i and Δ_i is the degree of vertex v_i. Although using these inhibitor arcs decreases the number of markings in the state-space of APN-LA-IA, it increases the complexity of the structure of APN-LA-IA. To generate a legal coloring by APN-LA-IA, the lowest priority is considered for transitions $t_{i,m+1}, \ldots, t_{i,2m}, (i = 1, \ldots, n)$. This way, the assigned color to each vertex (i.e., appeared token in place $P_{i,1}|P_{i,2}|\ldots|P_{i,m}$) stays the same until the next coloring assignment to the vertex. The algorithm represented by APN-LA-IA is referred to as APN6. In APN6, the number of actions of each vertex v_i is equal to $\Delta_i + 1$ where Δ_i is the degree of vertex c_i. When the action probability of one of these actions, corresponding to $LA_i, (i = 1, \ldots, n)$, reaches a predetermined threshold τ, the color

Table 6.16 Results of experiment for the algorithms represented by the CAPN-LA-VC

| Graph | APN | | | | | | | | | | | | | | |
| | APN1 | | | APNC2 | | | APN3 | | | APN4 | | | APN5 | | |
	CN	$\mathcal{U}$	$M_E(\mathbb{K})$	CN	$\mathcal{U}$	$M_E(\mathbb{K})$	CN	$\mathcal{U}$	$M_E(\mathbb{K})$	CN	$\mathcal{U}$	$M_E(\mathbb{K})$	CN	$\mathcal{U}$	$M_E(\mathbb{K})$
DSJC125_1	5	1,044,679	.439	5	116,002	.561	5	35,679	.801	5	26,876	.842	5	12,593	.897
DSJC125_5	19	1,067,235	.494	18	124,642	.558	17	46,410	.760	17	39,055	.827	17	27,263	.851
DSJC125_9	48	1,308,101	.486	45	128,119	.579	44	55,224	.758	44	46,407	.802	44	32,004	.827
DSJC250_1	10	641,605	.563	8	84,145	.689	8	28,757	.752	8	24,925	.857	8	14,661	.874
DSJC250_5	31	4,418,776	.532	29	421,963	.672	28	130,756	.739	28	100,635	.838	28	45,645	.839
DSJC250_9	76	6,741,839	.510	75	626,042	.671	73	233,568	.699	72	150,091	.833	72	59,200	.847
DSJC500_1	14	1,110,579	.502	13	144,954	.602	12	86,799	.721	12	67,603	.870	12	50,270	.889
DSJC500_5	52	2,712,399	.492	51	345,971	.588	49	222,161	.706	49	156,358	.813	49	83,597	.814
DSJC500_9	139	6,735,701	.482	131	648,780	.578	128	324,640	.685	127	285,403	.773	127	158,908	.782
le450_15a	15	4,279,467	.561	15	480,839	.687	15	162,445	.775	15	107,575	.892	15	78,335	.910
le450_15b	15	4,629,513	.556	15	497,638	.635	15	183,004	.734	15	138,001	.798	15	59,340	.864
le450_15c	16	4,672,330	.562	15	589,748	.663	15	311,569	.762	15	265,139	.815	15	102,384	.849
le450_15d	17	5,284,660	.585	15	502,709	.663	15	299,948	.771	15	203,358	.809	15	11,965	.831
le450_25c	26	5,516,730	.469	26	621,156	.607	25	313,236	.790	25	255,099	.825	25	103,542	.830
le450_25d	27	5,737,808	.448	26	636,840	.592	25	334,672	.699	25	275,394	.786	25	157,722	.798

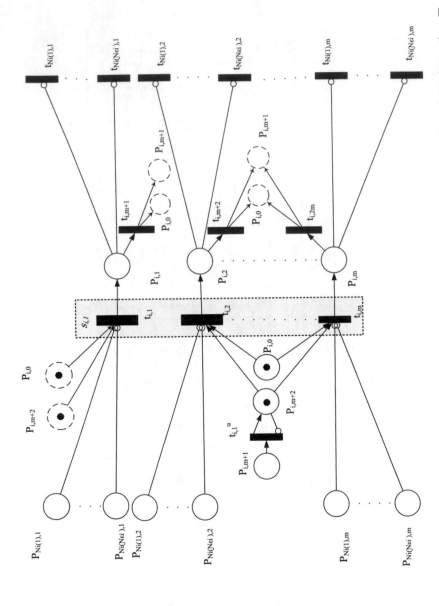

Fig. 6.28 An APN-LA is associated to vertex v_i, in which inhibitor arcs are utilized to disable illegal coloring of the neighboring vertices. The vertex v_i has $N\mathcal{E}_i$ neighbors indicated by $N_i(1), \ldots, N_i(N\mathcal{E}_i)$. For simplicity several places have been duplicated (indicated by dash-line)

Table 6.17 Results of experiment for APN6 represented by the APN-LA-IA. #IA is the required number of inhibitor arcs for applying APN6

| Graph | APN | | | | | | | |
| | APN5 | | | | APN6 | | | |
	CN	$\mathcal{U}$	$M_E(\mathbb{K})$	#IA	CN	$\mathcal{U}$	$M_E(\mathbb{K})$	#IA
DSJC125_1	5	12,593	.897	125	5	35,722	.782	20,279
DSJC125_5	17	27,263	.851	125	18	46,415	.745	495,871
DSJC125_9	44	32,004	.827	125	44	55,231	.756	1,565,939
DSJC250_1	8	14,661	.874	250	8	28,765	.750	178,900
DSJC250_5	28	45,645	.839	250	28	130,771	.734	3,974,526
DSJC250_9	72	59,200	.847	250	73	233,584	.698	1,253,204
DSJC500_1	12	50,270	.889	500	12	86,814	.719	1,289,234
DSJC500_5	49	83,597	.814	500	49	222,180	.704	31,560,758
DSJC500_9	127	158,908	.782	500	128	324,658	.685	101,382,226
le450_15a	15	78,335	.910	450	15	162,443	.777	737,068
le450_15b	15	59,340	.864	450	15	182,986	.736	730,290
le450_15c	15	102,384	.849	450	15	311,587	.761	2,709,830
le450_15d	15	119,652	.831	450	16	299,983	.767	2,730,334
le450_25c	25	103,542	.830	450	26	313,249	.773	3,091,758
le450_25d	25	157,722	.798	450	26	334,671	.699	3,098,528

of the vertex v_i will be fixed to the corresponding color. Table 6.17 compares APN5 (applied on the CAPN-LA-VC) and APN6 (applied on the APN-LA-IA) in terms of $CN, \mathcal{U}$, and $M_E(\mathbb{K})$ criteria. In addition, this table also reports the required number of inhibitor arcs for representing two mentioned scenarios. Consider a graph $\mathbb{G}$ with n vertices and $|\mathbb{E}|$ edges, in which the degree of vertex $v_i, (i = 1, \ldots, n)$ is Δ_i and the number of possible colors for vertex v_i is $m_i = \Delta_i + 1$. Regarding the size of PNs (in terms of the number of places, transitions, arcs, and inhibitor arcs) used in these two scenarios, one can mention the following points:

- In the CAPN-LA-VC, for each vertex v_i, there exists a Petri net (Fig. 6.22) in which there are $m_i + 3$ places, $2m_i$ transitions, $6m_i + 2$ arcs, one updating transition, and one inhibitor arc. Therefore, we have n Petri nets in which, there are totally $|\mathbb{E}| + 4n$ places, $2|\mathbb{E}| + 2n$ transitions, $6|\mathbb{E}| + 8n$ arcs, n updating transitions, and n inhibitor arcs.
- In the APN-LA-VC, there exists only one Petri net for the whole graph (Fig. 6.28) in which there are $|\mathbb{E}| + 4n$ places, $2|\mathbb{E}| + 2n$ transitions, $6|\mathbb{E}| + 8n$ arcs, n updating transitions, and $n + \sum_{i=1}^{n} m_i\Delta_i$ inhibitor arcs.

From the Table 6.17, the following remarks may be concluded:

- APN5 outperforms APN6 in terms of $CN, \mathcal{U}$, and $M_E(\mathbb{K})$ criteria.
- When the cooperation between the neighboring cells is handled within the controlling mechanisms, the complexity of the firing rules in Petri net is decreased.

6.6 Conclusion and Future Work

The aim of this chapter is to propos hybrid machines by the fusion of Petri nets and learning automata, which are appropriate for problem solving in dynamic environments. For this purpose, several Adaptive Petri nets based on learning automata and their applications for solving several problems in different areas were presented. Some of problems which have arisen in the course of research described in this chapter for further studies are summarized in the following paragraphs:

- Notion of expediency was introduced for the proposed CAPN-LA. A similar concept can be defined for APN-ICLA, and conditions under which, this hybrid machine become expedient can be studied. In addition another concept, which has been already defined for a learning automaton, is absolute expediency; a learning automaton is said to be absolutely expedient if its average input (i.e., reward) is an absolutely monotonically increasing function of the time. A similar notation can be defined for CAPN-LA and APN-ICLA, and conditions under which, these adaptive Petri nets become absolutely expedient can be studied.
- In this chapter, an algorithm with two phases has been designed for solving the shortest path problem in stochastic graphs. In this algorithm, each node of the graph is equipped with a learning automaton and these LAs, in cooperation with each other, solve the problem. One may say that this is an interconnected model of LAs, in which a number of LAs is organized into a graph. Hence, it is possible to propose an interconnected structure of LAs like DLA or eDLA, and then use it for solving different problems in the stochastic graphs. Based on this novel complex structure, the proposed algorithm in this chapter can be restated.
- The adaptive controlling mechanism in this chapter can be used in other types of Petri nets such as Time Petri net (TPN) and Stochastic Reward Net (SRN). Hence, adaptive TPN based on LA (ATPN-LA) and adaptive SRN based on LA (ASRN-LA) can be proposed.
- In all proposed adaptive Petri nets, we can say that learning automata are fused with transitions. To form other types of adaptive Petri net, fusion of LA with other elements of Petri nets such as tokens, places, arcs, and inhibitor arcs may be pursued.
- In this chapter, the CAPN-LA system was used for designing adaptive algorithms to solve vertex coloring problem. Applications of this model for designing algorithms to solve other types of graph coloring problem such as multi-coloring, bandwidth coloring, and bandwidth multi-coloring problems can be taken into consideration.

Chapter 7
Summary and Future Directions

Learning automaton (LA) is an effective decision making agent, especially within unknown stochastic environments. This book aims at collecting recent theoretical advancements as well as applied experiences of LAs in different research areas of the computer science discipline. The book starts with a brief explanation of the LA and its baseline variations. Then, a number of recently developed complex structures atop of the LA are introduced and their steady-state behaviors are studied. These complex structures have been developed considering the fact that an LA is by design a simple unit by which simple things can be done. The full potential of the LA is realized when a cooperative effort is made by a set of interconnected LAs to achieve the group synergy. These models are five versions of the *CLAs*, namely irregular cellular learning automaton (*ICLA*), dynamic *ICLA* (*DICLA*), heterogeneous *DICLA* (*HDICLA*), closed asynchronous dynamic *CLA* (*CADCLA*), and *CADCLA* with varying number of LAs in each cell (*CADCLA-VL*).

The next part of the book is devoted to provide a range of LA-based applications in different computer science domains, from wireless sensor networks, to peer-to-peer networks, then to complex social networks, and finally to Petri nets.

Putting all these matters together, the reader will go through a comprehensive journey, starting from basic concepts, continuing to recent theoretical findings, and ending in appliances of LA in vast varieties of problems from different research domains. The book is well suited to all computer engineers, scientists, and students, especially those who study/work in reinforcement learning and artificial intelligence domains.

Justifying the suitability of the learning automaton model in designing algorithms and protocols for *WSNs* are investigated. There are strong evidences which support this idea. First, *LA* is proved to perform well in distributed environments like the environments of *WSNs*, where the number of distributed elements is very large and the overhead of using centralized algorithms is very high (Lakshmivarahan and Narendra 1982; Mason and Gu 1986; Economids and Silvester 1988; Economides 1997; Atlasis et al. 1998). Second, *LA* has a very low computational and communicational overhead which makes it an outstanding

© Springer International Publishing AG 2018
A. Rezvanian et al., *Recent Advances in Learning Automata*, Studies in
Computational Intelligence 754, https://doi.org/10.1007/978-3-319-72428-7_7

model to be used in resource limited environments as of *WSN*s. Third, *LA* model is highly adaptive to the environmental changes, and hence, is well-suited to highly dynamic environments like the environments of *WSN*s. Finally, the reinforcement signal used by the *LA* is considered as a random variable and hence, its instant values do not affect the performance of the *LA* in the long run.

A framework for cognitive peer-to-peer networks was introduced and then an approach based on *DCLAs* for designing cognitive engines in the cognitive peer-to-peer networks was proposed. The proposed approach was used to suggest three cognitive engines for solving topology mismatch and super-peer selection problems. In order to design the structure updating rule of the *DCLAs*, the restructuring rules of Schelling segregation model, fungal growth model, and Voronoi diagrams model were borrowed. Experimental results showed that the suggested cognitive engine can compete with existing algorithms.

We introduced several learning automata based algorithms for solving stochastic graph problems, social network analysis, community detection and network sampling algorithms for deterministic and stochastic graphs. The learning automata based algorithms for stochastic graphs on network of learning automata tried to estimate the unknown distribution of the weigh associated with an edge by taking samples from the edges of stochastic graphs. For the sake of social network analysis when stochastic graph are used as graph model of network, the learning automata can be used by the network as a means for observing the time varying parameters of the network for the purpose of analyzing network. The aim of the LA-based algorithms are to collect information from the social network in order to find good estimates for the network parameters and measurements using fewer numbers of samples than that of standard sampling methods. We believe that the learning automata approach introduced in this book for modeling and analysis of social networks can provide a better avenue for studying online social networks by taking into consideration the continuum of behavioral parameters of network occurring over time.

Finaly several hybrid machines by the fusion of Petri nets and learning automata were proposed, which are appropriate for problem solving in dynamic environments. For this purpose, several Adaptive Petri nets based on learning automata and their applications for solving several problems in different areas were presented.

Bibliography

Agache M, Oommen BJ (2001) Continuous and discretized pursuit learning schemes: various algorithms and their comparison. IEEE Trans Syst Man Cybern Part B Cybern 31:277–287

Agache M, Oommen BJ (2002) Generalized pursuit learning schemes: new families of continuous and discretized learning automata. IEEE Trans Syst Man Cybern PartBibliography

Agache M, Oommen BJ (2001) Continuous and discretized pursuit learning schemes: various algorithms and their comparison. IEEE Trans Syst Man Cybern Part B Cybern 31:277–287

Agache M, Oommen BJ (2002) Generalized pursuit learning schemes: new families of continuous and discretized learning automata. IEEE Trans Syst Man Cybern Part B Cybern 32:738–749

Agerwala T (1979) Putting Petri nets to work. Computer 12:85–94

Ahangaran M, Taghizadeh N, Beigy H (2017) Associative cellular learning automata and its applications. Appl Soft Comput 53:1–18

Ahmed NK, Neville J, Kompella R (2014) Network sampling: from static to streaming graphs. ACM Trans Knowl Discov Data (TKDD) 8:7

Aliakbary S, Habibi J, Movaghar A (2015) Feature extraction from degree distribution for comparison and analysis of complex networks. Comput J 58:2079–2091

Alt FB (1982) Bonferroni inequalities and intervals. Encycl Stat Sci

Ambagis J (2002) Census and monitoring techniques for leach's storm petrel (Oceanodroma Leucorhoa). Msc thesis, College of the Atlantic. Bar Harbor

Amiri F, Yazdani N, Faili H, Rezvanian A (2013) A novel community detection algorithm for privacy preservation in social networks. In: Abraham A (ed) Intelligent informatics, pp 443–450

Amis A, Prakash R, Vuong T, Huynh and DT (2000) Max-min d-cluster formation in wireless ad hoc networks. In: Proceedings of the IEEE INFOCOM, pp 32–41

Arenas A, Duch J, Gómez S, Danon L, Díaz-Guilera A (2010) Communities in complex networks: identification at different levels. Encyclopedia of life support system (EOLSS) EOLSS Publishers, Oxford, UK, Developed under the auspices of the Unesco

Arora A, Dutta P, Bapat S, Kulathumani V, Zhang H, Naik V, Mittal V, Cao H, Demirbas M, Gouda M, Choi Y, Herman T, Kulkarni S, Arumugam U, Nesterenko M, Vora A, Miyashita and M (2004) A line in the sand: a wireless sensor network for target detection, classification, and tracking. Comput Netw 46:605–634

Asmuni H, Burke EK, Garibaldi JM, McCollum B, Parkes and AJ (2009) An investigation of fuzzy multiple heuristic orderings in the construction of university examination timetables. Comput Oper Res 36:981–1001

Atlasis A, Saltouros MP, Vasilakos and AV (1998) The use of a stochastic estimator learning algorithms to the ATM routing problem: a methodology. Comput Commun 21:538–546

Aurenhammer F (1991) Voronoi diagrams—a survey of a fundamental geometric data structure. ACM Comput Surv (CSUR) 23:345–405

© Springer International Publishing AG 2018
A. Rezvanian et al., *Recent Advances in Learning Automata*, Studies in Computational Intelligence 754, https://doi.org/10.1007/978-3-319-72428-7

Badie R, Aleahmad A, Asadpour M, Rahgozar M (2013) An efficient agent-based algorithm for overlapping community detection using nodes' closeness. Phys A Stat Mech Appl 392:5231–5247

Bagheri M, Hefeeda M (2007) Randomized k-coverage algorithms for dense sensor networks. In: Proceedings of IEEE INFOCOM 2007 Mini-symposium. Anchorage, AK, pp 2376–2380

Bail JL, Alia H, David R (1991) Hybrid Petri nets. In: The first European control conference. Grenoble, pp 1472–1477

Bandeira N, Lobo VJ, Moura-Pires F (1998) Training a self-organizing map distributed on a PVM network. In: Proceedings of neural networks conference of the IEEE, pp 457–461

Bandyopadhyay S, Coyle E (2003) An energy efficient hierarchical clustering algorithm for wireless sensor networks. In: Proceedings of the 22nd annual joint conference of the IEEE computer and communications societies. IEEE, San Francisco, California

Banerjee S, Khuller S (2001) A clustering scheme for hierarchical control in multi-hop wireless networks. In: Proceedings of the 20th annual joint conference of the IEEE computer and communications societies. Anchorage, AK, USA, pp 1028–1037

Barabási AL, Albert R (1999) Emergence of scaling in random networks. Science 286:509–512

Barnier N, Brisset and P (2002) Graph coloring for air traffic flow management. In: Proceedings of the fourth international workshop on integration of AI and OR techniques, Le Croisic. France, pp 133–147

Barto AG, Sutton RS, Anderson CW (1983) Neuronlike adaptive elements that can solve difficult learning control problems. IEEE Trans Syst Man Cybern SMC-13:834–846

Barzegar S, Davoudpour M, Meybodi MR, Sadeghian A, Tirandazian M (2011) Formalized learning automata with adaptive fuzzy colored Petri net; an application specific to managing traffic signals. Sci Iranica 18:554–565

Bash BA, Desnoyers PJ (2007) Exact distributed Voronoi cell computation in sensor networks. In: Proceedings of the 6th international conference on Information processing in sensor networks. ACM, New York, NY, USA, pp 236–243

Baumgart I, Heep B, Krause S (2009) OverSim: a scalable and flexible overlay framework for simulation and real network applications. Peer-to-Peer Computing. IEEE Computer Society, Seattle, Washington, USA, pp 87–88

Bause F (1996) On the analysis of Petri net with static priorities. Acta Informatica 33:669–685

Bause F (1997) Analysis of Petri nets with a dynamic priority method. In: Application and theory of Petri Nets, pp 215–234

Beaver J, Sharaf MA, Labrinidis A, Chrysanthis PK (2003) Location-aware routing for data aggregation in sensor networks. In: Proceedings of the 2nd hellenic data management symposium

Beigy H, Meybodi MR (2002) A new distributed learning automata based algorithm for solving stochastic shortest path problem. In: Proceedings of the sixth international joint conference on information Science (JCIS2002), Triangle Park. Durham, NC, USA, pp 339–343

Beigy H, Meybodi MR (2004) A mathematical framework for cellular learning automata. Adv Complex Syst 3:295–319

Beigy H, Meybodi MR (2006) Utilizing distributed learning automata to solve stochastic shortest path problems. Int J Uncertainty Fuzziness Knowl-Based Syst 14:591–615

Beigy H, Meybodi MR (2007) Open synchronous cellular learning automata. Adv Complex Syst 10:527–556

Beigy H, Meybodi MR (2008) Asynchronous cellular learning automata. Automatica 44:1350–1357

Beigy H, Meybodi MR (2009) Cellular learning automata based dynamic channel assignment algorithms. Int J Comput Intell Appl 8:287–314

Beigy H, Meybodi MR (2010) Cellular learning automata with multiple learning automata in each cell and its applications. IEEE Trans Syst Man Cybern Part B Cybern 40:54–65

Bentz C, Costa M-C, Picouleau C, Ries B, de Werra D (2012) d-Transversals of stable sets and vertex covers in weighted bipartite graphs. J Discrete Algorithms 17:95–102

Bettstetter C, Krause O (2001) On border effects in modeling and simulation of wireless ad hoc networks. In: Proceedings of the IEEE international conference on mobile and wireless communication networks. Recife, Brazil, pp 20–27

Bettstetter C, Hartenstein H, Costa and XP (2004) Stochastic properties of the random waypoint mobility model. ACM/Kluwer Wireless Networks. Special Issue on Modeling and Analysis of Mobile Networks 10:555–567

Beyens P, Peeters M, Steenhaut K, Nowe A (2005) Routing with compression in wireless sensor networks: a q-learning approach. In: 5th European workshop on adaptive agents and multi-agent systems. Paris, France

Bhattacharya S, Xing G, Lu C, Roman GC, Chipara O, Harris B (2005) Dynamic wake-up and topology maintenance protocol with spatiotemporal guarantees. In: Proceedings of the 4th international symposium on information processing in sensor networks, pp 28–34

Bild DR, Liu Y, Dick RP, Mao ZM, Wallach DS (2015) Aggregate characterization of user behavior in Twitter and analysis of the retweet graph. ACM Trans Internet Technol (TOIT) 15:4

Billard EA (1994) Instabilities in learning automata playing games with delayed information. In: 1994 IEEE international conference on systems, man, and cybernetics, pp 1160–1165

Billard EA (1996) Stability of adaptive search in multi-level games under delayed information. IEEE Trans Syst Man Cybern Part A Syst Hum 26:231–240

Blagus N, Šubelj L, Weiss G, Bajec M (2015) Sampling promotes community structure in social and information networks. Phys A Stat Mech Appl 432:206–215

Blough DM, Santi P (2002) Investigating upper bounds on network lifetime extension for cell-based energy conservation techniques in stationary ad hoc networks. In: Proceedings of the ACM/IEEE international conference on mobile computing and networking

Bolch G, Greiner S, de Meer H, Trivedi KS (2006) Queuing system and markov chain, 2nd edition. Wiley Publication

Bonferroni CE (1936) Teoria statistica delle classi e calcolo delle probabilita. Libreria internazionale Seeber

Borgatti SP (2005) Centrality and network flow. Soc Networks 27:55–71

Bruno G, Biglia P (1985) Performance evaluation and validation of tool handling in flexible manufacturing systems using Petri nets. In: International workshop on timed Petri Nets. 1, Torino, Italy, pp 64–71

Burkhard HD (1981) Ordered firing in Petri nets. In: EIK (Journal of information processing and cybernetics), pp 71–86

Cai Y, Li M, Shu W, Wu M-Y (2007) ACOS: an area-based collaborative sleeping protocol for wireless sensor networks. Ad Hoc Sens Wireless Netw 3:77–97

Cardoso RV, Dubois D (1999) Possibilistic Petri nets. IEEE Trans Syst Man Cybern Part B Cybern 29:573–582

Carmo R, Züge A (2012) Branch and bound algorithms for the maximum clique problem under a unified framework. J Braz Comput Soc 18:137–151

Carter M, Laporte G, Lee S (1996) Examination timetabling: algorithmic strategies and applications. J Oper Res Soc 47:373–83

Chakrabarty K, Iyengar SS, Qi H, Cho E (2002) Grid coverage for surveillance and target location in distributed sensor networks. IEEE Trans Comput 51:1448–1453

Chawathe Y, Ratnasamy S, Breslau L, Lanham N, Shenker S (2003) Making gnutella-like P2P systems scalable. In: Proceedings of the conference on applications, technologies, architectures, and protocols for computer communications. ACM, Karlsruhe, Germany, pp 407–418

Chen MI, Lin KJ (1990) Dynamic priority ceilings: a concurrency control protocol for real-time systems. Real-Time Syst 2:325–346

Chen M, Ke JS, Chang JF (1990) Knowledge representation using fuzzy Petri nets. IEEE Trans Knowl Data Eng 2:311–319

Chen WP, Hou JC, Sha L (2004) Dynamic clustering for acoustic target tracking in wireless sensor networks. IEEE Trans Mob Comput 3:258–271

Chen S, Ge QW, Shao QM, Zhu Q (2008) Modelling and performance analysis de wireless sensor network systems using Petri nets. In: International technical conference on circuits systems, computers and communications, pp 1689–1692

Chen Y, Li Z, Al-Ahmari A (2013) Non-pure Petri net supervisors for optimal deadlock control of flexible manufacturing systems. IEEE Trans Syst Man Cybern Syst 43:252–265

Cheng J, Ke Y, Fu AWC, Yu JX, Zhu L (2011) Finding maximal cliques in massive networks. ACM Trans Database Syst (TODS) 36:21

Cherkassky V, Goldberg AV, Radzik T (1996) Shortest paths algorithms: theory and experimental evaluation. Math Program 73:129–174

Chiola G, Ferscha A (1993) Distributed simulation of Petri nets. IEEE Concurrency 3:33–50

Clauset A, Newman MEJ, Moore C (2004) Finding community structure in very large networks. Phys Rev E 70:066111

Clauset A, Shalizi CR, Newman ME (2009) Power-law distributions in empirical data. SIAM Rev 51:661–703

Cobham A (1954) Priority assignment in waiting line problems. Oper Res 2:70–76

Costa LF, Rodrigues FA, Travieso G, Boas PRV (2007) Characterization of complex networks: a survey of measurements. Adv Phy 56:167–242

Crowell B (2004) Newtonian Physics. Light and matter

Damerchilu B, Norouzzadeh MS, Meybodi MR (2016) Motion estimation using learning automata. Mach Vis Appl 27:1047–1061

Danon L, Diaz-Guilera A, Duch J (2005) Comparing community structure identification. J Stat Mech Theory Exp 2005:P09008

Dasgupta K, Namjoshi KKP (2003) An efficient clustering-based heuristic for data gathering and aggregation in sensor networks. In: Proceedings of the IEEE wireless communications and networking conference, pp 1948–1953

David R, Alla H (1994) Petri nets for modeling of dynamic systems. Automatica 30:175–202

Demetrescu B, Italiano GF (2004) A new approach to dynamic all Pairs shortest paths. J ACM (JACM) 51:968–992

Demirbas M, Arora A, Mittal V (2004) FLOC: a fast local clustering service for wireless sensor networks. In: Proceedings of workshop on dependability issues in wireless ad hoc networks and sensor networks, Palazzo dei Congressi. Florence, Italy

Dharmaraja S, Trivedi KS, Logothetis D (2003) Performance modeling of wireless networks with generally distributed handoff interarrival times. Comput Commun 26:1747–1755

Dhillon IS, Guan Y, Kulis B (2004) Kernel k-means: spectral clustering and normalized cuts. In: Proceedings of the tenth ACM SIGKDD international conference on knowledge discovery and data mining. ACM, pp 551–556

Dhillon SS, Chakrabarty K, Iyengar and SS (2002) Sensor placement for grid coverage under imprecise detections. In: Proceedings of international conference on information fusion. pp 1581–1587

Di Mascolo M, Frein Y, Dallery Y, David R (1991) A unified modeling of Kanban systems using Petri nets. Int J Flex Manuf Syst 3:275–307

Dietrich I, Dressler F (2009) On the lifetime of wireless sensor networks. ACM Trans Sens Netw (TOSN) 5:1–5

Dijkstra W (1959) A note on two problems in connection with graphs. Numer Math 1:269–271

Ding P, Holliday J, Celik A (2005) Distributed energy efficient hierarchical clustering for wireless sensor networks. In: Proceedings of the IEEE international conference on distributed computing in sensor systems, Marina Del Rey. CA, USA

Ding Z, Zhou Y, Zhou MC (2014) Modeling self-adaptive software systems with learning Petri nets. In: Proceedings of international conference on software engineering. Hyderabad, India, pp 464–467

Dingxing Z, Ming X, Yingwen C, Shulin W (2006) Probabilistic coverage configuration for wireless sensor networks. In: Procedings of the international conference on wireless communications, networking, and mobile computing. Wuhan, pp 1–4

Domic NG, Goles E, Rica S (2011) Dynamics and complexity of the schelling segregation model. Phys Rev E 83:96–111

Economides AA (1997) Real-time traffic allocation using learning automata. In: Proceedings of the IEEE international conference on computational cybernetics and simulation, pp 3307–3312

Economids AA, Silvester JA (1988) Optimal routing in networks with unreliable links. In: Proceedings of the computer networking symposium, pp 288–297

Elamvazuthi I, Vasant P, Ganesan T (2013) Hybrid optimization techniques for optimization in a fuzzy environment. In: Handbook of Optimization. Springer, pp 1025–1046

Elyasi M, Meybodi M, Rezvanian A, Haeri MA (2016) A fast algorithm for overlapping community detection. In: 2016 eighth international conference on information and knowledge technology (IKT), pp 221–226

Emiris IZ, Fragoudakis C, Markou E (2006) Maximizing the guarded interior of an art gallery. In: Proceedings of the 22nd European workshop on computational geometry, Delphi, pp 165–168

Enayatzare M, Meybodi MR (2009) Solving graph coloring problem using cellular learning automata. In: Proceedings of 14th annual CSI computer conference of Iran, Amirkabir University of Technology. Tehran, Iran

Epema DHJ (1991) Mean waiting times in a general feedback queue with priorities. Perform Eval 13:45–58

Eraghi AE, Torkestani JA, Meybodi MR (2009) Cellular learning automata-based graph coloring problem. In: Proceedings of 2009 international conference on machine learning and computing. Perth, Australia, pp 163–167

Erdos P, Rényi A (1960) On the evolution of random graphs. Publ Math Inst Hung Acad Sci 5:17–61

Esnaashari M, Meybodi MR (2008) A cellular learning automata based clustering algorithm for wireless sensor networks. Sens Lett 6:723–735

Esnaashari M, Meybodi MR (2010) Dynamic point coverage problem in wireless sensor networks: a cellular learning automata approach. Ad Hoc Sens Wirel Netw 10:193–234

Esnaashari M, Meybodi MR (2010) Data aggregation in sensor networks using learning automata. Wirel Netw 16:687–699

Esnaashari M, Meybodi MR (2010) A learning automata based scheduling solution to the dynamic point coverage problem in wireless sensor networks. Comput Netw 54:2410–2438

Esnaashari M, Meybodi M (2011) A cellular learning automata-based deployment strategy for mobile wireless sensor networks. J Parallel Distrib Comput 71:988–1001

Esnaashari M, Meybodi M (2013) Deployment of a mobile wireless sensor network with k-coverage constraint: a cellular learning automata approach. Wirel Netw 19:945–968

Esnaashari M, Meybodi MR (2015) Irregular cellular learning automata. IEEE Trans Cybern 45:1622–1632

Esnaashari M, Meybodi MR (2017) Dynamic irregular cellular learning automata. J Comput Sci. https://doi.org/10.1016/j.jocs.2017.08.012

Floyd RW (1962) Algorithm 97: shortest path. Commu ACM 5:345

Fortunato S (2010) Community detection in graphs. Phys Rep 486:75–174

Frank O (2011) Survey sampling in networks. In: The SAGE handbook of social network analysis. SAGE publications, pp 370–388

Freeman LC (1979) Centrality in social networks conceptual clarification. Soc Netw 1:215–239

Freiheit I, Billington J (2003) New developments in closed-form computation for GSPN aggregation. Int Conf Formal Eng Meth. Springer, Berlin Heidelberg, pp 471–490

Frigioni D, Marchetti-Spaccamela A, Nanni U (2000) Fully dynamic algorithms for maintaining shortest paths trees. J Algorithms 34:251–281

Fujimoto RM (2001) Parallel simulation: parallel and distributed simulation systems. In: Proceedings of the 33nd conference on winter simulation, IEEE Computer Society. pp 147–157

Gama J, Gaber MM (2007) Learning from data streams: processing techniques in sensor networks. Springer-Verlag, Berlin, Heidelberg

Ganesan T, Vasant P, Elamvazuthi I (2014) Hopfield neural networks approach for design optimization of hybrid power systems with multiple renewable energy sources in a fuzzy environment. J Intel Fuzzy Syst 26:2143–2154

Gao MZ, Wu Z, Huang X (2003) Fuzzy reasoning Petri nets. In: IEEE transactions on systems, man and cybernetics, Part A: systems and humans. pp 314–324

Gao L, Sun P-G, Song J (2009) Clustering algorithms for detecting functional modules in protein interaction networks. J Bioinf Comput Biol 7:217–242

Geetha S, Jayaparvathy R (2010) Modeling and analysis of bandwidth allocation in IEEE 802.16 MAC: a stochastic reward net approach. Int J Commun Netw Syst Sci 3:631–637

Ghavipour M, Meybodi MR (2016) An adaptive fuzzy recommender system based on learning automata. Electron Commer Res Appl 20:105–115

Ghavipour M, Meybodi MR (2017) Irregular cellular learning automata-based algorithm for sampling social networks. Eng Appl Artif Intel 59:244–259

Ghavipour M, Meybodi MR (2017b) Trust propagation algorithm based on learning automata for inferring local trust in online social networks. Knowledge based systems in-press. https://doi.org/10.1016/j.knosys.2017.06.034

Gholami S, Meybodi MR, Saghiri AM (2014) A learning automata-based version of SG-1 protocol for super-peer selection in peer-to-peer networks. In: Proceedings of the 10th international conference on computing and information technology. Springer, Angsana Laguna, Phuket, Thailand, pp 189–201

Gile KJ, Handcock MS (2010) Respondent-driven sampling: an assessment of current methodology. Sociol Methodol 40:285–327

Girvan M, Newman MEJ (2002) Community structure in social and biological networks. Proc Nat Acad Sci 99:7821–7826

Giusti A, Murphy AL, Picco GP (2007) Decentralized scattering of wake-up times in wireless sensor networks. In: Proceedings of the 4th European conference on wireless sensor networks, Lecture Notes in Compute Science. pp 245–260

Gjoka M, Butts CT, Kurant M, Markopoulou A (2011) Multigraph sampling of online social networks. IEEE J Sel Areas Commun 29:1893–1905

Glockner A, Pasquale J (1993) Co-adaptive behavior in a simple distributed job scheduling system. IEEE Trans Syst Man Cybern 23:902–907

Goldstein ML, Morris SA, Yen GG (2004) Problems with fitting to the power-law distribution. Eur Phys J B-Condens Matter Complex Syst 41:255–258

Gui C, Mohapatra P (2004) Power conservation and quality of surveillance in target tracking sensor networks. In: Proceeding of the 10th annual international conference on mobile computing and networking (MOBICOM 2004). Philadelphia, PA, USA, Sep–Oct 2004

Gupta R, Das SR (2003) Tracking moving targets in a smart sensor network. In: Proceedings of the 2003 IEEE 58th vehicular technology conference. pp 3035–3039

Gupta A, Pál M, Ravi R, Sinha A (2011) Sampling and cost-sharing: approximation algorithms for stochastic optimization problems. SIAM J Comput 40:1361–1401

Haines RJ, Clemo GR, Munro ATD (2007) Petri-nets for formal verification of MAC protocols. IET Softw 1:39–47

Hao F, Park D-S, Min G, Jeong Y-S, Park J-H (2016) k-Cliques mining in dynamic social networks based on triadic formal concept analysis. Neurocomputing 209:57–66

Hasanzadeh M, Meybodi MR (2014) Grid resource discovery based on distributed learning automata. Computing 96:909–922

Hasanzadeh M, Meybodi MR (2015) Distributed optimization grid resource discovery. J Supercomput 71:87–120

Hasanzadeh M, Meybodi MR, Ebadzadeh MM (2013) Adaptive cooperative particle swarm optimizer. Appl Intel 39:397–420

Hasanzadeh-Mofrad M, Rezvanian A (2017) Learning automata clustering. J Comput Sci. https://doi.org/10.1016/j.jocs.2017.09.008

Hatler M (2004) Wireless sensor networks: mass market opportunities. ON World, San Diego, CA, USA

He T, Vicaire P, Yan T, Luo L, Gu L, Zhou G (2006) Achieving real-time target tracking using wireless sensor networks. In: Proceedings of the 12th IEEE real-time and embedded technology and applications symposium

Heinzelman W (2000) Application-specific protocol architecture for wireless networks. Institute of Technology, Massachusetts

Heinzelman W, Chandrakasan A, Balakrishnan H (2000) Energy efficient communication protocol for wireless microsensor networks. In: Proceedings of the international conference on system sciences. Hawaii

Heinzelman W, Chandrakasan A, Balakrishnan H (2002) An application-specific protocol architecture for wireless microsensor networks. IEEE Trans Wirel Commun 1:660–670

Heo N, Varshney PK (2005) Energy-efficient deployment of intelligent mobile sensor networks. IEEE Trans Syst Man Cybern Part A Syst Hum 35:78–92

Holliday MA, Vernon MK (1987) A generalized timed Petri net model for performance analysis. IEEE Trans Softw Eng 12:1297–1310

Holloway LE, Krogh BH (1994) Controlled Petri nets: a tutorial survey. In: 11th international conference on analysis and optimization of systems discrete event systems. Springer, pp 158–168

Holloway LE, Krogh B, Giua and A (1997) A survey of Petri net methods for controlled discrete event systems. Discrete Event Dyn Syst 7:151–190

Howard A, Mataric MJ, Sukhatme GS (2002) Mobile sensor network deployment using potential fields: a distributed scalable solution to the area coverage problem. In: Proceedings of the international symposium on distributed autonomous robotics systems. Fukuoka, Japan, pp 299–308

Howell MN, Frost GP, Gordon TJ, Wu QH (1997) Continuous action reinforcement learning applied to vehicle suspension control. Mechatronics 7:263–276

Huang H, Wu J (2005) A probabilistic clustering algorithm in wireless sensor networks. In: Proceedings of the IEEE 62nd semiannual vehicular technology conference

Huffaker B, Plummer D, Moore D, Claffy KC (2002) Topology discovery by active probing. In: Proceedings of symposium on applications and the internet workshops. Washington, DC, USA, pp 90–96

Hutson KR, Shier DR (2006) Minimum spanning trees in networks with varying edge weights. Ann Oper Res 146:3–18

Hwang HJ, Velázquez JJ (2013) Bistable stochastic biochemical networks: large chemical networks and systems with many molecules. J Math Chem 51:2074–2103

Ilyas M, Mahgoub I (2005) Handbook of sensor networks: compact wireless and wired sensing systems. CRC Press, London, Boca Raton, New York, Washington D.C

Intanagonwiwat C, Govindan R, Estrin D (2003) Directed diffusion for wireless sensor networks. IEEE/ACM Trans Netw 11:2–16

Irit D, Safra S (2005) On the hardness of approximating minimum vertex cover. Ann Math 162:439–485

Isaacson DL, Madsen RW (1976) Markov chains: theory and applications. Wiley, New York

Isella L, Stehlé J, Barrat A, Cattuto C, Pinton J-F, Van den Broeck W (2011) What's in a crowd? Analysis of face-to-face behavioral networks. J Theor Biol 271:166–180

Jadliwala M, Bilogrevic I, Hubaux J-P (2013) Optimizing mix-zone coverage in pervasive wireless networks. J Comput Secur 21:317–346

Jain LC, Martin N (1999) Fusion of neural networks, fuzzy sets and genetic algorithms: industrial applications. CRC press

Jaiswal K (1968) Priority queues. Elsevier

Jalali ZS, Rezvanian A, Meybodi MR (2015) A two-phase sampling algorithm for social networks. In: 2015 2nd international conference on knowledge-based engineering and innovation (KBEI). IEEE, pp 1165–1169

Jalali ZS, Rezvanian A, Meybodi MR (2016) Social network sampling using spanning trees. Int J Mod Phys C 27:1650052

James-Romero B, Munoz-Rodriguez D, Molina C, Tawfik H (1997) Modeling resource management in cellular systems using Petri nets. IEEE Trans Veh Technol 46:298–312

Jelasity M, Kowalczyk W, Van Steen M (2003) Newscast computing. Vrije Universiteit Amsterdam, Department of Computer Science, Amsterdam, Netherlands

Jensen K (1981) Colored Petri nets and the invariant-method. Theor Comput Sci 14:317–336

Jensen TR, Toft B (1995) Graph coloring problems. Wiley, New York

Jeong J, Sharafkandi S, Du DHC (2006) Energy-aware scheduling with quality of surveillance guarantee in wireless sensor networks. In: Proceedings of the international conference on mobile computing and networking. Los Angeles, CA, USA

Jeong J, Hwang T, He T, Du D (2007) MCTA: target tracking algorithm based on minimal contour in wireless sensor networks. In: Proceedings 26th IEEE international conference on computer communications (INFOCOM 2007)

Jiang Y, Tham CK, Ko CC (2002) A probabilistic priority scheduling discipline for multi-service networks. Comput Commun 25:1243–1254

Jiang Y, You J, He X (2006) A particle swarm based network hosts clustering algorithm for peer-to-peer networks. International conference on computational intelligence and security. IEEE Computer Society, Guangzhou, China, pp 1176–1179

Jiang B, Ravindran B, Cho H (2008) Energy efficient sleep scheduling in sensor networks for multiple target tracking. Lect Notes Comput Sci 498–509

Jin L, Chen Y, Wang T, Hui P, Vasilakos AV (2013) Understanding user behavior in online social networks: A survey. IEEE Commun Mag 51:144–150

Jin-Hua Z, Hai-Jun Z (2014) Statistical physics of hard combinatorial optimization: vertex cover problem. Chin Phys B 23:078901

Joshi-Tope G, Gillespie M, Vastrik I, D'Eustachio P, Schmidt E, de Bono B, Jassal B, Gopinath GR, Wu GR, Matthews L, Lewis S, Birney E, Stein L (2005) Reactome: a knowledgebase of biological pathways. Nucl Acids Res 33:D428–D432. https://doi.org/10.1093/nar/gki072

Ju H-J, Du L-J (2012) Nodes clustering method in large-scale network. 8th international conference on wireless communications, networking and mobile computing. IEEE Computer Society, Shanghai, China, pp 1–4

Jurczyk M (2000) Traffic control in wormhole-routing multistage interconnection networks. In: International conference on parallel and distributed computing and systems. pp 157–162

Kadjinicolaou MG, Abdelrazik MBE, Musgrave G (1990) Structured analysis for neural networks using Petri nets. In: Proceedings of the 33rd midwest symposium on circuits and systems, IEEE. pp 770–773

Karp RM (1972) Reducibility among combinatorial problems. In: Complexity of computer computations. pp 85–103

Khalifa YMA, Okoene E, Al-Mourad MB (2007) Autonomous intelligent agent-based tracking systems. ICGST-ACSE J 7:21–31

Khomami MMD, Rezvanian AR, Meybodi MR (2014) Irregular cellular automata for multiple diffusion. In: Proceedings of the 22th Iranian conference on electrical engineering. Shahid Beheshti University, Tehran, Iran

Khomami MMD, Bagherpour N, Sajedi H, Meybodi MR (2016) A new distributed learning automata based algorithm for maximum independent set problem. Artificial intelligence and robotics (IRANOPEN), 2016. IEEE, Qazvin, Iran, pp 12–17

Khomami MMD, Rezvanian A, Meybodi MR (2016) Distributed learning automata-based algorithm for community detection in complex networks. Int J Mod Phys B 30:1650042

Khomami MMD, Rezvanian A, Bagherpour N, Meybodi MR (2017a) Irregular cellular automata based diffusion model for influence maximization. In: 2017 5th Iranian joint congress on fuzzy and intelligent systems (CFIS). IEEE, pp 69–74

Khomami MMD, Rezvanian A, Meybodi MR (2017b) A new cellular learning automata-based algorithm for community detection in complex social networks. J Comput Sci. https://doi.org/10.1016/j.jocs.2017.10.009

Khomami MMD, Rezvanian AR, Bagherpour N, Meybodi MR (2017c) Minimum positive influence dominating set and its application in influence maximization: a learning automata approach. Appl Intel 1–24

Khosla R, Dillon T (1997) Intelligent hybrid multi-agent architecture for engineering complex systems. In: International conference on neural networks, IEEE

Kilhwan K, Chae KC (2010) Discrete-time queues with discretionary priorities. Eur J Oper Res 473–485

Kim I-J, Barthel BP, Park Y, Tait JR, Dobmeier JL, Kim S, Shin D (2014) Network analysis for active and passive propagation models. Networks 63:160–169

Kittsteiner T, Moldovanu B (2005) Priority auctions and queue disciplines that depend on processing time. Manage Sci 51:236–248

Kleinberg J (2000) The small-world phenomenon: an algorithmic perspective. In: Proceedings of the thirty-second annual ACM symposium on Theory of computing. ACM, pp 163–170

Kleinrock L (1975) Queueing systems. Wiley, New York

Kohonen T (1998) The self-organizing map. Neurocomputing 21:1–6

Kolchin V, Sevast'yanov B, Chistyakov V (1978) Random allocations. Winston and Sons, Washington D.C

KONECT (2016) Linux kernel mailing list replies network dataset. In: KONECT. http://konect.uni-koblenz.de/networks

Kounty M (1992) Modeling systems with dynamic priorities. Springer-Verlag, Advances in Petri nets

Krishna K (1993) Cellular learning automata: a stochastic model for adaptive controllers. Master's thesis, Department of Electrical Engineering, Indian Institue of science, Banglore, India

Krishna P, Vaidya NH, Chatterjee M, Pradhan DK (1997) A cluster-based approach for routing in dynamic networks. ACM SIGCOMM Comput Commun Rev 49–65

Krishna PV, Misra S, Joshi D, Gupta A, Obaidat MS (2014) Secure socket layer certificate verification: a learning automata approach. Secur Commun Netw 17:1712–1718

Krishnamurthy B, Wang J (2000) On network-aware clustering of web clients. In: ACM SIGCOMM Computer Communication Review. ACM, pp 97–110

Krogh H, Magott J, Holloway and LE (1991) On the complexity of forbidden state problems for controlled marked graphs. In: Proceedings of the 30th IEEE conference on decision and control, IEEE. pp 85–91

Kuhn F, Mastrolilli M (2013) Vertex cover in graphs with locally few colors. Inf Comput 222:265–277

Kumar RS (1986) A simple learning scheme for priority assignment at a single-server queue. IEEE Trans Syst Man Cybern 16:751–754

Kumar N, Lee J-H (2015) Collaborative-learning-automata-based channel assignment with topology preservation for wireless mesh networks under QoS constraints. IEEE Syst J 9:675–685

Kumar S, Lai TH, Balogh J (2004) On k-Coverage in a mostly sleeping sensor network. In: Proceedings of the 10th annual international conference on mobile computing and networking. Philadelphia, Pennsylvania, USA, pp 144–158

Kumar N, Misra S, Obaidat M, Rodrigues J, Pati B (2014) Networks of learning automata for the vehicular environment: a performance analysis study. IEEE Wirel Commun 21:41–47

Kumar N, Lee J-H, Rodrigues JJ (2015) Intelligent mobile video surveillance system as a bayesian coalition game in vehicular sensor networks: learning automata approach. IEEE Trans Intel Trans Syst 16:1148–1161

Kumar N, Misra S, Obaidat MS (2015) Collaborative learning automata-based routing for rescue operations in dense urban regions using vehicular sensor networks. IEEE Syst J 9:1081–1090

Kumpula J, Saramaki J, Kaski K, Kertesz J (2007) Limited resolution and multiresolution methods in complex network community detection. Fluctuation Noise Lett 7:L209–L214

Kurant M, Markopoulou A, Thiran P (2010) On the bias of BFS (Breadth First Search). In: 2010 22nd International Teletraffic Congress (ITC). pp 1–8

Lakshmivarahan S, Narendra KS (1982) Learning algorithms for two-person zero-sum stochastic games with incomplete information: a unified approach. SIAM J Control Optim 20:541–552

Lancichinetti A, Fortunato S, Radicchi F (2008) Benchmark graphs for testing community detection algorithms. Phys Rev E 78:046110

Lanctot JK, Oommen BJ (1992) Discretized estimator learning automata. IEEE Trans Syst Man Cybern 22:1473–1483

Lee H, Dong H, Aghajan H (2006) Robot-assisted localization techniques for wireless image sensor networks. In: Proceedings of the IEEE international conference on sensor, mesh, and ad hoc communications and networks. pp 383–392

Lee CH, Xu X, Eun DY (2012) Beyond random walk and metropolis-hastings samplers: why you should not backtrack for unbiased graph sampling. In: Proceedings of the 12th ACM SIGMETRICS/PERFORMANCE joint international conference on measurement and modeling of computer systems. pp 319–330

Leitão J, Marques JP, Pereira J, Rodrigues L (2012) X-bot: a protocol for resilient optimization of unstructured overlay networks. IEEE Trans Parallel Distrib Syst 23:2175–2188

Leskovec J, Faloutsos C (2006) Sampling from large graphs. In: Proceedings of the 12th ACM SIGKDD international conference on knowledge discovery and data mining. ACM, Philadelphia, pp 631–636

Leskovec J, Kleinberg J, Faloutsos C (2007) Graph evolution: densification and shrinking diameters. ACM Trans Knowl Discov Data (TKDD) 1:1–41

Lewis I, Paechter B (2007) Finding feasible timetables using group based operators. IEEE Trans Evol Comput 11:397–413

Li JS, Kao HC (2010) Distributed k-Coverage self-location estimation scheme based on voronoi diagram. IET Commun 4:167–177

Li Y, Wonham WM (1994) Control of vector discrete-event systems. II. controller synthesis. IEEE Trans Autom Control 39:512–531

Li Y, Wu C, Wang X, Luo P (2014) A network-based and multi-parameter model for finding influential authors. J Informetrics 8:791–799

Li Y, Yu Z (2011) An improved genetic algorithm for network nodes clustering. In: Proceedings of the second international conference on information computing and applications. Springer Berlin Heidelberg, Qinhuangdao, China, pp 399–406

Li Z, Zhou MC (2009) Deadlock resolution in automated manufacturing systems: a novel Petri net approache. Springer Science & Business Media

Lima PU, Saridis GN (1999) Intelligent controllers as hierarchical stochastic automata. IEEE Trans Syst Man Cybern Part B (Cybernetics) 29:151–63

Lin SY, Chan TY (2009) A Petri-net-based automated distributed dynamic channel assignment for cellular network. IEEE Trans Veh Technol 58:4540–4553

Lin CR, Gerla M (1997) Adaptive clustering for mobile wireless networks. IEEE J Sel Areas Commun 15:1265–1275

Lin Y-K, Huang C-F (2013) Backup reliability of stochastic imperfect-node computer networks subject to packet accuracy rate and time constraints. Int J Comput Math 90:457–474

Lindsey S, Raghavendra CS (2002) PEGASIS: power-efficient gathering in sensor information systems. In: Proceedings of the IEEE Aerospace Conference. pp 1125–1130

Lindsey S, Raghavendra C, Sivalingam KM (2002) Data gathering algorithms in sensor networks using energy metrics. IEEE Trans Parallel Distrib Syst 13:924–935

Linial N (1992) Locality in distributed graph algorithms. SIAM J Comput 21:193–201

Liu D (2007) Resilient cluster formation for sensor networks. In: Proceedings of the 27th international conference on distributed computing systems

Liu Y (2008) A two-hop solution to solving topology mismatch. IEEE Trans Parallel Distrib Syst 19:1591–1600

Liu C, Zhang Z-K (2014) Information spreading on dynamic social networks. Commun Nonlinear Sci Numer Simul 19:896–904

Liu M, Harjula E, Ylianttila M (2013) An efficient selection algorithm for building a super-peer overlay. J Internet Serv Appl 4:1–12

Liu J, Ye X, Zhang J, Li J (2008) Security verification of 802.11i 4-way handshake protocol. In: International Conference on Communications (ICC'08). pp 1642–1647

Llinas J, Liggins ME, Hall D (2001) Handbook of multisensor data fusion. CRC Press

Lo V, Zhou D, Liu Y, GauthierDickey C, Li J (2005) Scalable supernode selection in peer-to-peer overlay networks. Hot topics in peer-to-peer systems. IEEE Computer Society, Washington, DC, USA, pp 18–25

Looney B (1988) Fuzzy Petri nets for rule-based decision making. IEEE Trans Syst Man Cybern 18:178–183

Luo P, Li Y, Wu C, Zhang G (2015) Toward cost-efficient sampling methods. Int J Mod Phys C 26:1550050

Lusseau D, Schneider K, Boisseau OJ, Haase P, Slooten E, Dawson SM (2003) The bottlenose dolphin community of doubtful Sound features a large proportion of long-lasting associations. Behav Ecol Sociobiol 54:396–405

Ma Y, Han JJ, Trivedi KS (2002) Call admission control for reducing dropped calls in CDMA cellular systems. Comput Commun 25:689–699

Mabrouk B, Hasni H, Mahjoub Z (2009) On a parallel genetic-tabu search based algorithm for solving the graph coloring problem. Eur J Oper Res 197:1192–1201

Mahadevan P, Krioukov D, Fomenkov M, Huffaker B, Dimitropoulos X, Vahdat A (2005) Lessons from three views of the internet topology. University of California, San Diego, CA, USA

Mahdaviani M, Kordestani JK, Rezvanian A, Meybodi MR (2015) LADE: learning automata based differential evolution. Int J Artif Intel Tools 24:1550023

Mahmoud QH (2007) Cognitive networks. Wiley Online Library

Malaguti A, Toth P (2008) An evolutionary approach for bandwidth multi-coloring problems. Eur J Oper Res 189:638–651

Marsan MA, Balbo G, Conte G, Donatelli S, Franceschinis G (1994) Modeling with generalized stochastic Petri net. In: Wiley Series in Parallel Computing. Wiley

Mason LG, Gu XD (1986) Learning automata models for adaptive flow control in packet-switching networks. In adaptive and learning systems, Springer, New York

Massaro A, Pelillo M, Bomze IM (2002) A complementary pivoting approach to the maximum weight clique problem. SIAM J Optim 12:928–948

Mehta D, Lopez M, Lin L (2003) Optimal coverage paths in ad hoc sensor networks. In: Proceedings of the IEEE international conference on communications. pp 507–511

Merabet A (1986) Synchronization of operations in a flexible manufacturing cell: the Petri net approach. J Manufac Syst 5:161–169

Merlin PM, Farber DJ (1976) Recoverability of communication protocols: implications of a theoretical study. IEEE Trans Commun 24:1036–1043

Meškauskas A, Fricker MD, Moore D (2004) Simulating colonial growth of fungi with the Neighbour-Sensing model of hyphal growth. Mycol Res 108:1241–1256

Meybodi MR (1983) Learning automata and its application to priority assignment in a queuing system with unknown characteristics. Ph.D. dissertation, University of Oklahoma

Meybodi MR, Beigy H (2001) Solving stochastic path problem using distributed learning automata. In: Proceedings of the sixth annual international CSI computer conference (CSICC2001). Iran, pp 70–86

Meybodi MR, Beigy H (2003) Solving stochastic shortest path problem using monte carlo sampling method: a distributed learning automata approach. In: Neural networks and soft computing, physica-verlag HD. pp 626–632

Meybodi MR, Lashmivarahan S (1983) A learning approach to priority assignment in a two class M/M/1 queuing system with unknown parameters. In: The yale workshop on adaptive system theory. pp 106–109

Meybodi MR, Beigy H, Taherkhani M (2003) Cellular learning automata and its applications. J Sci Technol 54–77

Mirsaleh MR, Meybodi MR (2015) A learning automata-based memetic algorithm. Genet Program Evolvable Mach 16:399–453

Mirsaleh MR, Meybodi MR (2016) A Michigan memetic algorithm for solving the community detection problem in complex network. Neurocomputing 214:535–545

Mirsaleh MR, Meybodi MR (2016) A new memetic algorithm based on cellular learning automata for solving the vertex coloring problem. Memetic Comput 8:211–222

Misra S, Oommen B (2004a) Stochastic learning automata-based dynamic algorithms for the single source shortest path problem. In: Innovations in applied artificial intelligence. pp 239–248

Misra S, Oommen BJ (2004) GPSPA: a new adaptive algorithm for maintaining shortest path routing trees in stochastic networks. Int J Commun Syst 17:963–984

Misra S, Oommen BJ (2006) An efficient dynamic algorithm for maintaining all-pairs shortest paths in stochastic networks. IEEE Trans Comput 55:686–702

Misra S, Chatterjee SS, Guizani M (2015) Stochastic learning automata-based channel selection in cognitive radio/dynamic spectrum access for WiMAX networks. Int J Commun Syst 28:801–817

Mittal V, Demirbas M, Arora A (2003) LOCI: Local clustering service for large scale wireless sensor networks. Ohio State University

Mofrad MH, Sadeghi S, Rezvanian A, Meybodi MR (2015) Cellular edge detection: Combining cellular automata and cellular learning automata. AEU-Int J Electron Commun 69:1282–1290

Mofrad MH, Jalilian O, Rezvanian A, Meybodi MR (2016) Service level agreement based adaptive Grid superscheduling. Future Gener Comput Syst 55:62–73

Mollakhalili Meybodi MR, Meybodi MR (2008) Link prediction in adaptive web sites using distributed learning automata. In: Proceeding of the 13th annual CSI computer conference of Iran, Kish Island. Iran

Mollakhalili Meybodi MR, Meybodi MR (2014) Extended distributed learning automata: an automata-based framework for solving stochastic graph. Appl Intel 41:923–940

Montresor A (2004) A robust protocol for building superpeer overlay topologies. In: Proceedings of the 4th international conference on peer-to-peer computing. IEEE Computer Society, Zurich, Switzerland, pp 202–209

Moradabadi B, Beigy H (2014) A new real-coded Bayesian optimization algorithm based on a team of learning automata for continuous optimization. Genet Program Evolvable Mach 15:169–193

Moradabadi B, Meybodi MR (2016) Link prediction based on temporal similarity metrics using continuous action set learning automata. Phys A Stat Mech Appl 460:361–373

Moradabadi B, Meybodi MR (2017) A novel time series link prediction method: learning automata approach. Phys A Stat Mech Appl 482:422–432. https://doi.org/10.1016/j.physa.2017.04.019

Moradabadi B, Ebadzadeh MM, Meybodi MR (2016) A new real-coded stochastic Bayesian optimization algorithm for continuous global optimization. Genet Program Evolvable Mach 17:145–167

Morshedlou H, Meybodi MR (2014) Decreasing impact of SLA violations: a proactive resource allocation approach for cloud computing environments. IEEE Trans Cloud Comput 2:156–167

Morshedlou H, Meybodi MR (2017) A new local rule for convergence of ICLA to a compatible point. IEEE Trans Syst Man Cybern Systems in-press:1–12

Motzkin TS, Straus EG (1965) Maxima for graphs and a new proof of a theorem of Turán. Can J Math 17:533–540

Mousavian A, Rezvanian A, Meybodi MR (2013) Solving minimum vertex cover problem using learning automata. In: 13th Iranian conference on fuzzy systems (IFSC 2013). pp 1–5

Mousavian A, Rezvanian A, Meybodi MR (2014) Cellular learning automata based algorithm for solving minimum vertex cover problem. In: 2014 22nd Iranian conference on electrical engineering (ICEE). IEEE, pp 996–1000

Moustakas V, Akcan H, Roussopoulos M, Delis A (2016) Alleviating the topology mismatch problem in distributed overlay networks: A survey. J Syst Softw 216–245

Mozafari M, Shiri ME, Beigy H (2015) A cooperative learning method based on cellular learning automata and its application in optimization problems. J Comput Sci 11:279–288

Murai F, Ribeiro B, Towsley D, Wang P (2013) On set size distribution estimation and the characterization of large networks via sampling. IEEE J Sel Areas Commun 31:1017–1025

Murata I (1989) Petri nets: properties, analysis and applications. In: Proceedings of the IEEE. pp 541–580

Musman S, Lehner PE, Elsaesser C (1997) Sensor planning for elusive targets. J Comput Math Model 25:103–115

Narendra KS, Thathachar MA (1989) Learning automata: an introduction. Prentice-Hall

Newman MEJ (2005) A measure of betweenness centrality based on random walks. Soc Netw 27:39–54

Newman ME (2006) Modularity and community structure in networks. Proc Nat Acad Sci 103:8577–8582

Newman MEJ (2015) Newman dataset. In: Network data. http://www-personal.umich.edu/~mejn/netdata/

Ni Y (2012) Minimum weight covering problems in stochastic environments. Inf Sci 214:91–104

Ni Y, Shi Q (2013) Minimizing the complete influence time in a social network with stochastic costs for influencing nodes. Int J Uncertainty Fuzziness Knowl Based Syst 21:63–74

Nicopolitidis P (2015) Performance fairness across multiple applications in wireless push systems. Int J Commun Syst 28:161–166

Noe JD, Nutt GJ (1973) Macro e-nets representation of parallel systems. IEEE Trans Comput 31:718–727

Norman MF (1968) Some convergence theorems for stochastic learning models with distance diminishing operators. J Math Psychol 5:61–101

O'kelly ME (1987) A quadratic integer program for the location of interacting hub facilities. Eur J Oper Res 32:393–404

Oommen BJ, Lanctot JK (1990) Discretized pursuit learning automata. IEEE Trans Syst Man Cybern 20:931–938

Opsahl T, Panzarasa P (2009) Clustering in weighted networks. Soc Netw 31:155–163

Packard NH, Wolfram S (1985) Two-dimensional cellular automata. J Stat Phys 38:901–946

Palmer E (1985) Graphical Evolution. Wiley, New York

Papadimitriou I (1994) A new approach to the design of reinforcement schemes for learning automata: stochastic estimator learning algorithms. IEEE Trans Knowl Data Eng 6:649–654

Papadimitriou I, Pomportsis AS, Kiritsi S, Talahoupi E (2002) Absorbing stochastic estimator learning algorithm. Inf Sci 147:193–199

Papagelis M, Das G, Koudas N (2013) Sampling online social networks. IEEE Trans Knowl Data Eng 25:662–676

Paximadis AV, Paximadis GT (1994) Fault-tolerant algorithms using estimator discretized learning automata for high-speed packet-switched networks. IEEE Trans Reliab 43:582–593

Pedraza F, Medaglia AL, Garcia A (2006) Efficient coverage algorithms for wireless sensor networks. In: Proceedings of the IEEE systems and information engineering design symposium. Charlottesville, VA, USA, pp 78–83

Pengand Y, Zhanting Y, Jizeng W (2007) Petri net model of session initiation protocol and its verification. In: International Conference on Communications (ICC). pp 1861–1864

Peterson L, Net P (1981) Theory and the modeling of systems. Prentice-Hall

Petri CA Fundamentals of a theory of asynchronous information flow. In: Proceedings of the 1962 IFIP Congress. pp 386–390

Petriu EM, Georganas ND, Petriu D, Makrakis D, Groza VZ (2000) Sensor-based information appliances. IEEE Instrum Measur Mag 31–35

Polastre JR (2003) Design and implementation of wireless sensor networks for habitat monitoring

Pons P, Latapy M (2005) Computing communities in large networks using random walks. Comput Inf Sci ISCIS 2005:284–293

Popova-Zeugmann L (2013) The classic Petri net. In: Time and Petri nets. Springer Berlin Heidelberg

Qiu T, Chan E, Ye M, Chen G, Zhao BY (2009) Peer-exchange schemes to handle mismatch in peer-to-peer systems. J Supercomput 48:15–42

Rabbany R, Takaffoli M, Fagnan J, Zaïane OR, Campello RJ (2013) Communities validity: methodical evaluation of community mining algorithms. Soc Netw Anal Min 3:1039–1062

Raghavan UN, Albert R, Kumara S (2007) Near linear time algorithm to detect community structures in large-scale networks. Phys Rev E 76:036106

Raghunathan V, Schurgers C, Park S, Srivastava MB (2002) Energy-aware wireless microsensor networks. IEEE Signal Proces Mag 19:40–50

Rajiv G, Halperin E, Khuller S, Kortsarz G, Srinivasan A (2006) An improved approximation algorithm for vertex cover with hard capacities. J Comput Syst Sci 72:16–33

Ramadurai V, Sichitiu ML (2003) Localization in wireless sensor networks: a probabilistic approach. In: Proceedings of the International conference on wireless networks. pp 275–281

Ramalingam G, Reps T (1996) On the computational complexity of dynamic graph problems. Theor Comput Sci 158:233–277

Ramamoorthy V, Ho GS (1980) Performance evaluation of asynchronous concurrent systems using Petri nets. IEEE Trans Softw Eng 6:440–449

Ramchandani A (1974) Analysis of asynchronous concurrent systems by Timed Petri Nets, Ph.D. dissertation. MIT, Cambridge, MA

Ranjbar A, Maheswaran M (2014) Using community structure to control information sharing in online social networks. Comput Commun 41:11–21

Rasch PJ (1970) A queuing theory study of round-robin scheduling of time-shared computer systems. J ACM (JACM) 17:131–145

Rastegar R, Meybodi MR (2004) A new evolutionary computing model based on cellular learning automata. In: IEEE conference on cybernetics and intelligent systems. Singapore, pp 433–438

Rastegar R, Arasteh AR, Hariri A, Meybodi MR (2004) A fuzzy clustering algorithm using cellular learning automata based evolutionary algorithm. In: Fourth international conference on hybrid intelligent systems. IEEE, pp 310–314

Rastegar R, Meybodi MR, Hariri A (2006) A new fine-grained evolutionary algorithm based on cellular learning automata. Int J Hybrid Intel Syst 3:83–98

Ratnasamy S, Handley M, Karp R, Shenker S (2002) Topologically-aware overlay construction and server selection. The 21st Annual Joint Conference of the IEEE computer and communications societies. IEEE Computer Society, New York, USA, pp 1190–1199

Ravichandran R, Chakravarty AK (1986) Decision support in flexible manufacturing systems using Timed Petri Nets. J Manufac Syst 5:89–101

Reichardt J, Bornholdt S (2004) Detecting fuzzy community structures in complex networks with a Potts model. Phys Rev Lett 93:218701

Reisig W (2013) The synthesis problem. Transactions on Petri Nets and other models of concurrency VII. Springer, Berlin Heidelberg, pp 300–313

Reisig W (2013) Understanding Petri nets: modeling techniques, analysis methods. Case Studies, Springer, Berlin, Heidelberg

Rezvanian A, Meybodi MR (2010) LACAIS: learning automata based cooperative artificial immune system for function optimization. Contemporary computing. Springer, Berlin Heidelberg, pp 64–75

Rezvanian A, Meybodi MR (2010b) An adaptive mutation operator for artificial immune network using learning automata in dynamic environments. In: Proceedings of the 2010 second world congress on nature and biologically inspired computing (NaBIC). pp 479–483

Rezvanian A, Meybodi MR (2015) Finding minimum vertex covering in stochastic graphs: a learning automata approach. Cybern Syst 46:698–727

Rezvanian A, Meybodi MR (2015) Finding maximum clique in stochastic graphs using distributed learning automata. Int J Uncertainty Fuzziness Knowl Based Syst 23:1–31

Rezvanian A, Meybodi MR (2015) Sampling social networks using shortest paths. Phys A Stat Mech Appl 424:254–268

Rezvanian A, Meybodi MR (2016) Stochastic graph as a model for social networks. Comput Hum Behav 64:621–640

Rezvanian A, Meybodi MR (2016) Sampling algorithms for weighted networks. Soc Netw Anal Min 6:1–22

Rezvanian A, Meybodi MR (2016c) Stochastic social networks: measures and algorithms. LAP LAMBERT Academic Publishing

Rezvanian A, Meybodi MR (2017) Sampling algorithms for stochastic graphs: a learning automata approach. Knowl Based Syst 127:126–144

Rezvanian A, Meybodi MR (2017) A new learning automata-based sampling algorithm for social networks. Int J Commun Syst 30:e3091

Rezvanian A, Rahmati M, Meybodi MR (2014) Sampling from complex networks using distributed learning automata. Phys A Stat Mech Appl 396:224–234

Robson GD, Van West P, Gadd G (2007) Exploitation of fungi. Cambridge University Press

Ronhovde P, Nussinov Z (2009) Multiresolution community detection for megascale networks by information-based replica correlations. Phys Rev E 80:016109

Rostami H, Habibi J (2007) Topology awareness of overlay P2P networks. Concurrency Comput Pract Experience 19:999–1021

Rubenstein M, Ahler C, Nagpal R (2012) Kilobot: a low cost scalable robot system for collective behaviors. International Conference On Robotics and Automation. IEEE Computer Society, Saint Paul, Minnesota, pp 3293–3298

Rubenstein M, Ahler C, Hoff N, Cabrera A, Nagpal R (2014) Kilobot: a low cost robot with scalable operations designed for collective behaviors. Robot Autono Syst 62:966–975

Rummery GA, Niranjan M (1994) On-line Q-learning using connectionist systems. University of Cambridge, Department of Engineering

Rutkowski L, Cpalka K (2003) Flexible neuro-fuzzy systems. IEEE Trans Neural Netw 14:554–574

Rysz M, Pajouh FM, Krokhmal P, Pasiliao EL (2016) Identifying risk-averse low-diameter clusters in graphs with stochastic vertex weights. Ann Oper Res 1–20

Saber RO (2007) Distributed tracking for mobile sensor networks with information—driven mobility. In: Proceedings of the American Control Conference. New York, NY, pp 4606–4612

Safavi SM, Meybodi MR, Esnaashari M (2014) Learning automata based face-aware mobicast. Wirel Pers Commun 77:1923–1933

Saghiri AM, Meybodi MR (2015) A distributed adaptive landmark clustering algorithm based on mOverlay and learning automata for topology mismatch problem in unstructured peer-to-peer networks. Int J Commun Syst. https://doi.org/10.1002/dac.2977

Saghiri AM, Meybodi MR (2016) An approach for designing cognitive engines in cognitive peer-to-peer networks. J Netw Comput Appl 70:17–40

Saghiri AM, Meybodi MR (2016) A self-adaptive algorithm for topology matching in unstructured peer-to-peer networks. J Netw Syst Manage 24:393–426. https://doi.org/10.1007/s10922-015-9353-9

Saghiri AM, Meybodi MR (2017) A distributed adaptive landmark clustering algorithm based on mOverlay and learning automata for topology mismatch problem in unstructured peer-to-peer networks. Int J Commun Syst 30:e2977. https://doi.org/10.1002/dac.2977

Saghiri AM, Meybodi MR (2017) A closed asynchronous dynamic model of cellular learning automata and its application to peer-to-peer networks. Genet Program Evolvable Mach 18:313–349

Saghiri AM, Meybodi MR (2017c) On expediency of closed asynchronous dynamic cellular learning automata. J Comput Sci. https://doi.org/10.1016/j.jocs.2017.08.015

Saghiri AM, Meybodi MR (2017d) An adaptive super-peer selection algorithm considering peers capacity utilizing asynchronous dynamic cellular learning automata. Appl Intell. https://doi.org/10.1007/s10489-017-0946-8

Salah A, Mustafa K (2004) Protocol verification and analysis using colored Petri nets. DePaul University, Chicago USA

Salehi M, Rabiee HR, Rajabi A (2012) Sampling from complex networks with high community structures. Chaos: An Interdisciplinary. J Nonlinear Sci 22:023126

Salhieh A, Weinmann J, Kochhal M, Schwiebert L (2001) Power efficient topologies for wireless sensor network. In: Proceedings of the international conference on parallel processing. Spain, pp 156–163

Sanli HO, Cam H, Cheng X (2004) EQoS: an energy efficient QoS protocol for wireless sensor networks. In: Proceedings of 2004 western simulation multiconference. San Diego, CA, USA

Santharam G, Sastry PS, Thathachar MAL (1994) Continuous action set learning automata for stochastic optimization. J Franklin Inst 331:607–628

Sato T (1999) On some asymptotic properties of learning automaton networks

Scheidegger M, Braun T (2007) Improved locality-aware grouping in overlay networks. Kommunikation in verteilten systemen. Springer, Bern, Switzerland, pp 27–38

Schelling TC (1971) Dynamic models of segregation†. J Math Sociol 1:143–186

Schoenen R, Sediq AB, Yanikomeroglu H, Senarath G, Chao Z (1983) Fairness analysis in cellular networks using stochastic Petri nets. In: Personal indoor and mobile radio communications (PIMRC)

Schwagre M, McLurkin J, Rus D (2006) Distributed coverage control with sensory feedback for networked robots. In: Proceedings of robotics: science and systems. Philadelphia, PA, USA

Schwartz A (1993) A reinforcement learning method for maximizing undiscounted rewards. In: Proceedings of the tenth international conference on machine learning. pp 298–305

Schwiebert L, Gupta S, Weinmann J (2001) Research challenges in wireless sensor networks of biomedical sensors. In: Proceedings of the 7th annual international conference on mobile computing and networking. pp 151–165

Shannon CE (1948) A mathematical theory of communication. Bell Syst Tech J 27:379–423

Sharifzadeh M, Shahabi C (2004) Supporting spatial aggregation in sensor network databases. In: Proceedings of the 12th annual ACM international workshop on geographic information systems. ACM, Washington DC, USA, pp 166–175

Shi J, Malik J (2000) Normalized cuts and image segmentation. IEEE Trans Pattern Anal Mach Intell 22(8):888–905

Shucker B, Bennett JK (2005) Target tracking with distributed robotic macrosensors. In: Proceedings of the military communications conference. Atlantic City, New Jersey

Sibley GT, Rahimi MH, Sukhatme GS (2002) Robomote: a tiny mobile robot platform for large-scale sensor networks. In: Proceedings of the IEEE international conference on robotics and automation

Sifakis J (1978) Structural properties of Petri nets. In: Mathematical foundations of computer science. pp 474–483

Simha R, Kurose JF (1989) Relative reward strength algorithms for learning automata. IEEE Trans Syst Man Cybern 19:388–398

SNAP (2014) Network datasets: Gnutella peer-to-peer network. In: SNAP: Network datasets: Gnutella peer-to-peer network. http://snap.stanford.edu/data/p2p-Gnutella04.html. Accessed 23 Jan 2014

Snyder PL, Greenstadt R, Valetto G (2009) Myconet: a fungi-inspired model for superpeer-based peer-to-peer overlay topologies. Third IEEE international conference on self-adaptive and self-organizing systems. IEEE Computer Society, San Francisco, CA, pp 40–50

Sobeih A, Chen WP, Hou JC, Kung LC, Li N, Lim H, Tyan HY, Zhang H (2006) J- Sim: a simulation and emulation environment for wireless sensor networks. IEEE Wirel Commun 13:104–119

Soe KT (2008) Increasing lifetime of target tracking wireless sensor networks. In: Proceedings of the world academy of science, engineering and technology. pp 2070–3740

Sohrabi K, Minoli D, Znati T (2007) Wireless Sensor Networks. In: Technology, protocols, and applications. Wiley, Hoboken, New Jersey

Solcimani-Pouri M, Rezvanian A, Meybodi MR (2012) Solving maximum clique problem in stochastic graphs using learning automata. In: 2012 fourth international conference on computational aspects of social networks (CASoN). pp 115–119

Sutton RS, Barto AG (1998) Reinforcement learning: an introduction. MIT press, Cambridge

Talavan M, Yanez J (2008) The graph coloring problem: a neuronal network approach. Eur J Oper Res 191:100–111

Teng SH, Black JT (1990) Cellular manufacturing systems modeling: the Petri net approach. J Manufac Syst 9:45–54

Terán-Villanueva D, Huacuja HJF, Valadez JMC, Rangel RAP, Soberanes HJP, Flores JAM (2013) Cellular processing algorithms. In: Soft computing applications in optimization, control, and recognition, Springer Berlin Heidelberg, pp 53–74

Thathachar M (1987) Learning automata with changing number of actions. IEEE Trans Syst Man Cybern 17:1095–1100

Thathachar MAL, Oommen BJ (1979) Discretized reward-inaction learning automata. J Cybern Inf Sci 2:24–29

Thathachar MAL, Sastry PS (1985) A class of rapidly converging algorithms for learning automata. IEEE Trans Syst Man Cybern 15:168–175

Thathachar MAL, Sastry PS (1986) Estimator algorithms for learning automata. In: Platinum jubilee conference on systems and signal processing. Bangalore, India

Thathachar MAL, Sastry PS (1997) A hierarchical system of learning automata that can learn die globally optimal path. Inf Sci 42:143–166

Thathachar MAL, Sastry PS (2004) Networks of learning automata: techniques for online stochastic optimization. Springer, Boston, MA

Tian R, Xiong Y, Zhang Q, Li B, Zhao BY, Li X (2005) Hybrid overlay structure based on random walks. In: Peer-to-peer systems IV. Springer, pp 152–162

Tilak S, Abu–Ghazaleh NB, Heinzelman W (2002) Infrastructure trade-offs for sensor networks. ACM WSNA 49–58

Tong C, Lian Y, Niu J, Xie Z, Zhang Y (2016) A novel green algorithm for sampling complex networks. J Netw Comput Appl 59:55–62

Torkestani JA (2010) A new vertex coloring algorithm based on variable action-set learning automata. J Comput Inform 29:1001–1020

Torkestani JA (2012) Degree-constrained minimum spanning tree problem in stochastic graph. Cybern Syst 43:1–21

Torkestani JA (2013) A new approach to the vertex coloring problem. Cybern Syst 44:444–466

Torkestani JA (2013) A learning automata-based algorithm to the stochastic min-degree constrained minimum spanning tree problem. Int J Found Comput Sci 24:329–348

Torkestani JA, Meybodi MR (2009) Graph coloring problem based on learning automata. In: Information management and engineering, 2009. ICIME'09. International Conference on. IEEE, pp 718–722

Torkestani J, Meybodi MR (2010) Learning automata-based algorithms for finding minimum weakly connected dominating set in stochastic graphs. Int J Uncertainty Fuzziness Knowl Syst 18:721–758

Torkestani JA, Meybodi MR (2010) Mobility-based multicast routing algorithm for wireless mobile Ad-hoc networks: a learning automata approach. Comput Commun 33:721–735

Torkestani JA, Meybodi MR (2011) A cellular learning automata-based algorithm for solving the vertex coloring problem. Expert Syst Appl 8:9237–9247

Torkestani JA, Meybodi MR (2012) A learning automata-based heuristic algorithm for solving the minimum spanning tree problem in stochastic graphs. J Supercomput 59:1035–1054

Torkestani JA, Meybodi MR (2012) Finding minimum weight connected dominating set in stochastic graph based on learning automata. Inf Sci 200:57–77

Tsetlin M (1973) Automation theory and modeling of biological systems. In: Mathematics in science and engineering. p 102

ul Asar A, Zhou MC, Caudill RJ (2005) Making Petri nets adaptive: a critical review. In: Proceedings networking, sensing and control. pp 644–649

Vafashoar R, Meybodi MR (2016) Multi swarm bare bones particle swarm optimization with distribution adaption. Appl Soft Comput 47:534–552

Vahidipour SM, Esnaashari M (2017) Priority assignment in queuing systems with unknown characteristics using learning automata and adaptive stochastic petri nets. J Comput Sci. https:// doi.org/10.1016/j.jocs.2017.08.009

Vahidipour SM, Meybodi MR, Esnaashari M (2015) Learning automata based adaptive Petri net and its application to priority assignment in queuing systems with unknown parameters. IEEE Trans Syst Man Cybern 45:1373–1384 . doi:10.1109/TSMC.2015.2406764

Vahidipour SM, Meybodi MR, Esnaashari M (2017) Adaptive Petri net based on irregular cellular learning automata with an application to vertex coloring problem. Appl Intell 46:272–284

Vahidipour SM, Meybodi MR, Esnaashari M (2017) Finding the shortest path in stochastic graphs using learning automata and adaptive stochastic Petri nets. Int J Uncertainty Fuzziness Knowl Based Syst 25:427–455

Vahidipour SM, Meybodi MR, Esnaashari M (2017) Cellular adaptive Petri net based on learning automata and its application to the vertex coloring problem. Discrete Event Dyn Syst 27:609–640

van der Aalst WMP (1998) Three good reasons for using a Petri-net-based workflow management system. In: Information and process integration in enterprises: rethinking documents, Kluwer international series in engineering and computer science. pp 161–182

van der Aalst C, Stahl W (2011) Modeling business processes: a Petri net-oriented approach. MIT Press, Cambridge, London

Varshavskii VI, Meleshina MV, Tsetlin ML (1968) Priority organization in queueing systems using a model of the collective behavior of automata. Probl Peredachi Informatsii 4:73–76

Verbeeck K, Nowe A (2002) Colonies of learning automata. IEEE Trans Syst Man Cybern Part B Cybern 32:772–780

Veremyev A, Sorokin A, Boginski V, Pasiliao EL (2014) Minimum vertex cover problem for coupled interdependent networks with cascading failures. Eur J Oper Res 232:499–511

Viswanath B, Mislove A, Cha M, Gummadi KP (2009) On the evolution of user interaction in facebook. In: Proceedings of the 2nd ACM workshop on online social networks, pp 37–42

Viswanathan R, Narendra KS (1973) Stochastic automata models with applications to learning systems. IEEE Trans Syst Man Cybern 1:107–111

Volz E, Heckathorn DD (2008) Probability based estimation theory for respondent driven sampling. J Official Stat Stockholm 24:79

Wan C, Eisenman S, Campbelt A (2003) CODA: Congestion detection and avoidance in sensor networks. In: Proceedings of the ACM SenSys. CA, pp 266–279

Wang L, Xiao Y (2006) A survey of energy-efficient scheduling mechanisms in sensor networks. Mob Netw Appl 11:723–740

Wang B, Chua KC, Srinivasan V, Wang W (2006a) Sensor density for complete information coverage in wireless sensor networks. In: Proceedings of the EWSN. pp 69–82

Wang B, Chua KC, Srinivasan V, Wang W (2006b) Scheduling sensor activity for point information coverage in wireless sensor networks. In: Proceedings of the 4th international symposium on modeling and optimization in mobile, ad hoc and wireless networks. pp 1–8

Wang G, Cao G, Porta TFL (2006c) Movement-assisted sensor deployment. IEEE Trans Mob Comput 5:640–652

Wang X, Ma JJ, Ding L, Bi DW (2007a) Robust forecasting for energy efficiency of wireless multimedia sensor networks. Sens J 7:2779–2807

Wang X, Ma JJ, Wang S, Bi DW (2007) Cluster-based dynamic energy management for collaborative target tracking in wireless sensor networks. Sens J 7:1193–1215

Wang M, Zuo W, Wang Y (2016) An improved density peaks-based clustering method for social circle discovery in social networks. Neurocomputing 179:219–227

Watfa MK (2006) An energy efficient approach to dynamic coverage in wireless sensor networks. J Netw 1:10–20

Watfa MK, Commuri S (2006) A reduced cover approach to energy efficient tracking using wireless sensor networks. In: World congress in computer science, computer engineering and applied computing, Las Vegas. Nevada, USA

Watkins CJCH (1989) Learning from delayed rewards. University of Cambridge England

Watts DJ, Strogatz SH (1998) Collective dynamics of 'small-world' networks. Nature 393:440–442

Williams RJ (1988) Toward a theory of reinforcement-learning connectionist systems. Northeastern University

Willig A, Shah R, Rabacy J, Wolisz A (2002) Altruists in the pico-radio sensor network. In: Proceedings of the 4th IEEE international workshop on factory communication systems. Sweden, pp 175–184

Wong Y, Dillon TS, Forward and KE (1985) Timed Petri nets with stochastic representation of place time. In: International workshop on timed Petri nets, IEEE Computer Society. Torino, Italy, pp 96–103

Wolf S, Merz P (2007) Evolutionary local search for the super-peer selection problem and the p-hub median problem. In: Proceedings of the 4th international conference on hybrid metaheuristics. Springer-Verlag, Berlin, Heidelberg, pp 1–15

Wu Q, Hao J-K (2015) A review on algorithms for maximum clique problems. Eur J Oper Res 242:693–709

Wu Y, Fahmy S, Shroff NB (2006) Optimal QoS-aware sleep/wake scheduling for time-synchronized sensor networks. Proceeding of the 40th annual conference on information sciences and systems. IEEE. Princeton, NJ, pp 924–930

Xu Y, Heidemann J, Estrin D (2001) Geography informed energy conservation for ad hoc routing. In: Proceedings of the ACM mobile computing and networking. pp 70–84

Xu Z, Tang C, Zhang Z (2003) Building topology-aware overlays using global soft-state. 23rd international conference on distributed computing systems. IEEE, Providence, RI, USA, pp 500–508

Yang H, Sidkar B (2005) Lightweight target tracking protocol using ad hoc sensor networks. In: Proceedings of the IEEE VTC. Stockholm, Sweden

Yap TN, Shelton CR (2008) Simultaneous learning of motion and sensor model parameters for mobile robots. In: Proceedings of the IEEE international conference on robotics and automation. pp 2091–2097

Ye F, Zhong G, Cheng J, Lu S, Zhang L (2003) PEAS: a robust energy conserving protocol for long-lived sensor networks. In: Proceedings of the 23rd international conference on distributed computing systems. pp 28–37

Ye F, Lu S, Zhang L (2006) A randomized energy-conservation protocol for resilient sensor networks. Wirel Netw 12:637–652

Yen LH, Yu CW, Cheng YM (2006) Expected k-coverage in wireless sensor networks. Ad Hoc Netw 4:636–650

Yoon S, Lee S, Yook SH, Kim Y (2007) Statistical properties of sampled networks by random walks. Phys Rev E 75:046114

Yoon S-H, Kim K-N, Hong J, Kim S-W, Park S (2015) A community-based sampling method using DPL for online social networks. Inf Sci 306:53–69

Zachary WW (1977) An information flow model for conflict and fission in small groups. J Anthropol Res 452–473

Zha XF, Fok SC, Lim SYE (1998) Integration of knowledge-based systems and neural networks: neuro-expert Petri net models and applications. In: Proceedings IEEE international conference robotics and automation, IEEE. pp 1423–1428

Zhang X-Y (2016) Simultaneous optimization for robust correlation estimation in partially observed social network. Neurocomputing 205:455–462

Zhang W, Cao G (2004) Optimizing tree reconfiguration for mobile target tracking in sensor network. In: Proceedings of the IEEE INFOCOM. pp 2434–2445

Zhang H, Hou JC (2005) Maintaining sensing coverage and connectivity in large sensor networks. In: Ad hoc & sensor wireless networks. pp 89–124

Zhang CW, Zhou MC (2014) Last-position elimination-based learning automata. IEEE Trans Cybern 44:2484–2492

Zhang XY, Zhang Q, Zhang Z, Song G, Zhu W (2004) A construction of locality-aware overlay network: mOverlay and its performance. IEEE J Sel Areas Commun 22:18–28

Zhao Z, Govindan R (2003) Understanding Packet delivery performance in dense wireless sensor networks. In: Proceedings of the 3rd ACM conference on embedded networked sensor systems. Los Angeles, CA, USA, pp 1–13

Zhao Y, Jiang W, Li S, Ma Y, Su G, Lin X (2015) A cellular learning automata based algorithm for detecting community structure in complex networks. Neurocomputing 151:1216–1226

Zhao Y, Li S, Jin F (2016) Identification of influential nodes in social networks with community structure based on label propagation. Neurocomputing 210:34–44

Zheng R, He G, Liu X (2005) Location-free coverage maintenance in wireless sensor networks. Department of Computer Science, University of Houston

Zheng X, Zeng Z, Chen Z, Yu Y, Rong C (2015) Detecting spammers on social networks. Neurocomputing 159:27–34

Zhou Y, Schembri J, Lamont L, Bird J (2009a) Analysis of stand-alone GPS for relative location discovery in wireless sensor network. In: Proceedings of the Canadian conference on electrical and computer engineering. Newfoundland, Canada, pp 437–441

Zhou Z, Das SR, Gupta H (2009) Variable radii connected sensor cover in sensor networks. ACM Trans Sens Netw (TOSN) 5:8

Zou Y, Chakrabarty K (2003) Sensor deployment and target localization based on virtual forces. In: Proceedings of the twenty-second annual joint conference of the IEEE computer and communications (INFOCOM 2003). pp 1293–1303

Zou Y, Chakrabarty K (2004) Sensor deployment and target localization in distributed sensor networks. ACM Trans Embed Comput Syst (TECS) 3:61–91

Zou Y, Chakrabarty K (2004) Uncertainty-aware and coverage-oriented deployment for sensor networks. J Parallel Distrib Comput 64:788–798

Zou Y, Chakrabarty K (2005) A distributed coverage-and connectivity-centric technique for selecting active nodes in wireless sensor networks. IEEE Trans Comput 54:978–991

Zuberek WM (1980) Timed Petri nets and preliminary performance evaluation. In: Proceedings of the 7th annual symposium on computer architecture, ACM, La Baule. France, pp 88–96

Zuniga M, Krishnamachari B (2004) Analyzing the transitional region in low power wireless links. In: Proceedings of the IEEE international conference on sensor and ad hoc communications and networks. Santa Clara, CA, pp 517–526

Zymolka M, Koster CA, Wessaly R (2003) Transparent optical network design with sparse wavelength conversion. In: Proceedings of the 7th IFIP working conference on optical network design a. Budapest, Hungary, pp 61–80

Printed in the United States
By Bookmasters